Reconstructing Earth's Climate History

To our families,
for their understanding and support
while we dedicated much time
and energy to the writing of this book

COMPANION WEBSITE

This book has a companion website:
www.wiley.com/go/stjohn/climatehistory
with Figures and Tables from the book and supplementary material for downloading

Reconstructing Earth's Climate History

Inquiry-Based Exercises for Lab and Class

Kristen St John, R Mark Leckie, Kate Pound,
Megan Jones and Lawrence Krissek

A John Wiley & Sons, Ltd., Publication

Wiley-Blackwell is an imprint of John Wiley & Sons, formed by the merger of Wiley's global Scientific, Technical and Medical business with Blackwell Publishing.

Registered office: John Wiley & Sons, Ltd, The Atrium, Southern Gate, Chichester, West Sussex, PO19 8SQ, UK

Editorial offices: 9600 Garsington Road, Oxford, OX4 2DQ, UK
The Atrium, Southern Gate, Chichester, West Sussex, PO19 8SQ, UK
111 River Street, Hoboken, NJ 07030-5774, USA

For details of our global editorial offices, for customer services and for information about how to apply for permission to reuse the copyright material in this book please see our website at www.wiley.com/wiley-blackwell.

Library of Congress Cataloging-in-Publication Data
Reconstructing Earth's climate history : inquiry-based exercises for lab and class / Kristen St John... [et al.].
p. cm.
Includes index.
Summary: "This project integrates scientific ocean drilling data and research (DSDP-ODP-IODPANDRILL) with education"– Provided by publisher.
ISBN 978-0-470-65805-5 (hardback)
ISBN 978-1-118-23294-1 (paper)
1. Paleoclimatology. 2. Climatic changes–Observations. 3. Climatic changes–History. I. St. John, Kristen.
QC884.R428 2012
551.609'01–dc23

2011039577

A catalogue record for this book is available from the British Library.

Wiley also publishes its books in a variety of electronic formats. Some content that appears in print may not be available in electronic books.

Set in 10/12pt Melior by Toppan Best-set Premedia Limited
Printed and bound in Singapore by Markono Print Media Pte Ltd

1 2012

Contents

COMPANION WEBSITE

This book has a companion website:
www.wiley.com/go/stjohn/climatehistory
with Figures and Tables from the book and supplementary material for downloading

The Authors

Dr Kristen St John is a Professor of Geology at James Madison University in Harrisonburg, VA. She is a marine sedimentologist, specializing in high latitude paleoclimate records and reconstructing ice-rafting histories. She has participated in four scientific ocean drilling expeditions with the Ocean Drilling Program (ODP) and the Integrated Ocean Drilling Program (IODP). She has co-led, with Mark Leckie, the science instruction of the IODP teacher professional development program "School of Rock" on several occasions. Kristen is the lead investigator on the NSF-funded project, Building Core Knowledge – Reconstructing Earth History, which was the catalyst for writing this book. She is currently the Editor of the *Journal of Geoscience Education*. Her undergraduate teaching responsibilities include Earth systems and climate change, oceanography for teachers, Earth science for teachers, geowriting and communication, paleoclimatology, and physical geology.

Dr R. Mark Leckie is a Professor of Geology and Department Head at the University of Massachusetts–Amherst. He co-led the scientific instruction of the IODP "School of Rock" expedition in 2005 and co-taught a number of subsequent shore-based short courses and workshops with Kristen St. John. Mark Leckie is a marine micropaleontologist, specializing in paleoceanography and ocean–climate history of the past 120 million years. He has participated in six Deep Sea Driling Projects (DSDP) and ODP scientific expeditions, and served as Co-Chief Scientist of ODP Leg 165. Mark Leckie is a co-author of a classroom activity book *Investigating the Oceans, an Interactive Guide to the Science of Oceanography*. He has been an instructor at the Urbino Summer School on Paleoclimatology since 2008. His teaching responsibilities include introductory oceanography, history of the Earth, introductory field methods, paleoceanography, and marine micropaleontology.

Dr Kate Pound is Professor of Geology and a member of the Science Education Group at St. Cloud State University. She has co-led the TIMES (Teaching Inquiry-based Minnesota Earth Science) project with Megan Jones. She has organized/co-convened NAGT-sponsored teacher workshops, associated conference sessions and associated "Hands-on Galleries" at American Geophysical Union (AGU) meetings and at the regional North-Central Geological Society of America (GSA) meeting. Kate Pound has a broad background in field geology, petrology, and provenance studies. Her teaching responsibilities include glacial geology, field geology, rocks and minerals, sedimentology, and the geological environment, as well as courses for pre-service teachers. She maintains a sedimentology lab for use in teaching and student-faculty research. She participated on-ice in ANDRILL ARISE (ANtarcticDRILLing, Andrill Research Immersion for Science Educators) during the fall of 2007.

Dr Megan Jones is a Professor of Geology at North Hennepin Community College. Her broad background and experience in marine micropaleontology, paleoceanography, sedimentology/stratigraphy, and field geology provides her students with a variety of options to pursue field experiences and undergraduate research at a two year institution. Megan Jones has been a co-facilitator, with Kate Pound, of the successful Twin Cities Metro Area TIMES Project Teaching Inquiry-based Minnesota Earth Science, a 10-day, field-based, summer institute for middle and high school, pre- and in-service teachers. Her teaching responsibilities include physical geology, historical geology, oceanography, and Minnesota field geology.

Dr Lawrence Krissek is a Professor and Associate Director in the School of Earth Sciences, Ohio State University. His primary scientific research is the study of the evolution of climates and ocean environments on the Earth during the past 65 million years. He has participated in nine field seasons of research in the Antarctic, including the 2006 and 2007 ANDRILL field seasons, as well as participation in seven cruises of DSDP, ODP, and IODP. His teaching responsibilities include oceanography, oceanography for educators, field geology for educators, natural hazards, physical and historical geology, and stratigraphy and sedimentation.

List of Contributors

Dr Kristen St John
Department of Geology &
Environmental Science
James Madison University
Msc 6903, 7125 Memorial Hall
395 High Street
Harrisonburg
VA 22807
USA
stjohnke@jmu.edu

Dr R Mark Leckie
Department of Geosciences
University of Massachusetts
611 N Pleasant Street
Amherst
MA 01003-0000
USA
mleckie@geo.umass.edu

Dr Kate Pound
Department of Atmospheric and
Hydrologic Sciences
St Cloud State University
WSB-44
St Cloud
MN 56301-4498
USA
kspound@stcloudstate.edu

Dr Megan Jones
Department of Geology
North Hennepin Community College
7411 85th Avenue North
Brooklyn Park
MN 55445
USA
megan.jones@nhcc.edu

Dr Lawrence Krissek
School of Earth Sciences
Ohio State University
275 Mendenhall Laboratory
125 South Oval Mall
Columbus
Ohio
OH 43210
USA
krissek.1@osu.edu

Foreword

Climate change has many manifestations, rising greenhouse gas concentrations, sea-level rise, abrupt climate change, ocean acidification, reduced Arctic sea ice, droughts, floods, hurricanes, melting glaciers and ice sheets, to mention a few. Few would doubt that climate is the environmental issue of our generation, but what scientific evidence causes so much concern about human influence on climate? Some might argue from the point of view of planetary physics; atmospheric greenhouse gases naturally affect the Earth's temperature and human carbon emissions have elevated carbon dioxide and methane concentrations and, as a consequence, global temperature. Others might claim that predictive climate models project future temperatures, rainfall patterns and sea levels that threaten society. The striking rise in global temperature observed from instruments over the past century also raises concern about future trends and impacts.

As important as these topics are, one field – paleoclimatology – is unique in providing the requisite baseline of natural climate variability against which human-induced climate change must be assessed. A rapidly growing discipline that draws on ocean, atmosphere, and earth sciences, paleoclimatology is today an essential foundation of climate science because it addresses climate history beyond the limited instrumental record and during climate states that the Earth may very well experience in the future. Consider these facts: Ice core records provide the primary evidence that modern greenhouse gas concentrations lie far outside the bounds of natural variability of the last 800,000 years. Thanks to tree rings, speleothems, and other records we now know that rising atmospheric and ocean temperatures during the last century cannot be explained by volcanic or solar activity but required forcing by elevated greenhouse gas concentrations. Lake and marine sediment records confirm what is suspected from satellite records – that polar climates are changing at unprecedented rates. Marine sediment records show us that ocean acidification – a major concern owing to human-induced perturbations of the global carbon cycle – typically accompanied massive increases in atmospheric carbon dioxide in the geological past.

Reconstructing Earth's Climate History – a novel classroom and laboratory educational guide by Kristen St John, R Mark Leckie, Kate Pound, Megan Jones and Lawrence Krissek – represents a major, long overdue effort to educate future generations about methods used to reconstruct climate history. From an academic perspective, the book exemplifies the authors' lifelong dedication to teaching. It includes practical discussions and exercises that teach students how climate history is reconstructed from "proxies" extracted from sediments, ice cores, speleothems, tree rings, coral skeletons, and other archives. It prepares students to engage in field and laboratory research to distinguish natural from anthropogenic climate change, evaluate computer model simulations of climatic under elevated greenhouse gas concentrations, and clarify the causes and impact of abrupt climate changes. Equally important, Kristen St. John and her co-authors also strive to explain why climate history is, and will continue to be, so relevant to policy debates about climate change. It is hoped that students of both natural and social sciences will use it for the benefit of the Earth's environments and future societies.

Thomas M. Cronin, Senior Research Geologist, US Geological Survey Reston Virginia

Acknowledgments

The authors and publisher gratefully acknowledge the permission granted to reproduce the copyright material in this book:

Figure 1.1A	http://www-saps.plantsci.cam.ac.uk/treerings/photos1/index.htm. Courtesy of Colin Bielby.
Figure 1.1B	Photo courtesy of John Haynes, James Madison University.
Figure 1.1B inset	S. Hollins, D. Cendon, J. Crawford, et al., "Climate Variability and Water Resources" (Australian Nuclear Science and Technology Organisation). http://www.ansto.gov.au/research/institute_of_environmental_research/science/isotopes_for_water/climate_variability_and_water_resources
Figure 1.1C	Courtesy of Lonnie Thompson, Ohio State University.
Figure 1.1D (top & bottom)	Janice Hough, Figure 2 from "Climate Records from Corals," http://onlinelibrary.wiley.com/doi/10.1002/wcc.39/full
Figure 1.1E	N. Marwan, M. H. Trauth, M. Vuille, et al., Figure 4 in "Comparing modern and Pleistocene ENSO-like influences in NW Argentina using nonlinear time series analysis methods," pp. 317–326 from *Climate Dynamics*, 21:3–4 (2003). doi:10.1007/s00382-003-0335-3.
Figure 1.2	Reproduced by permission of Lynn Fichter.
Figure 1.3A	Reproduced by permission of John Firth.
Figure 1.3B	WAIS Divide Ice Core Project, Photo ICDS (2006). http://www.atmos.washington.edu/~beckya/WAISDivide.htm
Figure 1.3C	Shipboard Scientific Party, Site 1220, pp. 1–93 from M. Lyle, P. A. Wilson, T. R. Janecek, et al., *Proceedings of the Ocean Drilling Program: Initial Reports* 199 (2002). doi:10.2973/odp.proc.ir.199.113.2002. http://www-odp.tamu.edu/publications/199_IR/chap_13/chap_13.htm. Courtesy of the Integrated Ocean Drilling Program.
Figure 1.4	Y. F. Li, "Global Population Distribution Database", from *A Report to the United Nations Environment Programme, under UNEP Sub-Project FP/1205-95-12*, (March 1996). http://cdiac.ornl.gov/newsletr/spring97/datas97.htm
Figure 1.5a	Reproduced by permission of Lonnie G. Thompson.
Figure 1.5b	Courtesy of Mark Leckie.
Figure 1.6	http://oregonstate.edu/dept/ncs/newsarch/2008/Sep08/icecore.html. Reproduced by permission of Edward J. Brooke.
Figure 1.7	http://bprc.osu.edu/Icecore/facilities.html#AnalyticalFacilities. Reproduced by permission of Ellen Mosley-Thompson.
Figure 1.8	William Crawford, IODP Expedition 321 Photos, IODP/TAMU. http://iodp.tamu.edu/scienceops/gallery/exp321/. Courtesy of the Integrated Ocean Drilling Program.
Figure 1.9	Drill Site Maps. http://iodp.tamu.edu/scienceops/maps.html. Courtesy of the Integrated Ocean Drilling Program.
Figure 1.10	Shipboard Scientific Party, Figure F1 in "Explanatory notes," pp. 1–70 from M. Lyle, P. A. Wilson, T. R. Janecek, et al., *Proceedings of the Ocean Drilling Program: Initial Reports* 199 (2002). doi:10.2973/odp.proc.ir.199.102.2002. Courtesy of the Integrated Ocean Drilling Program.

Figure 1.11	http://iodp.tamu.edu/database/. Courtesy of the Integrated Ocean Drilling Program.
Figure 2.1	Photo taken by Carlos Alvarez Zarikian, IODP Expedition 323 Photos (Week 3, July 2009 link) http://iodp.tamu.edu/scienceops/gallery/exp323/week3/ http://iodp.tamu.edu/scienceops/gallery/exp323/week3/. Courtesy of the Integrated Ocean Drilling Program.
Figure 2.2	Scanned image of Hubbard Scientific's product Sea Floor Physiography http://www.amep.com/standarddetail.asp?cid=1119. Reproduced by permission of American Educational Products.
Figure 2.3 A, B, G, I,	R. G. Rothwell, "Minerals and Mineraloids," p. 279 from *Marine Sediments: An Optical Identification Guide* (London: Elsevier Applied Science, 1989). Courtesy of Elsevier.
Figure 2.3 C, H, J	http://www.boscorf.org/curatorial/grain_id.html. Reproduced by permission of R. G. Rothwell.
Figure 2.3 D	http://www.boscorf.org/curatorial/ccd.html. Reproduced by permission of R. G. Rothwell.
Figure 2.3 E, F	http://www.boscorf.org/curatorial/pacific.html. Reproduced by permission of R. G. Rothwell.
Chapter 2 Appendix	Core photos courtesy of the Integrated Ocean Drilling Program.
Figure 3.1	Courtesy of Mark Leckie.
Figure 3.2	Photo micrographof planktic foraminifera from http://www.microscopy-uk.org.uk/mag/imgapr00/dwslide5.jpg. Reproduced by permission of David W. Walker. Other photomicrographs taken by teachers participating on the IODP Deep Earth Academy School of Rock, November 2005.
Figure 3.2	M. Leckie and K. St John, "How Old Is it? Part 1 - Biostratigraphy," written for the School of Rock Expedition, 2005. http://www.oceanleadership.org/education/deep-earth-academy/educators/classroom-activities/undergraduate/how-old-is-it-part-1-biostratigraphy/
Figure 3.3	F. Gradstein, p. 109 from *Report of the second Conference on Scientific Ocean Drilling (Cosod II)* (Strasbourg, France, European Science Foundation, 1987). Reproduced by permission of the European Science Foundation.
Figure 3.4	M. E. Katz, J. D. Wright, K. G. Miller, et al., "Biological Overprint of the Geological Carbon Cycle," pp. 323–338 from *Marine Geology* 217 (2005). Courtesy of Elsevier.
Figure 3.5	Reproduced by permission of Tina King.
Figure 3.6	Courtesy of Mark Leckie.
Figure 3.7 & 3.8	IODP Deep Earth Academy "School of Rock Expedition," (November 2005).
Figure 3.9	P. R. Bown, "Cenozoic Calcareous Nannofossil Biostratigraphy, ODP Leg 198 Site 1208 (Shatsky Rise, northwest Pacific Ocean)," pp. 1–44 from T. J. Bralower, I. Premoli Silva, and M. J. Malone (eds.), *Proceedings of the Ocean Drilling Program: Scientific Results* 198 (2005). doi:10.2973/odp.proc.sr.198.104.2005. http://www-odp.tamu.edu/publications/198_SR/104/104.htm. Courtesy of the Integrated Ocean Drilling Program.

Figure 3.10 P. R. Bown, Figure 1 in "Cenozoic calcareous nannofossil biostratigraphy, ODP Leg 198 Site 1208 (Shatsky Rise, northwest Pacific Ocean)," pp. 1–44 from T. J. Bralower, I. Premoli Silva, and M. J. Malone (Eds.), *Proceedings of the Ocean Drilling Program: Scientific Results* 198 (2005). doi:10.2973/odp.proc.sr.198.104.2005.
http://www-odp.tamu.edu/publications/198_SR/104/104.htm. Courtesy of the Integrated Ocean Drilling Program.

Figure 3.11 Courtesy of Mark Leckie.

Figure 3.12 Shipboard Scientific Party, Figure 4 in "Explanatory notes," pp. 1–63 from T. J. Bralower, I. Premoli Silva, M. J. Malone, et al., *Proceedings of the Ocean Drilling Program: Scientific Results* 198 (2002). doi:10.2973/odp.proc.ir.198.102.2002 http://www-odp.tamu.edu/publications/198_IR/chap_02/chap_02.htm. Courtesy of the Integrated Ocean Drilling Program.

Figure 3.13 Shipboard Scientific Party, Figure 22, pp. 256–333 from L. Mayer, N. Pisias,T. Janecek, et al., *Proceedings of the Ocean Drilling Program: Scientific Results* 198 (1992). doi:10.2973/odp.proc.ir.138.111.1992. http://www-odp.tamu.edu/publications/138_IR/VOLUME/CHAPTERS/ir138_11.pdf. Courtesy of the Integrated Ocean Drilling Program.

Table 3.1 & 3.2 P. R. Bown, Table 1 in "Cenozoic calcareous nannofossil biostratigraphy, ODP Leg 198 Site 1208 (Shatsky Rise, northwest Pacific Ocean)," pp. 1–44 from T. J. Bralower, I. Premoli Silva, and M. J. Malone (Eds.), *Proceedings of the Ocean Drilling Program: Scientific Results* 198 (2005). doi:10.2973/odp.proc.sr.198.104.2005.
http://www-odp.tamu.edu/publications/198_SR/104/104.htm. Courtesy of the Integrated Ocean Drilling Program.

Tables 3.3 P. R. Bown, Table 3, Calcareous nannofossil datums, ages and depths in "Cenozoic calcareous nannofossil biostratigraphy, ODP Leg 198 Site 1208 (Shatsky Rise, northwest Pacific Ocean)," pp. 1–44 from T. J. Bralower, I. Premoli Silva, and M. J. Malone (Eds.), *Proceedings of the Ocean Drilling Program: Scientific Results* 198 (2005). doi:10.2973/odp.proc.sr.198.104.2005
http://www-odp.tamu.edu/publications/198_SR/104/104.htm. Courtesy of the Integrated Ocean Drilling Program.

Table 3.4 K. Kameo and T. J. Bralower, Table 3 in "Neogene calcareous nannofossil biostratigraphy of Sites 998, 999, and 1000, Caribbean Sea," pp. 3–17 from R. M. Leckie, H. Sigurdsson, G. D. Acton et al. (eds.), *Proceedings of the Ocean Drilling Program, Scientific Results* 165 (2000). http://www-odp.tamu.edu/publications/165_SR/chap_01/chap_01.htm. Courtesy of the Integrated Ocean Drilling Program.

Figure 3.14 K. Kameo and T. J. Bralower, Figure 1 in "Neogene calcareous nannofossil biostratigraphy of Sites 998, 999, and 1000, Caribbean Sea," pp. 3–17 from R. M. Leckie, H. Sigurdsson, G. D. Acton, et al. (eds.), *Proceedings of the Ocean Drilling Program, Scientific Results* 165 (2000). http://www-odp.tamu.edu/publications/165_SR/chap_01/chap_01.htm. Courtesy of the Integrated Ocean Drilling Program.

Figure 4.1	http://sec.gsfc.nasa.gov/popscise.jpg and http://sohowww.nascom.nasa.gov/gallery/images/magfield.html. Courtesy of National Aeronautics and Space Administration.
Figure 4.2	http://www.askamathematician.com/?p=4129
Figure 4.3a	Shipboard Scientific Party, Figure F18 in "Leg 185 summary: inputs to the Izu-Mariana subduction system," pp. 1–63 from T. Plank, J. N. Ludden, C. Escutia, et al., *Proceedings of the Ocean Drilling Program, Initial Reports* 185 (2000). doi: 10.2973/odp.proc.ir.185.101.2000. http://www-odp.tamu.edu/publications/185_IR/chap_01/c1_f18.htm. Courtesy of the Integrated Ocean Drilling Program.
Figure 4.3b	Shipboard Scientific Party, Figure F16 in "Site 1172," pp.1–149 from N. F. Exon, J. P. Kennett, M. J. Malone, et al., *Proceedings of the Ocean Drilling Program, Initial Reports* 189 (2001). doi:10.2973/odp.proc.ir.189.107.2001 http://www-odp.tamu.edu/publications/189_IR/chap_07/chap_07.htm. Courtesy of the Integrated Ocean Drilling Program.
Figure 4.4	"Ocean Drilling Program Site Maps" http://www-odp.tamu.edu/sitemap/sitemap.html Courtesy of the Integrated Ocean Drilling Program.
Figure 4.5	http://www.es.ucsc.edu/~glatz/geodynamo.html. Reproduced by permission of Gary Glatzmaier.
Figure 4.6	http://www.odplegacy.org/operations/labs/paleomagnetism/
Figure 4.7	W. C. Pitman, III and J. R. Heirtzler, "Magnetic anomalies over the Pacific-Antarctic Ridge," pp. 1164–1171 from *Science* 154 (1966). Reprinted with permission from AAAS.
Figure 4.8	http://library.thinkquest.org/18282/lesson4.html. Reproduced from U.S. Geological Survey website, Department of the Interior/USGS. U.S. Geological Survey.
Figure 4.9	A. Cox, Figure 5 in "Geomagnetic reversals," pp. 239–245 from *Science* 163 (1969). Reproduced with permission from AAAS.
Figure 4.10	A. Cox, Figure 4 in "Geomagnetic reversals," pp. 239–245 from *Science* 163 (1969). Reproduced with permission from AAAS.
Figure 4.11	Shipboard Scientific Party, Figure F19 in " Site 1208," pp. 1–93 from T. J. Bralower, I. Premoli Silva, M. J. Malone, et al., *Proceedings of the Ocean Drilling Program, Initial Reports* 198 (2002). doi:10.2973/odp.proc.ir.198.104.2002. Courtesy of the Integrated Ocean Drilling Program.
Figure 4.12	A. E. Maxwell, R. P. Von Herzon, J. H. Andrews, et al., Figure 1 from *Initial Reports of the Deep Sea Drilling Project* 3 (Washington: U.S. Government Printing Office, 1970). http://www.deepseadrilling.org/03/dsdp_toc.htm. Courtesy of the Integrated Ocean Drilling Program.
Figure 4.13	F. J. Vine, Figure 3 in "Spreading of the ocean floor: New evidence," pp. 1405–1415 from *Science* 154 (1966). Reproduced with permission from AAAS.

Figure 4.14	E. A. Mankinen and C. M. Wentworth, Figure 2 from "Preliminary Paleomagnetic Results from the Coyote Creek Outdoor Classroom Drill Hole, Santa Clara Valley, California," USGS, (2003). http://pubs.usgs.gov/of/2003/of03-187/.
Figure 4.15	Shipboard Scientific Party, Figure F6 in T. Kanazawa, W. W. Sager, C. Escutia et al., pp. 1–49 from *Proceedings of the Ocean Drilling Program, Initial Reports* 191 (2001). doi:10.2973/odp.proc.ir.191.102.2001. Courtesy of the Integrated Ocean Drilling Program.
Figure 4.16	Shipboard Scientific Party, Figure F19 in " Site 1208," pp. 1–93 from T. J. Bralower, I. Premoli Silva, M., et al., *Proceedings of the Ocean Drilling Program, Initial Reports* 198 (2002). doi:10.2973/odp.proc.ir.198.104.2002. Courtesy of the Integrated Ocean Drilling Program.
Figure 4.17 (left)	W. Lowrie, "A revised magnetic polarity timescale for the Cretaceous and Cainozoic," pp. 129–136 from *Philosophical Transactions of the Royal Society of London* 306, (1982). Reproduced with permission.
Figure 4.17 (right)	S. C. Cande and D. V. Kent, "Revised calibration of the geomagnetic polarity timescale for the Late Cretaceous and Cenozoic," pp. 6093–6095 from *Journal of Geophysical Research* 100:B4 (1995). Reproduced with permission from the American Geophysical Union.
Figure 4.18	Shipboard Scientific Party, 2002, Explanatory Notes, Figure 4. In Proceedings ODP, Initial Reports of the Ocean Drilling Program, vol. 198, Bralower, T.J., et al., College Station, TX, Ocean Drilling Program, pp. 1–63. doi:10.2973/odp.proc.ir.198.102.2002. http://www-odp.tamu.edu/publications/198_IR/198TOC.HTM
Table 4.1	A. E. Maxwell, R. P. Von Herzon, J. E. Andrews, et al., Table 5a in Chapter 13 from *Initial Reports of the Deep Sea Drilling Project* 3 (Washington: U.S. Government Printing Office, 1970). http://www.deepseadrilling.org/03/dsdp_toc.htm. Courtesy of the Integrated Ocean Drilling Program.
Figure 5.1a	http://www.abc.net.au/science/articles/2007/04/27/1907876.htm. Reproduced by permission of Reuters/Bernado de Riz.
Figure 5.1b	http://www.alaska-in-pictures.com/mapco-oil-refinery-3329-pictures.htm. Reproduced by permission of AccentAlaska.com.
Figure 5.2	http://maps.grida.no/go/graphic/the-carbon-cycle1. Reproduced by permission of UNEP/GRID-Arendal.
Figure 5.3	P. Forster, V. Ramaswamy, P. Artaxo, et al., Figure 2 in "Changes in Atmospheric Constituents and in Radiative Forcing," from S. Solomon, D. Qin, M. Manning, et al., (eds.), *Climate Change 2007: The Physical Science Basis. Contribution of Working Group I to the Fourth Assessment Report of the Intergovernmental Panel on Climate Change* (Cambridge University Press, Cambridge, 2007). Reproduced with permission.
Figure 5.4	http://www.esrl.noaa.gov/gmd/ccgg/trends/

Figure 5.5	F. Joos and R. Spahni, Figure 1a in "Rates of change in natural and anthropogenic radiative forcing over the past 20,000 years," pp. 1425–1430 from *Proceedings of the National Academy of Sciences* 105:5 (2008). Copyright 2008, National Academy of Sciences, U.S.A.
Figure 5.6	R. A. Berner, "A new look at the long-term carbon cycle," pp. 2–6 from *GSA Today* 9 (1999). Reproduced by permission of the Geological Society of America. Originally drawn by Steve Petsch, Univ. of Massachuttes-Amherst.
Figure 5.7	Map courtesy of NOAA (http://www.ngdc.noaa.gov/mgg/ocean_age/data/2008/image/age_oceanic_lith.jpg).
Figure 5.8 & 5.9	D. L. Royer, R. A. Berner, I. P. Montanz, et al., "CO2 as a primary driver of Phanerozoic climate," pp. 4–10 from *GSA Today* 14 (2004). Reproduced by permission of the Geological Society of America.
Figure 5.10	http://www.ipcc-data.org/ddc_co2.html. Reproduced with permission from Intergovernmental Panel on Climate Change (IPCC) website.
Figure 5.11	http://www.epa.gov/climatechange/science/pastcc_fig1.html
Figure 6.1a, b & c	Courtesy of Megan Jones.
Figure 6.2 & 6.8	J. C. Zachos, G. R. Dickens and R. E. Zeebe, Figure 2 in "An early Cenozoic perspective on greenhouse warming and carbon-cycle dynamics," pp. 279–283 from *Nature* 451 (2008).
Figure 6.3	http://earthobservatory.nasa.gov/Features/Paleoclimatology_OxygenBalance/
Figure 6.4	Courtesy of Megan Jones.
Figure 6.5	Courtesy of Mark Leckie.
Figure 6.6 (diagram)	Drawn by Megan Jones.
Figure 6.6 (foram photomicro-graphs)	Courtesy of Mark Leckie.
Figure 6.7	Courtesy of Megan Jones.
Figure 7.1	http://www.marum.de/Alexius_Wuelbers.html
Figure 7.2	http://www.ipcc.ch/publications_and_data/ar4/syr/en/figure-spm-5.html
Figure 7.3	http://www.ipcc.ch/publications_and_data/ar4/syr/en/figure-spm-6.html
Figure 7.4	http://geology.com/world/arctic-ocean-map.shtml doi:10.2204/iodp.sp.302.2004. Courtesy of the Integrated Ocean Drilling Program.
Figure 7.5	M. Jakobsson, T. Flodén, and the Expedition 302 Scientists, Figure 14 in "Expedition 302 geophysics: integrating past data with new results," from J. Backman,, K. Moran, D. B. McInroy, et al., from *Proceedings of the Integrated Ocean Drilling Program* 302 (Edinburgh: Integrated Ocean Drilling Program Management International, Inc., 2006). doi:10.2204/iodp.proc.302.102.2006. Courtesy of the Integrated Ocean Drilling Program.

Figure 7.6	J. Zachos, M. Pagani, L. Sloan, et al., Figure 2 in "Trends, Rhythms, and Aberrations in Global Climate Change 65 Ma to Present," pp. 686–693 from *Science* 292 (2001). Reproduced with permission from AAAS.
Figure 7.7	A. Sluijs, S. Schouten, M. Pagani, et al., Figure 1 in "Subtropical Arctic Ocean temperatures during the Palaeocene/Eocene thermal maximum," pp. 610–613 from *Nature* 441 (2006).
Figure 7.8	J. Backman, K. Moran, D. B. McInroy, et al., Figure F19 from *Proceedings of the Integrated Ocean Drilling Program* 302 (2006). doi.10.2204/iodp.proc.302.104.2006. Courtesy of the Integrated Ocean Drilling Program.
Figure 8.1	http://www.nasa.gov/centers/goddard/news/topstory/2008/solar_variability.html
Figure 8.2	Courtesy of Kristen St. John.
Figure 8.3	Shipboard Scientific Party, Figure F9 in " Site 1208," from T. J. Bralower, I. Premoli Silva, M., et al., pp. 1–93 from *Proceedings of the Ocean Drilling Program, Initial Reports* 198 (2002). doi:10.2973/odp.proc.ir.198.104.2002. Courtesy of the Integrated Ocean Drilling Program.
Figure 8.4	F. Sangiorgi, E. E. van Soelen, D. J. A. Spofforth, et al., Figure 3 in "Cyclicity in the middle Eocene central Arctic Ocean sediment record: Orbital forcing and environmental response," p. PA1S08 from *Paleoceanography* 23 (2008). doi:10.1029/2007PA001487. Reproduced with permission from the American Geophysical Union.
Figure 8.5	H. A. Abels, H. A. Aziz, J. P. Calvo, et al., Figure 2 in "Shallow lacustrine carbonate microfacies document orbitally paced lake-level history in the Miocene Teruel Basin (North-East Spain)," pp. 399–419 from *Sedimentology* 56 (2009) doi: 10.1111/j.1365-3091.2008.00976.
Figure 8.6	H. A. Abels, H. A. Aziz, J. P. Calvo, et al., Figure 4 in "Shallow lacustrine carbonate microfacies document orbitally paced lake-level history in the Miocene Teruel Basin (North-East Spain)," pp. 399–419 from *Sedimentology* 56 (2009) doi: 10.1111/j.1365-3091.2008.00976.
Figure 8.7a	L. E. Lisiecki and M. E. Raymo, Figure 1 in "A Pliocene-Pleistocene stack of 57 globally distributed benthic δ18O records," p. PA1003 from *Paleoceanography* 20 (2005). doi:10.1029/2004PA001071. Reproduced with permission from the American Geophysical Union.
Figure 8.7b	L. E. Lisiecki and M. E. Raymo, Figure 2 in "A Pliocene-Pleistocene stack of 57 globally distributed benthic δ18O records," p. PA1003 from *Paleoceanography* 20 (2005). doi:10.1029/2004PA001071. Reproduced with permission from the American Geophysical Union.
Figure 8.8	L. E. Lisiecki and M. E. Raymo, Figure 4 in "A Pliocene-Pleistocene stack of 57 globally distributed benthic δ18O records," p. PA1003 from *Paleoceanography* 20 (2005). doi:10.1029/2004PA001071. Reproduced with permission from the American Geophysical Union.

Figure 8.9	J. R. Petit, J. Jouzel, D. Raynaud, et al., Figure 3 in "Climate and Atmospheric history of the past 420,000 years from the Vostok Ice core, Antarctica," pp. 429–36 from *Nature* 399 (1999). Courtesy of the Nature Publishing Group.
Figure 8.10	G. Kukla, F. Heller, L. X. Ming, et al., Figure 7 in "Pleistocene climates in china dated by magnetic susceptibility," pp. 811–814 from *Geology* 16 (1998). Courtesy of the Geological Society of America.
Figure 8.11	E. M. Pokras and A. C. Mix, Figure 1 in "Earth's Precession Cycle and Quaternary Climate Change in Tropical Africa," pp. 486–487 from *Nature* 326 (1987). Courtesy of the Nature Publishing Group.
Figure 8.12	Y. Sun, F. Wu, S. Clemens, et al., Figure 2 in "Processes controlling the geochemical composition of the South China Sea sediments during the last climatic cycle," p. 240–246 from *Chemical Geology* 257 (2008). Courtesy of Elsevier.
Figure 8.13, 8.14 & 8.15	http://en.wikipedia.org/wiki/Milankovitch_cycles (edited by wikipedia user Mysid). Original came from *On the Shoulders of Giants*, http://earthobservatory.nasa.gov/Features/Milankovitch/milankovitch_2.php. Image by Robert Simmon, NASA GSFC.
Figure 8.16	http://en.wikipedia.org/wiki/File:Carbon_Dioxide_400kyr.png. This figure was originally prepared by Robert A. Rohde from publicly available data and is incorporated into the Global Warming Art project.
Figure 9.1	Reproduced by permission of Jim Zachos.
Figure 9.2	J. Zachos, M. Pagani, L. Sloan, et al., Figure 2 in "Trends, Rhythms, and Aberrations in Global Climate Change 65 Ma to Present," pp. 686–693 from *Science* 292 (2001). Reproduced with permission from AAAS.
Figure 9.3	Kennett and Stott, 1991, Figure 1 in "Abrupt deep sea warming, paleoceanographic changes and benthic extinctions at the end of the Palaeocene," pp. 319–322 from *Nature* 353. Courtesy of the Nature Publishing Group.
Figure 9.4	Ocean Drilling Program Site Maps. http://www-odp.tamu.edu/sitemap/sitemap.html. Courtesy of the Integrated Ocean Drilling Program.
Figure 9.5	J. Zachos, M. Pagani, L. Sloan, et al., Figure 5 in "Trends, Rhythms, and Aberrations in Global Climate Change 65 Ma to Present," pp. 686–693 from *Science* 292 (2001). Reproduced with permission from AAAS.
Figure 9.6	T. J. Bralower, I. Premoli Silva, and M. J. Malone (eds.), pp. 1–47 from "Leg 198 synthesis: a remarkable 120-m.y. record of climate and oceanography from Shatsky Rise, northwest Pacific Ocean," *Proceedings of the Ocean Drilling Program, Initial Reports* 198 (2006). doi:10.2973/odp.proc.sr.198.101.2006 http://www-odp.tamu.edu/publications/198_SR/synth/synth.htm. Courtesy of the Integrated Ocean Drilling Program.

Figure 9.7	Shipboard Scientific Party, Figure F53 in T. J. Bralower, I. Premoli Silva, M. J. Malone, et al., pp. 1–148 from *Proceedings of the Ocean Drilling Program, Initial Reports* 198 (2002). doi:10.2973/odp.proc.ir.198.101.2002 http://www-odp.tamu.edu/publications/198_IR/chap_01/chap_01.htm Courtesy of the Integrated Ocean Drilling Program.
Figure 9.8	J. C. Zachos, M. W. Wara, S. Bohaty, et al., Figure 1 in "A transient rise in tropical sea surface temperature during the Paleocene-Eocene Thermal Maxium," pp. pp. 1551–1554 from *Science* 302 (2003). Reproduced with permission from AAAS.
Figure 9.9	K. Kaiho, K. Takeda, M. R. Petrizzo, et al., Figure 1 in "Anomalous shifts in tropical Pacific planktonic and benthic foraminiferal test size during the Paleocene-Eocene thermal maximum," pp. 456–464 from *Palaeogeography, Palaeoclimatology, Palaeoecology* 237 (2006). Courtesy of Elsevier.
Figure 9.10	Shipboard Scientific Party, Figure F1 in "Leg 199 summary," pp. 1–87 from M. Lyle, P. A. Wilson, T. R. Janecek, et al., *Proceedings of the Ocean Drilling Program, Initial Reports* 199 (2002). doi:10.2973/odp.proc.ir.199.101.2002 http://www-odp.tamu.edu/publications/199_IR/chap_01/chap_01.htm. Courtesy of the Integrated Ocean Drilling Program.
Figure 9.11	Shipboard Scientific Party, Figure F30 in "Leg 199 summary," pp. 1–87 from M. Lyle, P. A. Wilson, T. R. Janecek, et al., *Proceedings of the Ocean Drilling Program, Initial Reports* 199 (2002). doi:10.2973/odp.proc.ir.199.101.2002 http://www-odp.tamu.edu/publications/199_IR/chap_01/chap_01.htm. Courtesy of the Integrated Ocean Drilling Program.
Figure 9.12	Shipboard Scientific Party, Figures F8 and F24 in "Site 1220," pp. 1–93 from M. Lyle, P. A. Wilson, T. R. Janecek, et al., *Proceedings of the Ocean Drilling Program, Initial Reports* 199 (2002). doi:10.2973/odp.proc.ir.199.113.2002 http://www-odp.tamu.edu/publications/199_IR/chap_13/chap_13.htm. Courtesy of the Integrated Ocean Drilling Program.
Figure 9.13	D. J. Thomas, J. C. Zachos, T. J. Bralower, et al., Figures 1–3 in "Warming the fuel for the fire: Evidence for the thermal dissolution of methane hydrate during the Paleocene-Eocene thermal maximum," pp. 1067–1070 from *Geology* 30(12) (2002). Reproduced by permission of the Geological Society of America.
Figure 9.14	Shipboard Scientific Party, Figures F1 and F3 in "Leg 208 summary," pp. 1–112 from J. C. Zachos, D. Kroon, P. Blum, et al., *Proceedings of the Ocean Drilling Program, Initial Reports* 208 (2004). doi:10.2973/odp.proc.ir.208.101.2004 http://www-odp.tamu.edu/publications/208_IR/chap_01/chap_01.htm. Courtesy of the Integrated Ocean Drilling Program.
Figure 9.15	J. C. Zachos, U. Röhl, S. A. Schellenberg, et al., Figure 1 in "Rapid acidification of the ocean during the Paleocene-Eocene Thermal Maximum," pp. 1611–1615 from *Science* 308 (2005). Reproduced with permission from AAAS.

Figure 9.16 J. C. Zachos, U. Röhl, S. A. Schellenberg, et al., Figure 2 in "Rapid acidification of the ocean during the Paleocene-Eocene Thermal Maximum," pp. 1611–1615 from *Science* 308 (2005). Reproduced with permission from AAAS.

Figure 9.17 J. C. Zachos, U. Röhl, S. A. Schellenberg, et al., Figure 3 in "Rapid acidification of the ocean during the Paleocene-Eocene Thermal Maximum," pp. 1611–1615 from *Science* 308 (2005). Reproduced with permission from AAAS.

Figure 9.18 A. Sluijs, H. Brinkhuis, S. Schouten, et al., Figure 1 in "Environmental precursors to rapid light carbon injection at the Palaeocene/Eocene boundary," pp. 1218–1221 from *Nature* 450 (2007). Courtesy of the Nature Publishing Group.

Figure 9.19a A. Sluijs, H. Brinkhuis, S. Schouten, et al., Figure 2 in "Environmental precursors to rapid light carbon injection at the Palaeocene/Eocene boundary," pp. 1218–1221 from *Nature* 450 (2007). Courtesy of the Nature Publishing Group.

Figure 9.19b. A. Sluijs, H. Brinkhuis, S. Schouten, et al., Figure S1 in Supplemental Records of "Environmental precursors to rapid light carbon injection at the Palaeocene/Eocene boundary," pp. 1218–1221 from *Nature* 450 (2007). Courtesy of the Nature Publishing Group.

Figure 9.20 A. Sluijs, S. Schouten, M. Pagani, et al., Figure 1 in "Subtropical Arctic Ocean temperatures during the Palaeocene/Eocene thermal maximum," pp. 610–613 from *Nature* 441 (2006). Courtesy of the Nature Publishing Group.

Figure 9.21 A. Sluijs, S. Schouten, M. Pagani, et al., Figure 2 in "Subtropical Arctic Ocean temperatures during the Palaeocene/Eocene thermal maximum," pp. 610–613 from *Nature* 441 (2006). Courtesy of the Nature Publishing Group.

Figure 9.22 M. Storey, R. A. Duncan, and C. C. Swisher III, Figure 1 in "Paleocene-Eocene Thermal Maximum and the opening of the northeast Atlantic," pp. 587–589 from *Science* 316 (2007). Reproduced with permission from AAAS.

Figure 9.23 M. Storey, R. A. Duncan, and C. C. Swisher III, Figure 2 in "Paleocene-Eocene Thermal Maximum and the opening of the northeast Atlantic," pp. 587–589 from *Science* 316 (2007). Reproduced with permission from AAAS.

Figure 9.24 S. L. Wing, G. J. Harrington, F. A. Smith, et al., Figure 1 in "Transient floral change and rapid global warming at the Paleocene-Eocene boundary," pp. 993–996 from *Science* 310 (2005). Reproduced with permission from AAAS.

Figure 9.25 S. L. Wing, G. J. Harrington, F. A. Smith, et al., Figure 2 in "Transient floral change and rapid global warming at the Paleocene-Eocene boundary," pp. 993–996 from *Science* 310 (2005). Reproduced with permission from AAAS.

Figure 9.26 E. R. Clechenko, D. C. Kelly, G. J. Harrington, et al., Figure 2 in "Terrestrial records of a regional weathering profile at the Paleocene-Eocene boundary in the Williston Basin of North Dakota," pp. 428–442 from *Geological Society of America Bulletin* 119:3/4 (2007). Reproduced by permission of the Geological Society of America.

Figure 9.27 E. R. Clechenko, D. C. Kelly, G. J. Harrington, et al., Figures 7 and 8 in "Terrestrial records of a regional weathering profile at the Paleocene-Eocene boundary in the Williston Basin of North Dakota," pp. 428–442 from *Geological Society of America Bulletin* 119:3/4 (2007). Reproduced by permission of the Geological Society of America.

Figure 9.28 U. Röhl, T. J. Bralower, R. D. Norris, et al., Figure 1 in "A new chronology for the late Paleocene thermal maximum and its environmental implications," pp. 927–930 from *Geology* 28:10 (2000). Reproduced by permission of the Geological Society of America.

Figure 9.29 http://iodp.tamu.edu/database/. Courtesy of the Integrated Ocean Drilling Program.

Figure 9.30 U. Röhl, T. J. Bralower, R. D. Norris, et al., Figure 2 in "A new chronology for the late Paleocene thermal maximum and its environmental implications," pp. 927–930 from *Geology* 28:10 (2000). Reproduced by permission of the Geological Society of America.

Figure 9.31 U. Röhl, T. J. Bralower, R. D. Norris, et al., Figure 3 in "A new chronology for the late Paleocene thermal maximum and its environmental implications," pp. 927–930 from *Geology* 28:10 (2000). Reproduced by permission of the Geological Society of America.

Figure 9.32 U. Röhl, T. Westerhold, T. J. Bralower, et al., Figure 2 in "On the duration of the Paleocene-Eocene thermal maximum (PETM)," from *Geochemistry, Geophysics, Geosystems* 8:12 (2007). doi:10.1029/2007GC001784. Reproduced with permission from the American Geophysical Union.

Figure 9.33 http://www.esrl.noaa.gov/gmd/ccgg/trends/#mlo_full. Courtesy of NOAA Research.

Figure 9.34 http://en.wikipedia.org/wiki/Image:Instrumental_Temperature_Record.png. This figure was originally prepared by Robert A. Rohde from publicly available data and is incorporated into the Global Warming Art project.

Figure 9.35 http://en.wikipedia.org/wiki/File:Carbon_Dioxide_400kyr.png. This figure was originally prepared by Robert A. Rohde from publicly available data and is incorporated into the Global Warming Art project.

Figure 9.36 IPCC, "Summary for Policymakers,"from S. Solomon, D. Qin, M. Manning, et al. (eds.) *Climate Change 2007: The Physical Science Basis. Contribution of Working Group I to the Fourth Assessment Report of the Intergovernmental Panel on Climate Change* (Cambridge University Press, 2007).

Originally from M. E. Mann, R. S. Bradley, and M. K. Hughes, "Global-scale temperature patterns and climate forcing over the past six centuries," pp. 779–787 from *Nature* 392 (1998) and M. E. Mann, R. S. Bradley, and M. K. Hughes, "Northern hemisphere temperatures during the past millennium: Inferences, uncertainties, and limitations," pp. 759–762 from *Geophysical Research Letters* 26:6 (1999).

Table 9.1	W. F. Ruddiman, from *Earth's Climate Past and Future* (W.H. Freeman and Co., New York, 2008). Courtesy of the Nature Publishing Group.
Figure 10.1	http://earthobservatory.nasa.gov/IOTD/view.php?id=36839
Figure 10.2	J. C. Zachos, G. R. Dickens, R. E. Zeebe, et al., 2008. Figure 2 in "An early Cenozoic perspective on greenhouse warming and carbon-cycle dynamics," pp. 279–283 from *Nature* 451. Courtesy of the Nature Publishing Group.
Figure 10.3	J. C. Zachos, J. R. Breza, and S. W. Wise, Figure 1 in "Early Oligocene ice-sheet expansion on Antarctica: Stable isotope and sedimentological evidence from Kerguelen Plateau, southern Indian Ocean," pp. 569–573 from *Geology* 20 (1992). Reproduced by permission of the Geological Society of America.
Figure 10.4	J. C. Zachos, J. R. Breza, and S. W. Wise, Figure 2 in "Early Oligocene ice-sheet expansion on Antarctica: Stable isotope and sedimentological evidence from Kerguelen Plateau, southern Indian Ocean," pp. 569–573 from *Geology* 20 (1992). Reproduced by permission of the Geological Society of America.
Figure 10.5	Ocean Drilling Program Site Maps. http://www-odp.tamu.edu/sitemap/sitemap.html. Courtesy of the Integrated Ocean Drilling Program.
Figure 10.6	H. K. Coxall, P. A. Wilson, H. Pälike, et al., 2005, Figure 1 in "Rapid stepwise onset of Antarctic glaciation and deeper calcite compensation in the Pacific Ocean," pp. 53–57 from *Nature* 433 (2005). Courtesy of the Nature Publishing Group.
Figure 10.7	Shipboard Scientific Party, Figure F26 in "Leg 199 summary," pp. 1–87 from M. Lyle, P. A. Wilson, T. R. Janecek, et al., *Proceedings of the Ocean Drilling Program, Initial Reports* 199 (2002). doi:10.2973/odp.proc.ir.199.101.2002 http://www-odp.tamu.edu/publications/199_IR/chap_01/chap_01.htm. Courtesy of the Integrated Ocean Drilling Program.
Figure 10.8	Shipboard Scientific Party, Figure F50 in "Leg 208 summary," pp. 1–112 from J. C. Zachos, D. Kroon, P. Blum, et al., *Proceedings of the Ocean Drilling Program, Initial Reports* 208 (2004). doi:10.2973/odp.proc.ir.208.101.2004 http://www-odp.tamu.edu/publications/208_IR/chap_01/chap_01.htm. Courtesy of the Integrated Ocean Drilling Program.
Figure 10.9	Shipboard Scientific Party, Figure F7 in "Leg 189 summary," pp. 1–98 from N. F. Exon, J. P. Kennett, M. J. Malone, et al., *Proceedings of the Ocean Drilling Program, Initial Reports* 189 (2001). doi:10.2973/odp.proc.ir.189.101.2001 http://www-odp.tamu.edu/publications/189_IR/chap_01/chap_01.htm . Courtesy of the Integrated Ocean Drilling Program.
Figure 10.10	Shipboard Scientific Party, Figure F18 in "Leg 189 summary," pp. 1–98 from N. F. Exon, J. P. Kennett, M. J. Malone, et al., *Proceedings of the Ocean Drilling Program, Initial Reports* 189 (2001). doi:10.2973/odp.proc.ir.189.101.2001 http://www-odp.tamu.edu/publications/189_IR/chap_01/chap_01.htm. Courtesy of the Integrated Ocean Drilling Program.

Figure 10.11 Shipboard Scientific Party, Figure 1 in "Chapter 12, Preliminary results of subantarctic South Atlantic Leg 114 of the Ocean Drilling Program," pp. 797–804 from P. F. Ciesielski, Y. Kristoffersen, et al., *Proceedings of the Ocean Drilling Program, Initial Reports* 114 (1988). doi:10.2973/odp.proc.ir.114.112.1988 http://www-odp.tamu.edu/publications/114_IR/114TOC.HTM. Courtesy of the Integrated Ocean Drilling Program.

Figure 10.12 Shipboard Scientific Party, Figure 15 in "Chapter 6, Site 699," pp. 151–254 from P. F. Ciesielski, Y. Kristoffersen, et al., *Proceedings of the Ocean Drilling Program, Initial Reports* 114 (1988). doi:10.2973/odp.proc.ir.114.106.1988 http://www-odp.tamu.edu/publications/114_IR/114TOC.HTM. Courtesy of the Integrated Ocean Drilling Program.

Figure 10.13 Shipboard Scientific Party, Figure 2 in "Chapter 12, Site 699," pp. 151–254 from P. F. Ciesielski, Y. Kristoffersen, et al., *Proceedings of the Ocean Drilling Program, Initial Reports* 114 (1988). doi:10.2973/odp.proc.ir.114.106.1988 http://www-odp.tamu.edu/publications/114_IR/114TOC.HTM. Courtesy of the Integrated Ocean Drilling Program.

Figure 10.14 Shipboard Scientific Party, Figure 5 in "Chapter 5, Site 689," pp. 89–181 from P. E. Barker, J. P. Kennett, et al., *Proceedings of the Ocean Drilling Program, Initial Reports* 113 (1988). doi:10.2973/odp.proc.ir.113.106.1988 http://www-odp.tamu.edu/publications/113_IR/113TOC.HTM. Courtesy of the Integrated Ocean Drilling Program.

Figure 10.15 K. G. Miller, M. A. Kominz, J. V. Browning, et al., Figure 3 in "The Phanerozoic record of global sea-level change," pp. 1293–1298 from *Science* 310 (2005). Reproduced with permission from AAAS.

Figure 10.16 B. S. Wade and P. N. Pearson, Figure 5 in "Planktonic foraminiferal turnover, diversity fluctuations and geochemical signals across the Eocene/Oligocene boundary in Tanzania," pp. 244–255 from *Marine Micropalcontology* 68 (2008). Courtesy of Elsevier.

Figure 10.17 M. E. Raymo and W. F. Ruddiman, Figure 2a in "Tectonic forcing of late Cenozoic climate," pp. 117–122 from *Nature* 359 (1992). Courtesy of the Nature Publishing Group.

Figure 10.18 R. M. DeConto and D. Pollard, Figures 2 and 3 in "Rapid Cenozoic glaciation of Antarctica induced by declining atmospheric CO_2," pp. 245–249 from *Nature* 421 (2003). Courtesy of the Nature Publishing Group.

Figure 10.19 J. P. Kennett, Figures 4, 8, and 9 in "The development of planktonic biogeography in the Southern Ocean during the Cenozoic," pp. 301–345 from *Marine Micropaleontology* 3 (1978). Courtesy of Elsevier.

Figure 10.20 B. R. Jicha, D. W. Scholl, and D. K. Rea, Figure 4 in "Circum-Pacific arc flare-ups and global cooling near the Eocene-Oligocene boundary," pp. 303–306 from *Geology* 37 (2009). Courtesy of the Geological Society of America.

Figure 10.21 S. M. Savin, Figures 3b and c in "The history of the Earth's surface temperature during the past 100 million years," pp. 319–355 from *Annual Reviews of Earth and Planetary Science* 5 (1977).

Figure 10.22	Courtesy of Mark Leckie.
Figure 10.23	http://en.wikipedia.org/wiki/File:Thermohaline_Circulation_2.png. Originally found in http://earthobservatory.nasa.gov/Features/Paleoclimatology_Evidence/paleoclimatology_evidence_2.php
Figure 11.1	http://antarcticsun.usap.gov/science/contenthandler.cfm?id=2092. Provided by the ANDRILL Program
Figure 11.2	J. C. Zachos, G. R. Dickens and R. E. Zeebe, Figure 2 in "An early Cenozoic perspective on greenhouse warming and carbon-cycle dynamics," pp. 279–283 from *Nature* 451 (2008).
Figure 11.3	*Drill Site Maps.* http://iodp.tamu.edu/scienceops/maps.html. Courtesy of the Integrated Ocean Drilling Program.
Figure 11.4	http://www.classroom.antarctica.gov.au/6-climate/6-3-annual-ice-cycle. Reproduced with permission. Copyright © Australian Antarctic Division.
Figure 11.5	M. J. Hambrey, P.-N. Webb, D. M. Harwood, et al., "Neogene glacial record from the Sirius Group of the Shackleton Glacier region, central Transantarctic Mountains, Antarctica," pp. 994–1015 from *Geological Society of America Bulletin* 115:8 (August 1, 2003). Reproduced by permission of GSA.
Figure 11.6	http://www.unileipzig.de/~geologie/Forschung/ProjektSeiten/McMurdo/mcmurdo.htm. Reproduced by permission of Werner Ehrmann.
Figure 11.7	G. Warren, "Geology of the Terra Nova Bay-McMurdo Sound Area, Victoria Land," from *Antarctic Map Folio Series 12, Geology, Sheet 14* (American Geographical Society: New York, 1969). Reproduced by permission of the copyright holders: American Geographical Society, 1969.
Figure 11.8	D. M. Harwood, F. Florindo, F. Talarico, et al.,Figure 4 in "Background to the ANDRILL Southern McMurdo Sound Project, Antarctica," from *Studies from the ANDRILL, Southern McMurdo Sound Project, Antarctica, Initial Science Report on AND-2A* 15:1 (2008–9). Available at http://www.mna.it/english/Publications/TAP/volume14.html
Figure 11.9	D. M. Harwood, F. Florindo, F. Talarico, et al.,Figure 1 in "Background to the ANDRILL Southern McMurdo Sound Project, Antarctica," from *Studies from the ANDRILL, Southern McMurdo Sound Project, Antarctica, Initial Science Report on AND-2A* 15:1 (2008–9). Available at http://andrill.org/publications [Accessed 2008-07-15].
Figure 11.10	D. M. Harwood, L. Lacy, and R. H. Levy, (eds.), Figure 4.6-3 (p. 41) in "Future Antarctic Margin Drilling," from *Developing a Science Plan for McMurdo Sound* (ANDRILL Science Management Office (SMO), University of Nebraska–Lincoln, Lincoln, 2002). Reproduced by permission of Michael Hambrey.
Figure 11.11	D. M. Harwood, F. Florindo, F. Talarico, et al., Figure 4 in "Background to the ANDRILL Southern McMurdo Sound Project, Antarctica," from *Studies from the ANDRILL, Southern McMurdo Sound Project, Antarctica, Initial Science Report on AND-2A* 15:1 (2008–9). Available at http://andrill.org/publications [Accessed 2008-07-15].

Figure 12.1 http://www.andrill.org/iceberg/photos/theCore/index.html. Provided by the ANDRILL Program.

Figure 12.2 M. J. Hambrey, P. J. Barrett, and R. D. Powell, "Late Oligocene and early Miocene glacimarine sedimentation in the SW Ross Sea, Antarctica: the record from offshore drilling,"pp.105–108 from J. A. Dowdeswell and C.O. Cofaigh (eds.), *Glacier-Influenced Sedimentation on High Latitude Continental Margins*, Special Publications 203 (Geological Society, London, 2002). http://www.aber.ac.uk/en/iges/research-groups/centre-glaciology/research-intro/antarctica/. Reproduced with permission from The Geological Society.

Figure 12.3 and 12.5 M. J. Hambrey and B. C. McKelvey, Figure 2c in "Major Neogene fluctuations of the East Antarctic ice sheet: Stratigraphic evidence from the Lambert Glacier region," pp. 887–890 from *Geology* 28:10 (2000). Courtesy of the Geological Society of America.

Figure 12.4 http://www.coreref.org/projects/and1-1b/overview. Provided by the ANDRILL Program.

Figure 12.6 L. Krissek, G. Browne, L. Carter, et al., Figure 13 in "Sedimentology and Stratigraphy of the AND-1B Core, ANDRILL McMurdo Ice Shelf Project, Antarctica," pp. 185–222 from *Terra Antarctica* 14:3 (2007).

Figure 12.7 L. Krissek, G. Browne, L. Carter, et al., Figure 1 in "Sedimentology and Stratigraphy of the AND-1B Core, ANDRILL McMurdo Ice Shelf Project, Antarctica," pp. 185–222 from *Terra Antarctica* 14:3 (2007).

Figure 12.8 T. Naish, R. Powell, R. Levy, et. al., based on Figure 2 in "Obliquity-paced Pliocene West Antarctic ice sheet oscillations," p. 322–328 from *Nature* 458 (2009). doi: 10.1038/nature07867. Courtesy of the Nature Publishing Group.

Core Ligging Sheet http://www.csub.edu/geology/CSTA_core_log_sheet.jpg. Reproduced by permission of California State University, Bakersfield.

Figure 13.1 http://jan.ucc.nau.edu/~rcb7/nam.html. Reproduced by permission of Ron Blakey, Colorado Plateau Geosystems.

Figure 13.2 L. E. Lisiecki and M. E. Raymo, "A Pliocene-Pleistocene stack of 57 globally distributed benthic d18O records," from Paleoceanography, 20 (2005). doi:10.1029/2004PA001071. http://lorraine-lisiecki.com/stack.html. With permission of the American Geophysical Union.

Figure 13.3 http://geology.er.usgs.gov/eespteam/prism/index.html. Courtesy of United States Geological Survey.

Figure 13.4 H. J. Dowsett and M. M. Robinson, Figure 2 in "Mid-Pliocene equatorial Pacific sea surface temperature reconstruction: A multi-proxy perspective," pp. 109–125 from *Philosophical Transactions of the Royal Society A* 367 (2009). Reproduced with permission.

Figure 13.5 H. J. Dowsett and M. M. Robinson, Figure 4a in "Mid-Pliocene equatorial Pacific sea surface temperature reconstruction: A multi-proxy perspective," pp. 109–125 from *Philosophical Transactions of the Royal Society A* 367 (2009). Reproduced with permission.

Figure 13.6 H. J. Dowsett and M. M. Robinson, Figure 4b in "Mid-Pliocene equatorial Pacific sea surface temperature reconstruction: A multi-proxy perspective," pp. 109–125 from *Philosophical Transactions of the Royal Society A* 367 (2009). Reproduced with permission.

Figure 13.7 A. P. Ballantyne, D. R. Greenwood, J. S. Sinninghe, et al., Figure 3 in "Significantly warmer Arctic surface temperatures during the Pliocene indicated by multiple independent proxies," pp. 603–606 from *Geology* 38:7 (2010). Courtesy of the Geological Society of America.

Figure 13.8 http://earthobservatory.nasa.gov/IOTD/view.php?id=42392

Figure 13.9 M. Pagani, L. Zhonghui, J. LaRiviere, et al., Figure 2 in "High Earth-system climate sensitivity determined from Pliocene carbon dioxide concentrations," p. 27–30 from *Nature Geoscience* 3 (2009). doi:10.1038/ngeo724. Courtesy of the Nature Publishing Group.

Figure 13.10 L. E. Lisiecki and M. E. Raymo, adapted after Figure 4 in "A Pliocene-Pleistocene stack of 57 globally distributed benthic d18O records," from *Paleoceanography* 20 (2005). doi:10.1029/2004PA001071. http://lorraine-lisiecki.com/stack.html. With permission of the American Geeophysical Union.

Figure 13.11 http://geochange.er.usgs.gov/data/sea_level/ofr96000.html. Reproduced with the permission of Peter Schweitzer and Robert Thompson.

Figure 13.12 Figure 5 from http://www.gfdl.noaa.gov/climate-impact-of-quadrupling-co2. Courtesy of The National Geophysical Data Center.

Figure 13.13 http://sedac.ciesin.columbia.edu/gpw/. Reproduced by permission of Columbia University.

Figure 13.14 http://climatechange.rutgers.edu/images/nj_sl_rise_map.jpg. Reproduced by permission of Norbert Psuty. Graphs made by Serena Dameron from historical tide station data from http://tidesandcurrents.noaa.gov.

Table 13.1 http://pubs.usgs.gov/fs/fs2-00/. Courtesy of the U.S. Geological Survey.

Figure 14.1 Filename: IMG_5256.JPG. Album name: kleitzell / Sea Ice Knowledge Exchange.
http://nsidc.org/gallery/coppermine/displayimage.php?album=search&cat=0&pos=92.
Courtesy of The National Snow and Ice Data Center.

Figure 14.2 Courtesy of Kristen St John.

Figure 14.3 Courtesy of Kristen St John.

Figure 14.4 Ohio Division of Geological Survey, *Glacial map of Ohio* (Ohio Department of Natural Resources, Division of Geological Survey, 2005). http://www.dnr.ohio.gov/Portals/10/pdf/glacial.pdf. Courtesy of the Ohio Department of Natural Resources.

Figure 14.5 J. Elhers and P. L. Gibbard, Figure 6 in "The extent and chronology of Cenozoic global glaciations," p. 6–20 from *Quaternary International* 164-165 (2007). Courtesy of Elsevier.

Figure 14.6 http://iodp.tamu.edu/database/. Courtesy of the Integrated Ocean Drilling Program.

Figure 14.7 L. Polyak, R. B. Alley, J. T. Andrews, et al., 2010, Figure 6 in "History of Sea Ice in the Arctic, " from *Quaternary Science Review (QSR) Special Issue: Past Climate Variability and Change in the Arctic* (in press). doi:10.1016/j.quascirev.2010.02.010. Courtesy of Elsevier. Figure originally drawn by Matt O'Regan.

Figure 14.8 K. St. John, Figure 2 in "Cenozoic History of Ice-Rafting in the Central Arctic: Terrigenous Sands on the Lomonosov Ridge," pp. PA1S05 from *Paleoceanography*, 23 (2008). Reproduced by permission from the American Geophysical Union.

Figure 14.9 J. S. Eldrett, I. C. Harding, P. A. Wilson, et al., Figure 2 in "Continental ice in Greenland during the Eocene and Oligocene," pp. 176–179 from *Nature* 446 (2007). Courtesy of the Nature Publishing Group.

Figure 14.10 H. Flesche-Kleiven, E. Jansen, T. Fronval, et al., Figure 1 in "Intensification of Northern Hemisphere glaciations in the circum Atlantic region (3.5–2.4 Ma) – ice-rafted detritus evidence," p. 213–223 from *Palaeogeography, Palaeoclimatology, Palaeoecology* 184 (2002). Courtesy of Elsevier.

Figure 14.11 H. Flesche-Kleiven, E. Jansen, T. Fronval, et al., Figure 3 in "Intensification of Northern Hemisphere glaciations in the circum Atlantic region (3.5–2.4 Ma) – ice-rafted detritus evidence," p. 213–223 from *Palaeogeography, Palaeoclimatology, Palaeoecology* 184 (2002). Courtesy of Elsevier.

Figure 14.12 Shipboard Scientific Party, Figure 1 in "Introduction to Leg 145: North Pacific transect," in D. K. Rea, I. A. Basov, T. R. Janecek, et al., pp. 5–7 from *Proceedings of the Ocean Drilling Program, Initial Reports* 145 (1993). doi:10.2973/odp.proc.ir.145.103.1993. Courtesy of the Integrated Ocean Drilling Program.

Figure 14.13 L. A. Krissek, Figure 3 in "Late Cenozoic ice-rafting records from Leg 145 sites in the North Pacific: late Miocene onset, late Pliocene intensification, and Pliocene–Pleistocene events," in D. K. Rea, I. A. Basov, D. W. Scholl, (eds.), pp. 179–194 from *Proceedings of the Ocean Drilling Program, Scientific Results* 145 (1995). doi:10.2973/odp.proc.sr.145.118.1995. Courtesy of the Integrated Ocean Drilling Program.

Figure 14.14 L. E. Lisiecki and M. E. Raymo, Figure 4 in "A Pliocene-Pleistocene stack of 57 globally distributed benthic δ18O records," p. PA1003 from *Paleoceanography* 20 (2005). doi:10.1029/2004PA001071. Reproduced with permission from the American Geophysical Union.

Figure 14.15 S. J. Johnsen, D. Dahl-Jensen, N. Gundestrup, et al., 2001, Figure 1 in "Oxygen isotope and palaeotemperature records from six Greenland ice-core stations: Camp Century, Dye-3, GRIP, GISP2, Renland and NorthGRIP," pp. 299–307 from *Journal of Quaternary Science* 16 (2001).

Figure 14.16	S. J. Johnsen, D. Dahl-Jensen, N. Gundestrup, et al., 2001, Figure 3 in "Oxygen isotope and palaeotemperature records from six Greenland ice-core stations: Camp Century, Dye-3, GRIP, GISP2, Renland and NorthGRIP," pp. 299–307 from *Journal of Quaternary Science* 16 (2001).
Figure 14.17	http://www.geo.arizona.edu/palynology/geos462/07 nonmarin.html; Data compiled by O. Davis, University of Arizona.
Cover image	Scientific ocean drilling ship, the JOIDES Resolution (courtesy of John Beck, IODP), with inset photos (from left to right) of: (a) a stratified collection of four foraminifera (two planktics and two benthics): Sacculifer, a mixed layer species; Menardii, a thermocline species; Cibicidoides, an epifaunal benthic which is often used in stable isotope record; and Uvigerina – infaunal benthic (photomicrographs from Mark Leckie; (b) a core interval (133.07 to 133.88 mbsf) from ANDRILL 1-B showing glaciomarine sediments along the Antarctic margin; and (c) a core interval (Section 199-1220B-20X-2) from the central tropical Pacific of the Paleocene-Eocene boundary (from Lyle et al., 2002).
Geologic time line (inside front cover image)	J.D. Walker, and J.W. Geissman, compilers, 2009, Geologic Time Scale: Geological Society of America, doi: 10.1130/2009.CTS004R2C.
Inside back cover	The word cloud was produced by Kristen St. John using http://www.wordle.net/, and represents "in a picture" the main emphases and themes of our book. The bigger the word in the cloud, the more frequently that word was used in the Introduction and Table of Contents of our book.

Every effort has been made to trace copyright holders and to obtain their permission for the use of copyright material. The publisher apologizes for any errors or omissions in the above list and would be grateful if notified of any corrections that should be incorporated in future reprints or editions of this book.

Preface

A few years ago we first collaborated on a National Science Foundation (NSF) Course Curriculum and Laboratory Improvement (CCLI) grant (#0737335) with the goal of making the science of ocean drilling research accessible to educators. We set out to write seven data-rich exercises for the undergraduate classroom. However, we accomplished much more than we set out to do – we essentially developed an entire undergraduate course curriculum for paleoclimatology. We had written a book. This book would not have happened without the initial funding from NSF, as well as the support of the Deep Earth Academy (part of the Consortium of Ocean Leadership), the Integrated Ocean Drilling Program (IODP), and Antarctic Geological Drilling Program (ANDRILL). We are especially grateful to Leslie Peart, Education Director of Ocean Leadership's Deep Earth Academy, who is a leader in facilitating education and outreach for IODP. It was her vision of the "School of Rock" from which the seeds of this collaboration grew. We are thankful to Cathy Manduca, Director of the Science Education Resource Center (SERC) at Carleton College; her insight into the undergraduate curriculum, workshop development, and dissemination were probably more influential than she knows. We are grateful to Eric Pyle, Professor of Geology at James Madison University (JMU) and evaluator of our NSF grant; his input gave us much to consider about teaching and learning. This book could not have come together so smoothly without the skilled help of Serena Dameron to whom we give much thanks and admiration. Sarah Rangel was also helpful in assisting with book preparation. Jason Mallett of Ocean Leadership helped us by cleverly capturing the essence of our project in a graphic design: the Building Core Knowledge – Reconstructing Earth History logo, which appears at the beginning of each chapter in the book. We greatly appreciate the constructive comments from those who used and/or reviewed these exercise modules, including Stephen Schellenburg, Tom Cronin, Robert DeConto, Steve Petsch, Debbie Thomas, Jackie Hams, David Voorhees, Leilani Authors, Paul Holm, several anonymous reviewers, our workshop participants and, of course, our students. We would also like to thank the editors, and the development and production teams at Wiley-Blackwell, especially Ian Francis, Delia Sandford, Anna Bassett, Kevin Fung, Jessminder Kaur, Marilyn Grant, and Vivien Ward. What a fantastic, professional team to work with!

This material is based in large part upon scientific work of the Integrated Ocean Drilling Program (IODP). IODP is an international research program dedicated to advancing scientific understanding of the Earth through drilling, coring, and monitoring the subseafloor. The JOIDES Resolution is a scientific research vessel managed by the U.S. Implementing Organization (USIO) of IODP. Together, Texas A&M University, Lamont-Doherty Earth Observatory of Columbia University, and the Consortium for Ocean Leadership comprise the USIO. IODP is supported by two lead agencies: the U.S. National Science Foundation (NSF) and Japan's Ministry of Education, Culture, Sports, Science, and Technology. Additional program support comes from the European Consortium for Ocean Research Drilling (ECORD), the Australia-New Zealand IODP Consortium (ANZIC), India's Ministry of Earth Sciences, the People's Republic of China (Ministry of Science and Technology), and the Korea Institute of Geoscience and Mineral Resources.

This material is also based upon work supported by the National Science Foundation under Cooperative Agreement No. 0342484 to the ANDRILL Science Management Office at the University of Nebraska-Lincoln, which provides sub-awards to other US institutions as part of the ANDRILL US

Science Support Program. Any opinions, findings, and conclusions or recommendations expressed in this material are those of the author(s) and do not necessarily reflect the views of the National Science Foundation. The ANDRILL (ANtarctic geological DRILLing) Program is a multinational collaboration between the Antarctic Programs of Germany, Italy, New Zealand and the United States. Antarctica New Zealand is the project operator and has developed the drilling system in collaboration with Alex Pyne at Victoria University of Wellington and Webster Drilling and Enterprises Ltd. Scientific studies are jointly supported by the US National Science Foundation, NZ Foundation for Research Science and Technology, Royal Society of New Zealand Marsden Fund, the Italian Antarctic Research Programme, the German Research Foundation (DFG) and the Alfred Wegener Institute for Polar and Marine Research (Helmholtz Association of German Research Centres). Antarctica New Zealand supported the drilling team at Scott Base; Raytheon Polar Services supported the science team at McMurdo Station and the Crary Science and Engineering Laboratory. The ANDRILL Science Management Office at the University of Nebraska-Lincoln provided science planning and operational support.

Kristen St John, R Mark Leckie, Kate Pound, Megan Jones, Lawrence Krissek
January 2012

Book Introduction for Students and Instructors

Motivation and Purpose

There has never been a more critical time for students to understand how the Earth works (see the Geological Society of America position statement on *Teaching Earth Science*). Understanding the causes and potential consequences of Earth's changing climate are of particular importance because modern global warming is an issue that impacts economies, environments, and lifestyles. The context for understanding the global warming of today lies in the records of Earth's past. This is demonstrated by decades of paleoclimate research by scientists in organizations such as the Integrated Ocean Drilling Program, the Antarctic Geological Drilling Program, the Byrd Polar Ice Core Laboratory, and many others. The purpose of this book is to put key data and published case studies of past climate change at your fingertips, so that you can experience the nature of paleoclimate reconstruction. You will evaluate data, practice developing and testing hypotheses, and infer the broader implications of scientific results. It is our philosophy that addressing **how we know** is as important as addressing **what we know** about past climate.

Use of Real Data

All of the exercises in the book build upon **authentic data** from peer-reviewed scientific publications. One of the most effective means of conveying science data is through figures – graphs, photos, diagrams. Therefore, unlike many books where the figures supplement the text, **in this book the figures are the central elements of every exercise.** Questions and tasks within each exercise are constructed around the figures. This means that it is essential to examine carefully the figures and diligently read their captions. This may not be easy at first, but with practice (and this book will give you lots of it) it will become second nature.

Content Topics

The content topics chosen for this book support the **Essential Principles of Climate Science** (http://www.globalchange.gov/; **Table 1**). The relevance of the investigation of past climate change for understanding modern global warming and for predicting future climate change is evident throughout the book. Reconstructing past climate change relies on investigations of multiple archives, including tree rings, corals, speleothems, ice, sediment, and sedimentary rock. In this book we focus primarily on climate change during the Cenozoic Era (the last 65.5 million years of Earth history; see the timescale inside the front cover of the book). We therefore draw largely from sediment and ice core records, which are valuable archives of the past 200 million years and 500 thousand years, respectively. As you will see, obtaining detailed records of Earth's climate history from sediments and glacial ice is a challenging undertaking, often involving expeditions to remote locations, complex coring technology, careful planning and execution, and hard work. Once obtained, paleoclimate records must be systematically described, ages must be determined, and indirect evidence (i.e., **proxies**) of past climate must be analyzed. Much like the work of a detective, geoscientists and paleoclimatologists reconstruct **what happened** in the past, and **when** and **how it happened** based on the clues left behind by the events that took place.

The chapters in this book are organized to explore first the fundamental aspects of paleoclimate research and the "tools" used to conduct this research, including obtaining sediment and ice records (Chapter 1), describing them (Chapter 2), and determining their ages (Chapters 3 and 4). These initial chapters provide a foundation to explore further the nature of paleoclimate reconstruction, the important trends, events, and patterns of past climate change, and the particularly important paleoclimate proxies and models (Table 1). The next ten chapters of the book are case studies of past climate change. These chapters are organized chronologically. These case study chapters include: Phanerozoic CO_2 record (Chapter 5), the stable oxygen-isotope record of the Cenozoic (Chapter 6), scientific drilling in the Arctic (Chapter 7), climate cycles (Chapter 8), the Paleocene–Eocene thermal maximum (Chapter 9), the onset of Antarctic glaciation (Chapter 10), Antarctica and Neogene climate change (Chapter 11), interpreting Antarctic sediment cores (Chapter 12), Pliocene warmth (Chapter 13), and the onset of northern hemisphere glaciation (Chapter 14). A more detailed description of each chapter is provided in the expanded **Table of Contents**.

Transportable Skills

In addition to introducing and working with paleoclimate content, the exercises in this book are also designed to provide opportunities to develop and practice scientific and other life skills (Table 1). These include making observations, formulating hypotheses, practicing quantitative and problem-solving skills, making data-based interpretations, recognizing and dealing with uncertainty, working in groups, communicating (written and oral) with others, synthesizing data, and articulating evidence-based arguments. Many of these scientific skills are also valued by other disciplines, from business to the social sciences. Therefore whether your aspirations are to pursue a career in science or in another field, working with the data in this book in this inquiry-based way will help you develop valuable, transportable skill sets.

Practical Use of Multi-part Exercise Modules

Each chapter in this book is a multi-part exercise module. The first part within each module is typically designed to introduce a topic and gauge prior knowledge, and therefore to identify possible misconceptions. In-depth exploration of the topic follows in subsequent exercise parts, as does the synthesis of the important implications of the data. How you use these exercise modules in your course will depend on the focus of your course, time, prior knowledge (both of instructor and student), and class size. Therefore you may explore a chapter from beginning to end, or you may be extracting specific parts of exercises that support your curriculum and instructional goals, and course management decisions. Some exercise parts may be assigned as homework and others may serve as in-class activities that can be jumping-off points for lectures, or an entire exercise module may serve as a weekly lab activity. In all cases, the value of group discussion at different junctures within, and/or at the end of, an exercise cannot be underestimated. As undergraduate teachers, the authors of this book all practice a "do-talk-do" approach to teaching and learning, whereby we integrate both inquiry-based student learning and lecture in our classes. We encourage instructors using this book to do the same. For an instructor to adapt inquiry-based approaches successfully (as used in this book), it almost certainly is necessary to cover less material than would be covered in a semester (or quarter) of lecture-only classes. This is because inquiry takes more time than does lecture. However, the benefits of having students take active roles in the construction of their knowledge and the development of transportable skills is well worth this tradeoff. We recognize that instructors using this book are not necessarily experts themselves in paleoclimatology. Therefore, we have developed comprehensive instructor guides for each chapter to provide essential background information, detailed

TABLE 1 Summary of the scientific content and skills developed in the book, and the relationship to the essential principles of climate science.

Chapters →	1. Introduction to Paleoclimate Records	2. Seafloor Sediments	3. Microfossils and Biostratigraphy	4. Paleomagnetism and Magnetostratigraphy	5. CO2 as a Climate Regulator during the Phanerozoic and Today	6. The Benthic Foraminiferal Oxygen Isotope Record of Cenozoic Climate Change	7. Scientific Drilling in the Arctic Ocean: A Lesson on the Nature of Science	8. Climate Cycles	9. The Paleocene Eocene Thermal Maximum (PETM) Event	10. Glaciation of Antarctica: The Oi1 Event	11. Antarctica and Neogene Global Climate Change	12. Interpreting Antarctic Sediment Cores: A Record of Dynamic Neogene Climate	13. Pliocene Warmth: Are We Seeing Our Future?	14. Northern Hemisphere Glaciation
Scientific Content														
1. Science as Human Endeavor	X	X	X	X			X	X			X	X	X	X
2. Science as an Evolving Process / Nature of Science	X	X	X	X	X		X		X		X	X	X	X
3. Earth History Archives	X	X	X	X	X	X	X	X	X	X	X	X	X	X
4. Stratigraphic Principles	X		X	X		X	X	X	X	X	X	X	X	X
5. Fossils as age Indicators			X					X						
6. Paleomagnetism as an Age Indicator				X				X						
7. Stable Isotopes as Climate Proxies						X		X	X	X				
8. Climate Models					X		X			X			X	
9. Ocean Circulation and Sea Level Changes						X				X			X	
10. The Carbon Cycle and CO2 as a Climate Regulator					X			X	X					
11. Climate Changes (it is not static):					X	X	X	X	X	X	X	X		X
climate change can be cyclic					X	X		X			X	X		X
climate change can be gradual					X	X					X	X		
climate change can be abrupt									X	X	X	X		X
climate changes on human timescales					X			X	X				X	
regional to global scales of change					X	X	X	X	X	X	X	X	X	X

(*Continued*)

TABLE 1 *Continued*

Chapters →	1. Introduction to Paleoclimate Records	2. Seafloor Sediments	3. Microfossils and Biostratigraphy	4. Paleomagnetism and Magnetostratigraphy	5. CO2 as a Climate Regulator during the Phanerozoic and Today	6. The Benthic Foraminiferal Oxygen Isotope Record of Cenozoic Climate Change	7. Scientific Drilling in the Arctic Ocean: A Lesson on the Nature of Science	8. Climate Cycles	9. The Paleocene Eocene Thermal Maximum (PETM) Event	10. Glaciation of Antarctica: The Oi1 Event	11. Antarctica and Neogene Global Climate Change	12. Interpreting Antarctic Sediment Cores: A Record of Dynamic Neogene Climate	13. Pliocene Warmth: Are We Seeing Our Future?	14. Northern Hemisphere Glaciation
Scientific Skills														
1. Analysis of authentic data = thinking creatively, analytically, and critically thru inquiry & problem solving	X	X	X	X	X	X	X	X	X	X	X	X	X	X
2. Math Integration	X	X	X	X	X	X	X	X	X	X	X	X	X	
3. Collaboration (teamwork)	X	X	X	X	X	X	X	X	X	X			X	X
4. Communication	X	X	X	X	X	X	X	X	X	X	X	X	X	X
5. Uncertainty in Science	X	X	X	X	X	X	X	X	X	X	X	X	X	X
6. Information & Technological Literacy	X	X	X	X	X			X						
Essential Principles of Climate Science (http://www.globalchange.gov/)														
1. The Sun is the primary source of energy for Earth's climate system.					X									
2. Climate is regulated by complex interactions between components of the Earth system.						X		X	X	X	X	X	X	X
3. Life on Earth depends on, is shaped by, and affects climate.					X	X			X				X	
4. Climate varies over space and time through both natural and man-made processes.	X				X	X	X	X	X	X	X	X	X	X
5. Our understanding of the climate system is improved through observations, theoretical studies, and modeling.	X	X	X	X	X	X	X	X	X	X	X	X	X	X
6. Human activities are impacting the climate system.					X			X	X				X	
7. Climate change will have consequences for the Earth system and human lives.					X	X	X	X	X	X		X	X	X

answer keys and alternative implementation strategies, as well as to provide links to other supplementary materials and examples for assessment.

Audience

Because of the flexible design of these multi-part exercise modules, they can be (and have been) used at multiple levels and with multiple audiences. Collectively, the use of all the exercises in this book would strongly support an undergraduate course in paleoclimatology or global change. The select use of specific exercises, and exercise parts, can also support topics in many Earth science courses (e.g., historical geology, oceanography, stratigraphy, Earth science for teachers).

Classroom Tested

The flexible and effective use of these exercise modules with multiple audiences at multiple levels is demonstrated by our classroom testing; we have used these modules in our introductory geoscience courses for non-majors, K-12 teachers, as well as advanced courses for majors, and even in workshops for PhD students (e.g., Urbino Summer School on Paleoclimatology) and professional development workshops for undergraduate instructors. Our classroom experiences and the feedback from students and workshop participants informed the design and revisions of the exercises and the content included in the instructor guides.

Geologic Timescale

The geologic timescale (see inside front cover of book; Walker and Geissman, 2009a) is the common time reference for Earth's history. It is rich in information. Let's explore it a bit since you will be referring to it frequently as you work through the exercises in this book. There are several aspects of the geologic timescale that are important to understand. First is how it was constructed, second, is the nature of the units or divisions of geologic time, and third, is a subtle distinction between time units and rock units.

The early development of the geologic timescale from the 17th to 19th centuries resulted because many people examined rock outcrops in the areas where they lived and recognized an ordered arrangement of the different layers based on the included fossils and physical characteristics of the rocks. In most circumstances we can apply what was learned from these early observations and assume that: (1) older rocks are overlain by younger rocks (Principle of Superposition), (2) older rocks may be cut (or intruded) by young rocks or other geologic features like faults (Principle of Cross Cutting Relationships), and (3) layers of rock are laterally continuous over local to regional geographic areas (Principle of Lateral Continuity). Thus, the early geologic timescale was simply a representation of the **relative chronological order** (older to younger) represented by sequences of rock units, correlated across their region.

A method of accurately determining the **numerical ages** of rocks came much later (20th century) with the discovery of radioactive decay of unstable isotopes and its application in radiometric (i.e., isotopic) age dating. Modern geologic time, therefore, was developed by integrating radiometric age dating of key sequences of rock and sediment with the ability to correlate from one region to another using fossils (biostratigraphy, see Chapter 3) and the physical properties of rocks, the most important being paleomagnetic properties (magnetostratigraphy, see Chapter 4). One the geologic timescale the Cenozoic and Mesozoic history of **magnetic polarity** is indicated by the columns of alternating black and white stripes, which represent normal and reversed polarity, respectively. The corresponding magnetic anomalies and chrons are also included, which identify time units based on global paleomagnetic data.

The geologic timescale is a compilation of the most accurate geochronologic data to date. It is continually under review by professional geosciences organizations including the International Commission on Stratigraphy and the Timescale Advisory Committee of the Geological Society of America (e.g., Walker and Geissman, 2009b); these reviews ensure a high degree of quality control of the timescale. New datasets and improved age dating and correlation methods reduce timescale uncertainties and inform decisions about timescale revisions.

The primary divisions of geologic time are Eons, Eras, Periods, Epochs and Ages (listed across the top of the timescale), with Eons reflecting the longest time spans and Ages reflecting the shortest subdivisions of time. Notice that the time divisions on the geologic time scale do not represent equal increments of time. For example, the Precambrian (far right column) represents approximately 88% of Earth's history (4.6 billion years ago to 542 million years ago), whereas the first three columns represents 12% of Earth's history (the last 542 million years).

The boundaries between divisions of geologic time mark distinct shifts in fossil and/or geochemical records, which were typically caused by environmental and climatic changes in Earth's history. One of the most recent revisions of the geologic timescale was to shift the boundary between the Pliocene and Pleistocene Periods from 1.8 million years ago (Ma) to 2.6 Ma, based on the strongest evidence for the wide spread expansion of northern hemisphere glaciations (Gibbard and Cohen, 2008; explored in Chapter 14). Note that this timescale revision does not change the numerical age of any strata or geologic events, but reflects how new data (e.g, stable oxygen-isotope; explored in Chapter 6) and improved age dating methods constrain the timing of geologic events better.

It is important to comment on the relationship between time (**geochronologic**) units and rock (**chronostratigraphic**) units. For each geochronologic unit shown on the geologic timescale, there is a representative chronostratigraphic unit (a body of rock) located somewhere. In fact, the names of many of the time divisions reflect the location of a representative chronostratigraphic unit associated with that particular time span. In practice, when referring to geochronologic units we use the terms Early, Middle and Late (time). For example, the onset of Antarctic glaciation occurred during the Late Eocene and Early Oligocene Periods (see Chapters 10–12). In contrast, when referring to chronostratigraphic units, we use Lower, Middle, Upper (body of rock). For example, evidence for the onset of the Antarctic glaciation comes from studying Upper Eocene and Lower Oligocene rock outcrops and cores.

Finally, you will notice in this book (and elsewhere in geologic literature) that **dates** in the geologic past use a different abbreviation than **durations of time** in the geologic past. To indicate a date in the geologic past, the convention is to use the following abbreviations: ka (thousands of years *ago*), Ma (millions of years *ago*), and Ga (billions of years *ago*). For example, the K-P extinction occurred at 65.5 Ma. However, to describe the duration of time, the abbreviations ky (or kyr, thousands of years), my (or myr, millions of years), and gy (or gyr are used; also acceptable is by or byr, billions of years), are used. For example, the Cretaceous Period had a duration of 80 myr.

References

Gibbard, P. and Cohen, K.M., 2008, Global chronostratigraphical correlation table for the last 2.7 million years. Episodes, **31** (2), 243–47.

Walker, J.D. and Geissman, J.W. (compilers), 2009a, Geologic Time Scale. Geological Society of America. doi: 10.1130/2009.CTS004R2C.

Walker, J.D. and Geissman, J.W., 2009b, Commentary: 2009 GSA Geologic Time Scale. GSA Today, 60–1.

Chapter 1 Introduction to Paleoclimate Records

FIGURE 1.1. Cross sections (a view perpendicular to growth or accumulation) of (a) a tree (courtesy of Colin Bielby), (b) a cave deposit (speleothems; courtesy of John Haynes; inset photo from ANSTO.), (c) several tens of meters of glacial ice (courtesy of Lonnie Thompson), (d) a coral (Hough, 2010), and (e) a sedimentary sequence (Marwan et al., 2003).

Reconstructing Earth's Climate History: Inquiry-Based Exercises for Lab and Class,
First Edition. Kristen St John, R Mark Leckie, Kate Pound, Megan Jones and Lawrence Krissek.

NAME

SUMMARY

This exercise serves as an **introduction to paleoclimate records, with emphasis on sediment and ice cores.** In **Part 1.1,** you will compare and contrast the temporal and spatial scope of major paleoclimate archives: tree rings, speleothems, glacial ice, lake and marine sediments, and sedimentary rocks. In **Part 1.2** you will read about the 780,000 year old Owens Lake core record and create a summary figure to synthesize the paleoclimatic data and interpretations. In **Part 1.3** you will consider the challenges and strategies for obtaining cores from glacial ice and the sub-seafloor; in addition you will consider issues of sampling, reproducibility, resolution, and cost, which are common issues for all paleoclimate archive research.

Introduction to Paleoclimate Records

Part 1.1. Archives and Proxies

1 Think about how we know about past events in **human history** (e.g., the expansion of the Roman Empire, or the American Revolution). What types of records document these events?

2 Now think about **Earth's history**, specifically the past environmental or climatic conditions at times before recorded human history. What records might there be of such conditions? Make a list of your ideas.

3 Figure 1.1 shows an assemblage of five major types of **natural archive** of Earth's environmental and climatic history. What common feature(s) do each of these **paleoclimate** archives share?

NAME

The photographs in Figure 1.1 comprise an assemblage of five major types of natural records, or **archives**, of Earth's environmental and climatic history. Just like a diary or other historical document, the layers in these natural archives contain indirect evidence (i.e., **proxies**) about past conditions and events, recorded in a sequential order. The evidence is specific to a certain time period, and may be general or very detailed, depending on the rate that information was recorded. The faster the rate at which the recorder grew (trees and corals), accumulated (snow and ice), or was deposited (sedimentary sequences), the more detailed the record is and the higher its **resolution**. For example, a record in which an annual signal can observed has a very high resolution. In contrast, if the finest observable details are on the order of a million years, that record would have a low resolution.

The usefulness of a record can be affected by how regularly information was recorded. If information was recorded continuously the record would be more complete than if events were recorded only occasionally.

4 Consider again each of the archives in Figure 1.1. Mark an "X" on each figure to **indicate the oldest part** of the record. Explain your reasoning here:

5 Examine Figure 1.2. Place an "X" anywhere on the diagram where you would expect to find continuously recorded (i.e., relatively uninterrupted) paleoclimate information. Explain your answer.

Examine the photos of an outcrop, a **marine sediment core**, and an **ice core** (Figure 1.3). These could have originated from locations A, B_1, and B_3 respectively, in the block diagram (Figure 1.2).

6 How is the sedimentary outcrop similar to the sediment core? How are they different?

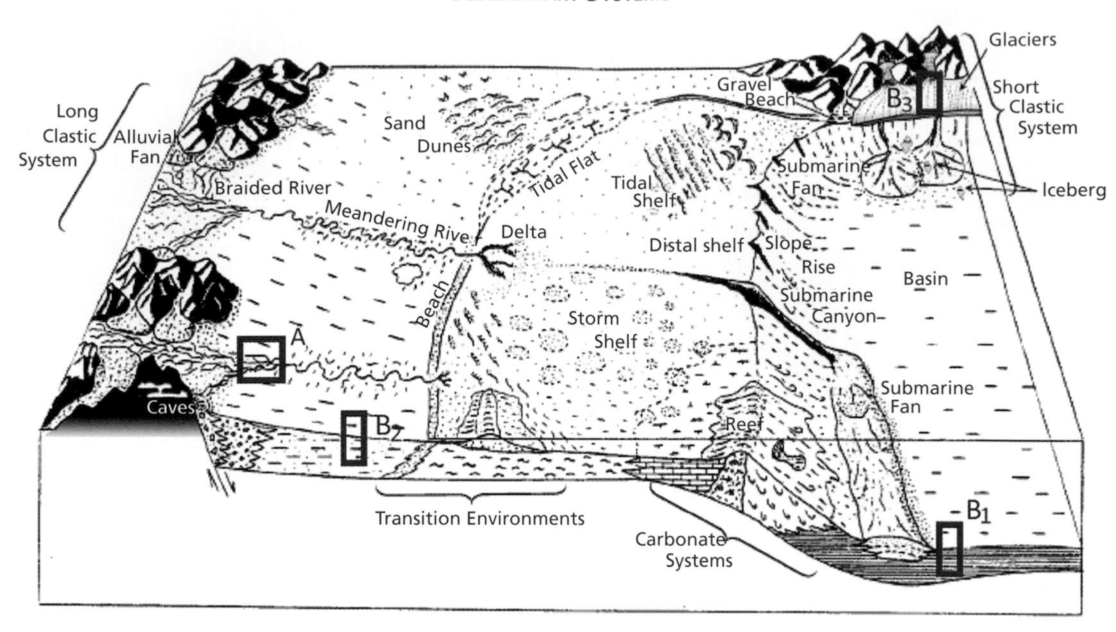

FIGURE 1.2. Major depositional systems. (Drawn by Lynn Fichter, James Madison University). A is an example outcrop setting, B_1 is an example of a core in a marine (ocean) setting, B_2 is an example of a core in a terrestrial (land) setting, B_3 is an example of an ice core in a terrestrial (land) setting.

Outcrop and Core Similarities	Outcrop and Core Differences

7 How are the two cores similar? How are they different?

Sediment and Ice Core Similarities	Sediment and Ice Core Differences

The quality of the record can also be affected by what has happened to the archive since the initial recording. For example, sediments can be lithified into sedimentary rock; rock and sediment, and the fossils they contain, can be weath-

FIGURE 1.3. (A) **Sedimentary outcrop** from Bastrop, TX (courtesy of John Firth). (B) Photo of an Antarctic ice core (Courtesy of WAIS Divide Ice Core Project). It is approximately 9 cm wide. (C) A 104-cm long section of a **marine core** recovered from drilling the Pacific Ocean sea floor. The core is approximately 7 cm wide; it was originally a cylinder that was cut in half lengthwise (Shipboard Scientific Party, 2002b).

ered and eroded; ice can melt; and organic matter can decay. All of these processes can degrade the information recorded and therefore increase uncertainty about the data and its interpretation. These processes also affect the **maximum time range** that an archive is useful in paleoclimate reconstructions.

The various types of proxy data must be described and interpreted by those who have learned how to "read" natural archives. This typically involves making observations and measurements using a wide range of analytical equipment (e.g., hand lens, microscopes, gas chromatographs, mass spectrometers). While we rely on the various proxy data to reconstruct paleoclimates, it is important to also recognize that each analytical method has its

NAME ______________________________

own limitations which can influence resolution, maximum time ranges, and scientific uncertainty.

Examine the data in **Table 1.1** which summarizes the typical (exceptions certainly exist!) spatial and temporal distribution of each of the paleoclimate archives in Figure 1.1. Also included in Table 1.1 are examples of the proxy data and climate parameters that can be reconstructed from these archives. **Supplemental information on each of these archives can be found on the website associated with this book.** In Table 1.1 notice that lake sediments, marine sediments, and sedimentary rocks, while they are all sedimentary sequences, are displayed in separate rows in this table because of their distinctly different timeframes and resolutions.

8 The oldest sediments in the modern ocean are approximately 200 **million** years old. What implication does this have for learning about ocean conditions prior to 200 million ago?

____

____

____

____

____

9 Which archive has the widest geographic distribution, and the greatest number of sites sampled?

____

10 Examine Figure 1.4, showing the distribution of human population vs. latitude. What latitudes are the most densely populated?

____

11 Propose likely physical and historical factors that underline the pattern in Figure 1.4.

____

____

____

____

____

NAME ____

TABLE 1.1. Selected major paleoclimate archives and their proxies*

Archive	Geographic Distribution; Approximate Number of Sites Sampled	Time Range of the Majority of the Records (years before present)	Resolution (years)	Types of Material Analyzed and/or Data Collected	Climate Parameter
Continental					
Tree Rings	Global, but most sites in the northern hemisphere; 1000s	<10,000	<1	Cellulose; isotopes; ring widths	Seasonal precipitation and temperature
Lake Sediments	Global, but most sites in the northern hemisphere; approximately 100s	<1,000,000	1–20	Microfossil (flora and fauna) types and abundances; stable isotopes; trace elements; alkenone biomarkers; sediment composition and geochemistry.	Regional changes in precipitation and evaporation; temperature; salinity; volcanic activity; glacial activity
Speleothems (cave deposits)	Global, but most sites in the northern hemisphere; >50	<500,000	1–2000	Stable isotopes; trace elements	Temperature; precipitation, atmospheric CO_2
Ice Sheets and Glaciers	Polar ice sheets, and high altitude (alpine) temperate and tropical glaciers; >80	<800,000	<1–1000	Stable isotopes; trace elements; dust and other particulates; atmospheric gas concentrations	Temperature; precipitation; wind/atmospheric circulation; atmospheric chemistry; volcanic activity; biomass burning
Sedimentary Rocks	Global; tens of thousands	<3,600,000,000	10,000–100,000	Microfossil (flora and fauna) types and abundances; sediment composition; stable isotopes; depositional patterns	Temperature; biological productivity; volcanic activity; glacial activity; sea level; ice volume; evolution
Oceanic					
Corals	Tropical and subtropical oceans, >60	<650,000	<1–1000	Stable isotopes; trace elements	Salinity; temperature; nutrients; sea level; ice volume
Marine Sediments	Global; >65,000	<200,000,000	100–1000	Microfossil (flora and fauna) types and abundances; stable isotopes; trace elements; alkenone biomarkers; sediment composition; depositional patterns	Ocean and atmospheric circulation; temperature; salinity; nutrients; aridity on land; biological productivity; volcanic activity; glacial activity; ice volume

*Adapted from Cronin (1999) with information from Ruddiman (2008), http://www.ncdc.noaa.gov/paleo/, and http://www.ngdc.noaa.gov/mgg/curator/curator.html.

NAME

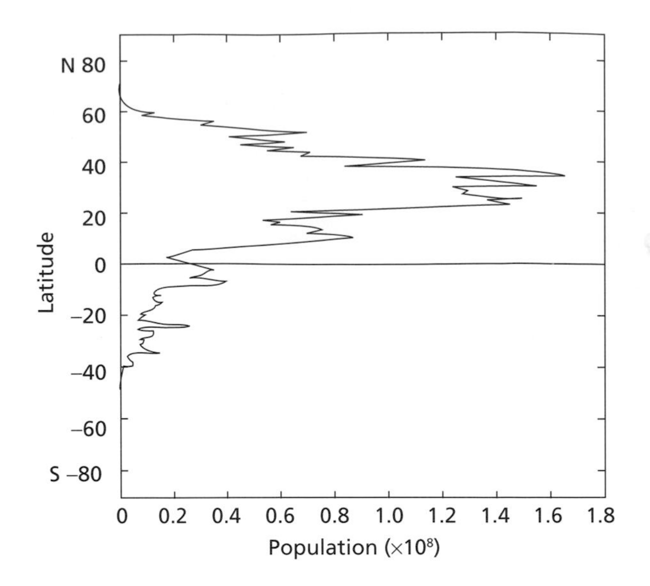

FIGURE 1.4. Latitudinal distribution of the 1990 human population. From Li, 1996.

12 How might the human population distribution (Figure 1.4) influence the geographic distribution of paleoclimate research sites? Use the summary data in Table 1.1 to support your hypothesis.

13 Imagine you have to reconstruct regional environmental and climatic conditions for the last **100,000** years in Europe as completely as possible. Which archive(s) would you choose to use and why?

NAME ______________________________

14 Imagine you have to reconstruct the global history of Earth's environmental and climatic conditions for the last **5 million** years as completely as possible. Which archive(s) would you choose to use and why?

____

____

____

____

15 Imagine you have to reconstruct the global history of environmental and climatic conditions of the Cretaceous period (**145–66 Ma ago**). What archive(s) would you choose to use and why?

____

____

____

____

NAME ___

Introduction to Paleoclimate Records

Part 1.2. Owens Lake – An Introductory Case Study of Paleoclimate Reconstruction

In Part 1.1 you were introduced to the range of archives and proxies of past climate change. In Part 1.2, the Owens Lake, CA sediment record will serve as an introductory paleoclimate case study; you will describe and interpret the clues that this archive and its proxies can offer us about climate in western North America over the last 800,000 years.

1 Use your library resources to **find and read** the following reference: Menking, K.M., 2000, A record of climate change from Owens Lake sediment, in Schneiderman, J.S. (ed.), *The Earth Around Us: Maintaining a Livable Planet*, New York: W.H. Freeman and Company, p. 322–335. **Use the work space at the end of Part 1.2** to (a) make a visual representation (a sketch) of the Owens Lake sediment **core**. Use different patterns or colors to represent the different layers described in the article, with the oldest at the bottom and the youngest at the top. (b) Adjacent to your sketch, list the **types of proxy data** (e.g., salt layer, pebbles, ash, microfossils, . . .) obtained from the different intervals of the core. (c) Next to the list of proxies, give an **interpretation** of the data with respect to past climatic and/or environmental conditions (e.g., dry, cold, volcanic eruption, . . .).

2 What methods were used to determine age within the core?

___

___

___

3 What evidence is there that part of the Owens Lake record is missing (i.e., that a hiatus exists)? What could cause this?

___

___

___

___

___

___

___

NAME ______________________________

4 What evidence would support the hypothesis that humans impacted the environment in and/or around Owens Lake?

5 Use Owens Lake to explain why a **multiproxy** approach is valuable in reconstructing past climatic and/or environmental change:

NAME ______________________________

Workspace for Question 1:

Sketch of Core	Type of Data Analyzed	Paleoclimatic/Environmental Interpretation

NAME

Introduction to Paleoclimate Records

Part 1.3. Coring Glacial Ice and Seafloor Sediments

Coring allows scientists to access a sequence of climatic and environmental conditions (i.e., a history) locked away in natural archives. Trees, corals, sediments, rock, and ice are all cored for paleoclimatic studies. **Cores** are narrow, cylindrical samples that extend through layer after layer of the material being investigated. Tree cores are thinnest, only approximately 0.5 cm diameter and <50 cm long. Coral, sediment, rock, and ice cores are often a few centimeters in diameter and of variable length (<50 cm to 10 m), depending on the coring technology used, the thickness of the material being cored, and the objectives of the study. Repeated drilling in the same hole can recover many successive meters of core from successively greater depths (Figure 1.5). The deepest cored hole in glacial ice (EPICA Dome C) is over 3.2 km deep and is located on the high polar plateau in East Antarctica. The deepest cored hole in the seafloor (Hole 504B) is over 2.1 km deep and is located in the eastern equatorial Pacific Ocean, on the Costa Rica Rift. This deep-sea drill site penetrated through the sedimentary layers and into the underlying igneous ocean crust.

FIGURE 1.5. Example of (left) an ice core from Huascaran, Peru (Courtesy of Lonnie Thompson), and (right) a sediment core from the western equatorial Pacific Ocean (Courtesy of Mark Leckie).

NAME ______________________________

In this exercise you will investigate how cores are obtained from arguably the most technologically challenging paleoclimate archives: glacial ice and seafloor sediments. We will also use these archives to explore issues common to all types of paleoclimate research: the need for teamwork, methods that keep samples uncontaminated and organized, reproducible results, and funding.

Coring Glacial Ice

1 Where does glacial ice (i.e., glaciers, ice sheets) exist today?

____

____

____

2 Watch the two on-line videos and read the short news article to gain an appreciation of the ice core drilling process. Then make a list of the challenges of obtaining ice cores for paleoclimate research. How do scientists and engineers overcome these challenges?

- 2-minute video on southern Alps ice core drilling: http://www.youtube.com/watch?v=T69_diWYbkQ
- 5-minute video on Antarctic ice core drilling: http://www.youtube.com/watch?v=kdfcNIFEnF8
- two-page news article (Stone, 2010) on the New Guinea ice core drilling: http://www.sciencemag.org/cgi/content/short/328/5982/1084 (Note that students will need to use their library resources to access this article.)

Challenges	Solutions

Bubbles of ancient air (Figure 1.6) found in glacial ice are unique and valuable indicators of past climate. Unlike most other climate indicators, which indirectly record climate parameters, the trapped air in glacial ice is a **direct** measure of atmospheric gases (e.g., CO_2 and CH_4) of the past. As snow recrystallizes into ice below the surface of a glacier, air is trapped in the pore spaces between ice crystals. The pore spaces are eventually closed off from the atmosphere by continued accumulation of new snow and by the recrystallization and fusing of individual ice crystals from layers of snow to firn (compacted snow) to ice. Because the pore spaces are open to the atmosphere until the ice forms, the age of the gases in the pore spaces is **younger** than the surrounding ice. Trapped gas comprises 10–15% of the volume of glacial ice at the "bubble close off depth" (firn-ice transition) (Bender et al., 1997).

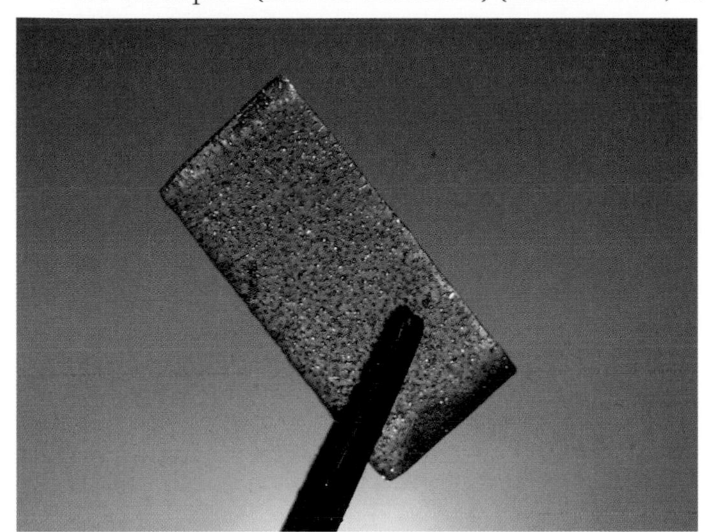

FIGURE 1.6. Piece of an Antarctic ice showing trapped air bubbles (Courtesy of Edward Brooke, Oregon State University).

FIGURE 1.7. Class 100 clean room at Bryd Polar Research Center. Class 100 means there are less than 100 particles (diameter >0.5 μm) per cubic foot of air. Photo courtesy of Ellen Mosley-Thompson and the Byrd Polar Research Center.

Examine Figure 1.7, which shows a lab where ice cores samples are analyzed, and read the following brief description of the sample preparation and gas analysis process:

> *Ice samples were cut with a band saw in a cold room (at about −15 °C) as close as possible to the center of the core in order to avoid surface contamination. Gas extraction and measurements were performed by crushing the ice sample (approximately 40 g) under vacuum in a stainless steel container without melting it, expanding the gas released during the crushing in a pre-evacuated sampling loop, and analyzing the CO_2 concentrations by gas chromatography. The analytical system, except*

NAME ______________________

for the stainless steel container in which the ice was crushed, was calibrated for each ice sample measurement with a standard mixture of CO_2 *in nitrogen and oxygen. Text is modified from: http://cdiac.ornl.gov/trends/co2/vostok.html*

3 Identify some specific conditions and methods from the above description and propose why they are necessary to produce robust gas concentration data from the ice core.

4 There are hundreds of analytical labs around the world that are capable of measuring CO_2 and CH_4 gas concentrations from ice core samples. How could investigators ensure that the results from different labs are comparable?

5 In 1992, the European Greenland Ice Core Project (GRIP) drilled down 3029 m to the base of the Greenland ice sheet at Summit, Greenland (72°N, 38°W). A year later the US Greenland Ice Sheet Project 2 (GISP2) completed drilling of a companion record through the ice sheet 30 km to the west. What value might there be in obtaining two parallel ice core records so close together?

NAME

6 The upper 2788 m of the GRIP ice core contain a Greenland paleoclimate record of the last 110,000 years. The European Project for Ice Coring in Antarctica (EPICA) recovered the deepest and oldest ice record to date at Dome C, Antarctica. This 3270.2 m-long ice core contains a paleoclimate record of the last 740,000 years.

(a) Calculate the average ice accumulation rates (cm/yr) for the GRIP and EPICA ice cores. Show your work, including the conversion from meters to centimeters.

(b) Which has a higher average ice accumulation rate: GRIP or EPICA?

(c) Which has a higher average resolution, GRIP or EPICA?

NAME ______________________________

(d) Which record spans a greater period of Earth history, GRIP or EPICA?

____

Coring Ocean Sediments

The *JOIDES Resolution* (Figure 1.8) is designed to obtain core from below the seafloor. Use the online resources at http://joidesresolution.org to learn about this ship. In particular, take a virtual tour of the ship by selecting **Meet the JR** and then **Research Vessel Tour**. Learn about the drilling process by selecting **Multimedia** and watching the three-minute video "**An Explanation of Deep Sea Coring**".

7 How are vessels used for scientific ocean drilling specially outfitted to enable them to recover cores from below the seafloor?

____

____

____

____

____

____

FIGURE 1.8. Scientific research vessel, *JOIDES Resolution*. Courtesy of Bill Crawford, IODP Expedition 321.

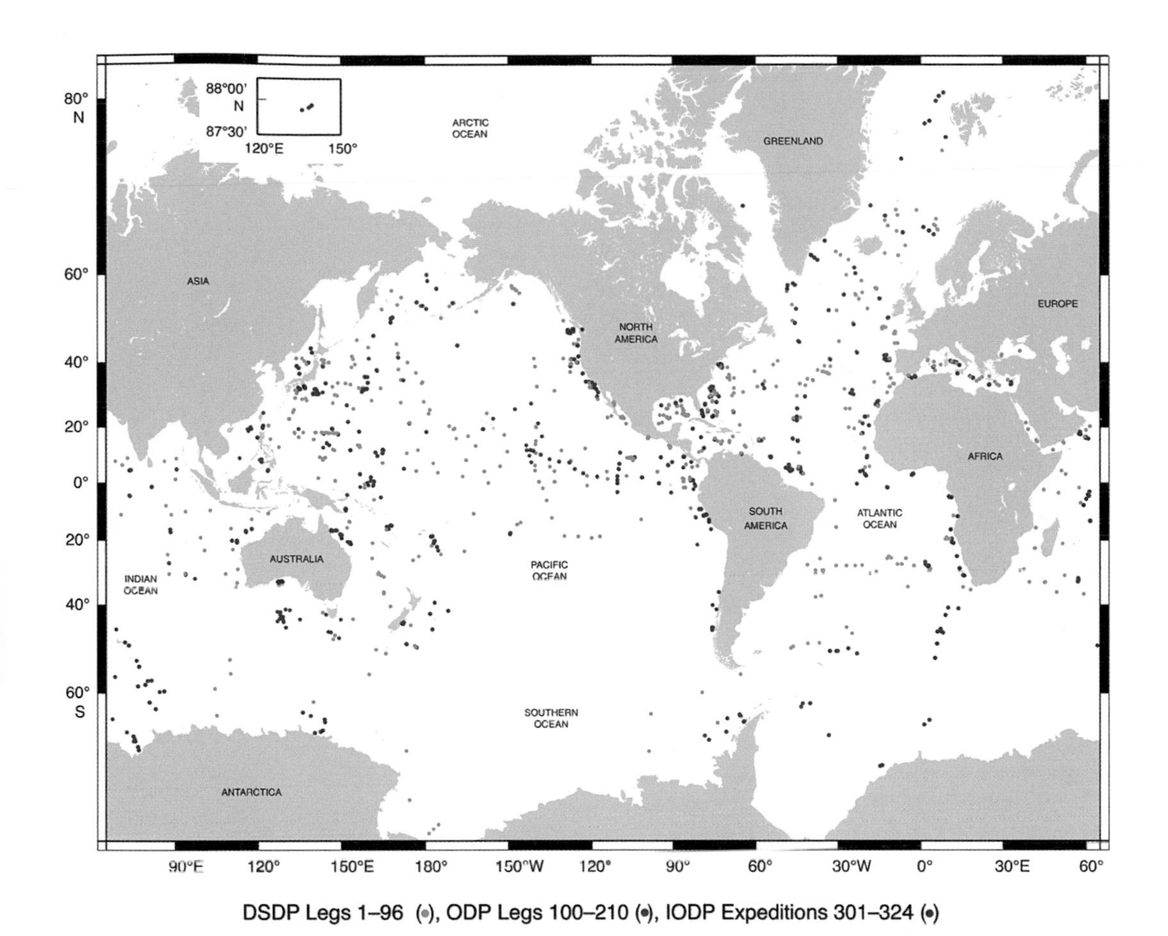

FIGURE 1.9. Scientific ocean drilling site locations of the Integrated Ocean Drilling Program (IODP; 2003–2013) and predecessor programs, the Ocean Drilling Program (ODP; 1983–2003) and the Deep Sea Drilling Project (DSDP; 1968–1983). Map courtesy of IODP.

The map above (Figure 1.9) shows all of the drill **site** locations from the over 40 year history of scientific ocean drilling **expeditions.** At **each** of these drill site locations several **holes** may have been drilled and tens to hundreds of 9.5-m length **cores** recovered. To carry and store the cores more easily, each is cut into 1.5-meter **sections** (Figures 1.5 & 1.10). The core sections are also split lengthwise into a two halves – a **working half**, which is used in sampling, and an **archive half**, which is used for non-destructive analyses and for core photography (Figure 1.11). In total there are currently >40,000 m of core recovered from below the seafloor and >2.3 million **samples** taken of the core sections in specific centimeter intervals. To keep so many samples organized for scientific research it is important that each sample has a unique and meaningful identification code.

8 The challenge of assigning unique identification codes to scientific samples is analogous to the challenge that libraries have in organizing and categorizing books. The US Library of Congress is the largest library in the world, with

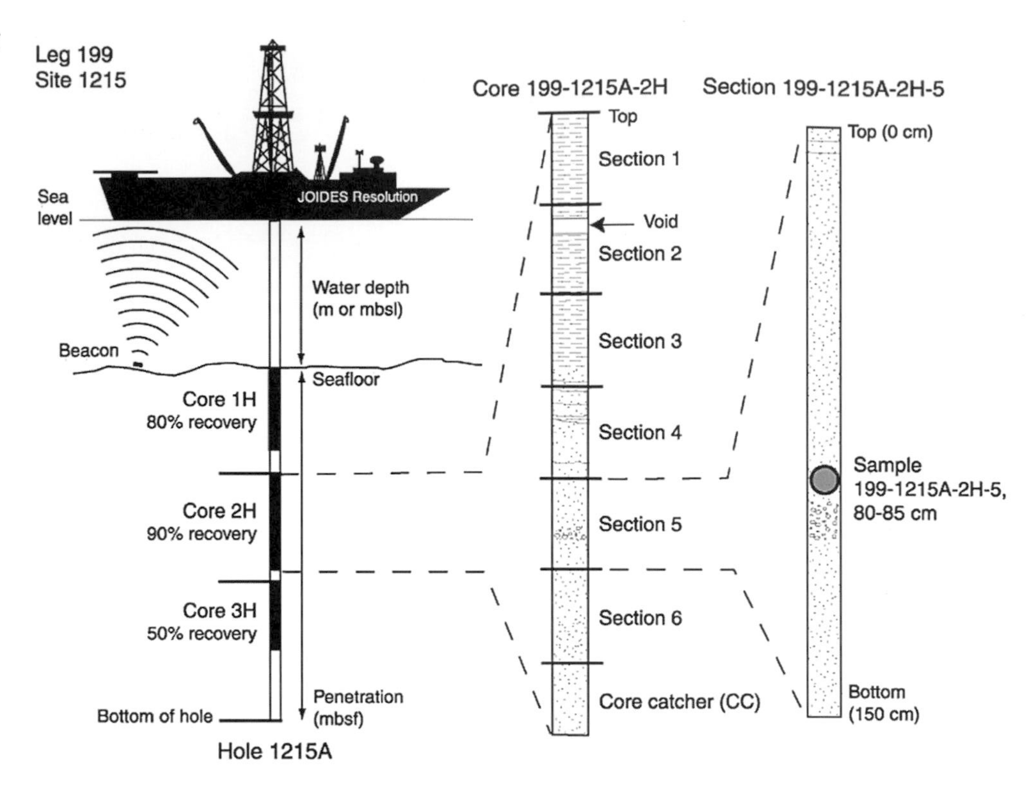

FIGURE 1.10. Example of coring and core terminology. Shipboard Scientific Party, 2002a.

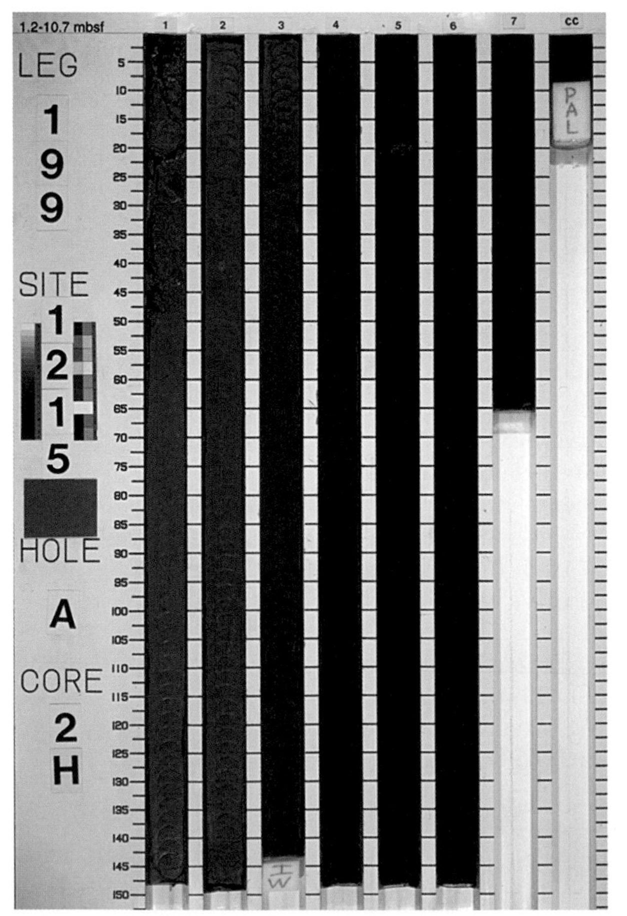

FIGURE 1.11. Photo of the archive-half of Core 2 from Hole 1215A, located in the central tropical Pacific Ocean. The sections of core are laid out next to each other, left to right. Section 1 of Core 2 is at the top of the drilled interval and the core catcher (CC) is at the bottom of the cored interval. The shipboard paleontologists took a sample (PAL) from the base of the core catcher to provide a preliminary age determination for Core 2. Site 1215 was cored during Ocean Drilling Program Leg (i.e., Expedition) 199. Note that an interstitial water sample (IW) was taken from the bottom of Section 3. Courtesy of IODP.

NAME ____

millions of books, recordings, photographs, maps, and manuscripts in its collections. Look up your favorite book in the US Library of Congress website (http://catalog.loc.gov/). What unique identification code distinguishes your book from all of the other books in the library?

____

9 Think about marine core samples. How could you ensure that each sample had a **unique** identification so that you knew exactly where in the sub-seafloor it came from? List your ideas.

____

____

____

____

____

10 The standard labeling for ocean drilling samples is shown in Figure 1.10.

(a) Deconstruct the sample identification "199-1215A-2H-5, 80-85" by filling in the blanks below:
Leg (or Expedition)____ Site_____ Hole____ Core____ Section____ Centimeter interval______.

(b) Place an X on Figure 1.11 marking the location of this sample.

Science requires both qualitative skills and quantitative skills. In questions 11–15 you will quantify some of the costs (with respect to time and money) involved in obtaining sediment cores from below the seafloor.

11 Expedition (Leg) 199 began in Honolulu, Hawaii on October 28, 2001. The *JOIDES Resolution* left port at 0830 hr on 28 October and transited 1158 km to the first drilling location, Site 1215, arriving at 2100 hr on 30 October, 2001. What was the average rate of travel (i.e., speed) during transit in km/hr? Convert this value to miles/hr. Show your work, including conversion.

____

____

____

____

____

NAME

12 Site 1215 was planned for 57.5 hr of drilling during Expedition 199. During this time the crew were able to drill 75.4 m below the seafloor at a single location (Hole 1215A). What was the average drilling rate (m/hr) for Hole 1215A? Show your work.

13 While the crew drilled 75.4 m below the seafloor at Hole 1215A, they only recovered 68.27 m of core. What was the percent core recovery for Hole 1215A? Propose a hypothesis to explain why core recovery would be less than the maximum drill depth.

14 A typical ocean drilling expedition lasts around 60 days and costs approximately US $6 million. At Hole 1215A on ODP Leg 199, 68.74–m of core were obtained during 57.5 hours drilling. What is the cost of 1–m of core from Hole 1215A? Show your work.

NAME

15 Sometimes scientific ocean drilling and NASA space exploration are compared because these are both large-scale, technologically dependent programs that are designed to help teams of scientists unravel the history of the Earth and our solar system by exploring in remote and challenging settings. Compare the cost of obtaining core from the seafloor to the cost of obtaining rocks from the moon in the following.

(a) The Apollo 11 mission cost US$355 million in 1969. Approximately 21.8 kg of moon rock were obtained on this successful and historic mission to the moon. What was the cost of 1 kg of moon rock in 1969? Show your work.

(b) The cost of a seafloor core that you calculated in question 14 was the cost per meter of core. To make a comparison to the cost of the moon rock we need to determine the cost per kg of core (so we will need to convert units). Use your skills in geometry to work this out, the average density of cores from Hole 1215A is approximately 1.3 g/cm^3. The core is a cylinder with a radius of 3.5 cm and a length (or height) of 1 m (100 cm). The volume of a cylinder is equal to $\pi r^2 h$. Recall from question 14 that a scientific ocean drilling expedi-

NAME ____

tion is typically 60 days and costs approximately US$6 million. What is the cost of 1 kg of core? How does this compare to the cost of 1 kg of moon rock?

16 In this book we will emphasize how and what we know about climate change in the Cenozoic Era (the most recent 65.5 million years of Earth's history). Consequently, the primary archive we will draw from is the ocean sediment record; it will be supplemented by data from other archives, especially the ice core record, depending on the particular timeframe and/or climatic event introduced. Go to http://recordings.wun.ac.uk/conf/nwo/oceandrilling2006 and select the 5-minute video titled "Why Ocean Drilling?" Based on this video and what you have learned in this exercise, list the benefits and limitations of utilizing the ocean sediment record to reconstruct Cenozoic climate change.

NAME

References

Alley, R.B., et al., 1995, Comparison of deep ice cores. Nature, **373**, 393–4.

Bender, M., et al., 1997, Gases in ice cores. Proceedings of the National Academy of Sciences, USA, 94, 8343–9.

Cronin, T., 1999. Principles of Paleoclimatology, Columbia University Press, 592 pp.

Li, Y.F., 1996, Global Population Distribution - 1990, Terrestrial Area, and Country Name Information on a One-by-One Degree Grid-Cell Basis. Compiled by A.L. Brenkert, CDIAC, DB-1016, (1996), http://cdiac.ornl.gov/newsletr/spring97/datas97.htm

Lough, J.M. 2010, Climate records from corals. Wiley Interdisciplinary Reviews: Climate Change.1: 318–331. doi: 10.1002/wcc.39

Lowel, J.M., 2010, Climate records from corals. Climate Change, **1** (3), 318–31, doi: 10.1002/wcc.39, http://wires.wiley.com/WileyCDA/WiresArticle/wisId-WCC39.html

Marwan, N., Trauth, M.H., Vuille, M., et al., 2003. Comparing modern and Pleistocene ENSO-like influences in NW Argentina using nonlinear time series analysis methods. *Climate Dynamics*, 21:3–4, p. 317–326. doi:10.1007/s00382-003-0335-3.

Menking, K.M., 2000, A record of climate change from Owens Lake sediment, The Earth Around Us: Maintaining a Livable Planet. Schneiderman, J.S. (ed.), W.H. Freeman and Company, New York, 322–35.

National Geophysical Data Center: http://www.ngdc.noaa.gov/mgg/curator/curator.html

NOAA Paleoclimatology database: http://www.ncdc.noaa.gov/paleo/

Ruddiman, W.F., 2008, Earth's Climate Past and Future, 2nd edition, Freeman, 388 pp.

Shipboard Scientific Party, 2002a. Explanatory notes. In Lyle, M., Wilson, P.A., Janecek, T.R., et al., Proc. ODP, Init. Repts., 199: College Station, TX (Ocean Drilling Program), 1–70. doi:10.2973/odp.proc.ir.199.102.2002

Shipboard Scientific Party, 2002b. Site 1220. In Lyle, M., Wilson, P.A., Janecek, T.R., et al., Proc. ODP, Init. Repts., 199: College Station, TX (Ocean Drilling Program), 1–93. doi:10.2973/odp.proc.ir.199.113.2002

Stone, R., 2010, Arduous expedition to sample last virgin tropical glaciers. Science, **328** (May 28) 1084–5.

Wilde, S.A., et al., 2001, Evidence from detrital zircons for the existence of continental crust and oceans on the Earth 4.4 Gyr ago. Nature, **409**, 175–8.

Chapter 2 Seafloor Sediments

FIGURE 2.1. Kelsie Dadd (Sedimentologist, Macquarie University, Australia) and Mea Cook (Sedimentologist, Williams College, USA) describe the sediment color of a core section from Site U1340 in the Bering Sea. Photo taken by Carlos Alvarez Zarikian, IODP.

Reconstructing Earth's Climate History: Inquiry-Based Exercises for Lab and Class, First Edition. Kristen St John, R Mark Leckie, Kate Pound, Megan Jones and Lawrence Krissek.

NAME ____________________

SUMMARY
This exercise set explores marine sediments using core photos and authentic datasets in an inquiry-based approach (Figure 2.1). Your prior knowledge of seafloor sediments is explored in **Part 2.1**. In **Part 2.2** you will **observe** and **describe the physical characteristics** of sediment cores. In **Part 2.3** you will **use composition** and **texture data** from smear slide samples taken from the cores to **determine** the **lithologic names** of the marine sediments. In **Part 2.4** you will **make a map** showing the distribution of the primary sediment lithologies of the Pacific and North Atlantic Oceans and **develop hypotheses** to explain the distribution of the lithologies shown on your map.

Seafloor Sediments

Part 2.1. Sediment Predictions

1 What kinds of materials do you expect to find on the seafloor?

____

____

____

____

2 Do you expect any geographic pattern for these materials in the global ocean? Explain your reasoning.

____

____

____

____

____

____

____

NAME

Seafloor Sediments

Part 2.2. Core Observations and Descriptions

Introduction

In this exercise your teacher will assign you one or more cores from **Table 2.1** below. **A corresponding photo of each core is included in an Appendix at the end of Chapter 2.** Note that all of the cores in Table 2.1 are either core number 1, 2, or 3. This means that these cores are **at** or **close to the top** of the sediment sequence on the seafloor. Therefore the sediment in these cores represents **modern** or very recent environmental conditions at that location in the ocean. For a review of the nomenclature for core identification see Chapter 1, Part 1.3.

TABLE 2.1. Seafloor cores

Core Identification: Expedition–Site & Hole–Core & Type*	Physiographic Site Location	Site Location (Latitude/ Longitude)	Water Depth (m)	Reference
Pacific Cores				
112-687A-2H	Peru continental shelf	12.9S/77.0W	316	Seuss et al., 1988
35-324-1	SE Pacific basin, North of Antarctica	69S/98.8W	4433	Hollister et al., 1976
28-269-1	Ross Sea, South of Australia, margin of Antarctica	61.7S/140.1E	4282	Hayes et al., 1975
145-886B-2H	Chinook Trough, North Pacific abyssal plain	44.7N/168.2W	5743	Rea et al., 1993
145-882A-2H	Detroit Seamount NW Pacific	50.36N/167.6E	3243.8	Rea et al., 1993
145-881A-1	NW Pacific, east of the Sea of Okhotsk	47.1N/161.5E	5531.1	Rea et al., 1993
145-887C-2H	Patton-Murray Seamount, NE Pacific	54.4N/148.5W	3633.6	Rea et al., 1993
19-188-2	Bering Sea	53.8N/178.7E	2649	Creager et al., 1973
18-182-1	Alaskan continental slope	57.9N/148.7W	1419	Klum et al., 1973
33-318-2	Line Islands Ridge, south central Pacific	14.8S/146.9W	2641	Schlanger et al., 1976
8-75-1	Marquesas Fracture Zone, central Pacific abyssal plain	12.5S/135.3W	4181	Tracey et al., 1971
92-597-1	SE Pacific abyssal plain	18.8S/129.8W	4166	Leinen et al., 1986
178-1101A-2H	Antarctic Peninsula continental rise	64.4S/70.3W	3279.7	Barker et al., 1991
178-1096A-1H	Antarctic Peninsula continental rise	67.57S/77.0W	3152	Barker et al., 1991
178-1097A-3R	Antarctic Peninsula shelf	66.4S/70.75W	551.7	Barker et al., 1991
29-278-3	South of New Zealand	56.6S/160.1E	3675	Kennett et al., 1974

NAME ______________________________

TABLE 2.1. *Continued*

Core Identification: Expedition–Site & Hole–Core & Type*	Physiographic Site Location	Site Location (Latitude/ Longitude)	Water Depth (m)	Reference
202-1236A-2H	Nazca Ridge, SE Pacific	21.4S/81.44W	1323.7	Mix et al., 2003
206-1256B-2H	Guatemala Basin, eastern tropical Pacific	6.7N/91.9W	3634.7	Wilson et al., 2003
8-74-1	Clipperton Fracture Zone, central Pacific abyssal plain	6.1N/136.1W	4431	Tracey et al., 1971
136-842A-1H	South of Hawaii	19.3N/159.1W	4430.2	Dziewonski et al., 1992
198-1209A-2H	Shatsky Rise, NW Pacific	32.7N/158.5E	2387.2	Bralower et al., 2002
199-1215A-2H	NE of Hawaii, North Pacific abyssal plain	26.0N/147.9W	5395.6	Lyle et al., 2002
86-576-2	West of Midway Island, North Pacific abyssal plain	32.4N/164.3E	6217	Heath et al., 1985
195-1201B-2H	Philippine Sea	19.3N/135.1E	5710.2	Salisbury et al., 2002
130-807A-2H	Ontong Java Plateau, western equatorial Pacific	3.6N/156.6E	2803.8	Kroenke et al., 1991
181-1125A-2H	Chatham Rise, east of New Zealand	42.6S/178.2W	1364.6	Carter et al., 1999
169-1037A-1H	Escanaba Trough, west of Oregon and N. California	41.0N/127.5W	3302.3	Fouquet et al., 1998
146-888B-2H	Cascadia margin, west of Vancouver, BC	48.2N/126.7W	2516.3	Westbrook et al., 1994
167-1010E-1H	West of Baja California	30.0N/118.1W	3464.7	Lyle et al., 1997
200-1224C-1H	North Pacific abyssal plain, south of the Murray Fracture Zone	27.9N/142.0W	4967.1	Stephen et al., 2003
127-795A-2H	Japan Sea	44.0N/139.0E	3300.2	Tamaki et al., 1990
28-274-2	North of Ross Ice Shelf, Antarctica	69.0S/173.4E	3305	Hayes et al., 1975
North Atlantic Cores				
37-333-2	Western flank of Mid-Atlantic Ridge	36.8N/33.7W	1666	Aumento et al., 1977
82-558-3	Western flank of Mid-Atlantic Ridge	33.8N/37.3W	3754	Bougault et al., 1995
172-1063A-2H	Northeast Bermuda Rise	33.7N/57.6W	4583.5	Keigwin et al., 1998
105-646A-2H	Labrador Sea, south of Greenland	58.2N/48.4W	3440.3	Srivastava et al., 1987
162-980A-2H	Rockall Bank, west of Ireland	55.5N/14.7W	2172.2	Jansen et al., 1996
152-919A-2H	SE Greenland, continental rise	62.7N/37.5W	2088.2	Larsen et al., 1994
174-1073-1H	New Jersey continental shelf	39.2N/72.3W	639.4	Austin et al., 1998
14-137-3H	Madeira abyssal plain	25.9N/27.1W	5361	Hayes et al., 1972

*The letter indicating the type of drilling (e.g., H for hydraulic piston coring) is not always included in the core identification (Column 1 of Table 2.1). This is because early on in the drilling program, there was only one type of coring (rotary), and thus no special notation was needed. Core identification in Table 2.1 matches the core identification on the related core photos (and the end of this chapter).

NAME ______________________________

To do:

1 Find your assigned core in Table 2.1 and on the base map (Figure 2.2). Examine the photo of your core (photos in the Appendix at end of Chapter 2) and make a list of observations and a list of questions about what you see:

2 **Lithology** refers to the visible physical characteristics of sediment or rock. Design a way to **organize** and **record** your observations of the lithology of the sediment in your core that could be used by all the students in the class for all the cores. You will need to think of **categories** (e.g., color, sedimentary structures) for your observations and also a means of recording them (i.e., all observations **written**, all observations **sketched**, a combination of the two?). Record your ideas below:

NAME

3 Use the space below to sketch and describe the lithology of your sediment core. Be sure to follow the template agreed upon by your class (or use the core description form provided by your instructor).

NAME

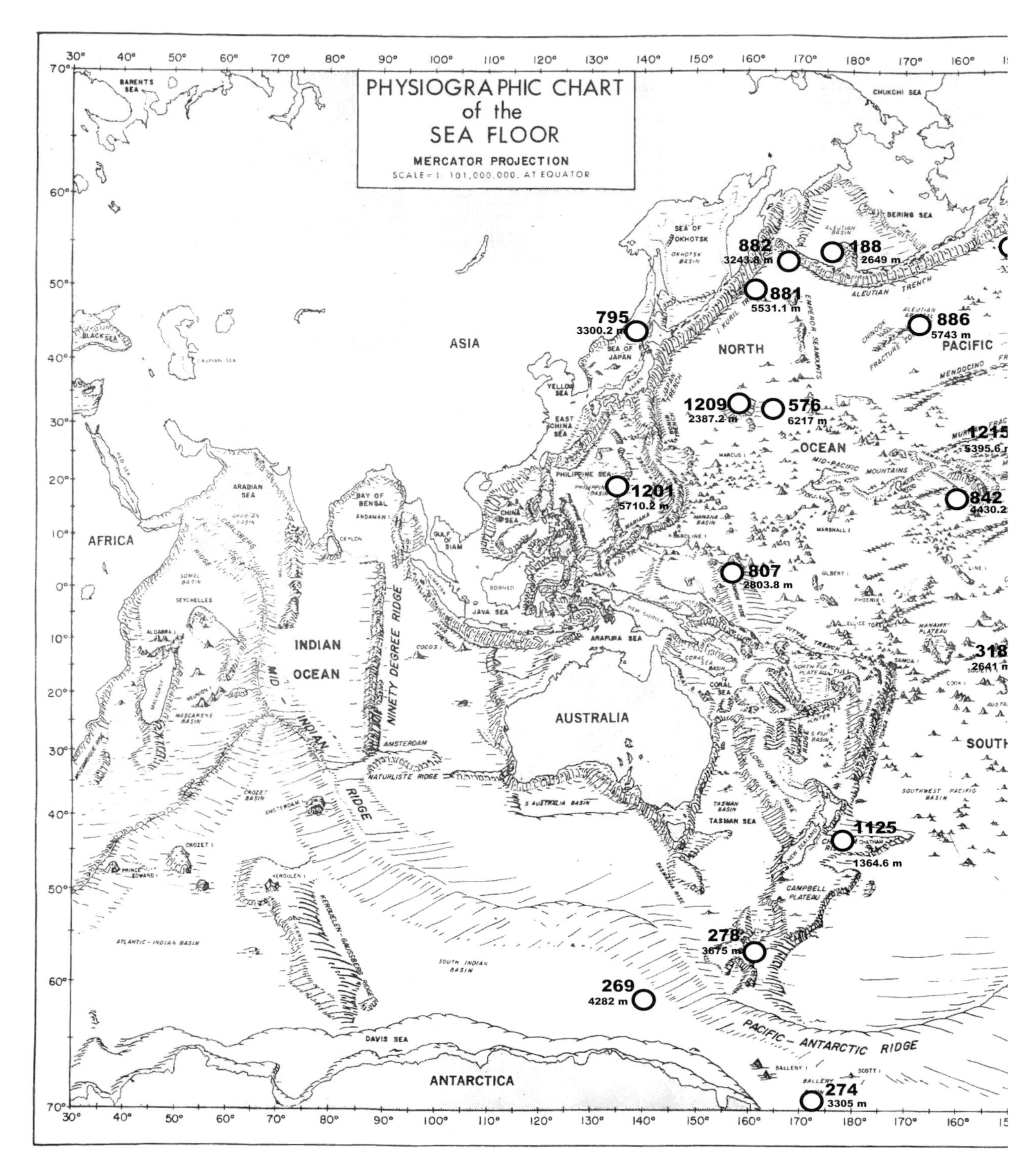

FIGURE 2.2. Physiographic map of the worlds oceans, showing the core sites and water depths used in this exercise. Map modified from Hubbard Scientific's Sea Floor Physiography map; American Education Products.

NAME

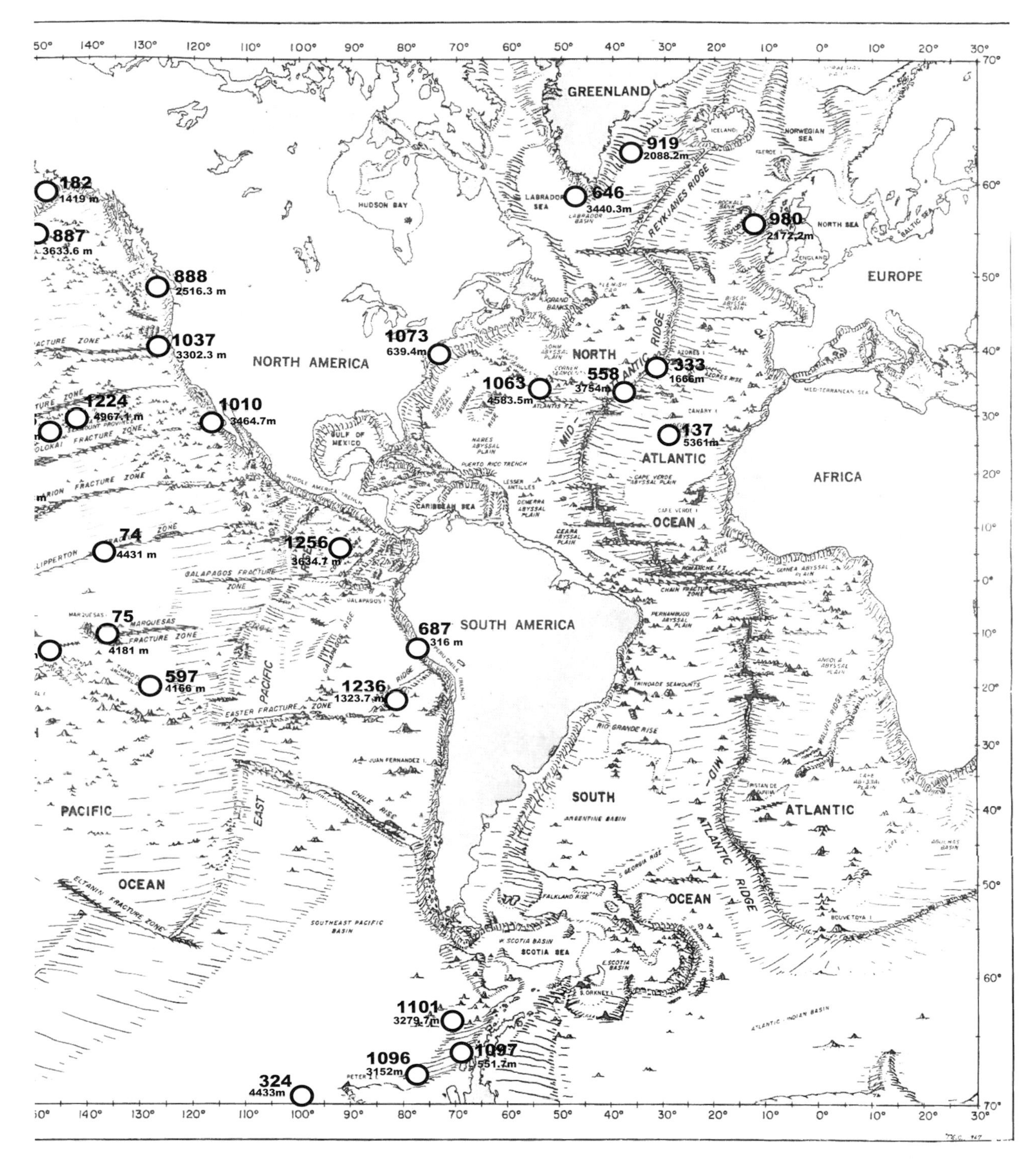

FIGURE 2.2. (*Continued*)

NAME ______________________________

4 Based on the work you have done in this exercise, explain the importance of a systematic, complete, and consistent method of recording scientific observations.

NAME

Seafloor Sediments
Part 2.3. Sediment Composition

Introduction

One lithologic category that typically **cannot** be determined from observation of the core alone is **sediment composition**. Composition can, however, be determined by examining a small (toothpick-tip sized) amount of sediment under a binocular microscope and matching the grain types observed to categories of known grain types. This method is called **smear slide analysis**. The main grain categories that compose marine sediment are **minerals**, **microfossils**, and **volcanic glass**. Some of the most common grain types are shown in Figure 2.3.

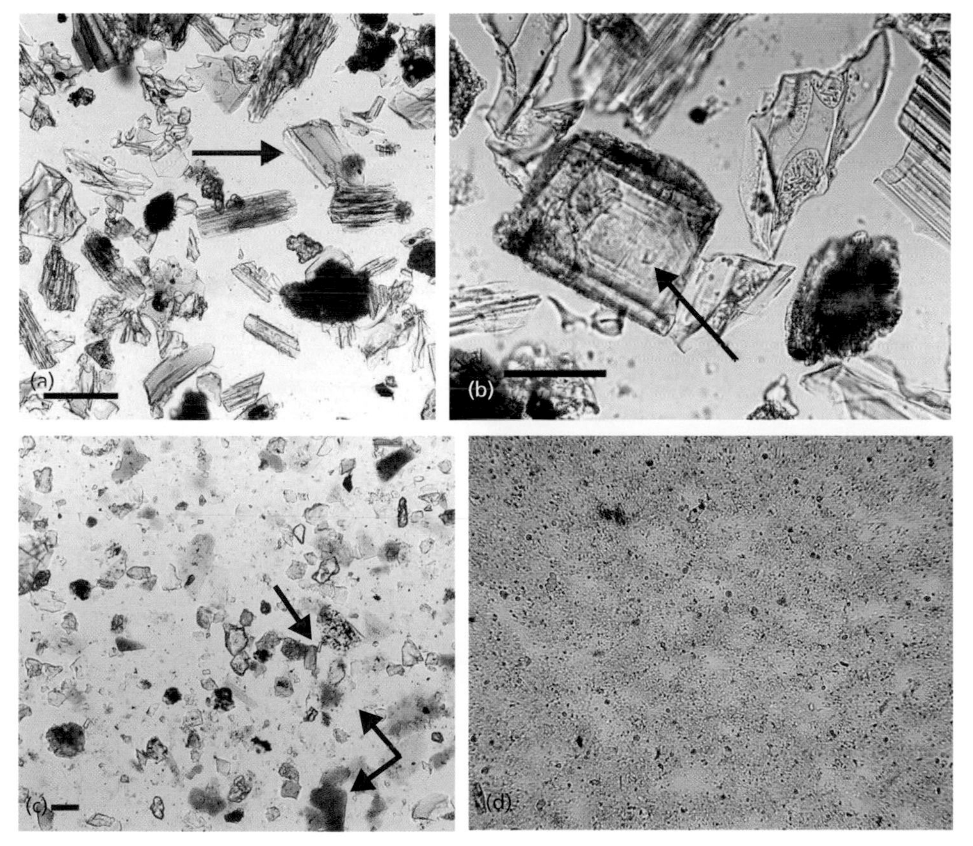

FIGURE 2.3. Volcanic glass, minerals, and mineral groups. (a) Volcanic glass, bar scale = 0.05 mm; (b) Feldspar mineral surrounded by volcanic glass. Bar scale = 0.05 mm; (c) Silt-size minerals including green and brown biotite (mica) flakes. Bar scale = 0.05 mm; (d) Clay minerals. Individual grains are under 4 μm (0.004 mm) in size; Photos reproduced with permission of R.G. Rothwell; available from the British Ocean Sediment Core Research Facility website: http://www.boscorf.org/

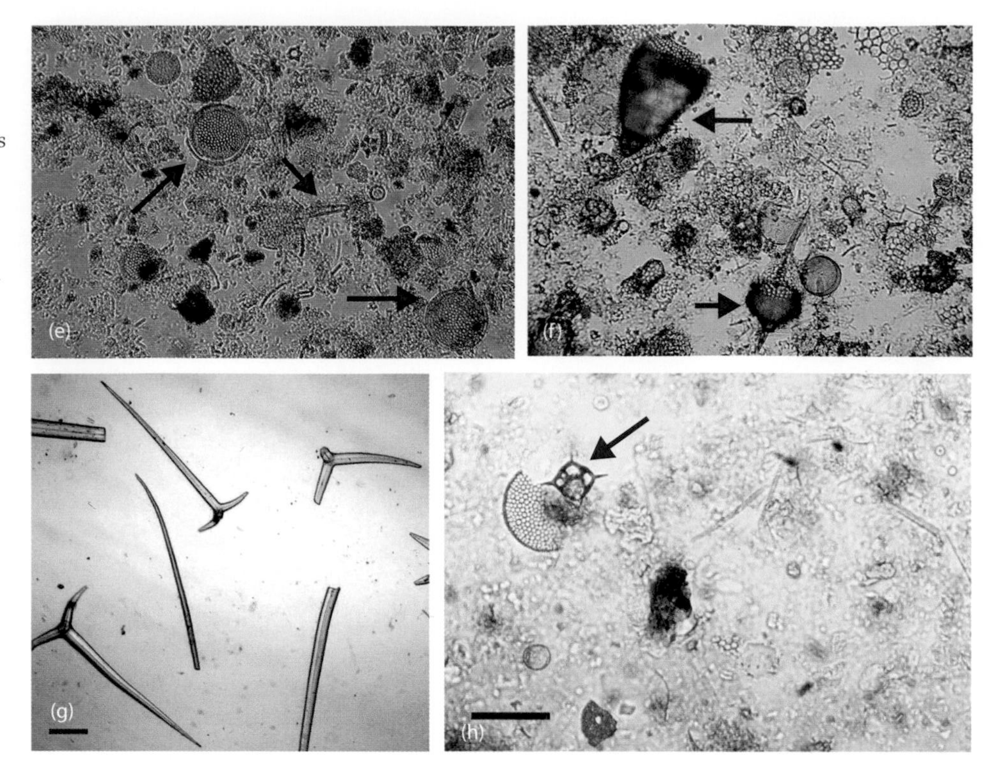

FIGURE 2.3. *Continued*: Siliceous (SiO_2) Microfossils. (e) Diatoms (and clay). High power (×100) view; (f) Radiolarians with some diatoms (and clay). High power (×100) view; (g) Sponge spicules. Scale bar = 0.05 mm; (h) Silicoflagellate (arrow) with diatom fragments (and clay). Scale bar = 0.05 mm. Photos are from the British Ocean Sediment Core Research Facility website: http://www.boscorf.org/

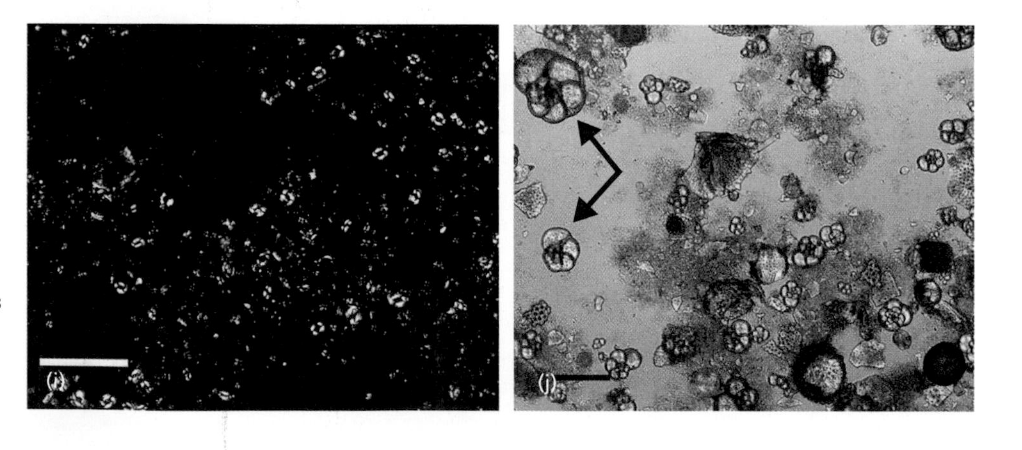

FIGURE 2.3. *Continued*: Calcareous ($CaCO_3$) Microfossils. (i) Scatter of calcareous nannofossils (coccolith plates) seen in cross-polarized light. Note the black interference crosses shown by each plate. Scale bar = 0.05 mm; (j) Foraminifera and clay. Scale bar = 0.05 mm. Photos are from the British Ocean Sediment Core Research Facility website: http://www.boscorf.org/

This exercise uses composition and texture data from smear slides from specific samples within the various cores you described in Part 2.2. The **composition** data are estimates of the relative abundances of specific grain types (i.e., minerals and microfossils), and the **texture** data are estimates of the relative percentages of sand, silt, and clay-sized grains. When considering

NAME ______________________________

these smear slide data, one should always consider their position within the core to assess whether they are representative of some broader interval or provide detailed information on a specific layer of interest (e.g., within light sediment, a dark layer that turns out to represent a volcanic ash layer when examined in smear slide).

To do:

1 Go to Table 2.2 (smear slide data) and find the row(s) of data that correspond to smear slide samples taken from your core(s). Note that this should be the same core(s) assigned to you in Part 2.2. What are the minerals and fossils in your core(s)? **List them below** and **circle the photos** in Figure 2.3 that show what these grains look like in a smear slide.

____

____

____

____

____

2 The composition and texture data from these smear slides will serve as the primary means of determining the **lithologic name** of the sediment. **Read the Box** below that describes important information about using the **decision tree**. Then use the decision tree (following pages) to determine the lithologic name of the sediment in your core(s):

Core: ______________________________

Lithologic Name of the Sediment: ______________________________

NAME

DETERMINING LITHOLOGIC NAMES WITH A DECISION TREE

The goal of the decision tree is to use composition and texture data from smear slide analysis, core photos, and knowledge of the water depth and distance from land of cored sites on the seafloor to determine the lithologic name of the sediment. The decision tree aims to capture the most distinctive (i.e., end-member) lithologies of marine sediments. These are:

1. **Calcareous ooze:** biogenic sediments composed of calcareous nannofossils and/or foraminifers;
2. **Siliceous ooze:** biogenic sediments composed of diatoms, radiolarians, sponge spicules, and/or silicoflagellates;
3. **Red clays:** very fine terrigenous sediment that often contains siliceous microfossils, fish teeth, Mn-Fe micronodules, and/or volcanic glass;
4. **Terrigenous sediment:** sediment from the weathering of continents or volcanic islands;
5. **Glaciomarine sediment:** sediment containing terrigenous sand, pebbles, or cobbles transported to the sea by icebergs.

► Note, while the above five marine sediment lithologies are the most distinctive sediment types, **mixed lithologies are common**. In addition, the **lithologies can also change within a core**. For example, the sediment could alternate between red clay and siliceous ooze, or gradually change from one to another.

► In any marine sediment lithology, but especially in biogenic oozes and deep sea red clays, layers of **volcanic ash** may be distinguishable as a minor lithology.

THE DECISION TREE CAN BE FOUND ON PAGES 49–51.

NAME

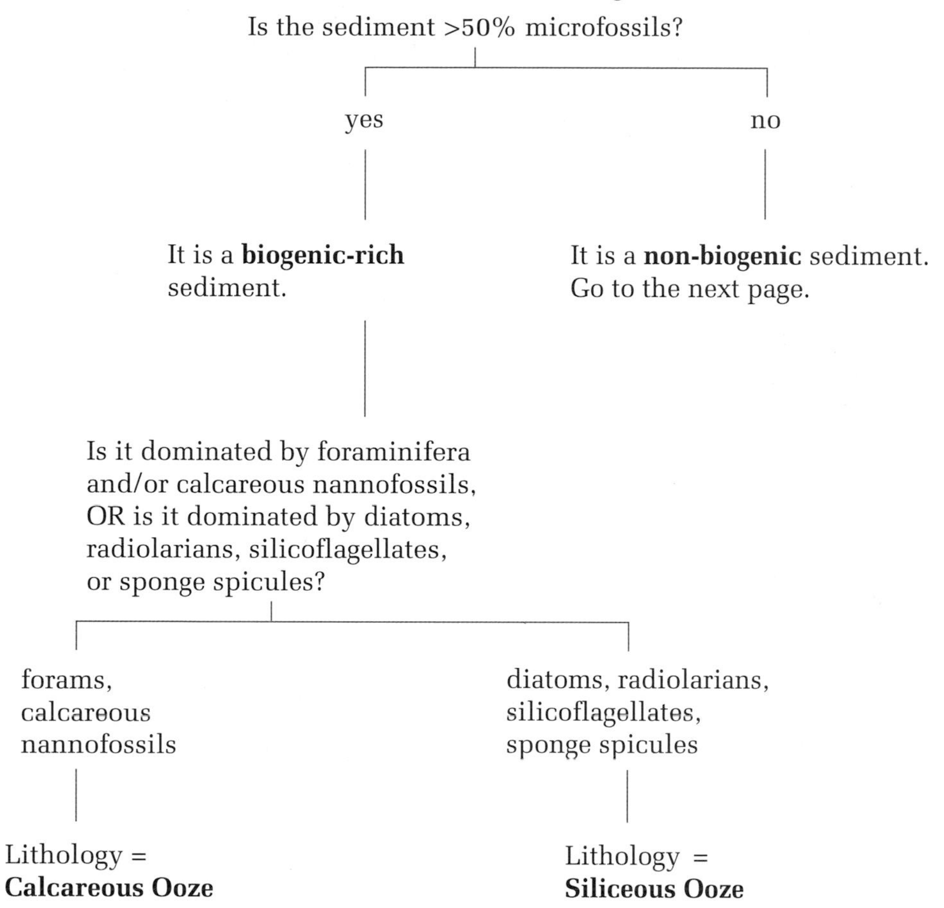

Decision Tree
To determine the dominant lithology of marine sediments based on smear slide data, visual core observations, and site characteristics use the following decision tree.
Is the sediment >50% microfossils?
yes
no
It is a **biogenic-rich** sediment.
It is a **non-biogenic** sediment. Go to the next page.
Is it dominated by foraminifera and/or calcareous nannofossils, OR is it dominated by diatoms, radiolarians, silicoflagellates, or sponge spicules?
forams, calcareous nannofossils
diatoms, radiolarians, silicoflagellates, sponge spicules
Lithology = **Calcareous Ooze**
Lithology = **Siliceous Ooze**

NAME

Non-biogenic Sediment

Is the texture and/or the mineral composition primarily clay (dust-size)?

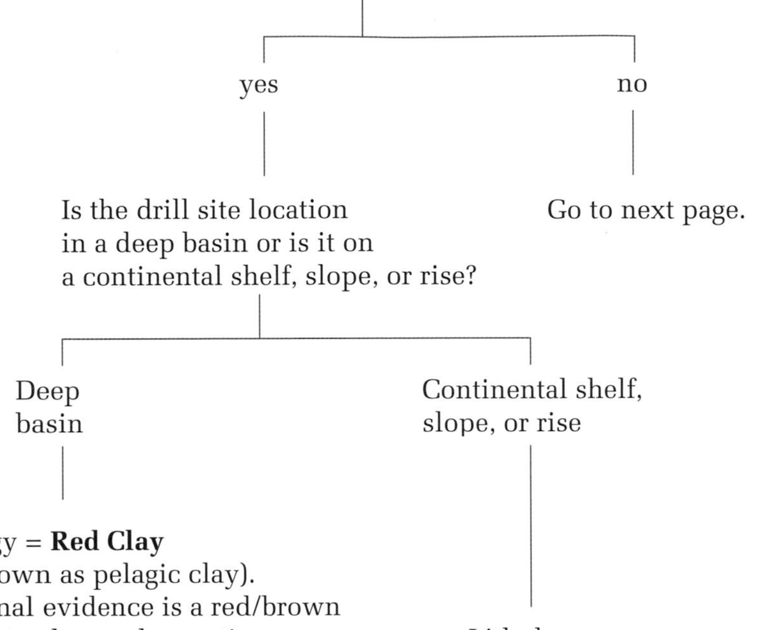

Lithology = **Red Clay** (also known as pelagic clay). Additional evidence is a red/brown sediment color, and sometimes black "spots" or nodules in the sediment, which are Mn & Fe mineral precipitates. Red Clays are a mix of wind-blown terrigenous dust, authigenic (precipitated) minerals, and may even include some siliceous microfossils or fish debris.

Lithology = **Terrigenous Sediment**. Silt and clay-rich, but may have thin layers of sand. Terrigneous sediment is largely transported by rivers to the oceans.

NAME

Mixed Grain Size, Non-biogenic Sediment

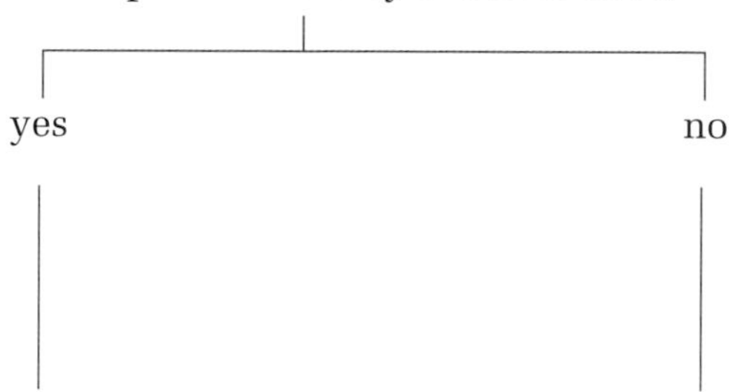

Lithology =
Glaciomarine Sediment
This sediment type contains material transported to the sea by floating ice (it is "ice-rafted debris").

Lithology =
Terrigenous Sediment
Additional evidence may include fining upward sequences with sharp bases. Shallow (shelf) terrigenous sediment may also contain abundant sand, sea shells (mollusks), and tilted (cross) bedding. Shallow, near shore terrigenous sediment is also known as neritic or margin sediment and is transported to the near shore ocean by rivers.

NAME ______________________________

TABLE 2.2. Smear slide data

Smear Slide Sample Identification	Texture (%)			Composition (%)										
	Grain Size			Mineral Grains									Microfossils	
Exp-Site&Hole-Core&Type-Section, Interval (cm)	Sand	Silt	Clay	Accessory Min.	Calcite/Dolomite	Clay Minerals	Fe Oxide	Feldspar	Other Minerals*	Mica	Quartz	Volcanic Glass	Calc. Nannos	Diatoms
Pacific samples														
112-687A-2H-1, 79	2	61	37	1	2	37		5			10		2	30
112-687A-2H-3, 69		80	20	2	1	19		23		2	35		5	
112-687A-2H-5, 61		45	55	10	2	17		5			13	1		
112-687A-2H-6, 36		25	75	1		75					1	1		20
35-324-1-1, 120	none given					40					10			50
35-324-1-2, 50	none given					85					10			5
35-324-1-3, 100	none given					35	2				3			60
35-324-1-6, 100	none given					97			1		2			
28-269-1-1, 134	none given				5	25					10			55
28-269-1-4, 37	none given					10					10		1	78
145-881A-1H-1, 50	30	60	10								1	1		98
145-881A-1H-2, 11	15	40	45			45			10					45
145-881A-1H-2, 116	46	60						4			2	94		
145-886B-2H-1, 39	2	10	88			81	2							4
145-886B-2H-5, 114	15	10	70			70	2	2			1			4
145-886B-2H-6, 16	25	10	65			64	3							5
145-882A-2H-2, 58	2	78	20			1						1	1	77
145-882A-2H-3, 34	2	85	13								1		1	78
145-882A-2H-4, 80	20	75	5									1	2	96
145-887C-2H-1, 75	10	82	8			7							1	72
145-887C-2H-3, 75	10	45	45			43					1			45
145-887C-2H-3, 85	3	92	5			5		2			1	92		

*Other minerals includes opaques, phillipsite, phroxene, hornblende, and others.
D = dominant, A = abundant, C = common, P = present, R = rare, T = trace.

NAME ____________________

									Lithologic Name of the Sediment	Reference
					Rock Frags./Other					
Foraminifers	Radiolarians	Silicoflagellates	Sponge Spicules	Skeletal Debirs	Carbonate Frags.	Organic matter	Nodules	Rock Fragments		
10								2		Seuss et al., 1988
								10		
							37	5		
										Hollister et al., 1976
	5									Hayes et al., 1975
		1								
										Rea et al., 1993
	3		10							Rea et al., 1993
	6		15							
	3		25							
			20							Rea et al., 1993
			20							
	1									
4			10		6					Rea et al., 1993
4	2				5					

(Continued)

NAME ______________________________

TABLE 2.2. *Continued*

Smear Slide Sample Identification	Texture (%)			Composition (%)										
	Grain Size			Mineral Grains									Microfossils	
Exp-Site&Hole-Core&Type-Section, Interval (cm)	Sand	Silt	Clay	Accessory Min.	Calcite/Dolomite	Clay Minerals	Fe Oxide	Feldspar	Other Minerals*	Mica	Quartz	Volcanic Glass	Calc. Nannos	Diatoms
Pacific samples														
19-188-2-1, 56	none given					25		5			5			65
19-188-2-1, 120	none given					10		2			3			85
19-188-2-2, 75	none given					10		5			5			80
19-188-2-3, 75	none given					10		5			5			80
18-182-1-5, 100	0	5	95	tr		95		1		1	3			
18-182-1-6, 100	0	8	92	1		92		1		2	5			
33-318-2-2, 67	none given						R						D	
8-75-1-1, 100	none given					90			9					
8-75-1-3, 10	none given					10							90	
8-75-1-5, 10	none given					5							94	
92-597-1-1, 35	none given					85			13			2		
92-597-1-2, 110	none given					10			5				85	
178-1101A-2H-1, 80	1	60	39	1	3	50		8	1	2	15		4	5
178-1101A-2H-2, 60	1	60	39	2	3	31		8	8	3	20			2
178-1101A-2H-2, 106	90	8	2		1	13		9		4	12			1
178-1101A-2H-4, 61	1	75	24			5		12	20	4	18			2
178-1101A-2H-6, 61		70	30			8		12	13	4	18			10
178-1096A-1H-1, 0		30	70			60		5		5				30
178-1096A-1H-1, 130		20	80			80		5	6	5				1
178-1096A-1H-4, 130		20	80			80		5	1	9				1
178-1096A-1H-6, 30	1	25	74			74		10		15				1

*Other minerals includes opaques, phillipsite, phroxene, hornblende, and others.
D = dominant, A = abundant, C = common, P = present, R = rare, T = trace.

NAME ______________________________

									Lithologic Name of the Sediment	Reference
					Rock Frags./Other					
Foraminifers	Radiolarians	Silicoflagellates	Spcnge Spicules	Skeletal Debirs	Carbonate Frags.	Organic matter	Nodules	Rock Fragments		
										Creager et al., 1973
										Klum et al., 1973
A				P	C					Schlanger et al., 1976
1										Tracey et al., 1971
1										
			2					14		Barker et al., 1991
2			1					20		
40			1					14		
			1					30		
			6					20		
1										Barker et al., 1991
1										
1										
1										

(Continued)

NAME

TABLE 2.2. *Continued*

Smear Slide Sample Identification	Texture (%)			Composition (%)											
	Grain Size			Mineral Grains										Microfossils	
Exp-Site&Hole-Core&Type-Section, Interval (cm)	Sand	Silt	Clay	Accessory Min.	Calcite/Dolomite	Clay Minerals	Fe Oxide	Feldspar	Other Minerals*	Mica	Quartz	Volcanic Glass	Calc. Nannos	Diatoms
Pacific samples														
178-1097A-3R-1	all gravel													
29-278-3-1, 127	none given					15							10	30
29-278-3-CC, 0	none given					2			2				3	75
202-1236A-2H,1, 75			100			14							57	
202-1236A-2H, 92			100						R				57	
202-1236A-2H, 123			100						R				70	
206-1256B-2H-2, 113	5	13	82			40		8			10		40	1
8-74-1-1, 2	none given												55	10
8-74-1-5, 10	none given												5	10
136-842A-1H-1, 27	10	50	40			40		17				23		
136-842A-1H-4, 90		25	75	1		75		1				10		
136-842A-1H-6, 68	25	60	15	10		15		20	20			30		
136-842A-1H-6, 130		52	48			18		1	1			1	30	6
136-842A-1H-7, 20		22	78	1		78		7	2			5		2
198-1209A-2H-1, 139	none given					21							70	2
198-1209A-2H-5, 138	none given				1	13							80	
199-1215A-2H-1, 60			100			90			8	2				
199-1215A-2H-3, 100			100			90			9	1				
199-1215A-2H-CC, 0			100			90			10					
Pacific samples														
86-576-2-1, 7		5	95			85		1			4			1
86-576-2-2, 80	1	7	92			87		1			6			
86-576-2-4, 74	30	68	2			2		1			3	93		
86-576-2-4, 110	1	1	98			94				1	1	1		1

*Other minerals includes opaques, phillipsite, phroxene, hornblende, and others.
D = dominant, A = abundant, C = common, P = present, R = rare, T = trace.

NAME ______________________________

									Lithologic Name of the Sediment	Reference
					Rock Frags./Other					
Foraminifers	Radiolarians	Silicoflagellates	Sponge Spicules	Skeletal Debirs	Carbonate Frags.	Organic matter	Nodules	Rock Fragments		
								100		Barker et al., 1991
10	25		10							Kennett et al., 1974
	3		15							
25										Mix et al., 2003
29										
30										
	1									Wilson et al., 2003
35	10		<1							Tracey et al., 1971
	75		10							
	2	1	1							Dziewonski et al., 1992
	10		1							
	5									
	30	10	2							
	3	1	1							
5	2									Bralower et al., 2002
5	1									
										Lyle et al., 2002
	3		1				5			Heath et al., 1985
	2						2			
							1			
	?						1			

(Continued)

NAME

TABLE 2.2. *Continued*

Smear Slide Sample Identification	Texture (%)			Composition (%)										
	Grain Size			Mineral Grains									Microfossils	
Exp-Site&Hole-Core&Type-Section, Interval (cm)	Sand	Silt	Clay	Accessory Min.	Calcite/Dolomite	Clay Minerals	Fe Oxide	Feldspar	Other Minerals*	Mica	Quartz	Volcanic Glass	Calc. Nannos	Diatoms
195-1201B-2H-1, 30		10	90			D		P	R		P	R		
195-1201B-2H-5, 73		5	95			D		R	P		R			
195-1201B-2H-7, 85		10	90			D		P	P	R	P			
130-807A-2H-2, 74	10	60	30	2									75	2
181-1125A-2H-1, 49	25	30	45		P	R							D	
169-1037A-1H-3, 80		C	A	R	R	A		R		R	C		R	R
169-1037A-1H-5, 62	A	C	R		R			R	R	C	A			R
146-888B-2H-5, 99	70	25	5			5		10	33		5	25		2
146-888B-2H-6, 145		75	25			25		20	30			10	1	2
167-1010E-1H-3, 143	5	15	80			80		2	1			9		
167-1010E-1H-4, 110		10	90			80					2	10		
200-1224C-1H-1, 70		10	90			90								
200-1224C-1H-2, 2		5	95			95								
200-1224C-1H-3, 70		20	80			65								
200-1224C-1H-5, 1		25	75			50								
Pacific samples														
127-795A-2H-1, 84		10	90		2	60		5			10	15		2
127-795A-2H-2, 146		40	60			30						10		60
127-795A-2H-3, 45		60	40			20		15			15	45		1
127-795A-2H-5, 81		40	60			60		10			20	10		
28-274-2-2,109	2	33	65		1	55		11	10		15			6
28-274-2-3, 86		25	75			60			1		1			35
28-274-2-6, 90		20	80		1	70		4	1		8			15

*Other minerals includes opaques, phillipsite, phroxene, hornblende, and others.
D = dominant, A = abundant, C = common, P = present, R = rare, T = trace.

NAME ______________________________

									Lithologic Name of the Sediment	Reference
					Rock Frags./Other					
Foraminifers	Radiolarians	Silicoflagellates	Sponge Spicules	Skeletal Debirs	Carbonate Frags.	Organic matter	Nodules	Rock Fragments		
										Salisbury et al., 2002
20										Kroenke et al., 1991
P			R	P						Carter et al., 1999
	R									Fouquet et al., 1998
			R							
							15			Westbrook et al., 1994
						2	10			
								1		Lyle et al., 1997
3								5		
	10									Stephen et al., 2003
			5							
	30		5							
	45		5							
		1	1			1				Tamaki et al., 1990
			2							Hayes et al., 1975
	1		2							
			1							

(Continued)

NAME ______________________

TABLE 2.2. *Continued*

Smear Slide Sample Identification	Texture (%)			Composition (%)											
	Grain Size			Mineral Grains										Microfossils	
Exp-Site&Hole-Core&Type-Section, Interval (cm)	Sand	Silt	Clay	Accessory Min.	Calcite/Dolomite	Clay Minerals	Fe Oxide	Feldspar	Other Minerals*	Mica	Quartz	Volcanic Glass		Calc. Nannos	Diatoms
North Atlantic samples															
37-333-2-1, 80	5	24	71											96	
82-558-3-3, 75	none given				2	9								87	
82-558-3-6, 75	none given				5	9								84	
172-1063A-2H-3, 62		40	60		R	D	T		T		A			C	R
172-1063A-2H-6, 66		25	75		R	D	T		T		C			A	C
105-646A-2H-1, 60	5	60	35		35	30					35				
105-646A-2H-2, 87	5	85	10	5	5	10		10	5		65				
105-646A-2H-5, 33	10	55	35		15	25		5			35			3	10
162-980A-2H-1, 90	20	30	50	2		8				1	11			50	4
162-980A-2H-3, 80	10	40	50			50			5	5	10			30	
162-980A-2H-6, 80	10	60	30	5		25		15	3	5	20			10	
North Atlantic samples															
152-919A-2H-1, 76	3	81	16			16		7	2		40	15			
152-919A-2H-3, 18		85	15								2	69			4
152-919A-2H-4, 50		62	38			38		6	2		25	5		7	4
174-1073A-1H-1, 10	none given			2		23		14		2	20			20	4
174-1073A-1H-1, 120	none given			2	1	39		12		4	20			8	1
14-137-3H-2, 90	none given					81	4		5	2		8			

*Other minerals includes opaques, phillipsite, phroxene, hornblende, and others.
D = dominant, A = abundant, C = common, P = present, R = rare, T = trace.

NAME ____________________

									Lithologic Name of the Sediment	Reference
					Rock Frags./Other					
Foraminifers	Radiolarians	Silicoflagellates	Sponge Spicules	Skeletal Debirs	Carbonate Frags.	Organic matter	Nodules	Rock Fragments		
4										Aumento et al., 1977
2										Bougault et al., 1995
2										
T	T	T	R							Keigwin et al., 1998
T	T	T	C							
										Srivastava et al., 1987
2			5							
13	5		3							Jansen et al., 1996
7			2					3		
	10		3					7		Larsen et al., 1994
			10							
			3					10		
5	5									Austin et al., 1998
8										

NAME

Seafloor Sediments
Part 2.4. Geographic Distribution and Interpretation

1 Use the following color scheme to **plot** the dominant lithologies of the sediment in your core (from Part 2.3) on **your base map** (Figure 2.2) and on the **class base map**. For example, sediment that is 100% calcareous ooze would be plotted as a blue circle on the class map. A good way to plot mixed or alternating sediment types is with a simple pie diagram (e.g., a sediment type that is an equal mix of calcareous and siliceous ooze would be a circle that is ½ blue and ½ yellow). Through group effort, the class map should ultimately contain all of the exercise core locations in the Pacific and North Atlantic Oceans and their sediment types.

- Blue = Calcareous Ooze
- Brown = Terrigenous Sediment
- Yellow = Siliceous Ooze
- Green = Glaciomarine Sediment
- Red = Red Clays

2 Make a list of your observations about the distribution of these marine sediment lithologies on the map your class constructed. Consider factors such as distance from the continents, water depth, and latitude/longitude, among others.

Sediment Lithologies	Observations about Distribution
Calcareous Ooze	

NAME

Siliceous Ooze	
Red Clay	
Terrigenous Sediment	
Glaciomarine Sediment	

3 Based on your observations (in question 2 above), **develop hypotheses to explain the distribution of each of these lithologies**. Since sediment lithologies (e.g., calcareous ooze) are named for their dominant grain types (e.g., calcareous microfossils), a good explanation will address what controls the relative abundances of different grain types that accumulate on the seafloor.

Sediment Lithologies	Hypotheses to Explain Lithologic Distributions
Calcareous Ooze	

NAME ____________________

Siliceous Ooze	
Red Clay	
Terrigenous Sediment	
Glaciomarine Sediment	

4 Compare your class map of seafloor sediment lithologies to a published map of sea floor sediment lithologies (e.g., see the map from Rothwell, 1989 distributed by your instructor). Are they generally similar? If not, where are the discrepancies? What might cause these discrepancies?

____

____

____

____

____

NAME

5 Compare the North Pacific and North Atlantic sediment distributions.

(a) In what basin are glaciomarine sediments more abundant? Propose a hypothesis to explain your observations.

(b) Is calcareous ooze in the North Atlantic found at the same depth, shallower depths, or deeper depths than in the North Pacific (see Table 2.1 or Figure 2.2 for water depths)? Propose a hypothesis to explain your observations.

6 The map you constructed represents the **modern** distribution of sediment lithologies in the Pacific Ocean. Do you think this map would also represent the distribution of marine sediment lithologies in the geologic past and in the geologic future? What factors might vary that could change the distribution of marine sediment lithologies?

NAME

NAME ___

References

Aumento, F., et al., 1977, Initial Reports of the Deep Sea Drilling Project, vol. 7, US Government Printing Office, Washington, DC.

Austin, J.A., Jr., et al., 1998, Proceedings of the Ocean Drilling Program, Initial Reports, vol. 174A, College Station, TX, Ocean Drilling Program, doi:10.2973/odp.proc.ir.174a.1998.

Barker, P.F., et al., 1999, Proceedings of the Ocean Drilling Program, Initial Reports, vol. 178, College Station, TX, Ocean Drilling Program, doi:10.2973/odp.proc.ir.178.1999.

Bougault, H., et al., 1985, Initial Reports of the Deep Sea Drilling Project, vol. 82, US Government Printing Office, Washington, DC.

Bralower, et al., 2002, Proceeding of the Ocean Drilling Program, Initial Reports, vol. 198, College Station, TX, Ocean Drilling Program, doi:10.2973/odp.proc.ir.198.2002.

Carter, R.M., et al., 1999, Proceedings of the Ocean Drilling Program, Initial Reports, vol. 181, doi:10.2973/odp.proc.ir.181.2000.

Creager, J. S., et al, 1973, Initial Reports of the Deep Sea Drilling Project, vol. 19, US Government Printing Office, Washington, DC.

Dziewonski, A., et al, 1992, Proceedings of the Ocean Drilling Program, vol. 136, College Station, TX, Ocean Drilling Program, doi:10.2973/odp.proc.ir.136.1992.

Fouquet, Y., et al., 1998, Proceedings of the Ocean Drilling Program, Initial Reports, vol. 169, doi:10.2973/odp.proc.ir.169.1998.

Hayes, D. E., et al, 1972, Initial Reports of the Deep Sea Drilling Project, vol. XIV, US Government Printing Office, Washington, DC.

Hayes, D. E., et al., 1975, Initial Reports of the Deep Sea Drilling Project, vol. 28, US Government Printing Office, Washington, DC.

Heath, G. R., et al., 1985, Initial Reports of the Deep Sea Drilling Project, vol. 86, US Government Printing Office, Washington, DC.

Hollister, C. D., et al., 1976, Initial Reports of the Deep Sea Drilling Project, vol. 35, US Government Printing Office, Washington, DC.

Jansen, E., et al., 1996, Proceedings of the Ocean Drilling Program, Initial Reports, vol. 162, College Station, TX, Ocean Drilling Program, doi:10.2973/odp.proc.ir.162.1996.

Keigwin, L.D., et al., 1998, Proceedings of the Ocean Drilling Program, Initial Reports, vol. 172, College Station, TX, Ocean Drilling Program, doi:10.2973/odp.proc.ir.172.1998.

Kennett, J. P., et al., 1974, Initial Reports of the Deep Sea Drilling Project, vol. 29, US Government Printing Office, Washington, DC.

Kroenke, et al., 1991, Proceedings of the Ocean Drilling Program, Initial Reports, vol. 130, College Station, TX, Ocean Drilling Program, doi:10.2973/odp.proc.ir.130.1991.

Kulm, L. D., et al., 1973, Initial Reports of the Deep Sea Drilling Project, vol. 18, US Government Printing Office, Washington, DC.

Larsen, H.C., et al., 1994, Proceedings of the Ocean Drilling Program, Initial Reports, vol. 152, College Station, TX, Ocean Drilling Program, doi:10.2973/odp.proc.ir.152.1994.

Leinen, M., et al., 1986, Initial Reports of the Deep Sea Drilling Project, vol. 92, US Government Printing Office, Washington, DC.

Lyle, M., et al., 1997, Proceedings of the Ocean Drilling Program, vol. 167, College Station, TX, Ocean Drilling Program, doi:10.2973/odp.proc.ir.167.1997.

Lyle, M., et al., 2002, Proceedings of the Ocean Drilling Program, Initial Reports, vol. 199, College Station, TX, Ocean Drilling Program, doi:10.2973/odp.proc.ir.199.2002.

Mazzullo and Graham, 1998, Handbook for Shipboard Sedimentologist, ODP Technical Notes No. 8, Texas A&M University.

Mix, A.C., et al., 2003, Proceedings of the Ocean Drilling Program, Initial reports, vol. 202, College Station, TX, Ocean Drilling Program, doi:10.2973/odp.proc.ir.202.2003.

Rea, D.K., et al., 1993, Proceedings of the Ocean Drilling Program, Initial Reports, vol. 145, College Station, TX, Ocean Drilling Program, doi:10.2973/odp.proc.ir.145.1993.

Rothwell, R. G., 1989, Minerals and Mineraloids in Marine Sediments: An optical identification guide, Elsevier Applied Science, London, 279 pp.

Salisbury, M.H., et al., 2002, Proceedings of the Ocean Drilling Program, Initial Reports, vol. 195, College Station, TX, Ocean Drilling Program, doi:10.2973/odp.proc.ir.195.2002.

Schlanger, S. O., et al., 1976, Initial Reports of the Deep Sea Drilling Project, vol. 33, US Government Printing Office, Washington, DC.

Srivastava, S. P., et al., 1987, Proceedings of the Ocean Drilling Program, Initial Reports, vol. 105, College Station, TX, Ocean Drilling Program, doi:10.2973/odp.proc.ir.105.1987.

Stephen, R.A., et al., 2003, Proceedings of the Ocean Drilling Program, Initial Reports, vol. 200, College Station, TX, Ocean Drilling Program, doi:10.2973/odp.proc.ir.200.2003.

Suess, E., et al., 1988, Proceedings of the Ocean Drilling Program, Initial Reports, vol. 112, College Station, TX, Ocean Drilling Program, doi:10.2973/odp.proc.ir.112.1988.

Tamaki, K., et al., 1990, Proceedings of the Ocean Drilling Program, Initial Reports, vol. 127, College Station, TX, Ocean Drilling Program, doi:10.2973/odp.proc.ir.127.1990.

Tracey, J. I. Jr., et al., 1971, Initial Reports of the Deep Sea Drilling Project, vol. 8, US Government Printing Office, Washington, DC.

Westbrook, G.K., et al., 1994, Proceedings of the Ocean Drilling Program, Initial Reports, vol. 146 (Part 1), College Station, TX, Ocean Drilling Program, doi:10.2973/odp.proc.ir.146-1.1994.

Wilson, D.S., et al., 2003, Proceedings of the Ocean Drilling Program, Initial Reports, vol. 206, College Station, TX, Ocean Drilling Program, doi:10.2973/odp.proc.ir.206.2003.

NAME

Appendix

Core Photos

Photos of each of the cores used in this exercise are included on the following pages. The order of the core photos matches the order of the cores listed in Table 2.1.

High resolution photos for each of the cores included in this exercise (and many others) can be easily accessed by going directly to the Integrated Ocean Drilling Program searchable database online of core photos. The steps are:

1. Go to: http://iodp.tamu.edu/database/coreimages.html.
2. Under "whole core photos" select the Janus database Core Photo query: http://iodp.tamu.edu/janusweb/imaging/photo.shtml.
3. This will bring up a short online form to fill out. Type in the core ID information (e.g., Leg-Site-Hole-Core). Note that you don't need to include all of the other information on the form. Then select "submit this request".
4. A new window should open with a list of files that should include a high resolution photo of the core of interest. Open or download that photo to view it.

NAME

Peru continental shelf, Pacific Ocean
112-687A-2

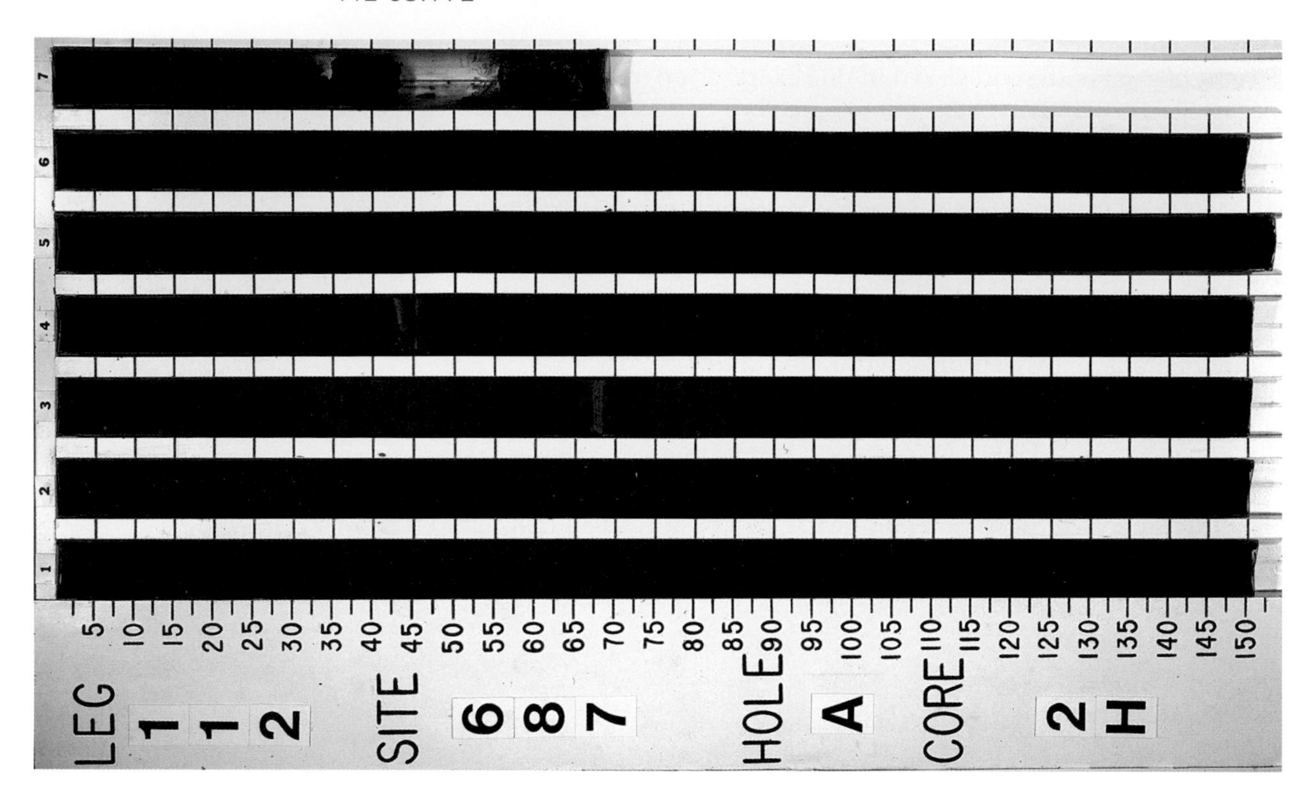

SE Pacific basin, North of Antarctica
35-324-1

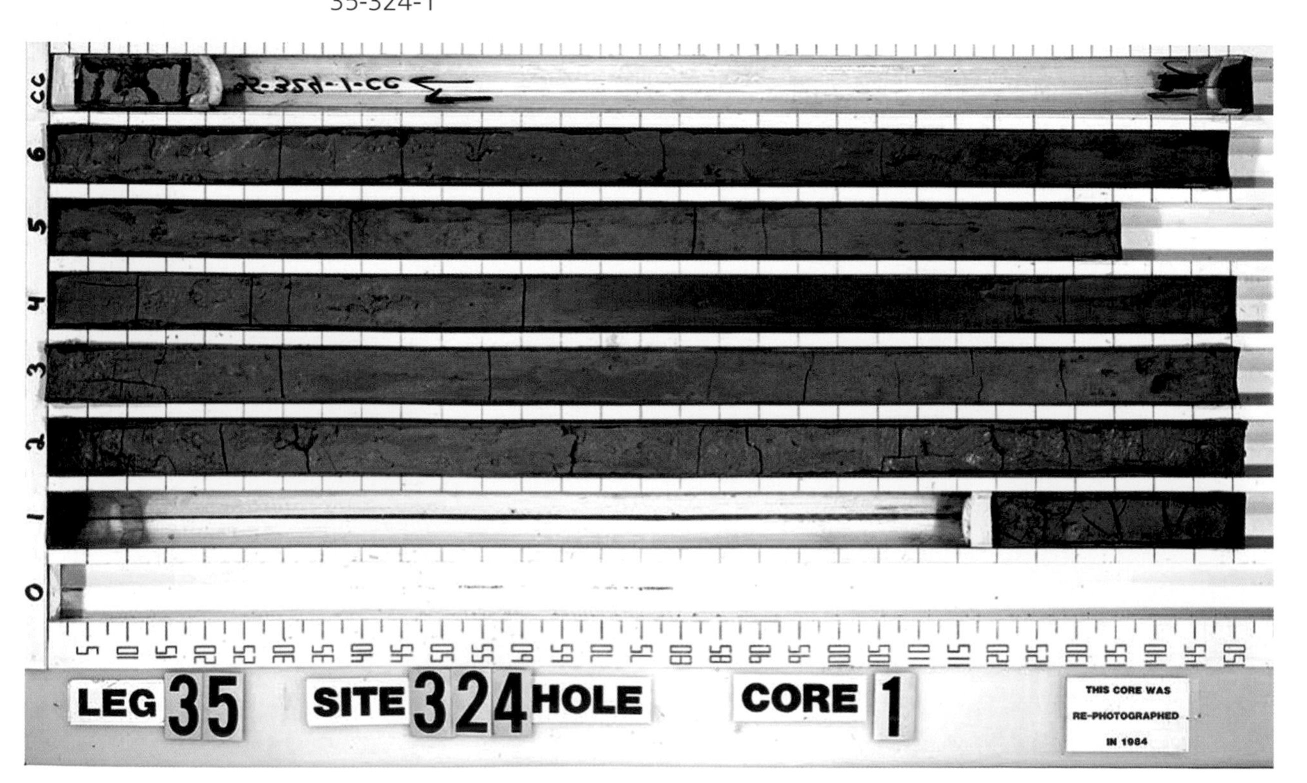

NAME

Ross Sea, South of Australia, Antarctic margin, Pacific Ocean
28-269-1

Chinook Trough, North Pacific abyssal plain
145-886B-2

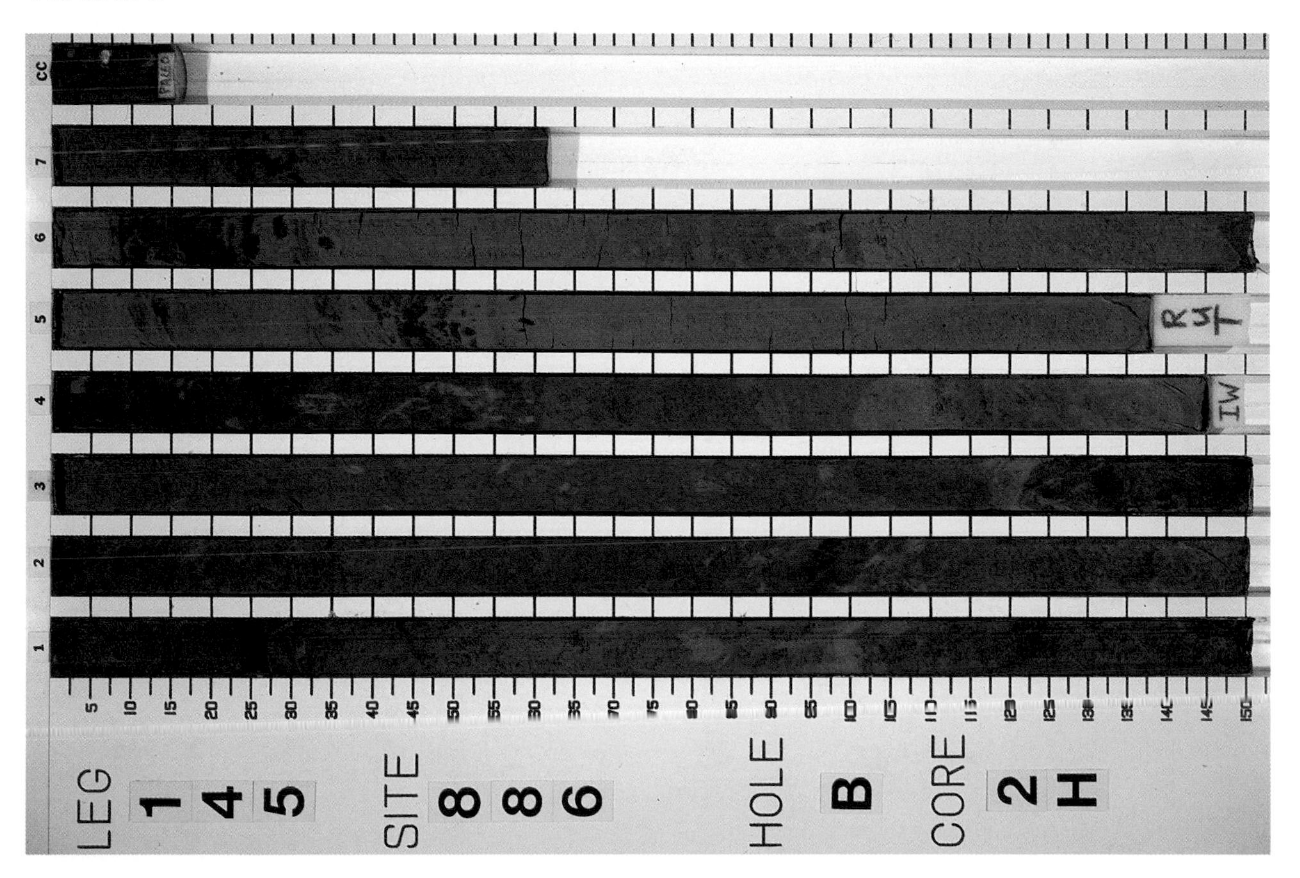

NAME

Detroit Seamount, NW Pacific
145-882A-2

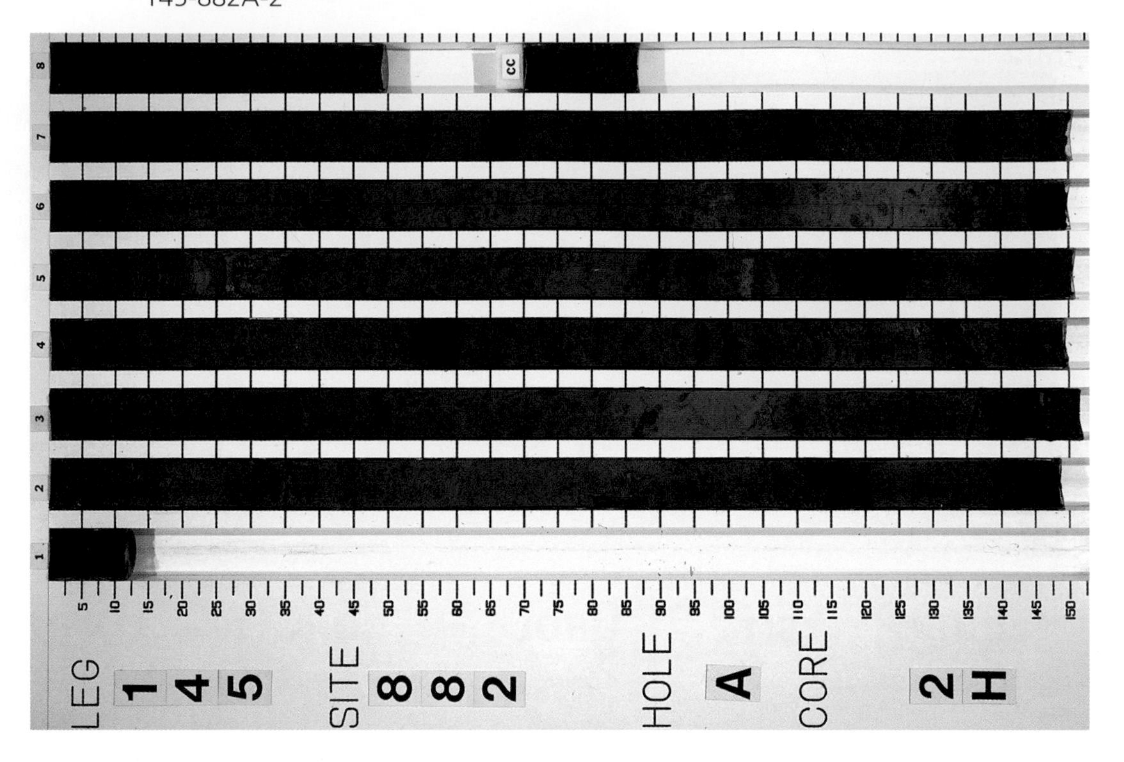

NW Pacific, east of the Sea of Okhotsk
145-881A-1

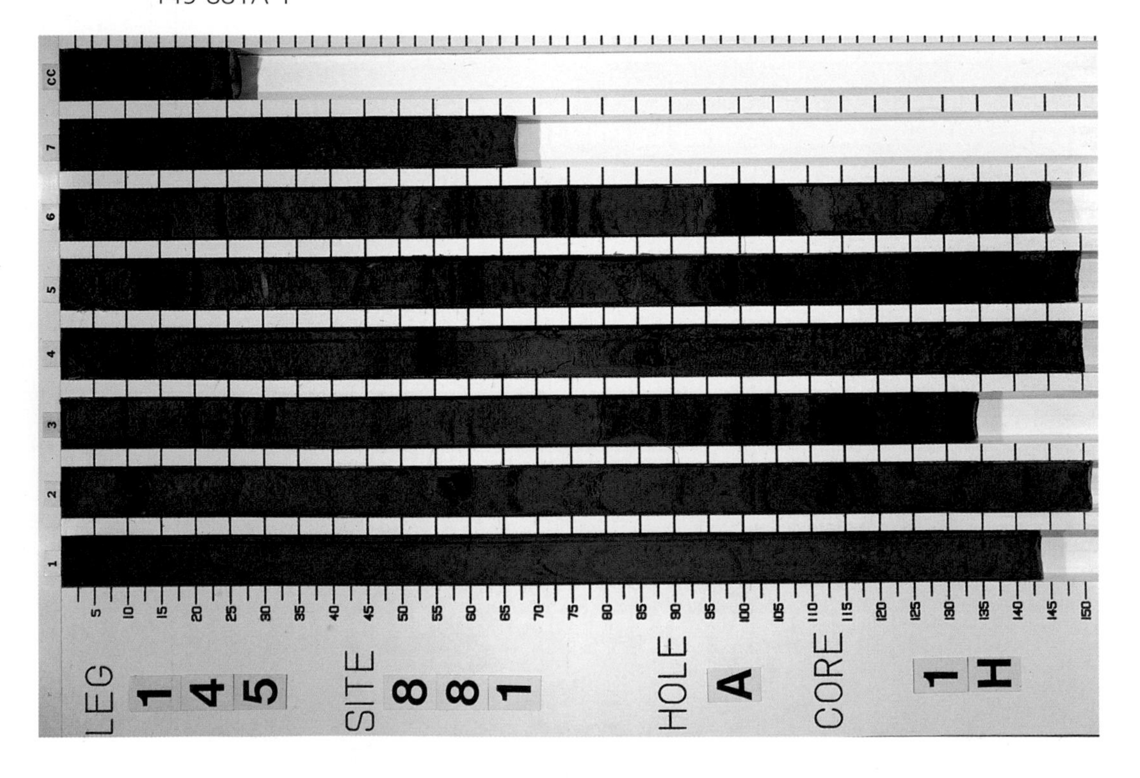

NAME

Patton-Murray Seamount, NE Pacific
145-887C-2

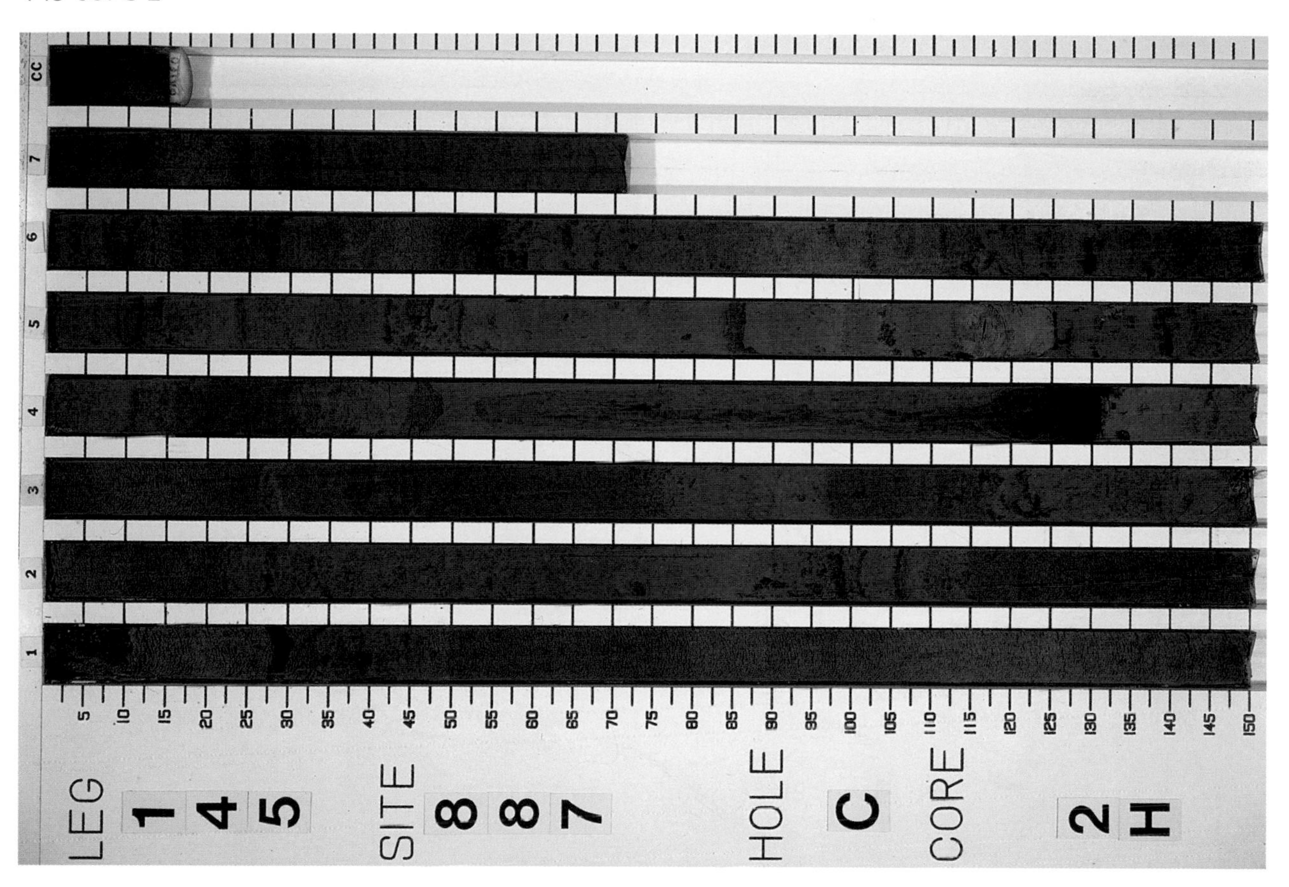

Bering Sea
19-188-2

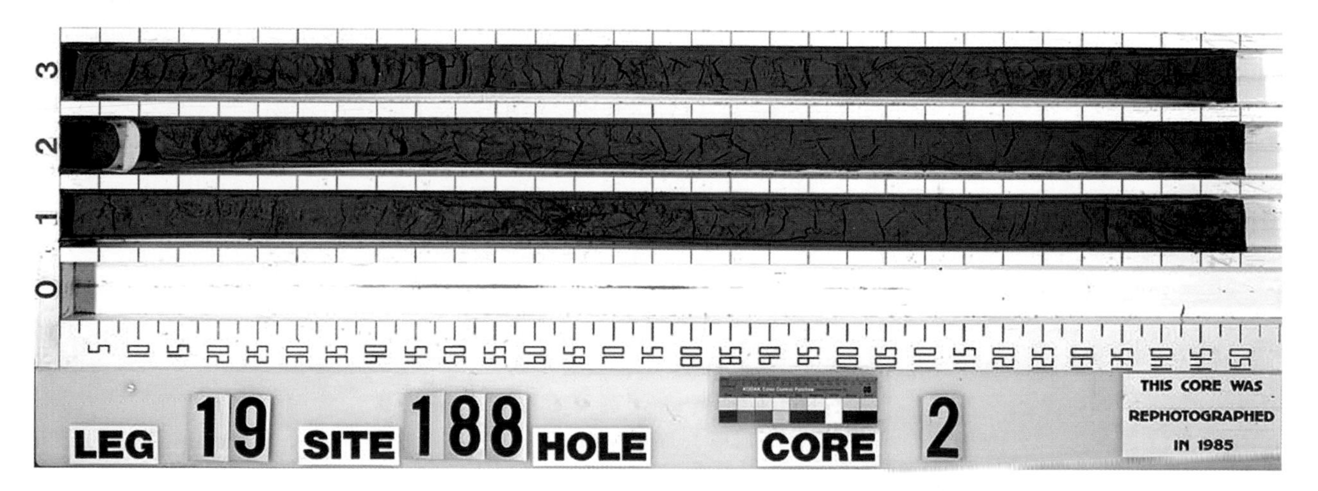

NAME

Alaskan continental slope, Pacific Ocean
18-182-1

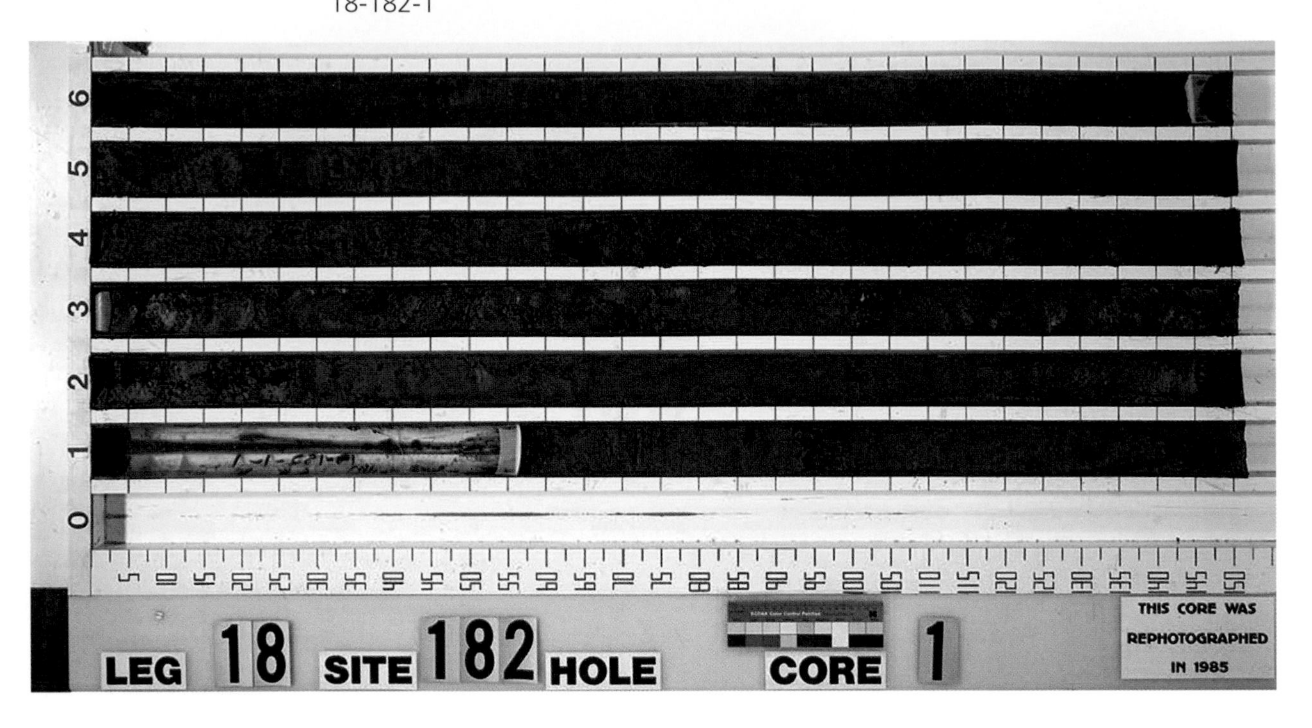

Line Islands Ridge, south central Pacific
33-318-2

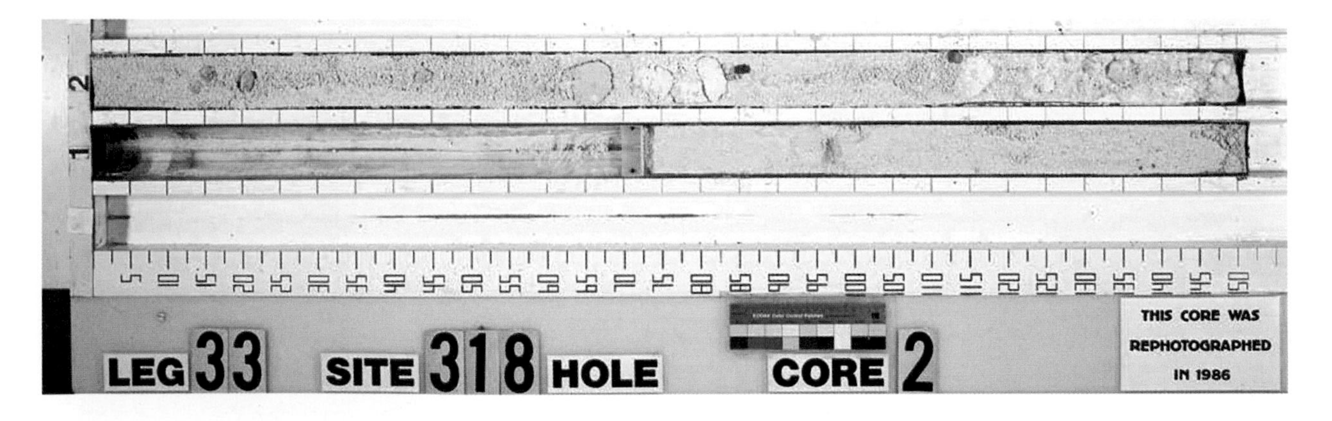

NAME

Marquesas Fracture Zone, central Pacific abyssal plain
8-75-1

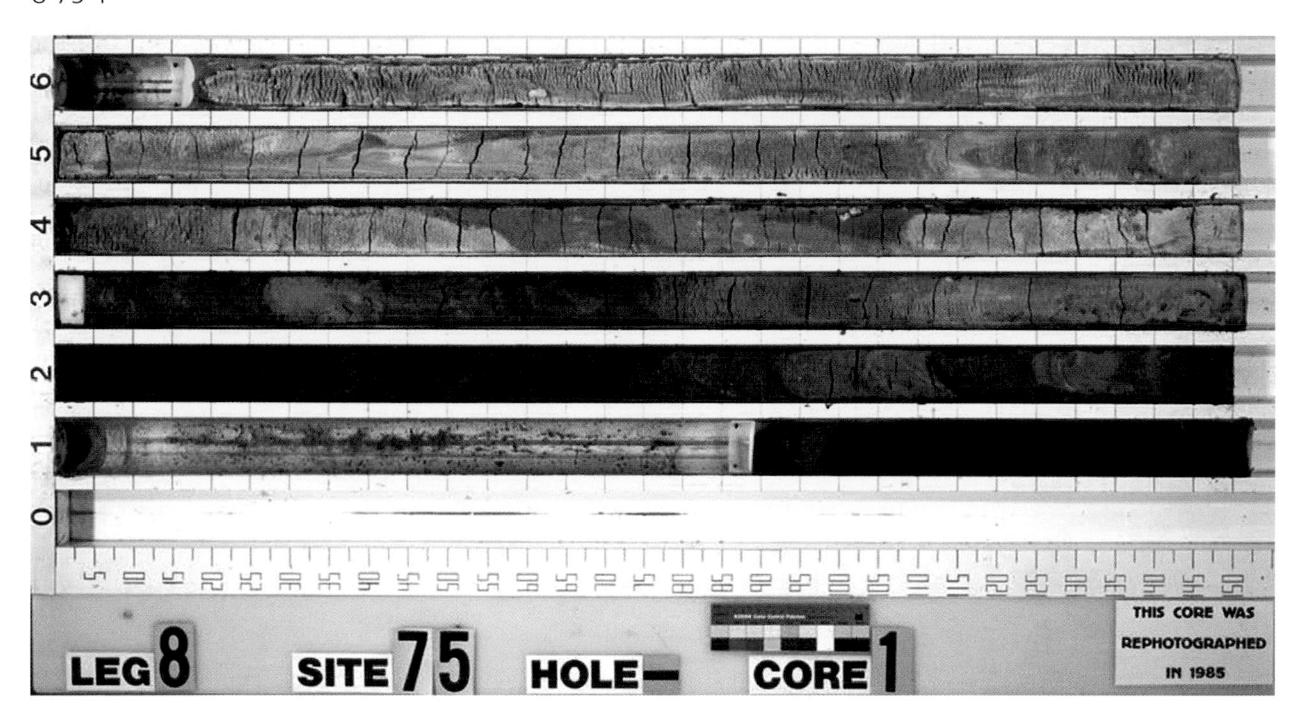

SE Pacific abyssal plain
92-597-1

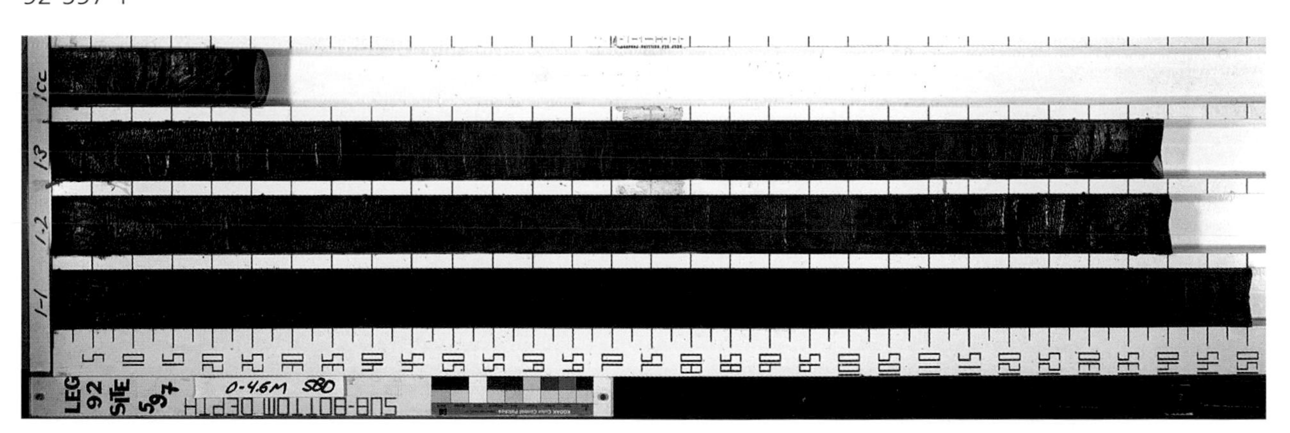

NAME

Antarctic Peninsula continental rise, Pacific Ocean
178-1101A-2

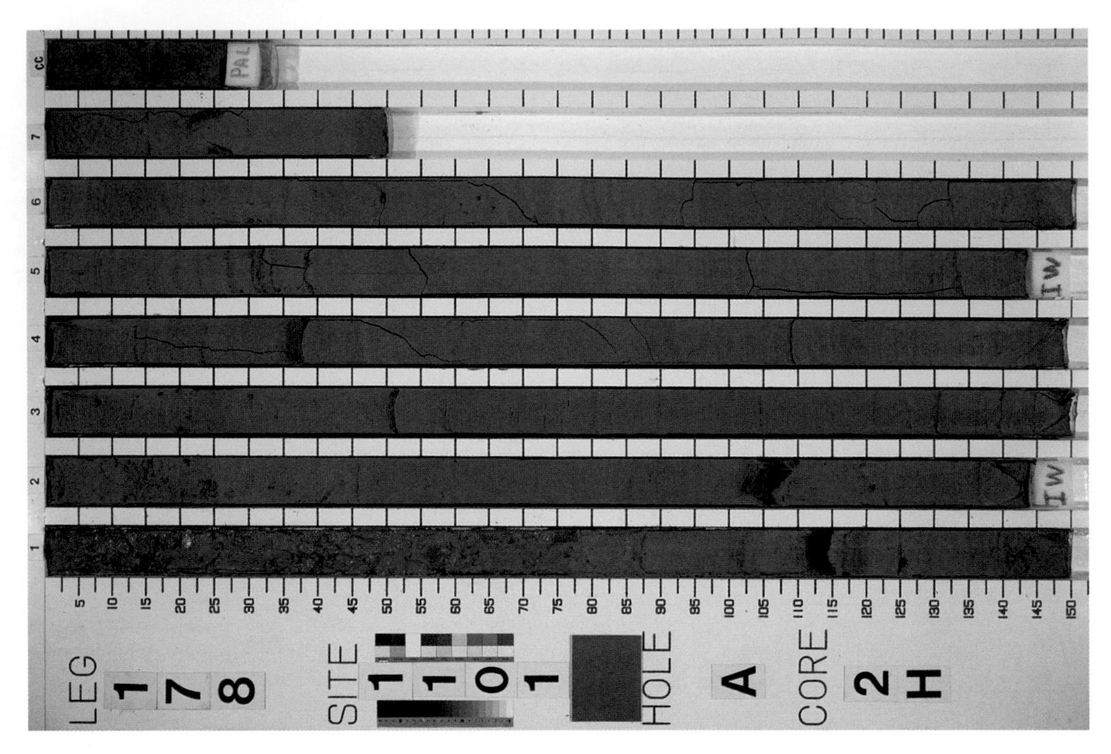

Antarctic Peninsula continental rise, Pacific Ocean
178-1096A-1

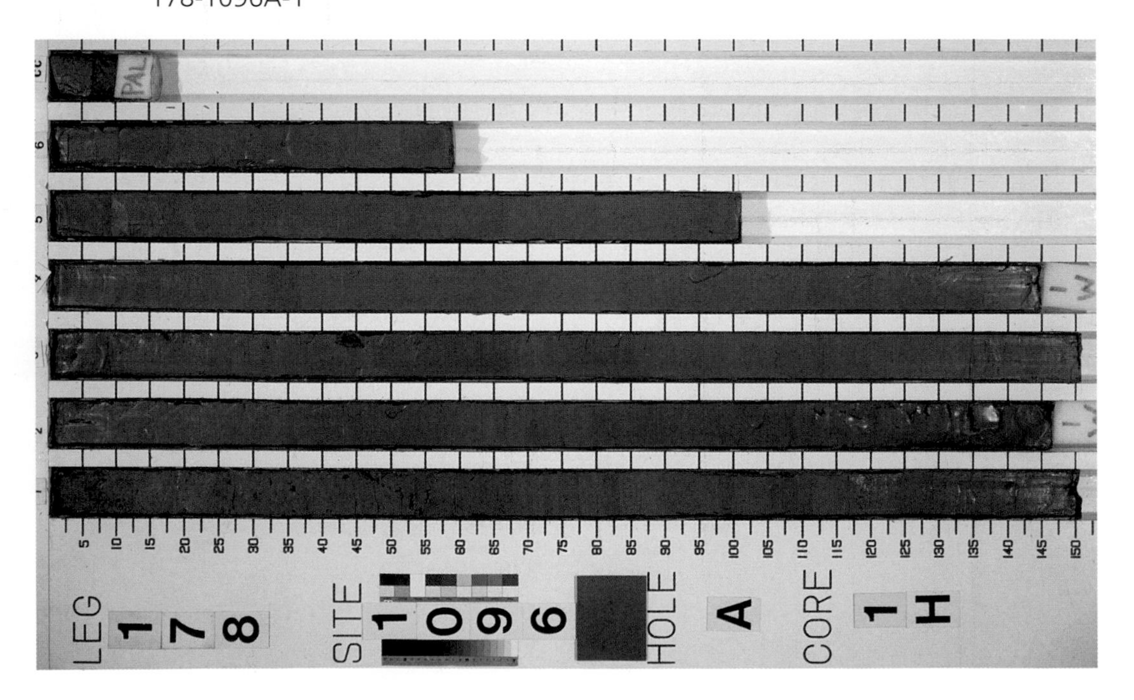

NAME

Antarctic Peninsula shelf, Pacific Ocean
178-1097A-3

South of New Zealand, Pacific Ocean
29-278-3

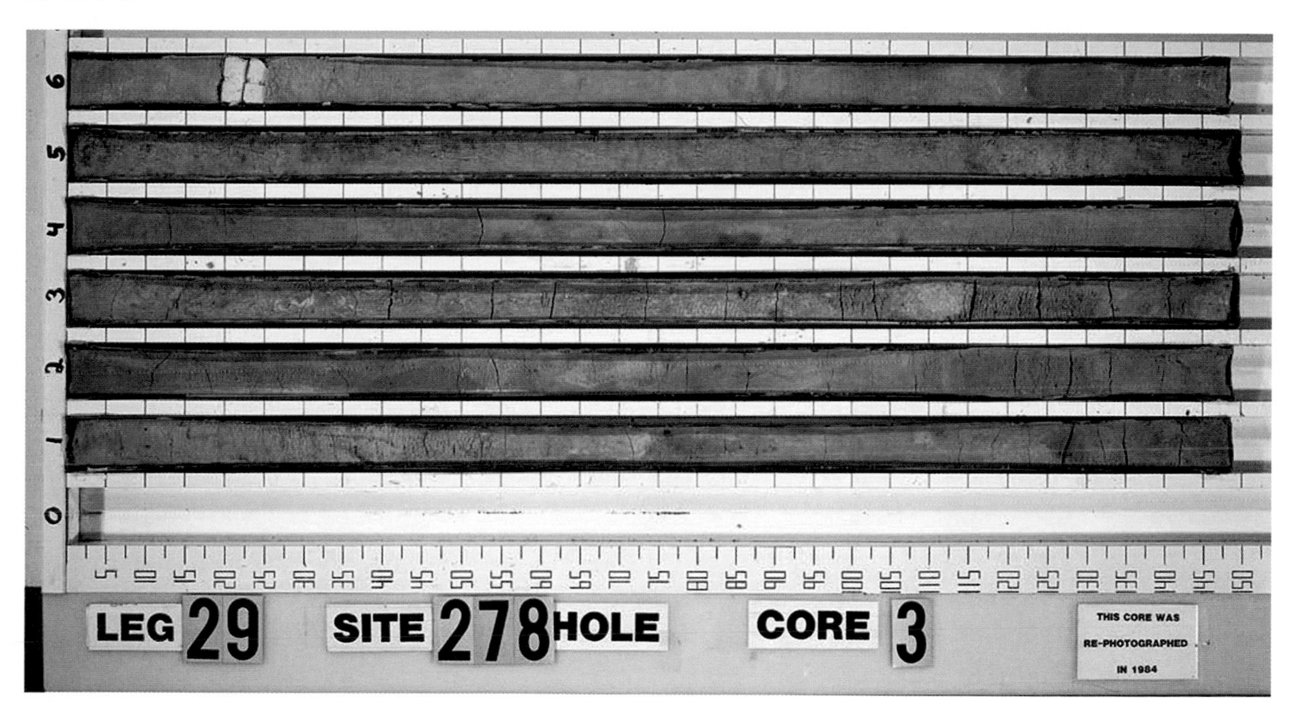

NAME

Nazca Ridge, SE Pacific
202-1236A-2

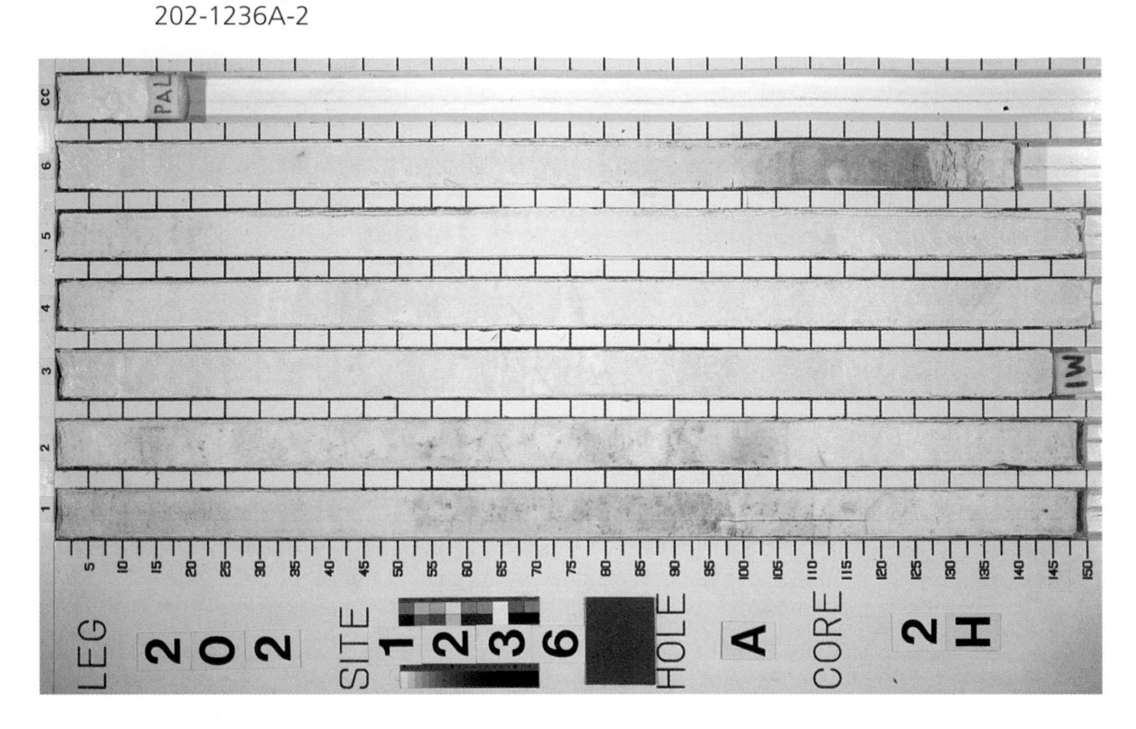

Guatemala Basin, eastern tropical Pacific
206-1256B-2

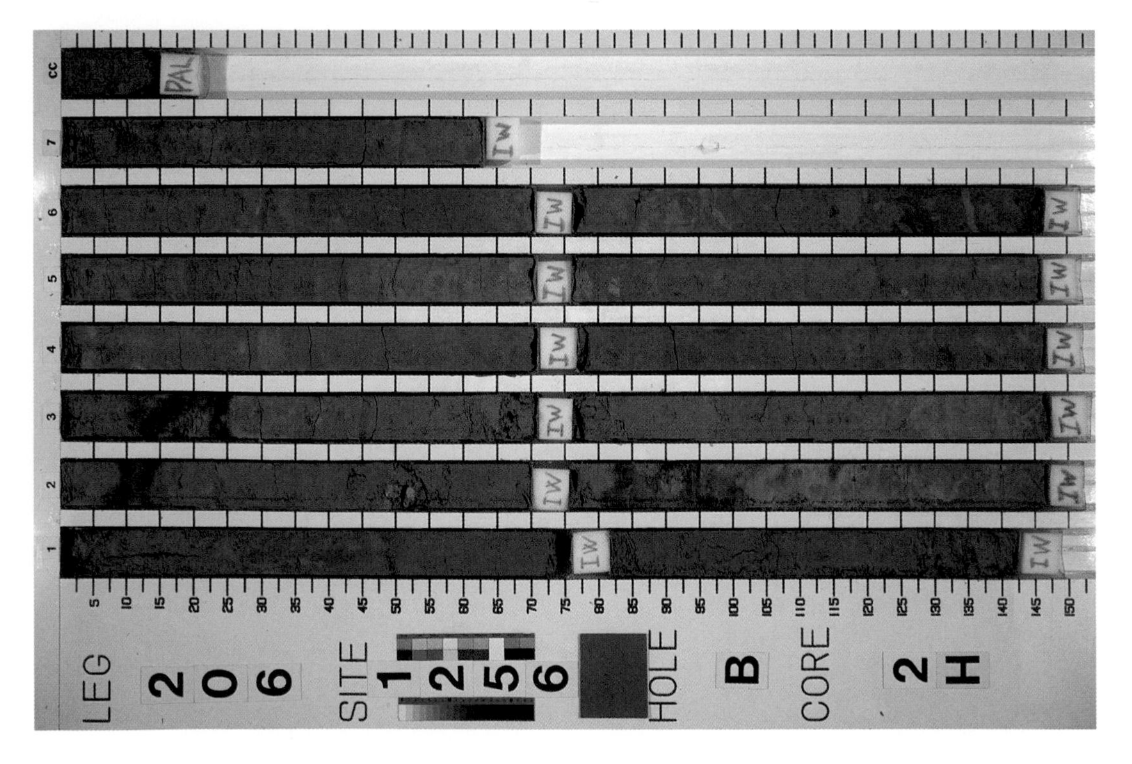

NAME

Clipperton Fracture Zone, central Pacific abyssal plain
8-74-1

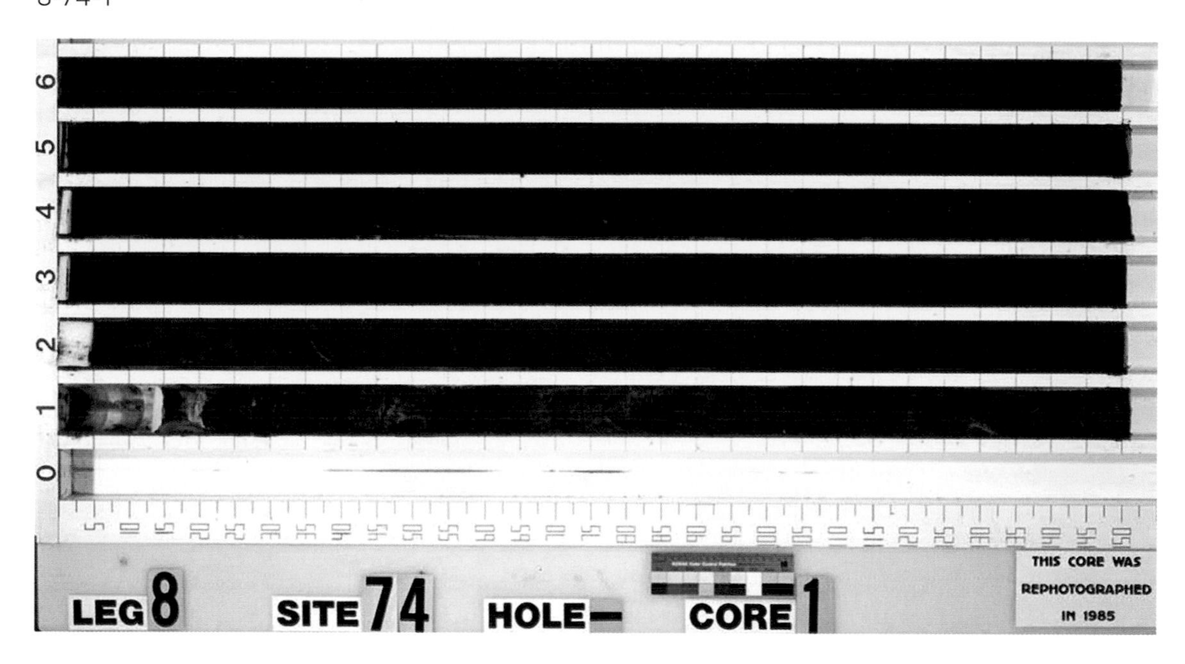

South of Hawaii, Pacific Ocean
136-842A-1

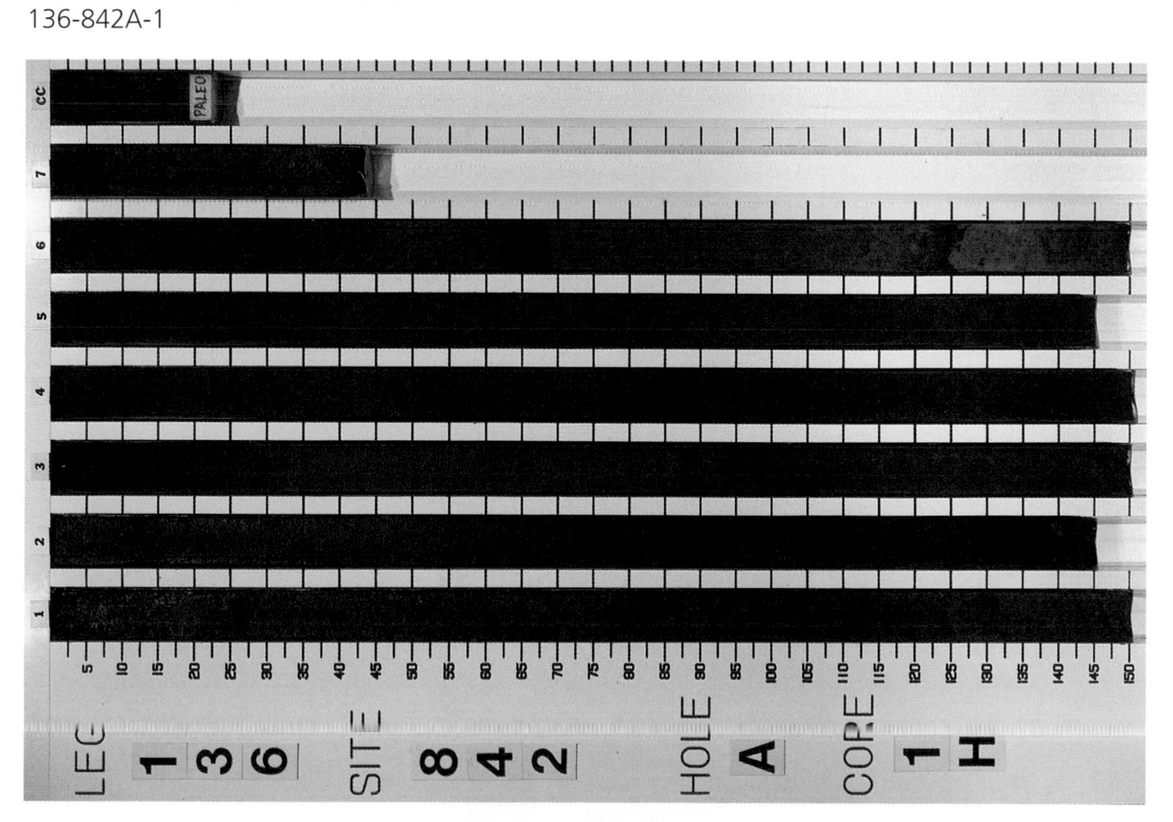

NAME

Shatsky Rise, NW Pacific
198-1209A-2

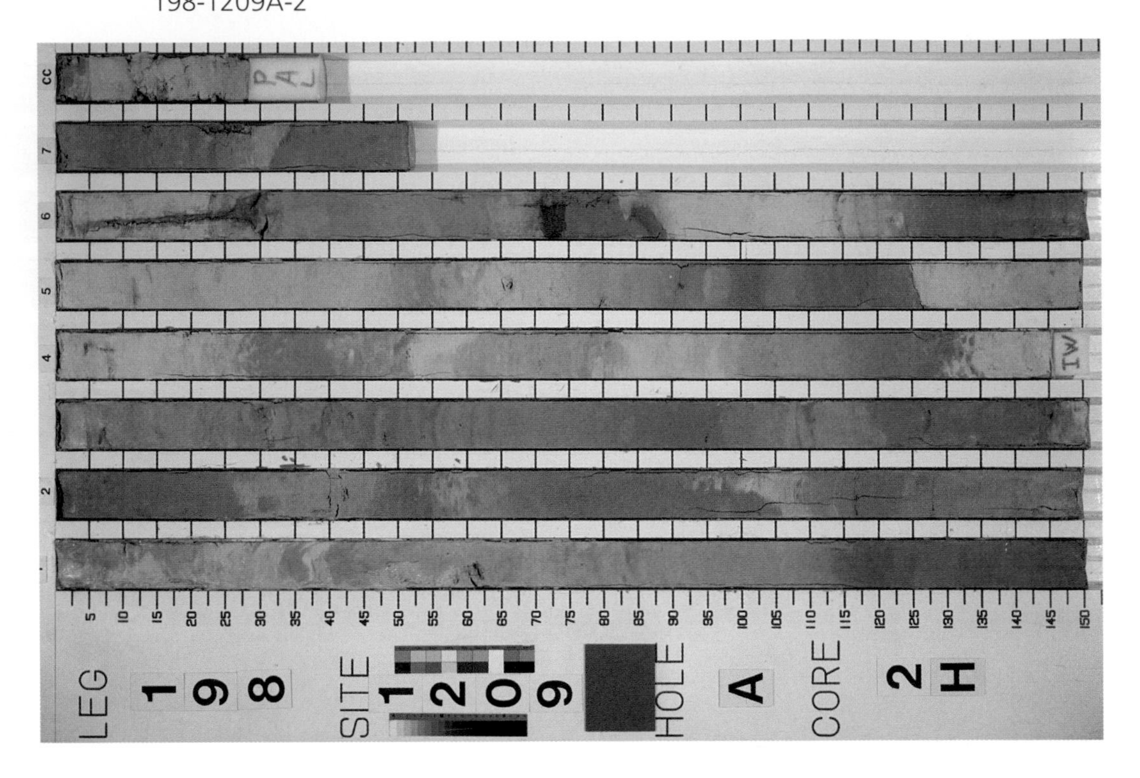

NE of Hawaii, North Pacific abyssal plain
199-1215A-2

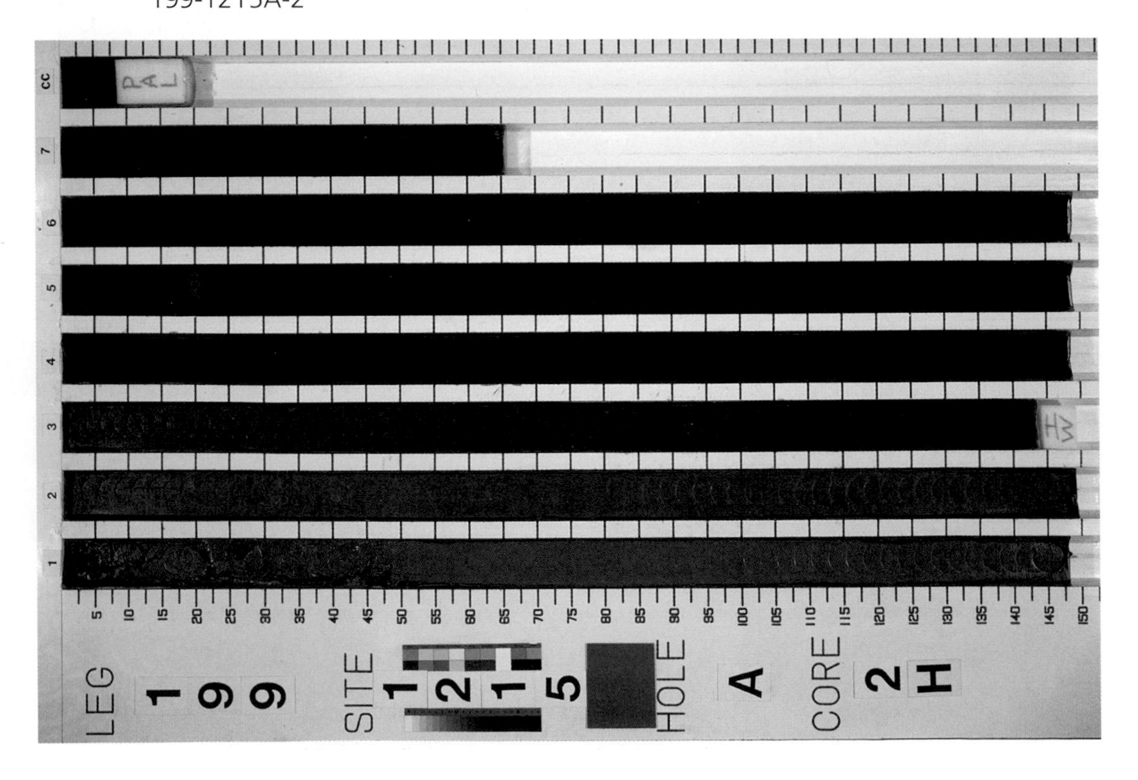

NAME ______________________________

West of Midway Island, North Pacific abyssal plain
86-576A-2

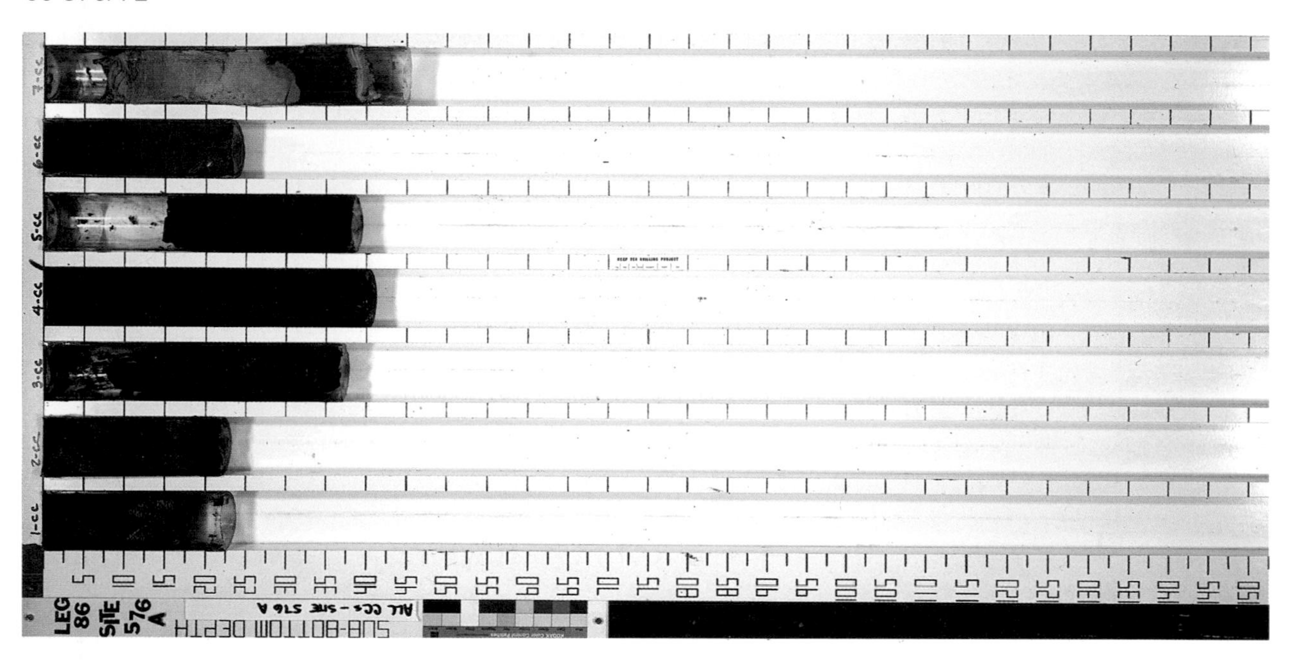

Philippine Sea
195-1201B-2

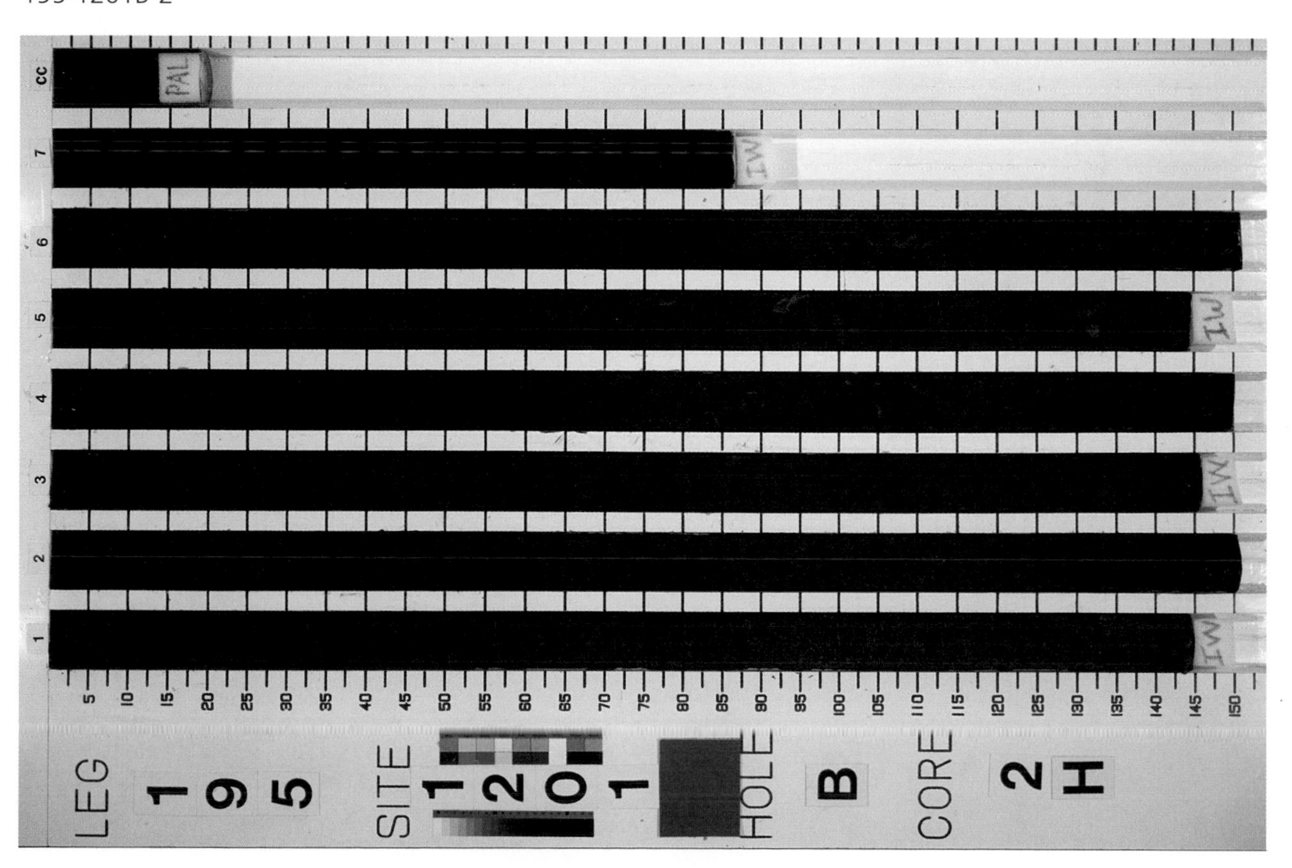

NAME

Ontong Java Plateau, western equatorial Pacific
130-807A-2

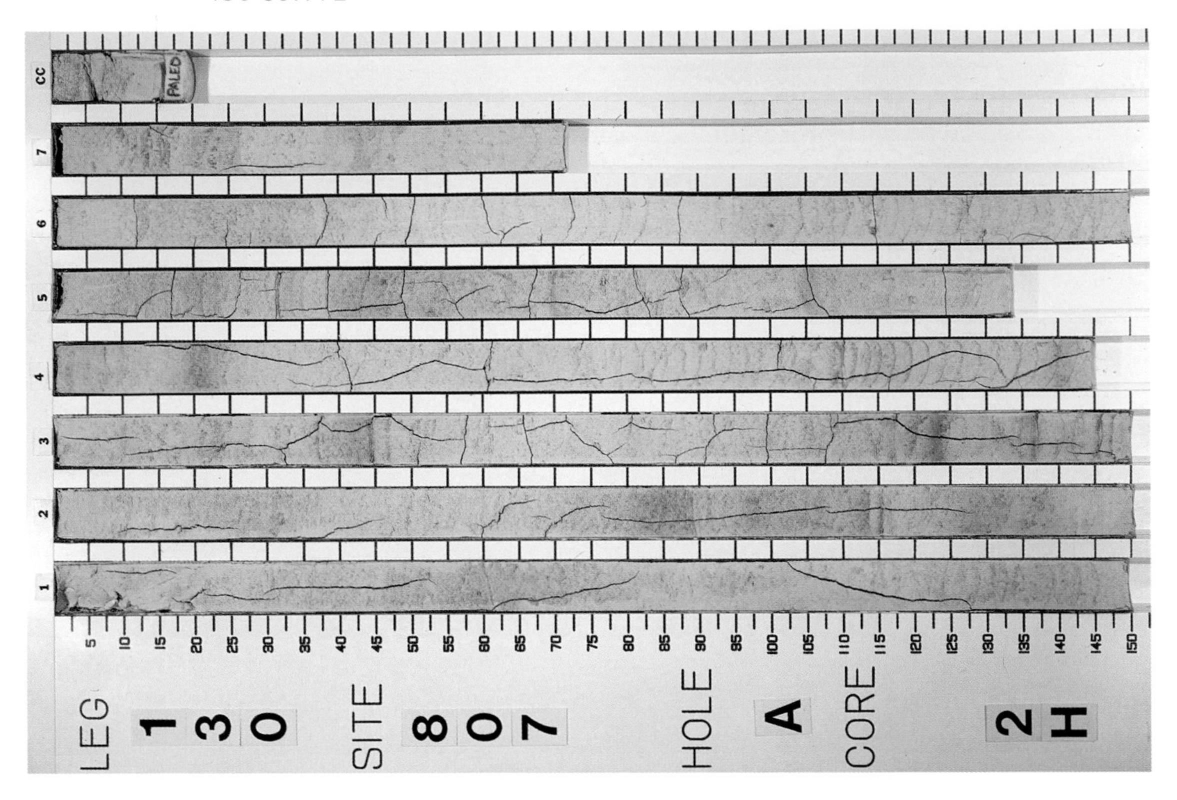

Chatham Rise, east of New Zealand
181-1125A-2

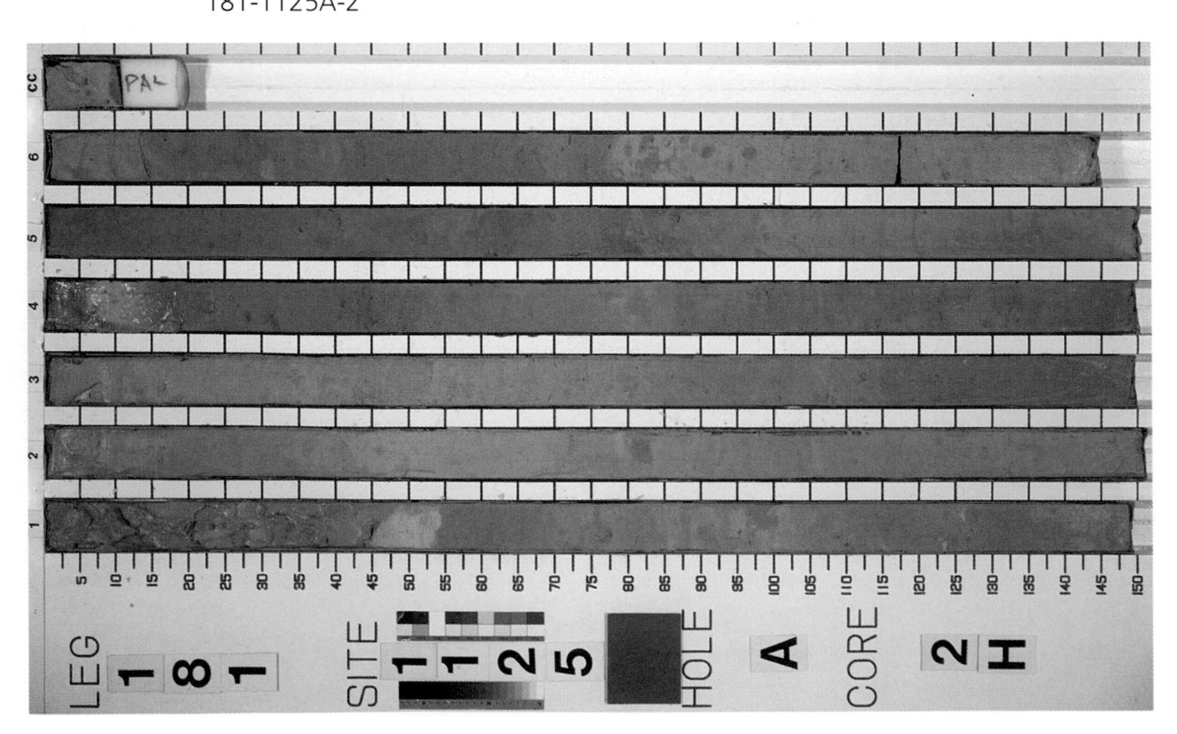

NAME

Escanaba Trough, west of Oregon and N. California
169-1037A-1

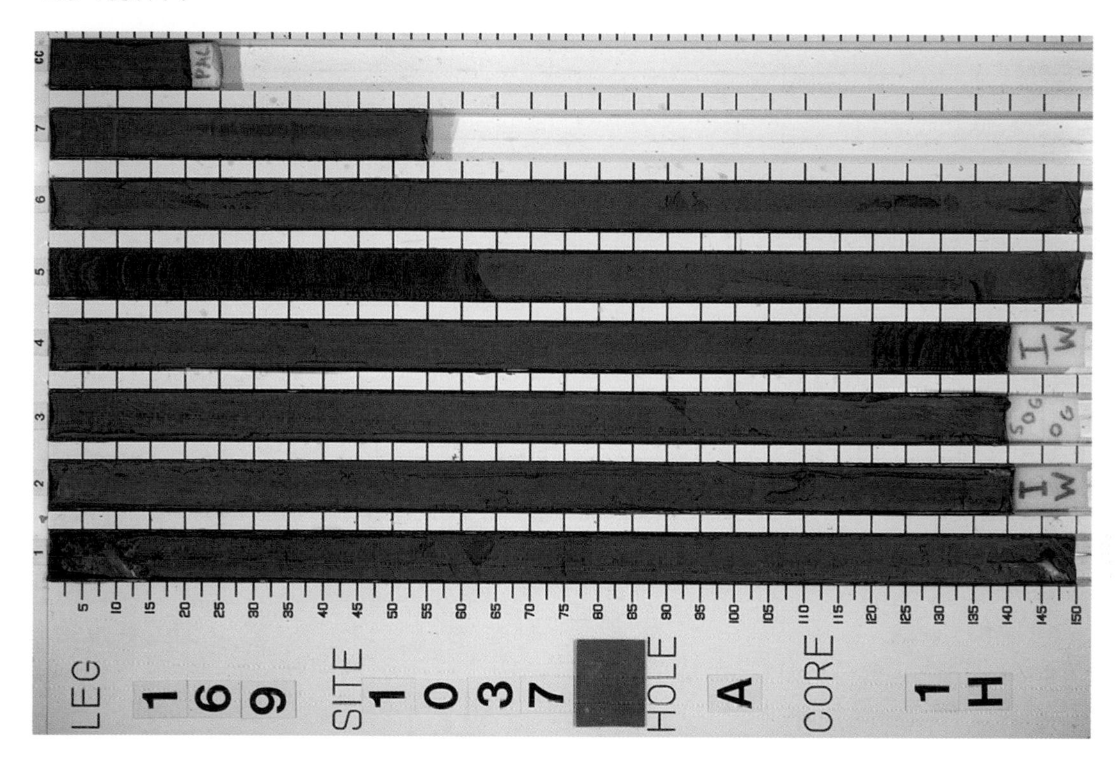

Cascadia margin, west of Vancouver, British Columbia
146-888B-2

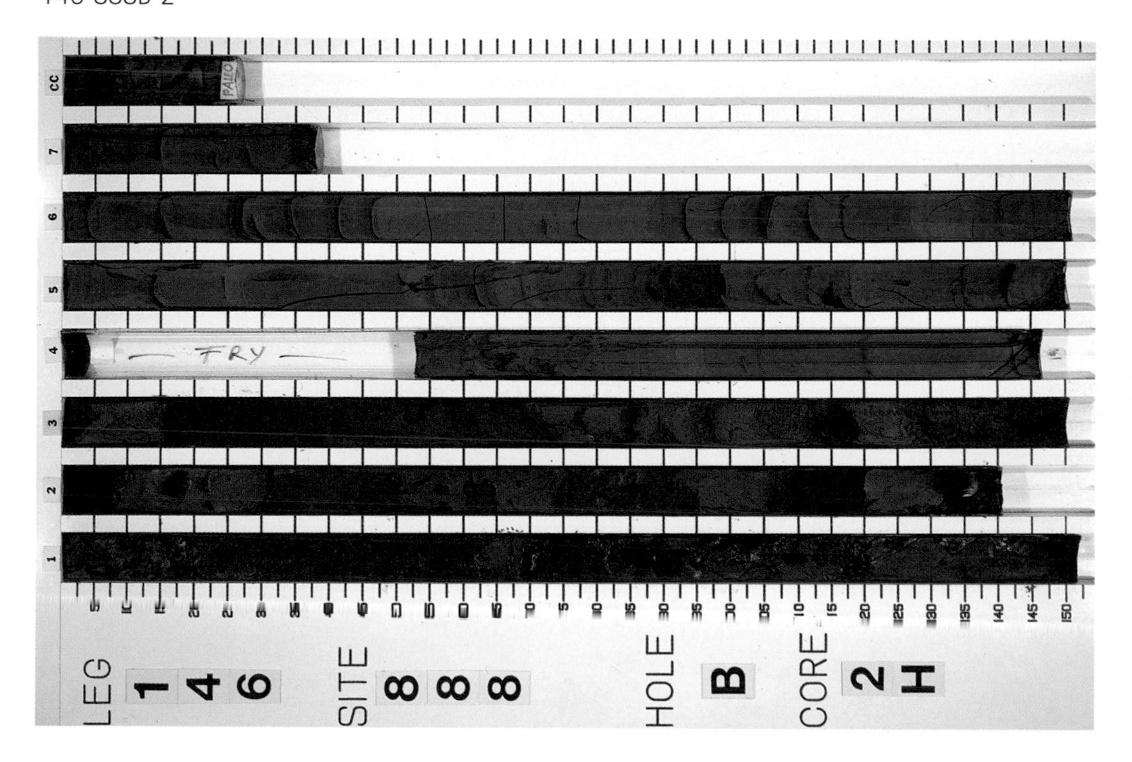

NAME

West of Baja California
167-1010E-1

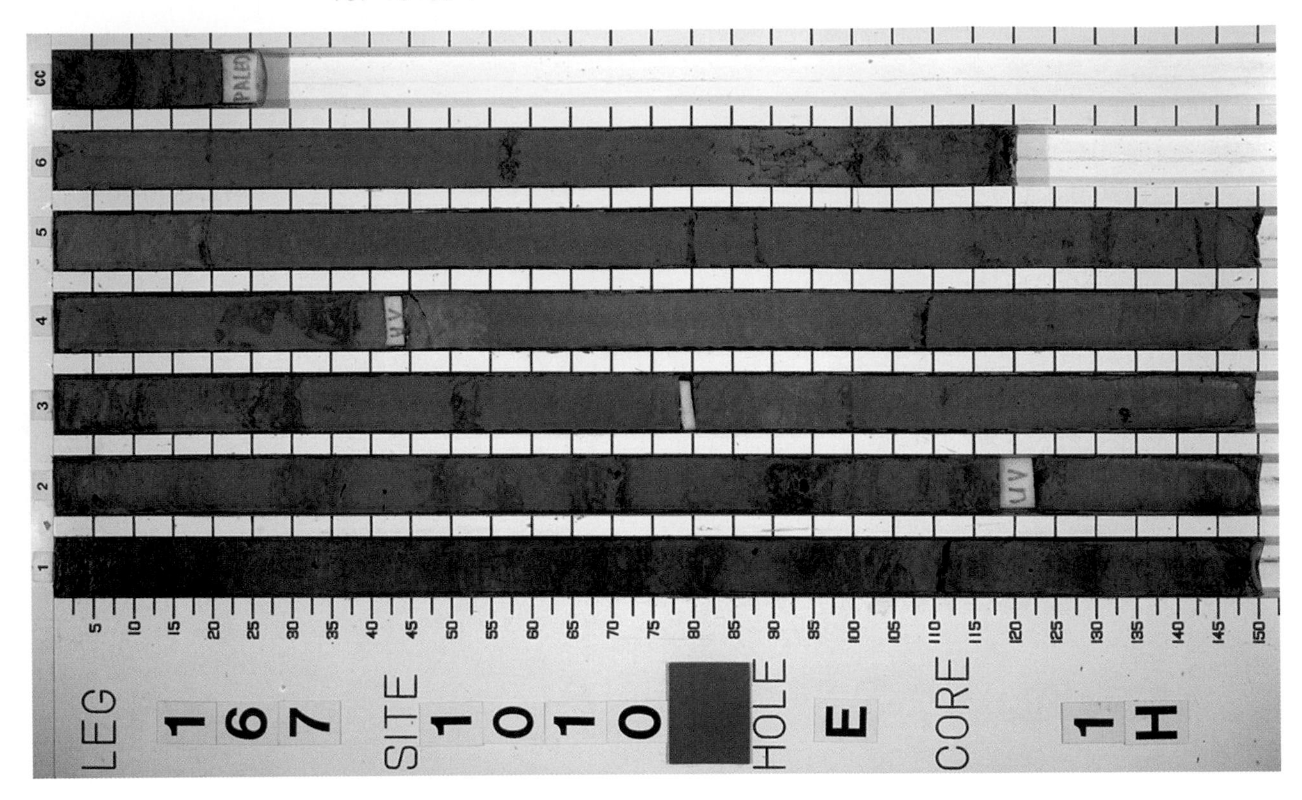

North Pacific abyssal plain, south of the Murray Fracture Zone
200-1224C-1

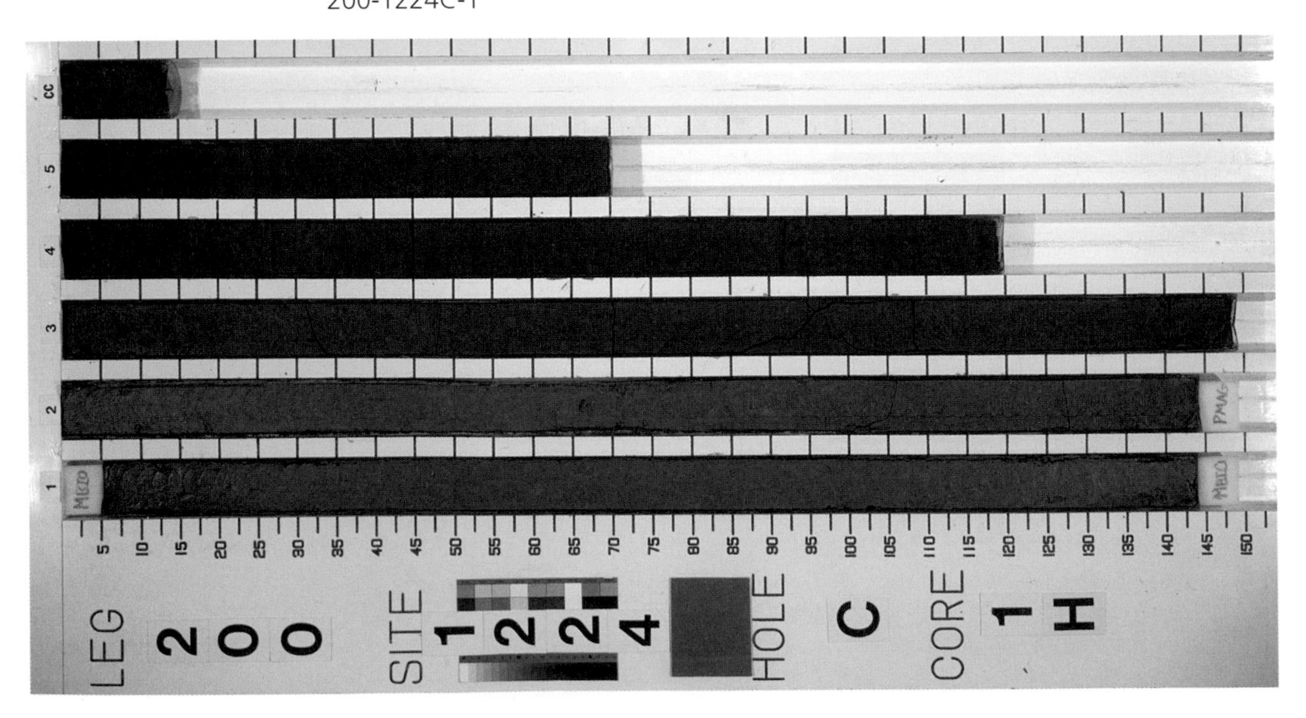

NAME

Japan Sea
127-795A-2

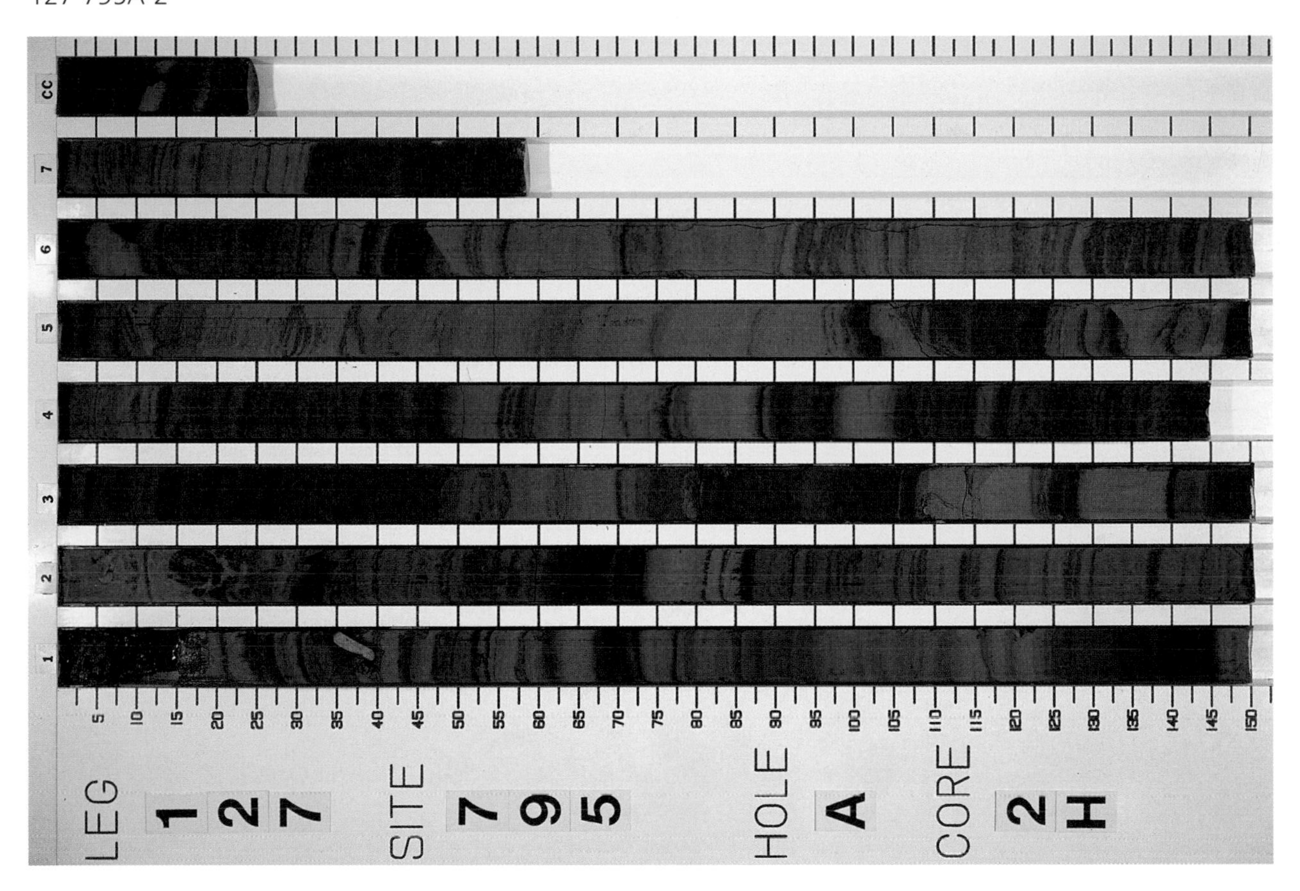

North of Ross Ice Shelf, Antarctic margin
28-274-2

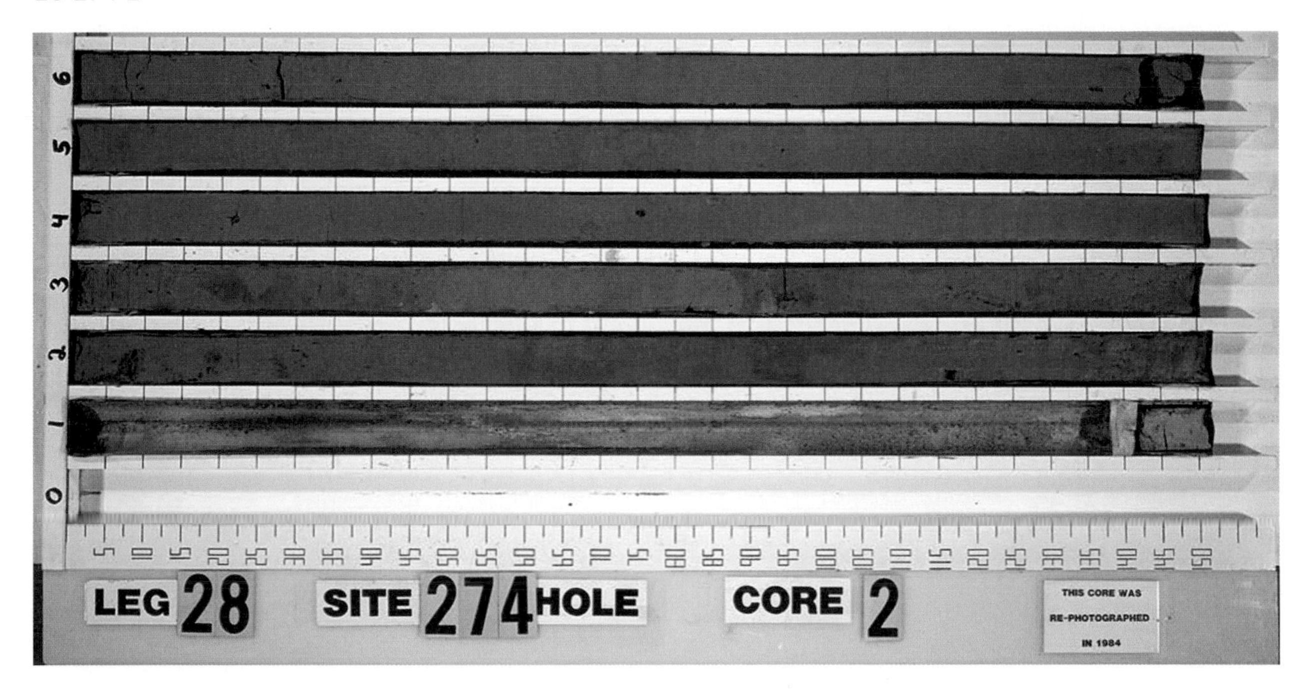

NAME

Western flank of Mid-Atlantic Ridge, North Atlantic Ocean
37-333-2

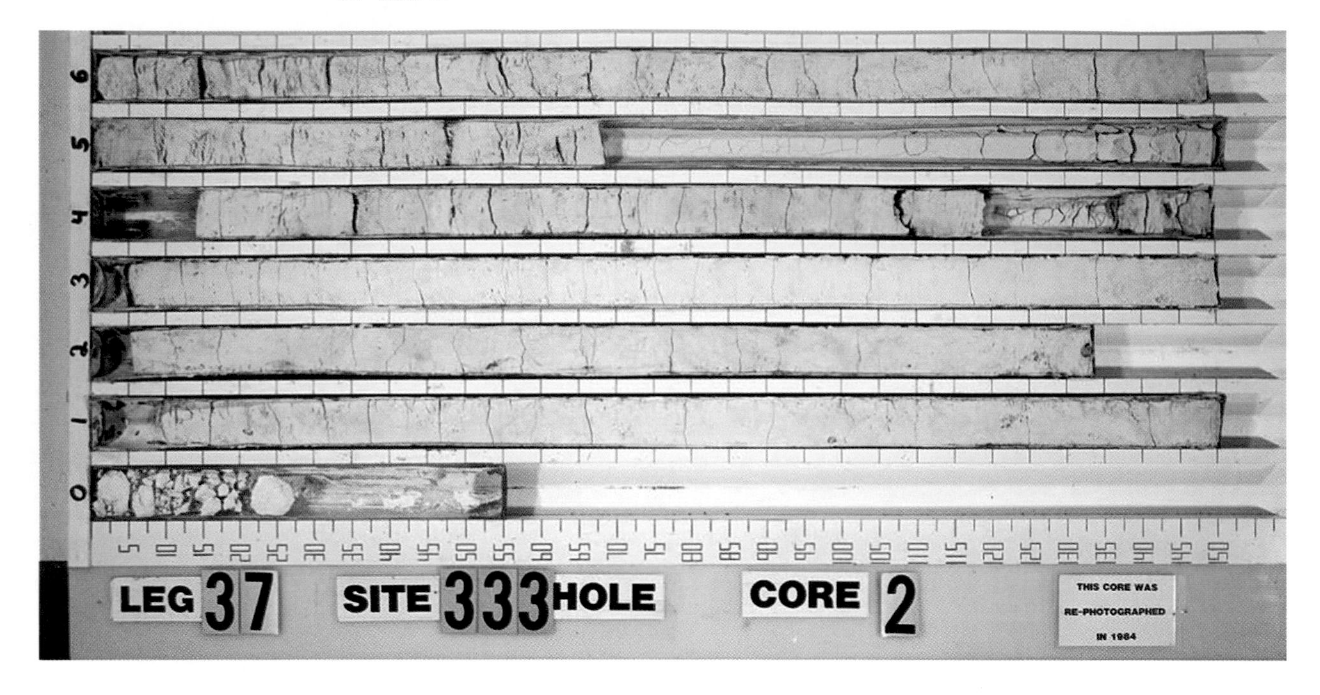

Western flank of Mid-Atlantic Ridge, North Atlantic Ocean
82-558-3

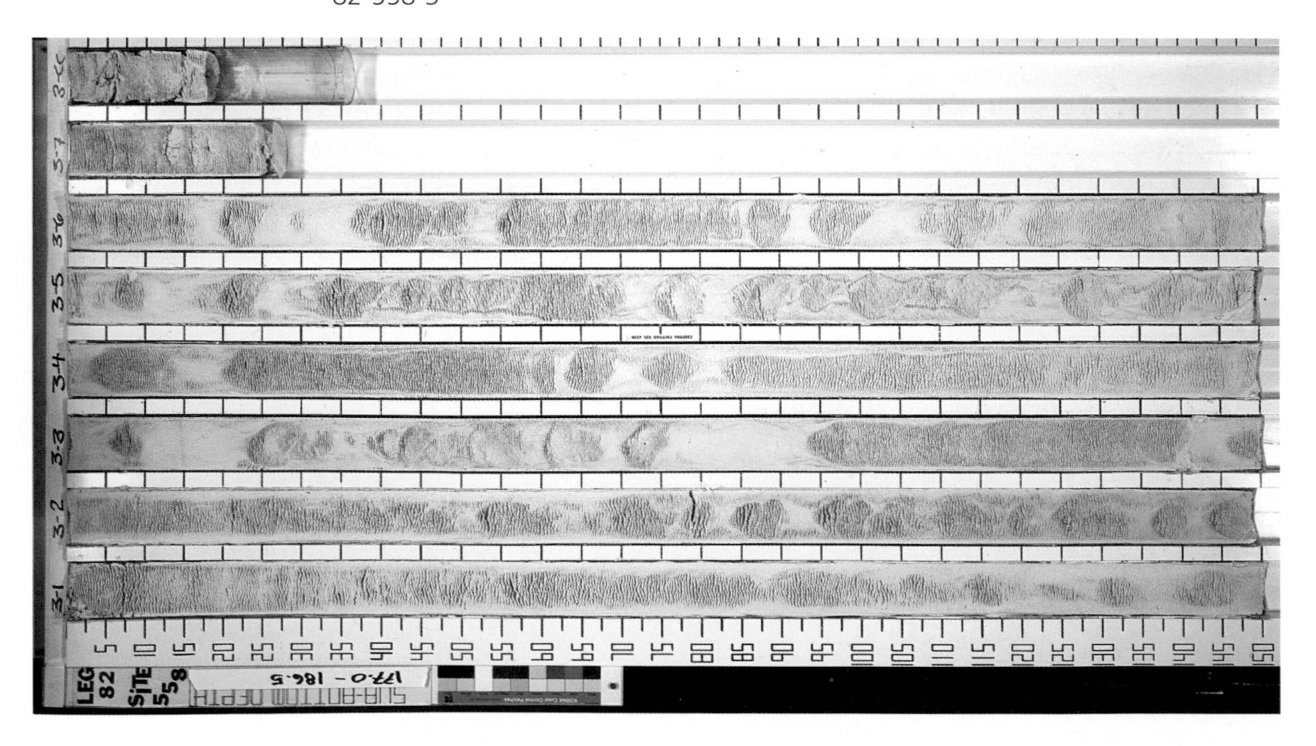

NAME

Northeast Bermuda Rise, North Atlantic Ocean
172-1063A-2

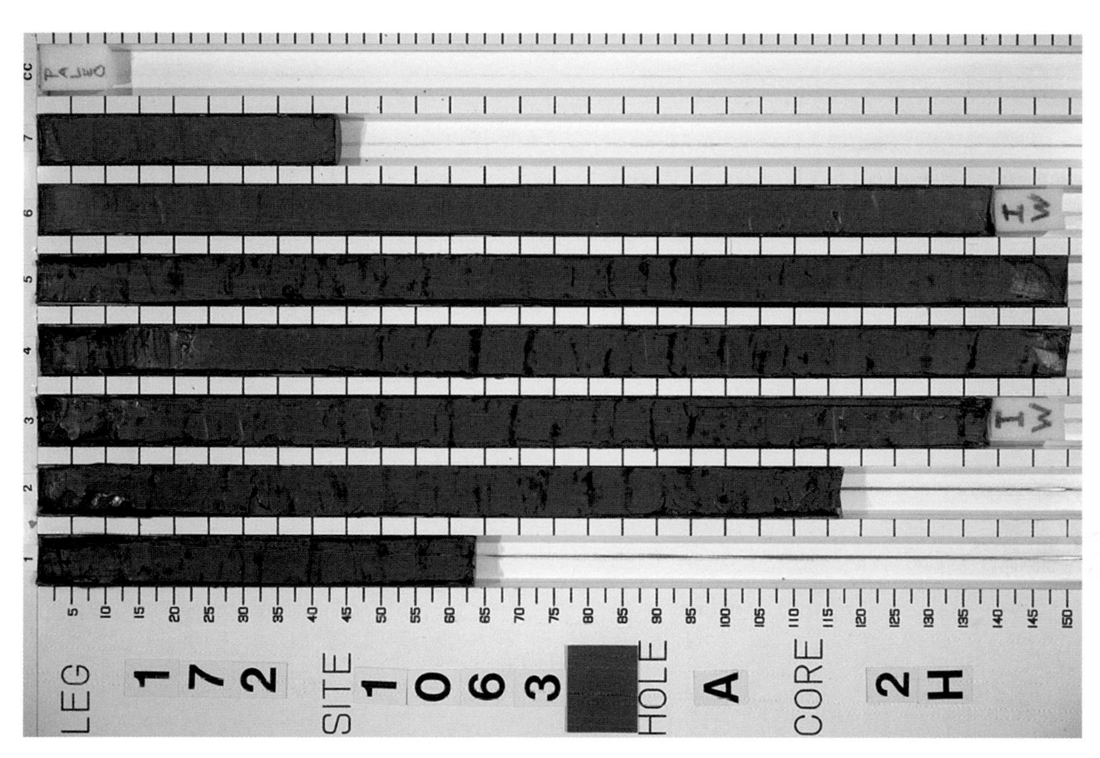

Labrador Sea, south of Greenland
105-646A-2

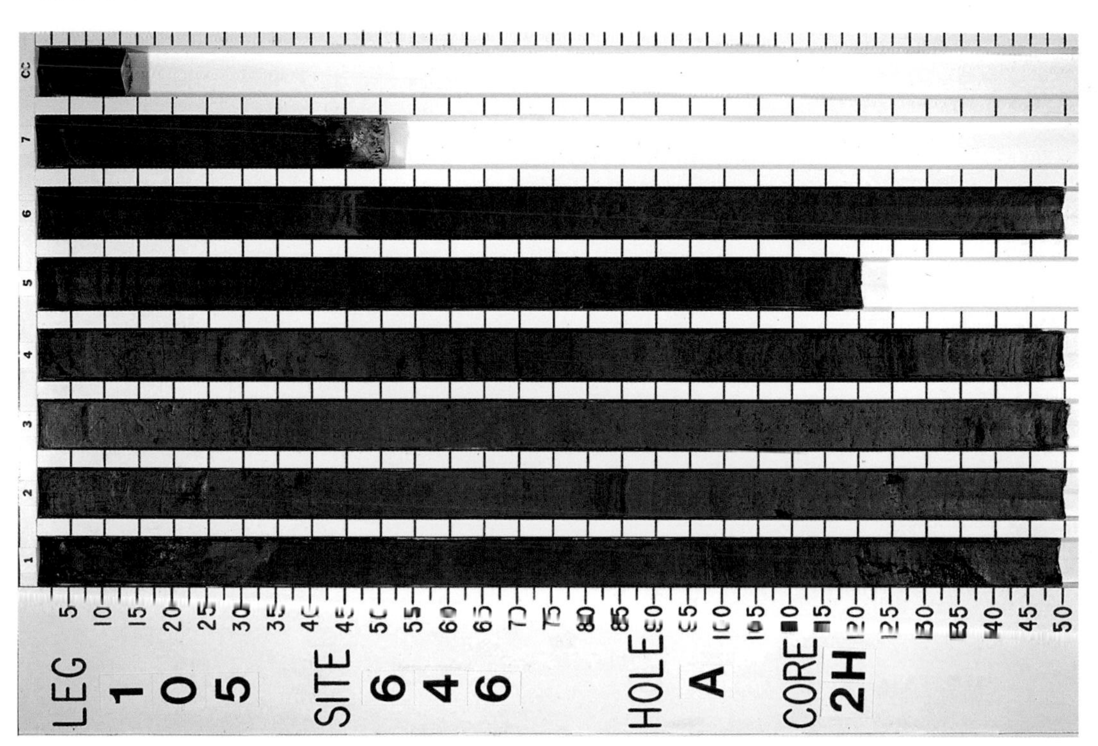

NAME

Rockall Bank, west of Ireland, North Atlantic Ocean
162-980A-2H

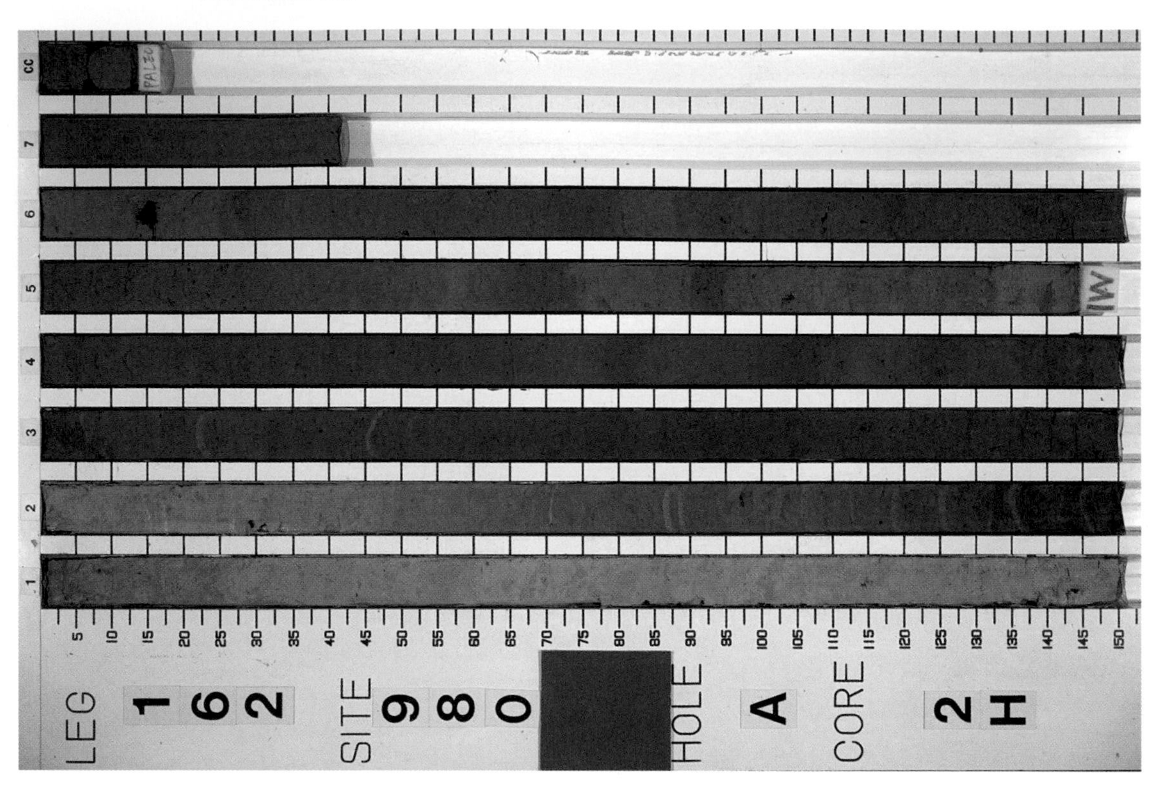

SE Greenland, continental rise, North Atlantic Ocean
152-919A-2

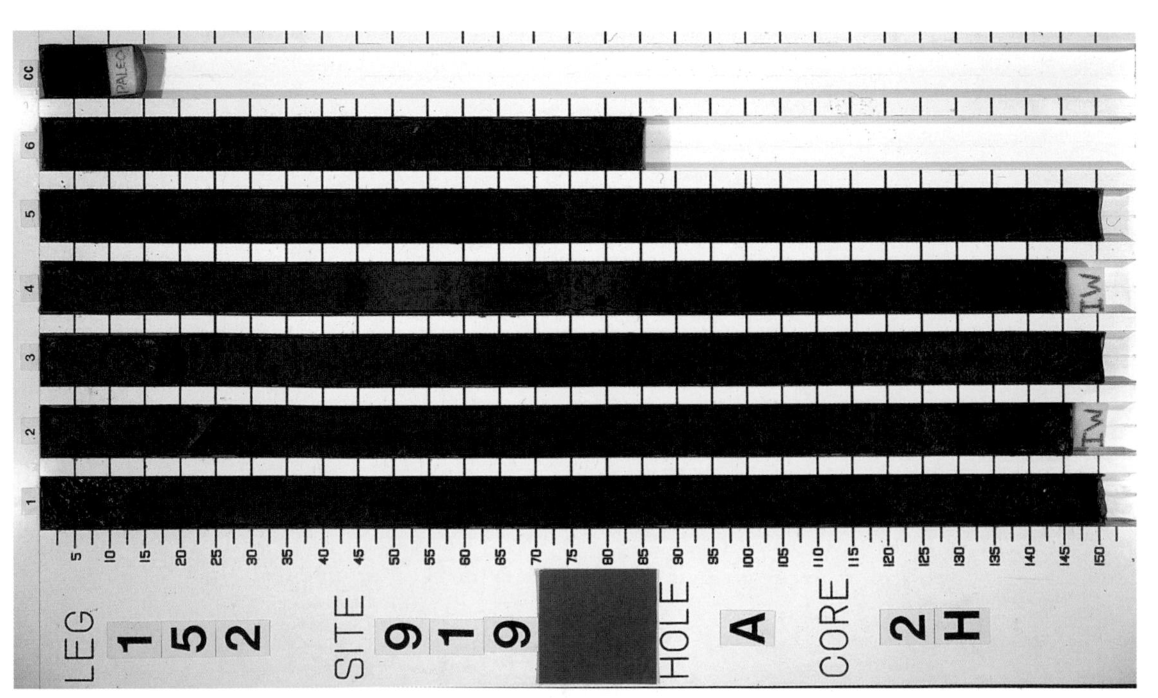

NAME

New Jersey continental shelf, North Atlantic Ocean
174-1073A-1

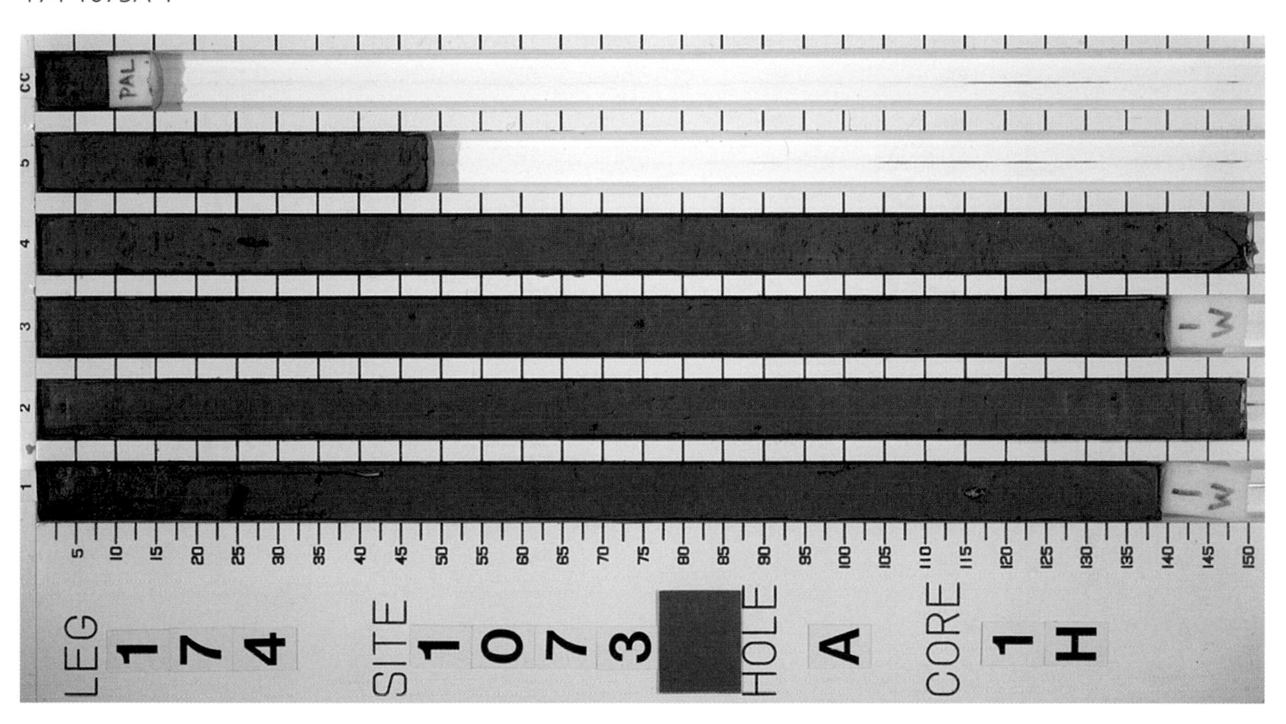

Madeira abyssal plain, North Atlantic Ocean
14-137-3

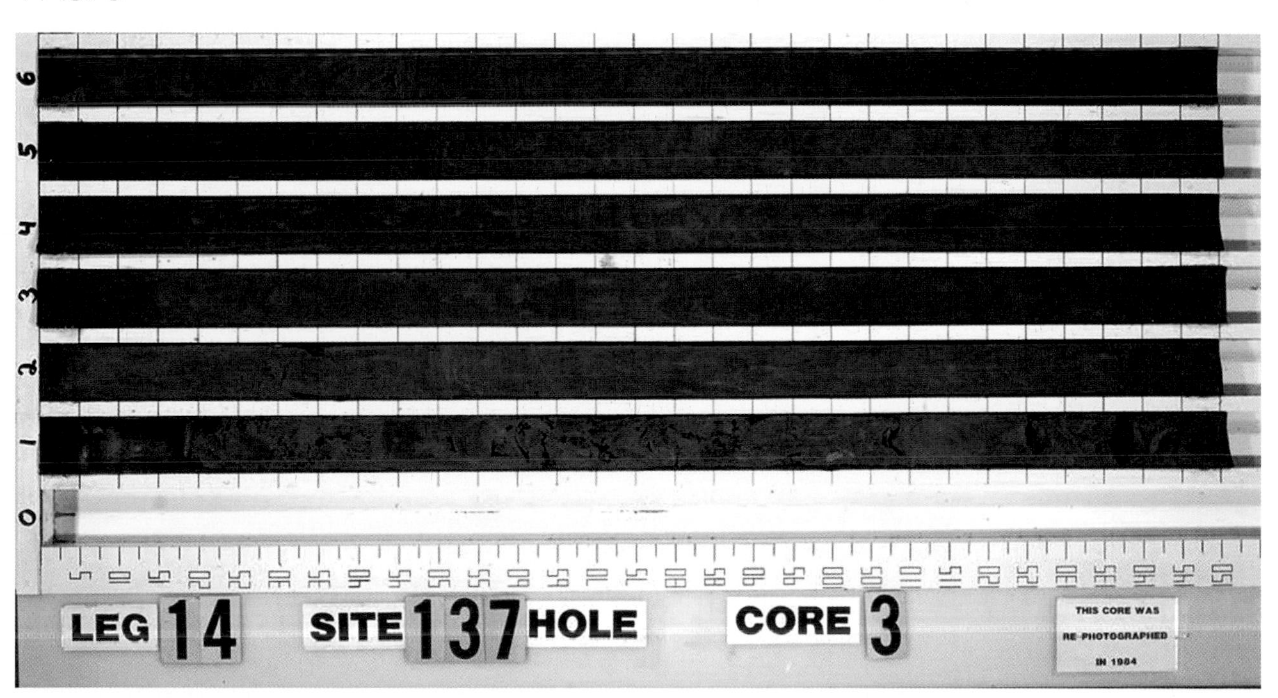

Chapter 3 Microfossils and Biostratigraphy

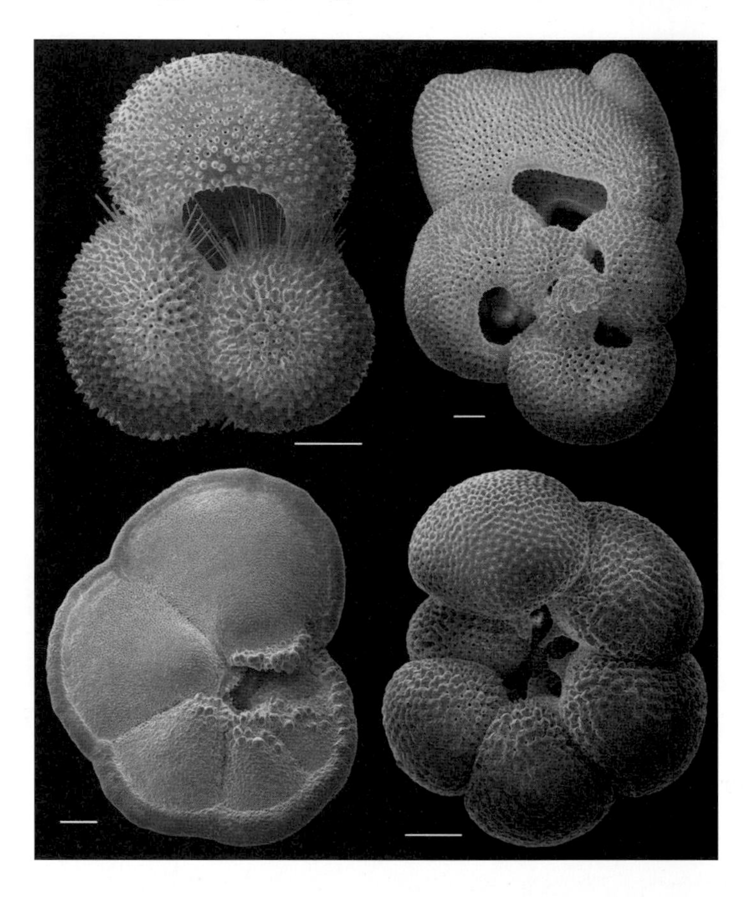

FIGURE 3.1. Four species of **planktic foraminifers** from the tropical western Atlantic: upper left: ***Globigerinoides ruber***, upper right: ***Globigerinoides sacculifer***, lower left: ***Globorotalia menardii***, lower right: ***Neogloboquadrina dutertrei***. Scale bar = 100 µm (100 µm = 0.1 mm). Photo courtesy of Mark Leckie.

SUMMARY

Microfossils are important, and in places the dominant constituents of deep sea sediments (see Chapter 2). The shells/hard parts of calcareous microfossils can be geochemically analyzed, for example by stable isotopes of oxygen and carbon, which are powerful **proxies (indirect evidence)** in paleoceanography and climate change studies (see Chapter 6). Microfossil **evolution**

Reconstructing Earth's Climate History: Inquiry-Based Exercises for Lab and Class, First Edition. Kristen St John, R Mark Leckie, Kate Pound, Megan Jones and Lawrence Krissek.

NAME

has provided a rich archive for establishing relative time in marine sedimentary sequences. In this chapter you will gain experience using microfossil distributions in deep-sea cores to apply a biostratigraphic zonation and **interpret relative age**, **correlate** from one region of the world ocean to another, and **calculate rates** of sediment accumulation. In **Part 3.1**, you will learn about the major types of marine microfossils and their **habitats**; you will consider their role in **primary productivity** in the world oceans, their geologic record and **diversity** through time, as well as the relationship between microfossil diversity and sea level changes through time. In **Part 3.2**, you will explore microfossil distribution in a deep-sea core and consider microfossils as age indicators. In **Part 3.3**, you will apply microfossil first and last occurrences (datum levels) in a deep-sea core to establish biostratigraphic zones. In **Part 3.4**, you will use these same microfossil datum levels to determine sediment accumulation rates. In **Part 3.5**, you will explore the reliability of microfossil datum levels in multiple locations.

Microfossils and Biostratigraphy

Part 3.1. What are Microfossils? Why are they Important in Climate Change Science?

While there are many types of tiny animals and microorganisms living in the sunlit surface waters of the ocean, at depth in the water column and on the seafloor, most of these organisms do not possess the hard parts necessary for them to be preserved on the sedimentary record as fossils. However, recalling from Chapter 2, there are some groups of organisms that have **mineralized shells** or other hard parts consisting of **carbonate** ($CaCO_3$) or **silica** (SiO_2), and others that possess a tough **organic wall**. The vast majority of the **"microfossils"** preserved in deep-sea sediments are **single-celled protists** belonging to the **Kingdom Protoctista** (i.e., not plants, not animals). Many of these protists reside in the upper water column and passively drift with other types of **plankton** at the whim of ocean currents (**planktic habitat**), mostly within or near the **photic zone** (the upper part of the water column where there is enough sunlight available to drive **photosynthesis**). Some of these shelled protists are able to photosynthesize and they are called **phytoplankton**, whereas those that do not are called **zooplankton**. Examples of phytoplankton include the calcareous **coccolithophorids (calcareous nannofossils)** and siliceous **diatoms**; examples of zooplankton include the calcareous **planktic foraminifers** ("planktic forams"; Figure 3.1) and the siliceous **radiolarians** (Figure 3.2).

In addition to planktic microfossils, there are a number of important microfossil groups that live on the seafloor (**benthic habitat**), including the **benthic foraminifera** and **ostracods**. Benthic "forams" live at all depths and all latitudes, from coastal salt marshes and estuaries to the deepest parts of the

NAME

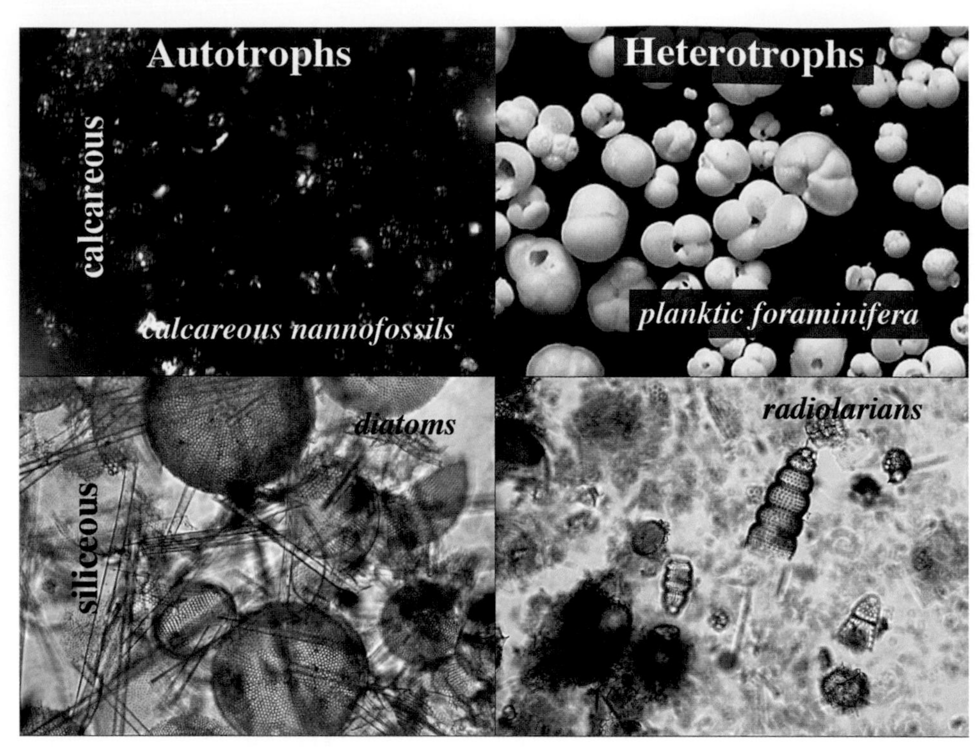

FIGURE 3.2. Calcareous microfossils (calcareous nannofossils and planktic forams) and **siliceous microfossils** (diatoms and radiolarians). All are **single-celled protists** with mineralized hard parts. The calcareous nannofossils (x400; polarized light) and diatoms (x100; plain light) are **photosynthetic primary producers** (i.e., **autotrophs**). The planktic foraminifera (x50; reflected light) and radiolarians (x100; plain light) are **consumers** (i.e., **heterotrophs**). Planktic foram photomicrograph courtesy of Dave Walker; all other photomicrographs from the IODP School of Rock 2005.

ocean, and from tropical coral reefs to frigid polar waters. Ostracods are tiny crustaceans (i.e., true animals) with two calcareous valves (shells) that superficially resemble microscopic clams.

Earth scientists study modern living organisms and the microfossils left behind. **Micropaleontologists** specialize in the study of microfossils, those fossils that are small enough to require the use of a microscope to identify the features of specimens and recognize species.

1 In Chapter 2, Seafloor Sediments, you learned that there are a variety of marine microfossils found in deep-sea sediments. Why are microfossils found in deep-sea sediments? Explain.

____

____

____

Trophic Levels and Productivity

2 What types of organisms make up the **base of the food chain** (i.e., food webs, food pyramids) **on land**?

NAME

3 Why are these organisms at the base of the food chain?

4 What types of organisms make up the **base of the food chain in the ocean**?

5 How might these organisms be similar, and how might they differ from those on land?

6 It is very cold and pitch black on the seafloor below 1000 m water depth. Speculate about the types of organisms that make up the **base of the food chain in the deep ocean.**

NAME

MARINE MICROFOSSILS AND PRODUCTIVITY

The siliceous **diatoms** and calcareous **coccolithophorids** (and other **calcareous nannofossils**) are similar to plants, in that they contain chlorophyll and, like plants, they are **autotrophs** ("self-feeding"; Figure 3.2). These autotrophs **synthesize organic molecules** (e.g., carbohydrates, proteins, and fats) **from inorganic molecules** (CO_2, water, and nutrients) using solar radiation. In other words, they make organic carbon from inorganic carbon. Such organisms are considered to be **primary producers** because they manufacture the organic materials that other types of organism require as food; thus, primary producers form the basis of many marine food chains/food pyramids. The process of **organic carbon production** in the sunlit surface waters is called **photosynthesis**. Free oxygen (O_2) is produced by the reduction of CO_2 and water during photosynthesis. The generalized reaction below describes the process of photosynthesis:

$$6CO_2 + 6H_2O + \text{inorganic nutrients} + \text{solar energy} \rightarrow C_6H_{12}O_6 + 6O_2$$

$6CO_2$	$6H_2O$	inorganic nutrients	$C_6H_{12}O_6$	$6O_2$
carbon dioxide	water	nitrates, phosphates, trace elements & vitamins	glucose (a simple sugar)	free oxygen

The siliceous **radiolarians** and calcareous **planktic foraminifers** (Figures 3.1 & 3.2) are similar to animals in that they are consumers, and like animals, they are **heterotrophs** ("other feeding"; Figure 3.2) and depend on other organisms, either autotrophs or other heterotrophs, for their food. The **dinoflagellates** are a diverse group of organic-walled protists that include both autotrophs and heterotrophs. The process of organic carbon consumption is called **respiration**. Carbon dioxide (CO_2) is released as a product during respiration. The generalized reaction for respiration is the reverse of that for photosynthesis:

$$C_6H_{12}O_6 + 6O_2 \rightarrow 6CO_2 + 6H_2O + \text{solid waste \& nutrients} + \text{chemical energy}$$

$C_6H_{12}O_6$	$6O_2$	$6CO_2$	$6H_2O$	solid waste & nutrients	chemical energy
'food' (carbohydrates, proteins, fats)	oxygen	carbon dioxide	water	to be broken-down further by bacteria	for metabolic processes

NAME

> *Note:* The oxidation, degradation, and **decomposition of organic matter by bacteria** in the water column and on the seafloor releases CO_2 and nutrients back to the marine environment where these raw materials are again available for use by autotrophs in the production of new organic carbon.

Read the box on Marine Microfossils and Productivity and then answer Questions 7–12.

7 How do **calcareous nannofossils** differ from **foraminifera**? Make a list.

8 What is the difference between an **autotroph** and a **heterotroph**?

9 Are **trees** autotrophs or heterotrophs? Explain.

10 Are **humans** autotrophs or heterotrophs? Explain.

NAME ______________________________

11 Why is **photosynthesis** a fundamentally important Earth systems process?

____

____

____

____

____

____

____

12 The **global carbon cycle** describes the exchange of different types of carbon compounds (e.g., CO_2, organic carbon, $CaCO_3$) between the various carbon reservoirs (e.g., atmosphere, biosphere, ocean, solid Earth) in the Earth system (see Chapter 5). Speculate how changes in photosynthesis might have an impact on the carbon cycle and therefore the **global climate.**

____

____

____

____

____

____

____

Marine Microfossils of the Mesozoic and Cenozoic Eras

The major mineralized microfossil groups have geologic records that extend back more than 100 millions years (Figures 3.3 & 3.4); in fact, the radiolarians are found in marine rocks dating back to 540 million years ago! Microfossils are very useful in **scientific ocean drilling** and in the **oil and gas industry** because their tiny size and great abundance in marine sediments allows geoscientists to reconstruct **ancient environmental conditions**, establish the **relative age** of the sedimentary layers (i.e., fossils found in lower or deeper strata are older than fossils found in higher or shallower strata; fossil occurrences can be used to identify the relative order of past events), and **correlate** the sedimentary layers with other localities around the world.

13 Consider **diversity** (i.e., numbers of species and numbers of genera) through the Mesozoic and Cenozoic Eras (Figures 3.3 & 3.4). In very general terms,

NAME ______________________________

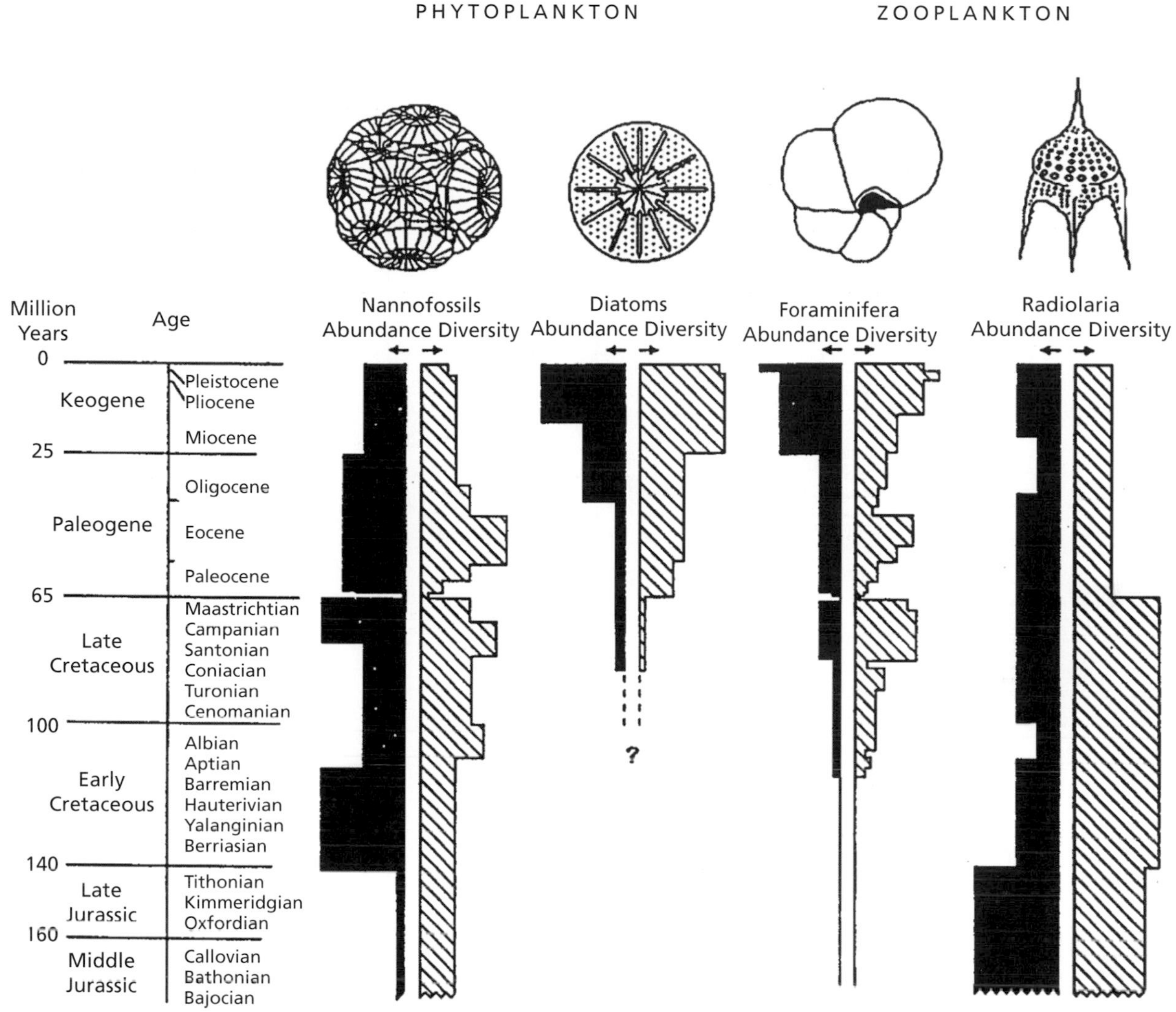

FIGURE 3.3. Generalized **abundance and taxonomic diversity** (numbers of species, genera, or families; this form of "diversity" is also known as taxonomic **richness**) of major mineralized **marine plankton groups** through time. From Gradstein, 1987.

describe the **pattern of evolution** (i.e., times of increasing or decreasing diversity) of each of the following groups of microfossils:

(a) Calcareous nannofossils

____

____

____

____

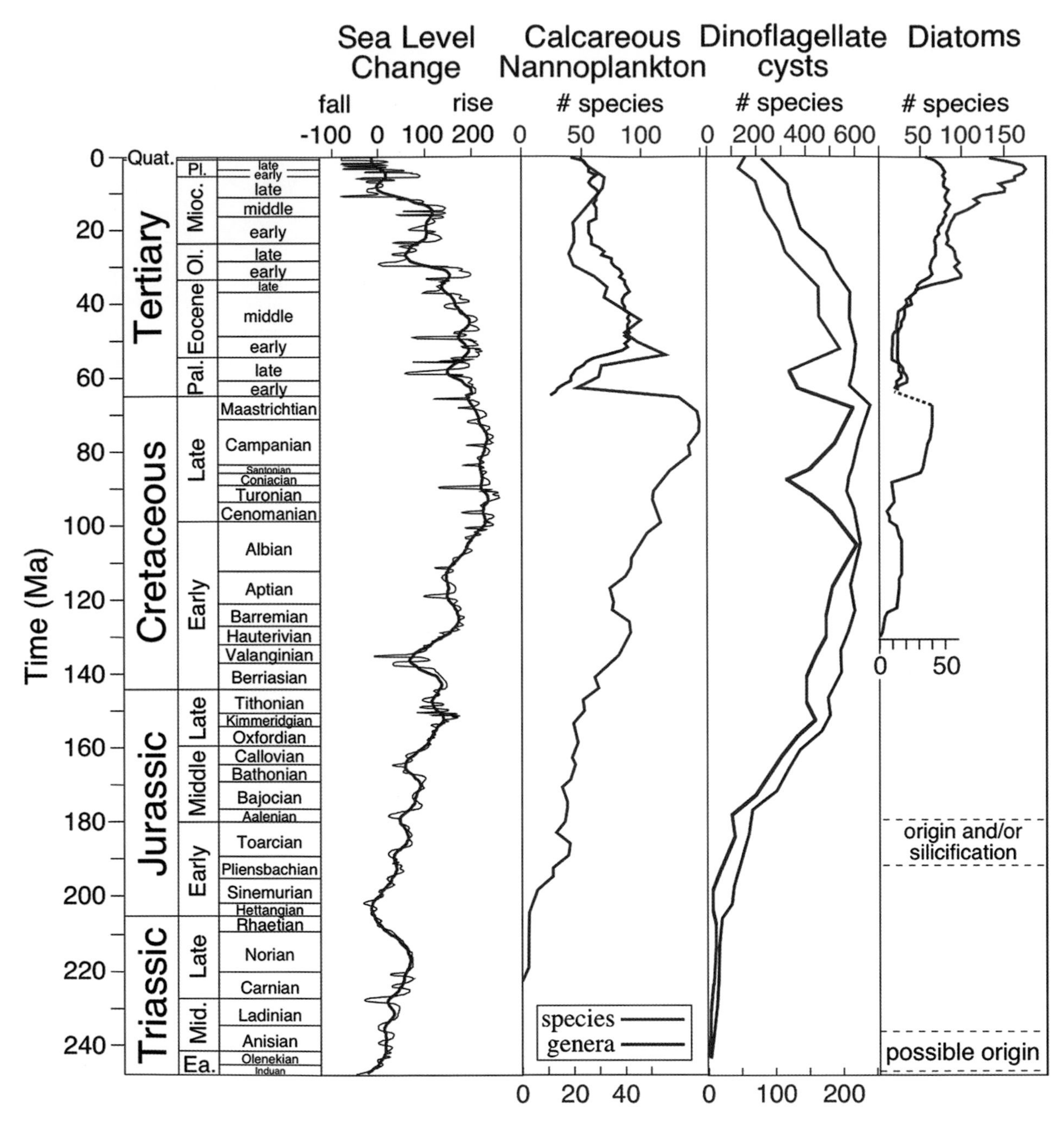

FIGURE 3.4. Comparison of **phytoplankton diversity** (numbers of species and genera, or richness, among the calcareous nannofossils, dinoflagellates, and diatoms) and **global sea level**. Note: when global sea level was much higher than today, for example during the **greenhouse climates** of the Cretaceous and early Tertiary periods, vast areas of the continents were flooded by shallow seas, called **epicontinental seas** or **epeiric seas**. From Katz et al., 2005.

NAME

(b) Diatoms

(c) Dinoflagellates

(d) Planktic foraminifers

NAME

(e) Radiolarians

14 Are there any similarities in the **diversity patterns** between any of the groups? If so, describe.

15 Do **phytoplankton** and **zooplankton groups** track each other? Explain.

NAME ____

16 Based on Figure 3.4, are there any similarities between the **pattern of plankton diversity** and **global sea level change**? If so, describe.

17 Hypothesize about why similarities between plankton diversity and global sea level might exist.

NAME ______________________________

Microfossils and Biostratigraphy
Part 3.2. Microfossils in Deep-Sea Sediments

Consider this scenario: We have a core of deep-sea sediment from a known location that has been described and sampled. What else do you want to know about this core?

A next logical step is to make an **age determination** in order to provide a temporal context for the core. Without an age, it is very difficult to tell a story, geologic or otherwise (e.g., "Once upon a time . . . ", or "Long, long ago in a galaxy far, far away . . . "). Geoscientists depend on reliable age assignments in order to (1) **calibrate** paleomagnetic records and other types of geologic data against the geologic time scale, (2) **calculate rates** of sediment accumulation, and (3) **correlate** core intervals representing a specific time from one location to another around the world oceans.

1 List your ideas on how geoscientists determine the **"age"** of a sequence of sedimentary layers.

____

____

____

____

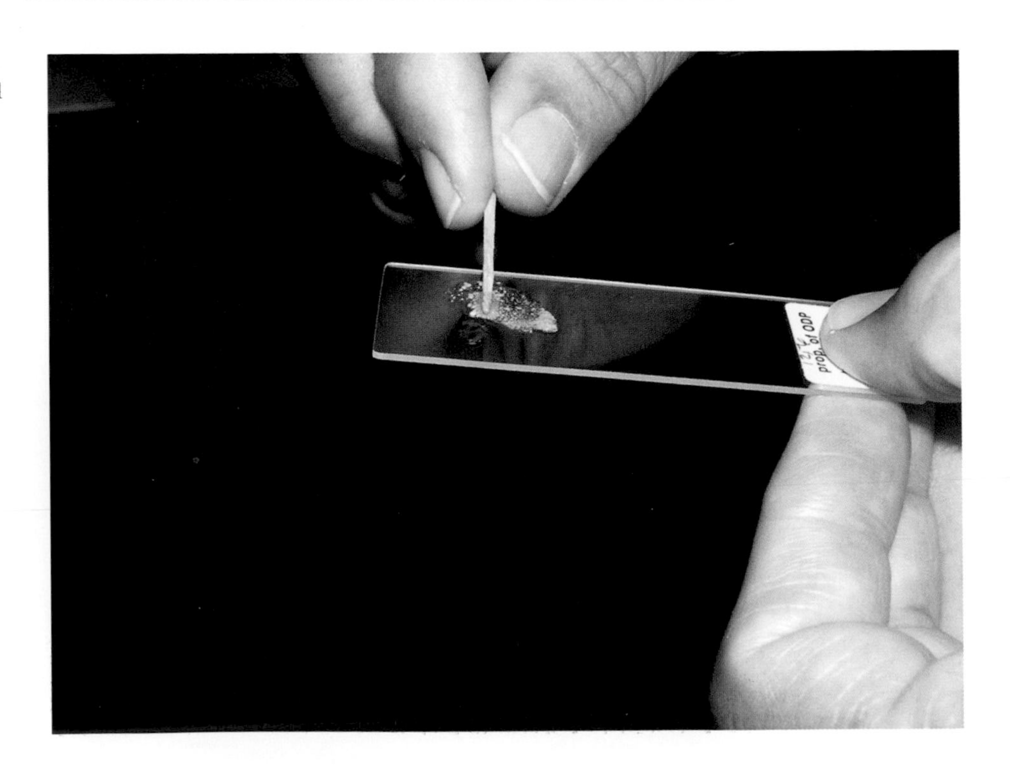

FIGURE 3.5. Making a smear slide. Note the sample label affixed to the glass slide describing exactly where this toothpick smear slide came from. Photo courtesy of Tina King.

NAME

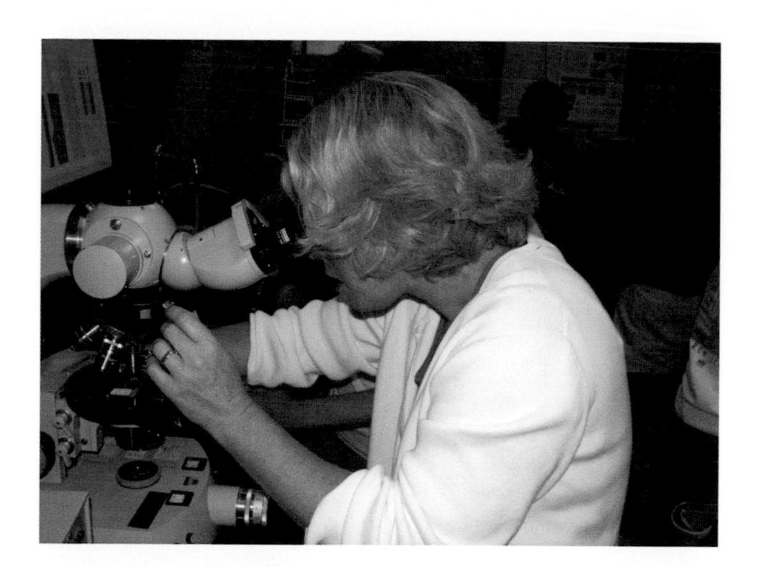

FIGURE 3.6. Geoscientist using a transmitted light microscope to examine microfossils and sediment on a smear slide. Photo courtesy of Mark Leckie.

SMEAR SLIDES AND CALCAREOUS NANNOFOSSILS

A **smear slide** is made to investigate the composition and texture of sediments (recall how you used smear slide data in Chapter 2 to determine sediment lithologies). To prepare a smear slide, a small amount of sediment from the core, collected with a toothpick, is mixed with a couple of drops of water to make a slurry (Figure 3.5). This slurry is smeared in a very thin layer across the glass slide, which is then dried on a hot plate. Adhesive and a cover slip are added. The adhesive is cured under a black light (UV radiation) for about 5 minutes. Now the slide is ready to be examined using a transmitted light microscope.

Micropaleontolgists specializing in **calcareous nannofossils** use a high-powered **transmitted light microscope** to study these very tiny microfossils (Figure 3.6). They typically work with magnifications of 1000–1400x and use different types of light to accentuate the subtle features of the calcite hard parts, including **plane-polarized light**, plane-polarized light with a phase contrast filter, and **cross-polarized light**.

2 Two photomicrographs of smear slides containing **calcareous nannofossils** are shown below, one using plane-polarized light (Figure 3.7) and the other using cross-polarized light (Figure 3.8). Can you see some particles that look like they could be of biological origin? Circle some of these microfossils.

---

3 Do your best to identify **one or two** of the tiny microfossils in the two smear slides (Figures 3.7 & 3.8) using the collection of calcareous nannofossil species shown in Figure 3.9. Put a box around the microfossil you are identifying (Figure 3.7 and/or Figure 3.8) and write the name of the species next to the box.

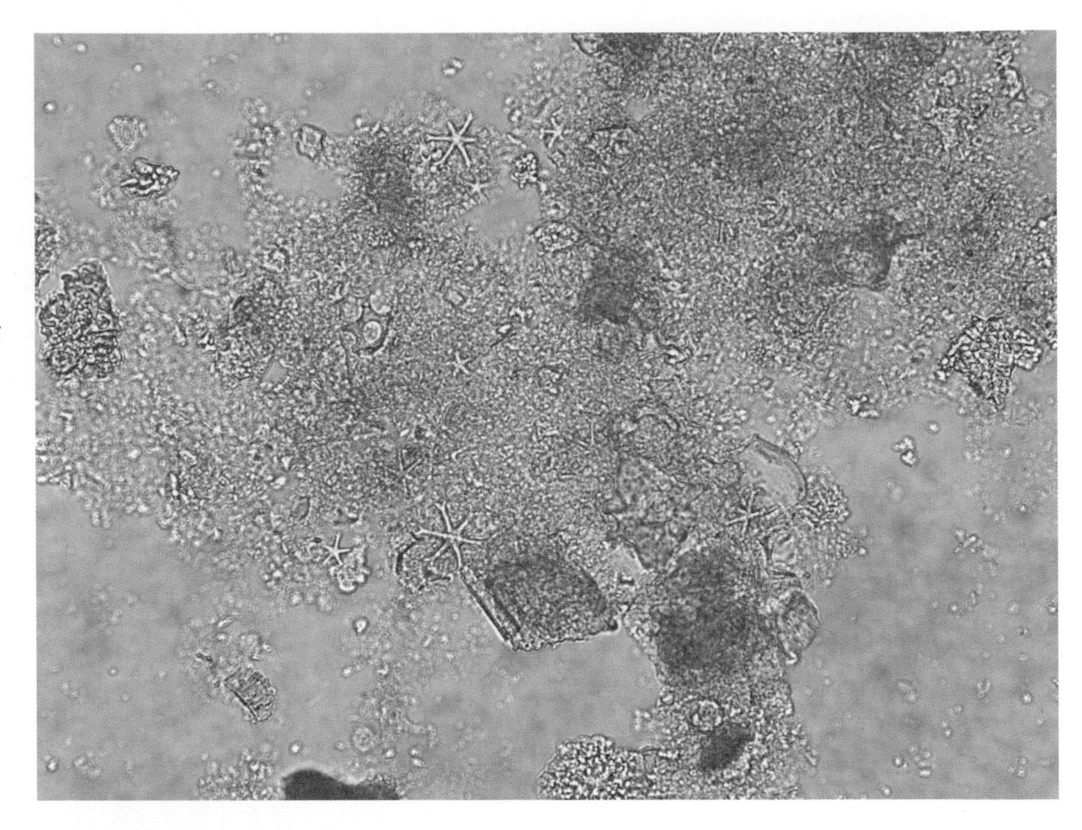

FIGURE 3.7. Smear slide of sample 807A-8H-5, 51 cm from Ontong Java Plateau in the western equatorial Pacific using **plane-polarized light**; 400× magnification. Photo courtesy of the IODP School of Rock, 2005.

FIGURE 3.8. Smear slide of sample 807A-8H-2, 70 cm from Ontong Java Plateau in the western equatorial Pacific using **crossed-polarized light**; 400× magnification. Photo courtesy of the IODP School of Rock, 2005.

NAME

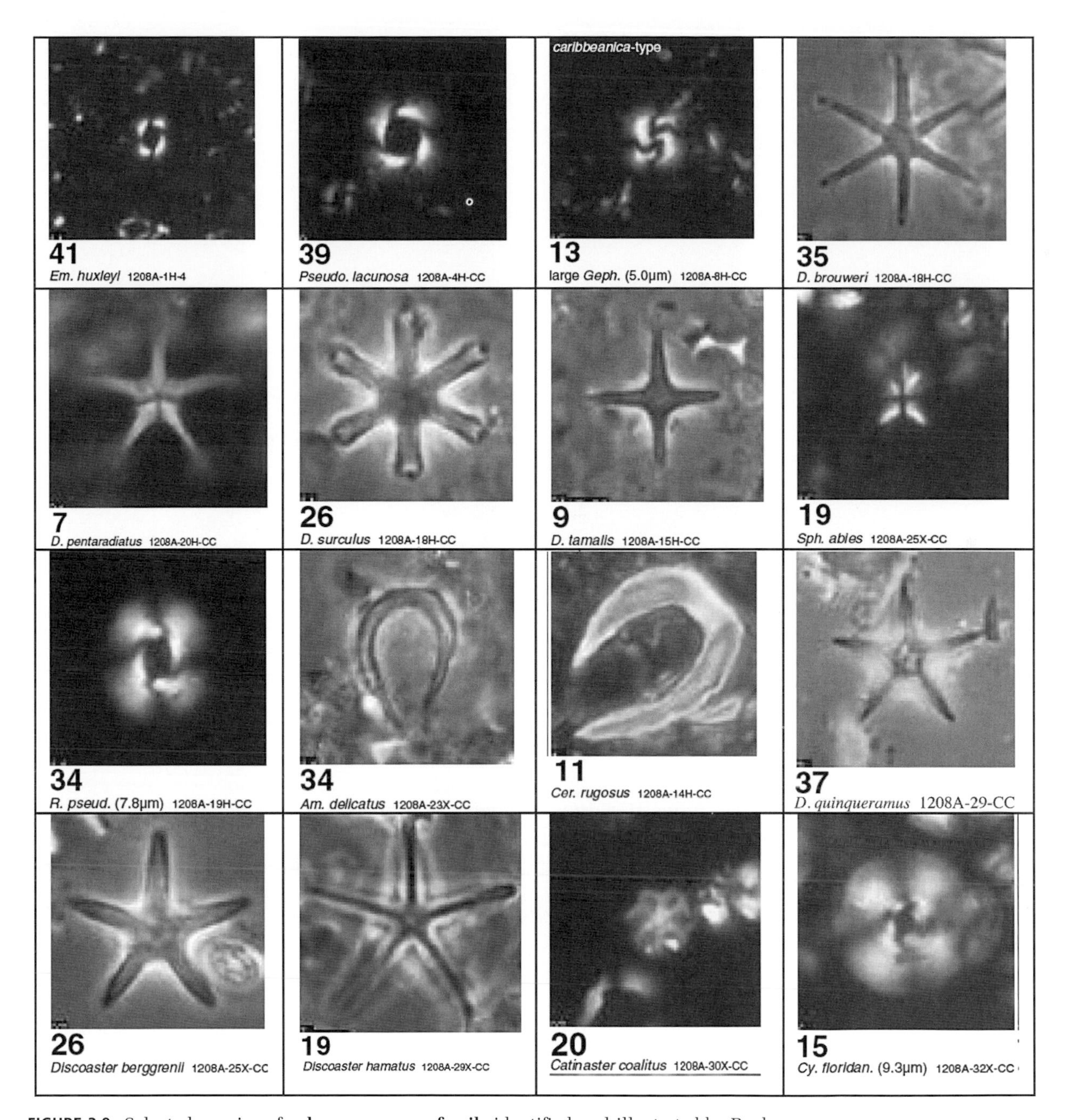

FIGURE 3.9. Selected species of **calcareous nannofossils** identified and illustrated by Paul Bown. These are **biostratigraphically useful zonal marker species**. All photomicrographs ×1000. Dark field = cross-polarized light; clear field = plane-polarized light with phase contrast filter. Bold numbers refer to specimen numbers on the original published plates. From Bown, 2005.

NAME

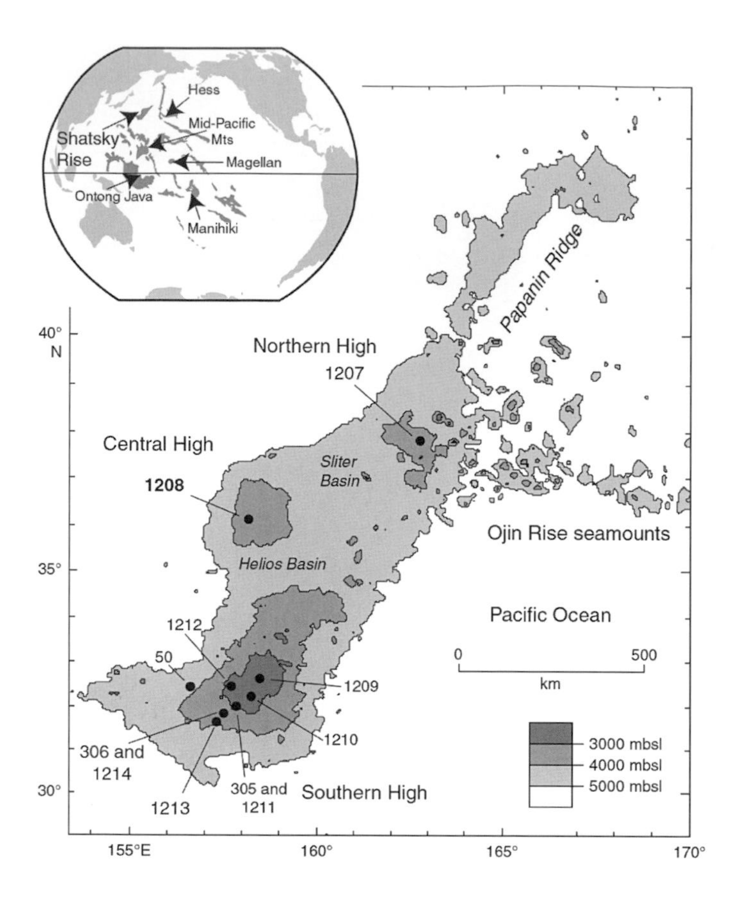

FIGURE 3.10. Location of ODP Site 1208 on Shatsky Rise in the northwest Pacific. Shaded bathymetric contours; mbsl = meters below sea level. From Bown, 2005.

Table 3.1 shows the **distribution of calcareous nannofossils** in Hole 1208A from the central high of Shatsky Rise in the northwest Pacific (Figure 3.10). Site 1208 was cored during Ocean Drilling Program (ODP) Leg 198.

4 Remember from Chapter 1 that sediments deeper in a cored interval accumulated before the sediments above; in other words, the sequence is oldest at the bottom and youngest at the top. Likewise, the species deeper/lower in the sedimentary sequence are older than the species that occur higher in the sequence. **Make a list of observations** about the distribution of species within the core as shown in Table 3.1. Also, **make a list of questions** that you may have about the distribution of species in this table. What **additional information** would you like to know about the species and their distribution in the Site 1208 drill site?

Observations:

NAME

Questions:

Additional Information:

5 Fossil occurrences can be used to identify the **relative order** of past events. Geoscientists "read" the sedimentary record from bottom to top (i.e., from older to younger). How might the species data in Table 3.1 be useful for determining **how old** the sediment layers are in **this** drill site? Explain.

NAME

TABLE 3.1. Rows of samples, the depth of those samples in the hole (mbsf = meters below seafloor), and columns of species (both *Genus* and *species* are italicized, e.g., *Emiliania huxleyi*).*

Hole 1208A																																						
SAMPLE: Core-Section, Interval (cm)	**DEPTH** (meters below sea floor: mbsf)	**PRESERVATION**	**ABUNDANCE**	*Amaurolithus* spp.	*Amaurolithus delicatus*	*Amaurolithus primus*	*Calcidiscus leptoporus*	*Calcidiscus macintyrei*	*Calcidiscus tropicus*	*Catinaster calyculus*	*Catinaster coalitus*	*Ceratolithus cristatus*	*Ceratolithus rugosus*	*Ceratolithus telesmus*	*Coccolithus miopelagicus*	*Coccolithus pelagicus*	*Coccolithus pelagicus* (w. bar)	*Cryptococcolithus mediaperforatus*	*Cyclicargolithus floridanus*	*Discoaster asymmetricus*	*Discoaster bellus*	*Discoaster bellus-berggrenii*	*Discoaster beggrenii*	*Discoaster blackstockiae*	*Discoaster bollii*	*Discoaster braarudii*	*Discoaster brouweri*	*Discoaster calcaris*	*Discoaster challengeri*	*Discoaster deflandrei*	*Discoaster exilis*	*Discoaster hamatus*	*Discoaster loeblichii*	*Discoaster pentaradiatus*	*Discoaster prepentaradiatus*	*Discoaster quinqueramus*	*Discoaster surculus*	*Discoaster tamalis*
1H-1, 15	0.15	M-G	A				A					F				A	A																					
1H-3, 60	3.6	M	A				C									C	C																					
1H-4, 50	4.4	G	A				C					C				A	A																					
1H-CC	4.70	M	A				C					•				A	A																					
2H-5, 120	11.90	P-M	A				C									C	C																					
2H-CC	14.24	P-M	A				A					R		R		A	A																					
3H-5, 107	21.27	M-G	A				C					•				C	C																					
3H-CC	23.82	M-G	A				F					•		•		A	A																					
4H-5, 60	30.3	G	A				F					•				C	C																					
4H-CC	33.47	M	A				C									A	A																					
5H-5, 120	40.40	M	A				C									A	C																					
5H-CC	43.11	P-M	A				F									A	A																					
6H-5, 120	49.90	M-G	A				C					•				F	F																					
6H-CC	52.62	M	A				A									C	C																					
7H-5, 120	59.40	M	A				C									C	F																					
7H-CC	62.15	M	A				A					•				C	C																					
8H-5, 120	68.90	M	A				C					•				F	F																					
8H-CC	71.60	M-G	A				C									C	C																					
9H-5, 120	78.40	G	A				C					•				F	F																					
9H-CC	81.09	M	A				A	•								C	C																					
10H-5, 120	87.90	M-G	A				C	F								C	F																					
10H-CC	90.62	M	A				C	F					•			C	C																					

*The letters correspond to the **relative abundance** of individual species in each sample based on a **visual estimate** (blank = not present; **A** = abundant; **C** = common; **F** = few; **R** = rare; • = very rare). The column labeled "Preservation" refers to the degree of breakage of nannofossils, calcite dissolution, or calcite overgrowths on the nannofossils in each sample (**G** = good; **M** = moderate; **P** = poor). The column labeled "Abundance" describes the relative proportion of calcareous nannofossils in each sample compared with other constituents such as other types of microfossils or mineral components such as clay (**A** = abundant). These data were produced by Paul Bown, a shipboard scientist during ODP Leg 198 (Bown, 2005; http://www-odp.tamu.edu/publications/198_SR/104/104.htm)

NAME ____

Discoaster triradiatus	Discoaster variabilis	Emiliania huxleyi	Florisphaera profunda	Gephyrocapsa aperta	Gephyrocapsa caribbeanica	Gephyrocapsa oceanica	Gephyrocapsa omega (= G. paraliela)	Gephyrocapsa spp. (medium)	Gephyrocapsa spp. (small)	**SAMPLE**
		A	A			C	R	A	C	1H-1, 15
		A							A	1H-3, 60
		A	A		F	F	R	A	A	1H-4, 60
		C	C		C	F		A	C	1H-CC
		C	C			C	C	C	C	2H-5, 120
		C	A	?A		R		A	A	2H-CC
			C		C	C	•	C	SA	3H-5, 107
			C		C	C		A	A	3H-CC
			C			•	1		SA	4H-5, 60
			C		C	C	C	A	A	4H-CC
			C			F	F	C	A	5H-5, 120
			A	C		F	F	A	A	5H-CC
			A					C	A	6H-5, 120
			C		C	F		C	A	6H-CC
			C						A	7H-5, 120
			C						A	7H-CC
						•		C		8H-5, 120
			C		C	F		A	A	8H-CC
			C		C	F		C	F	9H-5, 120
			A		C			C	A	9H-CC
			C		F				A	10H-5, 120
			A						C	10H-CC

Gephyrocapsa spp. (large)	Helicosphaera carteri	Helicosphaera inversa	Helicosphaera sellii	Helicosphaera wallichii	Hughesius gizoensis	Minylitha convalis	Pontosphaera discopora	Pontosphaera japonica	Pontosphaera multipora	**SAMPLE**
R	C						R			1H-1, 15
	C						?			1H-3, 60
R	C						R	R		1H-4, 60
F	F									1H-CC
C	F							•		2H-5, 120
R	R						•			2H-CC
	F	•						•		3H-5, 107
	R	R								3H-CC
	F	R					•			4H-5, 60
R	F						R	R		4H-CC
R	R			1			R	R		5H-5, 120
	F									5H-CC
	R							R		6H-5, 120
	F									6H-CC
	R							R		7H-5, 120
	F						R		R	7H-CC
R							•			8H-5, 120
R	F						F		R	8H-CC
	F		•	•			R			9H-5, 120
	F		•				R	R	R	9H-CC
	F		F	•			R	R		10H-5, 120
	F									10H-CC

Pseudoemiliania lacunosa	Pseudoemiliania ovata	Reticulofenestra asanoi	Reticulofenestra haqii	Reticulofenestra minuta	R. pseudoumbilicus (5–7 μm)	R. pseudoumbilicus (>7 μm)	R. pseudoumbilicus (>7 μm) w. grill	Reticulofenestra rotaria	Rhabdosphaera clavigera	**SAMPLE**
									F	1H-1, 15
										1H-3, 60
									F	1H-4, 60
										1H-CC
									R	2H-5, 120
										2H-CC
										3H-5, 107
				?A						3H-CC
C	F									4H-5, 60
C	C		F	A						4H-CC
C	F		C	C						5H-5, 120
C	C		C	A						5H-CC
C	F	R	A	C					•	6H-5, 120
C	C	C	A	A						6H-CC
C	F		A	C						7H-5, 120
C	A		A							7H-CC
C	F		A	C						8H-5, 120
C	C		R							8H-CC
C	C		SA	A						9H-5, 120
C			A	A						9H-CC
C	C		A	C						10H-5, 120
C			A	A						10H-CC

Sphenolithus abies	Syracosphaera pulchra	Triquetrorhabdulus rugosus	Umbilicosphaera jafari	Umbilicosphaera rotula	Umbilicosphaera sibogae	DIATOMS	SILICOFLAGELLATES	CALCISPHERES	**SAMPLE**
	F				F	C		R	1H-1, 15
	F								1H-3, 60
	C				R	C		F	1H-4, 60
	R				F	A	R	R	1H-CC
	R				•	A		R	2H-5, 120
					F	A	R		2H-CC
	•				R	C			3H-5, 107
	F				F	C			3H-CC
	•					C		R	4H-5, 60
	F				F	C	R		4H-CC
	•				F	R			5H-5, 120
					C	A			5H-CC
	•				•	C			6H-5, 120
	R				F	C		R	6H-CC
						A		R	7H-5, 120
	F				C	A	R		7H-CC
						A	R	R	8H-5, 120
	C			R	C	C	R	R	8H-CC
	F				R	A			9H-5, 120
				R	R	C			9H-CC
	F			R	R	C		R	10H-5, 120
	R				R	A	R		10H-CC

(Continued)

NAME

TABLE 3.1. *Continued*

Hole 1208A																																						
SAMPLE: Core-Section, interval (cm)	**DEPTH** (meters below sea floor: mbsf)	**PRESERVATION**	**ABUNDANCE**	*Amaurolithus* spp.	*Amaurolithus delicatus*	*Amaurolithus primus*	*Calcidiscus leptoporus*	*Calcidiscus macintyrei*	*Calcidiscus tropicus*	*Catinaster calyculus*	*Catinaster coalitus*	*Ceratolithus cristatus*	*Ceratolithus rugosus*	*Ceratolithus telesmus*	*Coccolithus miopelagicus*	*Coccolithus pelagicus*	*Coccolithus pelagicus* (w. bar)	*Cryptococcolithus mediaperforatus*	*Cyclicargolithus floridanus*	*Discoaster asymmetricus*	*Discoaster bellus*	*Discoaster bellus-berggrenii*	*Discoaster beggrenii*	*Discoaster blackstockiae*	*Discoaster bollii*	*Discoaster braarudii*	*Discoaster brouweri*	*Discoaster calcaris*	*Discoaster challengeri*	*Discoaster deflandrei*	*Discoaster exils*	*Discoaster hamatus*	*Discoaster loeblichii*	*Discoaster pentaradiatus*	*Discoaster prepentaradiatus*	*Discoaster quinqueramus*	*Discoaster surculus*	*Discoaster tamalis*
11H-2, 65	92.35	M	A				C	F								F	F																					
11H-5, 120	97.40	M-G	A				C	F								C	C																					
11H-CC	100.16	P-M	A				A	F	F							C	C										R											
12H-2, 120	102.40	M	A				C	R								C	C										F											
12H-5, 120	106.90	M	A				C	R								C	C										F											
12H-CC	109.50	M	A				C	C	C			•				C	C										F											
13H-2, 120	111.90	M-G	A				C	F	C							C	C										F											
13H-5, 120	116.40	M-G	A				C	F	C							C	C							1			F							R				
13H-CC	119.08	P-M	A				C	C	C				•	•		C	F										F							C			F	
14H-2, 120	121.40	M	A				C	F				F				F	F																	R			F	
14H-5, 120	125.90	M	A				C	R	C							C	F										R							F			F	
14H-CC	128.70	M-G	A				C	F	C				•			C	C			R							F							C			C	F
15H-2, 121	130.91	M	A				C	R								C				R							R							R				R
15H-5, 121	135.40	M	A				C	R								C				R							R							F			R	
15H-CC	138.20	M	A				A	F	F							C				C							C							F			F	C
16H-2, 120	140.40	M	A				C	F								C				F														F			F	F
16H-5, 120	144.90	M	A				C	F								F																		R			R	
16H-CC	147.77	M	A				C	C	C			•				F				F							F							F			F	F
17H-2, 120	149.90	M	A				C	F	F							C	F			F							F							R			F	F
17H-5, 120	154.50	M	A				C	F	C							F	F			F							F							R				F
17H-CC	157.21	M	A				A	C	C			F				F	F			C							F			?•				F			C	F
18H-2, 120	159.40	M	A				C	F	C			R				F				C							F							F			F	F
18H-5, 120	163.90	M	A				C	R	F			R				R	R			C							F							F			F	R
18H-CC	166.66	M	A				A	C	F			•	•			?F	F			F							F							F			F	•
19H-2, 118	168.88	P-M	A	•	•		C	R	F			R				F	F			1														F			F	

*The letters correspond to the **relative abundance** of individual species in each sample based on a **visual estimate** (blank = not present; **A** = abundant; **C** = common; **F** = few; **R** = rare; **•** = very rare). The column labeled "Preservation" refers to the degree of breakage of nannofossils, calcite dissolution, or calcite overgrowths on the nannofossils in each sample (**G** = good; **M** = moderate; **P** = poor). The column labeled "Abundance" describes the relative proportion of calcareous nannofossils in each sample compared with other constituents such as other types of microfossils or mineral components such as clay (**A** = abundant). These data were produced by Paul Bown, a shipboard scientist during ODP Leg 198 (Bown, 2005; http://www-odp.tamu.edu/publications/198_SR/104/104.htm)

NAME

Discoaster triradiatus	Discoaster variabilis	Emiliania huxleyi	Florisphaera profunda	Gephyrocapsa aperta	Gephyrocapsa caribbeanica	Gephyrocapsa oceanica	Gephyrocapsa omega (= G. paraliela)	Gephyrocapsa spp. (medium)	Gephyrocapsa spp. (small)	Gephyrocapsa spp. (large)	Helicosphaera carteri	Helicosphaera inversa	Helicosphaera sellii	Helicosphaera wallichii	Hughesius gizoensis	Minylitha convalis	Pontosphaera discopora	Pontosphaera japonica	Pontosphaera multipora	Pseudoemiliania lacunosa	Pseudoemiliania ovata	Reticulofenestra asanoi	Reticulofenestra haqii	Reticulofenestra minuta	R. pseudoumbilicus (5–7 μm)	R. pseudoumbilicus (>7 μm)	R. pseudoumbilicus (>7 μm) w. grill	Reticulofenestra rotaria	Rhabdosphaera clavigera	Sphenolithus abies	Syracosphaera pulchra	Triquetrorhabdulus rugosus	Umbilicosphaera jafari	Umbilicosphaera rotula	Umbilicosphaera sibogae	DIATOMS	SILICOFLAGELLATES	CALCISPHERES	**SAMPLE**
													F										C	C															11H-2, 65
			A								F		R				R			C	C		SA	A							F					A			11H-5, 120
			A								R						R			C			A	A							R			•	R	C	R		11H-CC
													F										C	C															12H-2, 120
											R		F				R			F	C		C	C															12H-5, 120
			A								R						R			C	C		A	A							R				R	C			12H-CC
			C								•						•			C	C		SA	C												A			13H-2, 120
			C								1		1				•			F	C		SA	A												A	R		13H-5, 120
			A														R			C			A	A							R				F	C			13H-CC
																							C	F															14H-2, 120
			C								•									F	C		A	A							•			R		A			14H-5, 120
	?R		C								F						R			C	C		A	A	?•						R			C		C			14H-CC
																					C		C	C							F			R					15H-2, 120
																					C		C	C										R					15H-5, 121
R			C														R			C	C		A	A							F			R		A	R		15H-CC
																				F	C		C	C										R					16H-2, 120
	F																				C		C	F										F					16H-5, 120
•			C																	C	F		A	A							R			F		A	R		16H-CC
			F														•				C		C	C												C			17H-2, 120
F			F						A		•						•				C		A	A	•									F		C			17H-5, 120
	R										F									F	C		A	A	1	•					R		R	C		A	F	R	17H-CC
•			C								1						•			F	F		C	F										F		A			18H-2, 120
1	1		C						R								•			F	F		C	C	R	C	C							F		C			18H-5, 120
			C						A		F						F		R	?F	C		A	A		C	C			F	F		C	C	?	A	R		18H-CC
	1		A								•						R				F		C	A	F	C	C			•	•			F		C			19H-2, 118

(Continued)

NAME

TABLE 3.1. *Continued*

Hole 1208A																																						
SAMPLE: Core-Section, Interval (cm)	**DEPTH** (meters below sea floor: mbsf)	**PRESERVATION**	**ABUNDANCE**	*Amaurolithus* spp.	*Amaurolithus delicatus*	*Amaurolithus primus*	*Calcidiscus leptoporus*	*Calcidiscus macintyrei*	*Calcidiscus tropicus*	*Catinaster calyculus*	*Catinaster coalitus*	*Ceratolithus cristatus*	*Ceratolithus rugosus*	*Ceratolithus telesmus*	*Coccolithus miopelagicus*	*Coccolithus pelagicus*	*Coccolithus pelagicus* (w. bar)	*Cryptococcolithus mediaperforatus*	*Cyclicargolithus floridanus*	*Discoaster asymmetricus*	*Discoaster bellus*	*Discoaster bellus-berggrenii*	*Discoaster beggrenii*	*Discoaster blackstockiae*	*Discoaster bollii*	*Discoaster braarudii*	*Discoaster brouweri*	*Discoaster calcaris*	*Discoaster challengeri*	*Discoaster deflandrei*	*Discoaster exilis*	*Discoaster hamatus*	*Discoaster loeblichii*	*Discoaster pentaradiatus*	*Discoaster prepentaradiatus*	*Discoaster quinqueramus*	*Discoaster surculus*	*Discoaster tamalis*
19H-5, 117	173.37	M-G	A	•	•		C	F	F			•				R	R																	F			F	
19H-CC	176.09	M	A	F	F		C	F				•	•			?F											F							F			F	
20H-2, 120	178.40	M	A	•	•		C	•	F			•				F	F										R							F			F	
20H-5, 136	183.06	M	A	•	•		F	•	F			•				•	•							1			R							F			R	
20X-CC	185.65	M	A	R	R		C	F				•															F							F			F	
21X-2, 120	187.90	M	A	•	•		F	F	F			•				R											•							F			F	
21X-4, 120	190.90	P-M	A	R	R	•	C	F	F							R											F							F			F	
21X-CC	191.07	M	A	F	F	R	A	F								C				R				•			C				?•			C			C	
22X-2, 120	193.10	P-M	A	•	•		F	F	F							R	R										F							F			F	
22X-5, 120	197.60	P	A	•	•		F	F	C							F	R										F							F			F	
22X-CC	199.99	P-M	A	F	F		C	F								C																		F			R	
23X-2, 121	202.81	P-M	A	•	•		F		F							F																		F				
23X-5, 90	207.00	P-M	A	•	•		F		F							F																		F		F		
23X-CC	208.18	P	A	•	•	•	F	F								R											F							F		R	R	
24X-2, 121	212.51	M	A				C		F							F	R										1							F		C		
24X-5, 120	217.00	P-M	A	•		•	F	•	F							F																		F		F		
24X-CC	218.74	P-M	A	•	F	•	A	C								C		R									F							C		C	R	
25X-2, 119	222.09	M	A	R	•	R	C		R							F		1																F		C		
25X-4, 90	224.50	M	A	•		•	F									C		1																R		C		
25X-CC	224.59	P-M	A	R	•	R	C	C								C				R			R				R							F		C	R	
26X-2, 119	231.79	M	A	1		1	F	R								F																		F		C		
26X-5, 82	235.37	M	A	1		1	F	R								F		F					F				•							F		F	•	
26X-CC	235.52	P-M	A	•		•	F	F								C		R					•				F		R					F		F	R	
27X-2, 120	241.40	M	A				F	F								F	1	F					F										1	•		R		
27X-5, 120	245.90	P-M	A				C	F	R							C		R			1		F				C							R		R		

*The letters correspond to the **relative abundance** of individual species in each sample based on a **visual estimate** (blank = not present; **A** = abundant; **C** = common; **F** = few; **R** = rare; • = very rare). The column labeled "Preservation" refers to the degree of breakage of nannofossils, calcite dissolution, or calcite overgrowths on the nannofossils in each sample (**G** = good; **M** = moderate; **P** = poor). The column labeled "Abundance" describes the relative proportion of calcareous nannofossils in each sample compared with other constituents such as other types of microfossils or mineral components such as clay (**A** = abundant). These data were produced by Paul Bown, a shipboard scientist during ODP Leg 198 (Bown, 2005; http://www-odp.tamu.edu/publications/198_SR/104/104.htm)

NAME ______________________________

Discoaster triradiatus	*Discoaster variabilis*	*Emiliania huxleyi*	*Florisphaera profunda*	*Gephyrocapsa aperta*	*Gephyrocapsa caribbeanica*	*Gephyrocapsa oceanica*	*Gephyrocapsa omega* (= *G. paraliela*)	*Gephyrocapsa* spp. (medium)	*Gephyrocapsa* spp. (small)	*Gephyrocapsa* spp. (large)	*Helicosphaera carteri*	*Helicosphaera inversa*	*Helicosphaera sellii*	*Helicosphaera wallichii*	*Hughesius gizoensis*	*Minylitha convalis*	*Pontosphaera discopora*	*Pontosphaera japonica*	*Pontosphaera multipora*	*Pseudoemiliania lacunosa*	*Pseudoemiliania ovata*	*Reticulofenestra asanoi*	*Reticulofenestra haqii*	*Reticulofenestra minuta*	*R. pseudoumbilicus* (5–7 μm)	*R. pseudoumbilicus* (>7 μm)	*R. pseudoumbilicus* (>7 μm) w. grill	*Reticulofenestra rotaria*	*Rhabdosphaera clavigera*	*Sphenolithus abies*	*Syracosphaera pulchra*	*Triquetrorhabdulus rugosus*	*Umbilicosphaera jafari*	*Umbilicosphaera rotula*	*Umbilicosphaera sibogae*	DIATOMS	SILICOFLAGELLATES	CALCISPHERES	**SAMPLE**
			C								1		?1				R						C	SA	R	C	C			R				F		C			19H-5, 117
	F		C														•				C		A	A		C	R			?F				C		A	R		19H-CC
	•		A								1						R				C			SA		C	R	1		F	•			F		C			20H-2, 120
	•		A																				F	SA	F	F				•				F		C			20H-5, 136
	F																						C		A	F				R				F		A			20X-CC
			C																				C	A	A	F	R			R				R	?1	C			21X-2, 120
•			A																				A	C	A	F				•				R		C			21X-4, 120
	R																						A	A	A	C				R				F		A	R		21X-CC
	F		A																				A	C	1			2		R				F		C			22X-2, 120
•	F		C																				A	F						R				F		C			22X-5, 120
	C																						A		C					F				R		C	R		22X-CC
	F		A																				C	F	F	C				F				R		C			23X-2, 121
	F		C																				A	C	R	C				R									23X-5, 90
	F																						A		?F	C				R			R	R		A	R		23X-CC
	C		A								1												A	C	F	A				R		?1		F		A			24X-2, 121
	F		A																				C	C	C	A				R				R		C			24X-5, 120
•	C																						A			A		F		F				F		A	R		24X-CC
	C		F																				C	A	C	C				R				F		C			25X-2, 119
1	C																						F	A	A	C		R		F				F		C			25X-4, 90
•	C																						A	?C	C	A		?		F						C	R		25X-CC
	C																						C	F	A	R		R		•				F		C			26X-2, 119
	C														•								C	C	C	R		1		•				F		C			26X-5, 82
R	C																						A	?C	C	C		F		•				R		C			26X-CC
	F																						A	A	A					•				F		C			27X-2, 120
1	C															F							C		F	F										C			27X-5, 120

(Continued)

TABLE 3.1. *Continued*

Hole 1208A																																						
SAMPLE: Core-Section, Interval (cm)	**DEPTH** (meters below sea floor: mbsf)	**PRESERVATION**	**ABUNDANCE**	*Amaurolithus* spp.	*Amaurolithus delicatus*	*Amaurolithus primus*	*Calcidiscus leptoporus*	*Calcidiscus macintyrei*	*Calcidiscus tropicus*	*Catinaster calyculus*	*Catinaster coalitus*	*Ceratolithus cristatus*	*Ceratolithus rugosus*	*Ceratolithus telesmus*	*Coccolithus miopelagicus*	*Coccolithus pelagicus*	*Coccolithus pelagicus* (w. bar)	*Cryptococcolithus mediaperforatus*	*Cyclicargolithus floridanus*	*Discoaster asymmetricus*	*Discoaster bellus*	*Discoaster bellus-berggrenii*	*Discoaster beggrenii*	*Discoaster blackstockiae*	*Discoaster bollii*	*Discoaster braarudii*	*Discoaster brouweri*	*Discoaster calcaris*	*Discoaster challengeri*	*Discoaster deflandrei*	*Discoaster exils*	*Discoaster hamatus*	*Discoaster loeblichii*	*Discoaster pentaradiatus*	*Discoaster prepentaradiatus*	*Discoaster quinqueramus*	*Discoaster surculus*	*Discoaster tamalis*
27X-CC	246.14	M	A				C	F								C		F		R			F		F		C		C		R		1	R		F		
28X-2, 120	250.80	M	A				C	F	R							C		F			R		R				C							R		R		
28X-5, 120	255.30	M	A				C	F	F							C		C			•					F	C						1	R	•			
28X-CC	256.71	P	A				R	F								F		R			•					R	C	R						S				
29X-2, 120	260.50	P-M	A				F	F	R							F		F			R						C	R					1	R	R			
29X-5, 120	265.00	M	A				F	F	R							C		R			F	R					C	R	F					•	R			
29X-CC	265.94	P	A				F	F		•						C		R		•	F	R					C	R			R	•		R	F			
30X-2, 120	270.10	P-M	A					R		R	R					F					R				F		R		C			•						
30X-4, 120	273.10	P-M	A				F	F	R		F					F					R						R		C		R	•						
30X-CC	274.20	P-M	A				F	C			F					C		F			R				R	R	R				F	•			R			
31X-2, 120	279.70	M	A				F	R	F		F					F									R		R		A		F							
31X-5, 97	283.97	P-M					F	F	F						2	F													A	R	F							
31X-CC	285.06	P-M					C	F	R						F	C		R							R				C	F	R							
32X-2, 120	289.30	P-M					F	F	F						F	C									F		•		C	R	F							
32X-5, 120	303.10	P-M					F		F						F	F		R							R		•		C	C	F							
32X-CC	295.41	P-M					C	?R	C						F	C		F	C						R		R		C	F	R							

*The letters correspond to the **relative abundance** of individual species in each sample based on a **visual estimate** (blank = not present; **A** = abundant; **C** = common; **F** = few; **R** = rare; • = very rare). The column labeled "Preservation" refers to the degree of breakage of nannofossils, calcite dissolution, or calcite overgrowths on the nannofossils in each sample (**G** = good; **M** = moderate; **P** = poor). The column labeled "Abundance" describes the relative proportion of calcareous nannofossils in each sample compared with other constituents such as other types of microfossils or mineral components such as clay (**A** = abundant). These data were produced by Paul Bown, a shipboard scientist during ODP Leg 198 (Bown, 2005; http://www-odp.tamu.edu/publications/198_SR/104/104.htm)

NAME

Discoaster triradiatus	*Discoaster variabilis*	*Emiliania huxleyi*	*Florisphaera profunda*	*Gephyrocapsa aperta*	*Gephyrocapsa caribbeanica*	*Gephyrocapsa oceanica*	*Gephyrocapsa omega* (= *G. paraliela*)	*Gephyrocapsa* spp. (medium)	*Gephyrocapsa* spp. (small)	*Gephyrocapsa* spp. (large)	*Helicosphaera carteri*	*Helicosphaera inversa*	*Helicosphaera sellii*	*Helicosphaera wallichii*	*Hughesius gizoensis*	*Minylitha convalis*	*Pontosphaera discopora*	*Pontosphaera japonica*	*Pontosphaera multipora*	*Pseudoemiliania lacunosa*	*Pseudoemiliania ovata*	*Reticulofenestra asanoi*	*Reticulofenestra haqii*	*Reticulofenestra minuta*	*R. pseudoumbilicus* (5–7 μm)	*R. pseudoumbilicus* (>7 μm)	*R. pseudoumbilicus* (>7 μm) w. grill	*Reticulofenestra rotaria*	*Rhabdosphaera clavigera*	*Sphernolithus abies*	*Syracosphaera pulchra*	*Triquetrorhabdulus rugosus*	*Umbilicosphaera jafari*	*Umbilicosphaera rotula*	*Umbilicosphaera sibogae*	DIATOMS	SILICOFLAGELLATES	CALCISPHERES	**SAMPLE**
R	C															F							C			C						?R		R		C			27X-CC
	C		C													C							F	F	F	R				R						C	R		28X-2, 120
•	C										•				•	C							F	A	F	C				F		•		C		C			28X-5, 120
•	C															C										C				F		R	R			C	R		28X-CC
	C															C							F		C	C						•		F		C			29X-2, 120
	C														•								C	C	C	C				F		F		F		C			29X-5, 120
	C																						A			A				R		•		F		C	R		29X-CC
	C																						C		C	C				R		•		R		A	R		30X-2, 120
	F																						F		C	C				F		F		R		C	R		30X-4, 120
	C																						C		A	A				F		F		F		A	R		30X CC
	F																						C		C	C				F		R				C			31X-2, 120
																							C		C	C						R				A			31X-5, 97
																									A	A				R				F		C	R		31X-CC
																							C		C	C						F		R		F			32X-2, 120
																							C		C	C						F	F	F		C			32X-5, 120
			?•																				A		A	A				F		F	C	F		C	R		32X-CC

NAME ______________________________

Microfossils and Biostratigraphy

Part 3.3. Application of Microfossil First and Last Occurrences

The presence of **microfossils** in many types of deep-sea sediments provides a **basis for determination of age**. The recurrent (and testable) pattern of fossil **first occurrences** (FOs) and **last occurrences** (LOs) recognized by paleontologists studying sedimentary sequences reveals the relative sequence of **evolutionary origination and extinction of species through time**. In other words, in a distribution table showing the occurrences of species in a deep-sea core, like Table 3.1, the **lowest occurrence of a species** represents the first time (i.e., **first occurrence, FO**) this particular taxon was found in this area of the northwest Pacific. This can be considered the evolutionary first appearance, or time of **origination** of this new species. By contrast, the **highest occurrence of a species** represents the last time (i.e., **last occurrence, LO**) this particular taxon was found in this area. This can be considered the evolutionary last appearance, or time of **extinction** of this species. Because of these evolutionary events, **microfossil species found in marine sediments are unique to a particular interval of geologic time** and this property makes them very useful for age determination (Figure 3.11).

1 How can geoscientists test whether a pattern or sequence of fossil first and last occurrences is consistent and repeatable?

____

____

____

____

____

____

Figure 3.12 shows **biozones** for two types of marine microfossils: **calcareous nannofossils (CN zones)** and **planktic foraminifers (N zones)** which have been correlated to the geomagnetic polarity time scale. Zonal marker species first and last occurrences (**B** = base or first occurrence; **T** = top or last occurrence) are shown on the right side of the figure and are used to recognize the base and top of the microfossil biozones similar to the example shown above. Notice that these **biostratigraphic datum levels** have **ages** assigned to them in Figure 3.12. For example, the base (**B**, or first occurrence) of ***Emiliania huxleyi*** is **0.26 Ma** (Ma = mega-annums, or millions of years ago), and the top (**T**, or last occurrence) of ***Discoaster hamatus*** is **9.63 Ma**. We will investigate

NAME

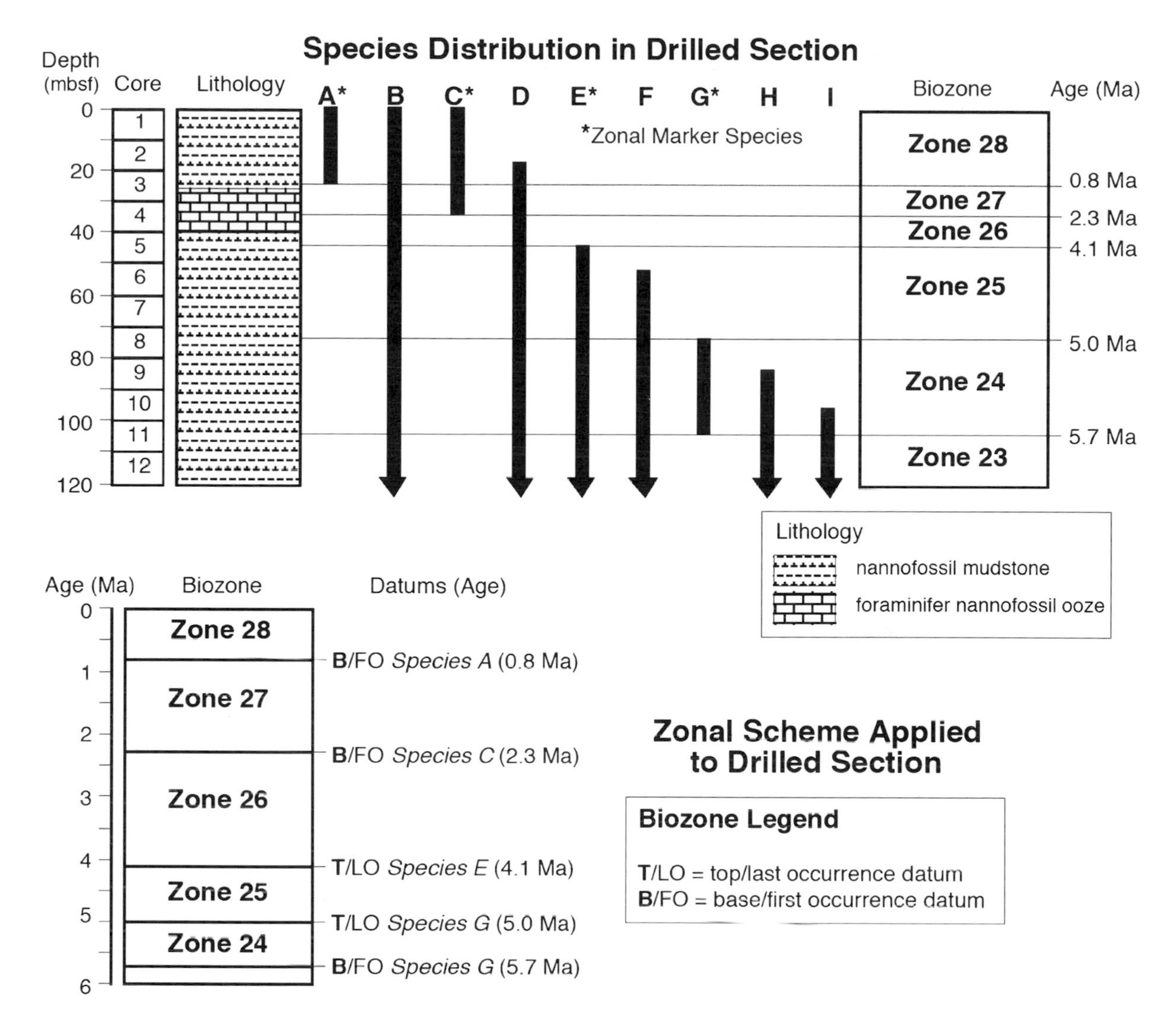

FIGURE 3.11. Two-part diagram illustrating how biostratigraphy is applied to a deep-sea sedimentary section or other fossiliferous sedimentary rocks. The **upper panel** shows a **hypothetical deep-sea sedimentary section** containing microfossils. Letters **A to I indicate species of microfossils** that occur in the cored interval. Vertical bars illustrate in which parts of the cored interval each species is found; this is also referred to as a species' **stratigraphic range**. The down arrows imply that the species also occurs in older sediments (i.e., this cored interval does not represent the total stratigraphic range of the species). The biozones on the right (Zones 23–28) show discrete intervals of the core based on microfossil first or last occurrences. The **lower panel** shows a hypothetical **biostratigraphic zonal scheme** based on a specific sequence of zonal marker species B/FOs and T/Los. The zonal scheme shown in the lower panel can be applied to the cored sedimentary section because some of the zonal marker species are present in the cored sedimentary section. Diagram courtesy of Mark Leckie.

NAME

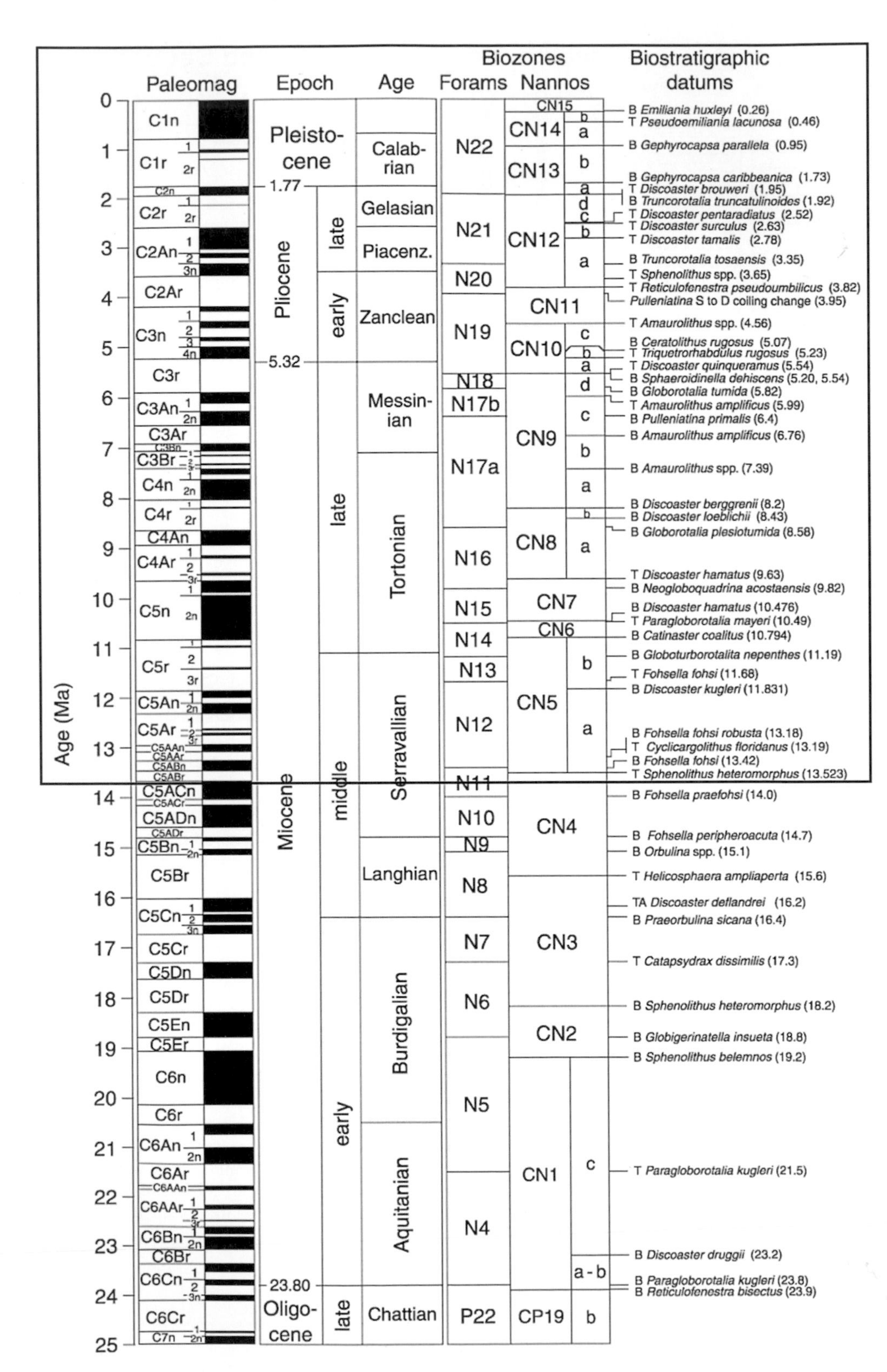

FIGURE 3.12. Biozones for planktic forams (N zones) and calcareous nannofossils (CN zones), along with the corresponding geomagnetic polarity time scale, geologic epochs, and ages. The boxed part of the timescale is the interval represented by the calcareous nannofossil data in Table 3.2. Shipboard Scientific Party, 2002, Explanatory notes.

NAME

BIOSTRATIGRAPHY

Biostratigraphy is the study of sedimentary layers based on fossil content. Paleontologists organize sedimentary layers into **biozones** based on the first and last occurrences of **selected species**; such **zonal marker species** (index fossils) are more suitable than other species for biostratigraphy, because they have short stratigraphic ranges, are geographically widespread, easily recognized, and well preserved. These levels of origination and extinction of marker species are called **biostratigraphic datum levels**. The **ages** of the biostratigraphic datum levels can be determined by multiple methods including paleomagnetic stratigraphy and radiometric age dating (this will be explored in Chapter 4, Paleomagnetism and Magnetostratigraphy).

Zonal marker species are used to **differentiate intervals of geologic time in a sedimentary section** cored from the deep sea, thus essentially determining the **age of successive layers of sediment**. In the case illustrated in Figure 3.11, the distribution of eight microfossil species (A to I) is shown in a 120-meter cored interval. Four of these species (A, C, E, and G) have been established previously as **zonal marker species** because their first and/or last occurrences have been shown to be reliable biostratigraphic datum levels.

how the ages of these biostratigraphic datum levels are determined in Chapter 4, Paleomagnetism and Magnetostratigraphy.

Table 3.2 shows is the **distribution of calcareous nannofossil species in Hole 1208A** on Shatsky Rise in the Northwest Pacific. This table is nearly identical with Table 3.1 and depicts the stratigraphic occurrence of species found in samples analyzed from Hole 1208A. The **first column** corresponds to a **sample identifier** (core-section, interval) and the **second column** is the **depth** in Hole 1208A (meters below seafloor, mbsf) for each of the samples. What is different about this version of the distribution table (compared to that shown in Table 3.1) is the addition of several columns: **Geochronology**, **Nannofossil Zone**, and **Nannofossil Datum**. These are to be filled in as part of this exercise.

In this exercise you will **interpret and apply the calcareous nannofossil zonal scheme (CN zones)** presented in Figure 3.12 **to the actual nannofossil distribution data** from Hole 1208A (Table 3.2).

Step 1: Find the species of calcareous nannofossils that correspond with the CN **zone boundaries** shown in Figure 3.12; these are the **zonal marker species** (i.e., species names that line up with zone boundaries). Each zonal marker species will either be a **base (B)** or a **top (T)**.

Step 2: Find each of these zonal marker species on the Hole 1208A distribution table on the following page. Be sure to note whether you need the base (first occurrence) or top (last occurrence) of the species range. Please note: some zonal species may not be present in the Hole 1208A data.

For example: The two youngest biozones are interpreted on Table 3.2:

NAME

TABLE 3.2. Distribution of calcareous nannofossil species in Hole 1208A on Shatsky Rise in the Northwest Pacific.*

Hole 1208A																														
SAMPLE: Core-Section, Interval (cm)	**DEPTH** (meters below sea floor: mbsf)	**GEOCHRONOLOGY**	**NANNOFOSSIL ZONE**	**NANNOFOSSIL DATUM**	**PRESERVATION**	**ABUNDANCE**	*Amaurolithus* spp.	*Amaurolithus delicatus*	*Amaurolithus primus*	*Calcidiscus leptoporus*	*Calcidiscus macintyrei*	*Calcidiscus tropicus*	*Catinaster calyculus*	*Catinaster coalitus*	*Ceratolithus cristatus*	*Ceratolithus rugosus*	*Ceratolithus telesmus*	*Coccolithus miopelagicus*	*Coccolithus pelagicus*	*Coccolithus pelagicus* (w. bar)	*Cryptococcolithus mediaperforatus*	*Cyclicargolithus floridanus*	*Discoaster asymmetricus*	*Discoaster bellus*	*Discoaster bellus-berggrenii*	*Discoaster berggrenii*	*Discoaster blackstockiae*	*Discoaster bollii*	*Discoaster braarudii*	*Discoaster brouweri*
1H-1, 15	0.15	**Pleistocene**			M-G	A				A					F				A	A										
1H-3, 60	3.6				M	A				C									C	C										
1H-4, 60	4.4				G	A				C					C				A	A										
1H-CC	4.70		**CN15**		M	A				C					•				A	A										
2H-5, 120	11.90				P-M	A				C									C	C										
2H-CC	14.24			**B** *E. huxleyi*	P-M	A				A					R		R		A	A										
3H-5, 107	21.27		**CN14b**		M-G	A				C					•				C	C										
3H-CC	23.82				M-G	A				F					•		•		A	A										
4H-5, 60	30.3			**T** *P. lacunosa*	G	A				F					•				C	C										
4H-CC	33.47				M	A				C									A	A										
5H-5, 120	40.40				M	A				C									A	C										
5H-CC	43.11				P-M	A				F									A	A										
6H-5, 120	49.90				M-G	A				C					•				F	F										
6H-CC	52.62				M	A				A									C	C										
7H-5, 120	59.40				M	A				C									C	F										
7H-CC	62.15				M	A				A					•				C	C										
8H-5, 120	68.90				M	A				C					•				F	F										
8H-CC	71.60				M-G	A				C									C	C										
9H-5, 120	78.40				G	A				C					•				F	F										
9H-CC	81.09				M	A				A	•								C	C										
10H-5, 120	87.90				M-G	A				C	F								C	F										
10H-CC	90.62				M	A				C	F					•			C	C										

Discoaster pentaradiatus	*Discoaster prepentaradiatus*	*Discoaster quinqueramus*	*Discoaster surculus*	*Discoaster tamalis*	*Discoaster triradiatus*	*Discoaster variabilis*	***Emiliania huxleyi***	*Florisphaera profunda*	*Gephyrocapsa aperta*	*Gephyrocapsa caribbeanica*	*Gephyrocapsa oceanica*	*Gephyrocapsa omega* (= *G. parallela*)	*Gephyrocapsa* spp. (medium)	*Gephyrocapsa* spp. (small)	*Gephyrocapsa* spp. (large)	*Helicosphaera carteri*	*Helicosphaera inversa*	*Helicosphaera sellii*	*Helicosphaera wallichii*	*Hughesius gizoensis*	*Minylitha convalis*	*Pontosphaera discopora*	*Pontosphaera japonica*	*Pontosphaera multipora*	***Pseudoemiliania lacunosa***	*Pseudoemiliania ovata*	*Reticulofenestra asanoi*	*Reticulofenestra haqii*	*Reticulofenestra minuta*	*R. pseudoumbilicus* (5–7 μm)	*R. pseudoumbilicus* (>7 μm)	*R. pseudoumbilicus* (>7 μm) w. grill	*Reticulofenestra rotaria*	*Rhabdosphaera clavigera*	*Sphenolithus abies*	*Syracosphaera pulchra*	*Triquetrorhabdulus rugosus*	*Umbilicosphaera jafari*	*Umbilicosphaera rotula*	*Umbilicosphaera sibogae*	DIATOMS	SILICOFLAGELLATES	CALCISPHERES	**SAMPLE**
							A	A			C	R	A	C	R	C						R												F		F				F	C		R	1H-1, 15
							A							A		C						?														F								1H-3, 60
							A	A		F	F	R	A	A	R	C						R	R											F		C				R	C		F	1H-4, 60
							C	C		C	F		A	C	F	F																				R				F	A	R	R	1H-CC
							C	C			C	C	C	C	C	F							•											R		R				•	A		R	2H-5, 120
							C	A	?A		R		A	A	R	R						•																		F	A	R		2H-CC
								C		C	C	•	C	SA		F	•						•													•				R	C			3H 5, 107
								C		C	C		A	A		R	R												?A							F				F	C			3H-CC
								C			•	1		SA		F	R					•			**C**	F										•					C		R	4H-5, 60
								C		C	C	C	A	A	R	F						R	R		**C**	C		F	A							F				F	C	R		4H-CC
								C			F	F	C	A	R	R			1			R	R		**C**	F		C	C							•				F	R			5H-5, 120
								A	C		F	F	A	A		F									**C**	C		C	A											C	A			5H-CC
								A					C	A		R							R		**C**	F	R	A	C					•		•				•	C			6H-5, 120
								C		C	F		C	A		F									**C**	C	C	A	A							R				F	C		R	6H-CC
								C						A		R							R		**C**	F		A	C												A		R	7H-5, 120
								C						A		F						R		R	**C**	A		A								F				C	A	R		7H-CC
											•		C		R							•			**C**	F		A	C												A	R	R	8H-5, 120
								C		C	F		A	A	R	F						F		R	**C**	C		R								C			R	C	C	R	R	8H-CC
								C		C	F		C	F		F		•	•			R			**C**	C		SA	A							F				R	A			9H-5, 120
								A		C			C	A		F		•				R	R	R	**C**			A	A										R	R	C			9H-CC
								C		F				A		F		F	•			R	R		**C**	C		A	C							F			R	R	C		R	10H-5, 120
								A						C		F									**C**			A	A							R				R	A	R		10H-CC

(Continued)

NAME

TABLE 3.2. *Continued*

Hole 1208A																														
SAMPLE: Core-Section, Interval (cm)	**DEPTH** (meters below sea floor: mbsf)	**GEOCHRONOLOGY**	**NANNOFOSSIL ZONE**	**NANNOFOSSIL DATUM**	**PRESERVATION**	**ABUNDANCE**	*Amaurolithus* spp.	*Amaurolithus delicatus*	*Amaurolithus primus*	*Calcidiscus leptoporus*	*Calcidiscus macintyrei*	*Calcidiscus tropicus*	*Catinaster calyculus*	*Catinaster coalitus*	*Ceratolithus cristatus*	*Ceratolithus rugosus*	*Ceratolithus telesmus*	*Coccolithus miopelagicus*	*Coccolithus pelagicus*	*Coccolithus pelagicus* (w. bar)	*Cryptococcolithus mediaperforatus*	*Cyclicargolithus floridanus*	*Discoaster asymmetricus*	*Discoaster bellus*	*Discoaster bellus-berggrenii*	*Discoaster berggrenii*	*Discoaster blackstockiae*	*Discoaster bollii*	*Discoaster braarudii*	*Discoaster brouweri*
11H-2, 65	92.35				M	A				C	F								F	F										
11H-5, 120	97.40				M-G	A				C	F								C	C										
11H-CC	100.16				P-M	A				A	F	F							C	C										R
12H-2, 120	102.40				M	A				C	R								C	C										F
12H-5, 120	106.90				M	A				C	R								C	C										F
12H-CC	109.50				M	A				C	C	C			•				C	C										F
13H-2, 120	111.90				M-G	A				C	F	C							C	C										F
13H-5, 120	116.40				M-G	A				C	F	C							C	C							1			F
13H-CC	119.08				P-M	A				C	C	C				•	•		C	F										F
14H-2, 120	121.40				M	A				C	F				F				F	F										
14H-5, 120	125.90				M	A				C	R	C							C	F										R
14H-CC	128.70				M-G	A				C	F	C				•			C	C			R							F
15H-2, 121	130.91				M	A				C	R								C				R							R
15H-5, 120	135.40				M	A				C	R								C				R							R
15H-CC	138.20				M	A				A	F	F							C				C							C
16H-2, 120	140.40				M	A				C	F								C				F							
16H-5, 120	144.90				M	A				C	F								F											
16H-CC	147.77				M	A				C	C	C			•				F				F							F
17H-2, 120	149.90				M	A				C	F	F							C	F			F							F

Discoaster pentaradiatus	*Discoaster prepentaradiatus*	*Discoaster quinqueramus*	*Discoaster surculus*	*Discoaster tamalis*	*Discoaster triradiatus*	*Discoaster variabilis*	***Emlliania huxleyi***	*Florisphaera profunda*	*Gephyrocapsa aperta*	*Gephyrocapsa caribbeanica*	*Gephyrocapsa oceanica*	*Gephyrocapsa omega* (= *G. parallela*)	*Gephyrocapsa* spp. (medium)	*Gephyrocapsa* spp. (small)	*Gephyrocapsa* spp. (large)	*Helicosphaera carteri*	*Helicosphaera inversa*	*Helicosphaera sellii*	*Helicosphaera wallichii*	*Hughesius gizoensis*	*Minylitha convalis*	*Pontosphaera discopora*	*Pontosphaera japonica*	*Pontosphaera multipora*	***Pseudoemiliania lacunosa***	*Pseudoemiliania ovata*	*Reticulofenestra asanoi*	*Reticulofenestra haqii*	*Reticulofenestra minuta*	*R. pseudoumbilicus* (5–7μm)	*R. pseudoumbilicus* (>7μm)	*R. pseudoumbilicus* (>7μm) w. grill	*Reticulofenestra rotaria*	*Rhabdosphaera clavigera*	*Sphenolithus abies*	*Syracosphaera pulchra*	*Triquetrorhabdulus rugosus*	*Umbilicosphaera jafari*	*Umbilicosphaera rotula*	*Umbilicosphaera sibogae*	DIATOMS	SILICOFLAGELLATES	CALCISPHERES	**SAMPLE**
																		F										C	C															11H-2, 65
								A								F		R				R			**C**	C		SA	A							F					A			11H-5, 120
								A								R						R			**C**			A	A							R			•	R	C	R		11H-CC
																		F										C	C															12H-2, 120
																R		F				R			**F**	C		C	C															12H-5, 120
								A								R						R			**C**	C		A	A							R				R	C			12H-CC
								C								•						•			**C**	C		SA	C												A			13H-2, 120
R								C								1		1				•			**F**	C		SA	A												A	R		13H-5, 120
C			F					A														R			**C**			A	A							R				F	C			13H-CC
R			F																									C	F															14H 2, 120
F			F					C								•									**F**	C		A	A							•			R		A			14H-5, 120
C			C	F		?R		C								F						R			**C**	C		A	A	?•						R			C		C			14H-CC
R				R																						C		C	C							F			R					15H-2, 121
F			R																							C		C	C										R					15H-5, 120
F			F	C	R			C														R			**C**	C		A	A							F			R		A	R		15H-CC
F			F	F																					**F**	C		C	C										R					16H-2, 120
R			R			F																				C		C	F										F					16H-5, 120
F			F	F	•			C																	**C**	F		A	A							R			F		A	R		16H-CC
R			F	F				F														•				C		C	C												C			17H-2, 120

(Continued)

NAME ______________________________

TABLE 3.2. *Continued*

Hole 1208A																														
SAMPLE: Core-Section, Interval (cm)	**DEPTH** (meters below sea floor: mbsf)	**GEOCHRONOLOGY**	**NANNOFOSSIL ZONE**	**NANNOFOSSIL DATUM**	**PRESERVATION**	**ABUNDANCE**	*Amaurolithus* spp.	*Amaurolithus delicatus*	*Amaurolithus primus*	*Calcidiscus leptoporus*	*Calcidiscus macintyrei*	*Calcidiscus tropicus*	*Catinaster calyculus*	*Catinaster coalitus*	*Ceratolithus cristatus*	*Ceratolithus rugosus*	*Ceratolithus telesmus*	*Coccolithus miopelagicus*	*Coccolithus pelagicus*	*Coccolithus pelagicus* (w. bar)	*Cryptococcolithus mediaperforatus*	*Cyclicargolithus floridanus*	*Discoaster asymmetricus*	*Discoaster bellus*	*Discoaster bellus-berggrenii*	*Discoaster berggrenii*	*Discoaster blackstockiae*	*Discoaster bollii*	*Discoaster braarudii*	*Discoaster brouweri*
17H-5, 120	154.50				M	A				C	F	C							F	F			F							F
17H-CC	157.21				M	A				A	C	C			F				F	F			C							F
18H-2, 120	159.40				M	A				C	F	C			R				F				C							F
18H-5, 120	163.90				M	A				C	R	F			R				R	R			C							F
18H-CC	166.66				M	A				A	C	F			•	•			?F	F			F							F
19H-2, 118	168.88				P-M	A	•	•		C	R	F			R				F	F			1							
19H-5, 117	173.37				M-G	A	•	•		C	F	F			•				R	R										
19H-CC	176.09				M	A	F	F		C	F				•	•			?F											F
20H-2, 120	178.40				M	A	•	•		C	•	F			•				F	F										R
20H-5, 136	183.06				M	A	•	•		F	•	F			•				•	•							1			R
20X-CC	185.65				M	A	R	R		C	F				•															F
21X-2, 120	187.90				M	A	•	•		F	F	F			•				R											•
21X-4, 120	190.90				P-M	A	R	R	•	C	F	F							R											F
21X-CC	191.07				M	A	F	F	R	A	F								C				R				•			C
22X-2, 120	193.10				P-M	A	•	•		F	F	F							R	R										F
22X-5, 120	197.60				P	A	•	•		F	F	C							F	R										F
22X-CC	199.99				P-M	A	F	F		C	F								C											
23X-2, 121	202.81				P-M	A	•	•		F		F							F											
23X-5, 90	207.00				P-M	A	•	•		F		F							F											

NAME ____

Discoaster pentaradiatus	*Discoaster prepentaradiatus*	*Discoaster quinqueramus*	*Discoaster surculus*	*Discoaster tamalis*	*Discoaster triradiatus*	*Discoaster variabilis*	***Emiliania huxleyi***	*Florisphaera profunda*	*Gephyrocapsa aperta*	*Gephyrocapsa caribbeanica*	*Gephyrocapsa oceanica*	*Gephyrocapsa omega (= G. parallela)*	*Gephyrocapsa* spp. (medium)	*Gephyrocapsa* spp. (small)	*Gephyrocapsa* spp. (large)	*Helicosphaera carteri*	*Helicosphaera inversa*	*Helicosphaera sellii*	*Helicosphaera wallichii*	*Hughesius gizoensis*	*Minylitha convalis*	*Pontosphaera discopora*	*Pontosphaera japonica*	*Pontosphaera multipora*	***Pseudoemiliania lacunosa***	*Pseudoemiliania ovata*	*Reticulofenestra asanoi*	*Reticulofenestra haqii*	*Reticulofenestra minuta*	*R. pseudoumbilicus* (5–7µm)	*R. pseudoumbilicus* (>7µm)	*R. pseudoumbilicus* (>7µm) w. grill	*Reticulofenestra rotaria*	*Rhabdosphaera clavigera*	*Sphenolithus abies*	*Syracosphaera pulchra*	*Triquetrorhabdulus rugosus*	*Umbilicosphaera jafari*	*Umbilicosphaera rotula*	*Umbilicosphaera sibogae*	DIATOMS	SILICOFLAGELLATES	CALCISPHERES	**SAMPLE**
R				F	F			F						A		•						•				C		A	A	•									F		C			17H-5, 120
F			C	F		R										F									**F**	C		A	A	1	•					R		R	C		A	F	R	17H-CC
F			F	F	•			C								1						•			**F**	F		C	F										F		A			18H-2, 120
F			F	R	1	1		C						R								•			**F**	F		C	C	R	C	C							F		C			18H-5, 120
F			F	•				C						A		F						F		R	**?F**	C		A	A		C	C			F	F		C	C	?	A	R		18H-CC
F			F			1		A								•						R				F		C	A	F	C	C			•	•			F		C			19H-2, 118
F			F					C								1		?1				R						C	SA	R	C	C			R				F		C			19H-5, 117
F			F			F		C														•				C		A	A		C	R			?F				C		A	R		19H-CC
F			F			•		A								1						R				C			SA		C	R	1		F	•			F		C			20H-2, 120
F			R			•		A																				F	SA	F	F				•				F		C			20H-5, 136
F			F			F																						C		A	F				R				F		A			20X-CC
F			F					C																				C	A	A	F	R			R				R	?1	C			21X-2, 120
F			F		•			A																				A	C	A	F				•				R		C			21X-4, 120
C			C			R																						A	A	A	C				R				F		A	R		21X-CC
F			F			F		A																				A	C	1			2		R				F		C			22X-2, 120
F			F		•	F		C																				A	F						R				F		C			22X-5, 120
F			R			C																						A		C					F				R		C	R		22X-CC
F						F		A																				C	F	F	C				F				R		C			23X-2, 121
F		F				F		C																				A	C	R	C				R									23X-5, 90

(Continued)

TABLE 3.2. *Continued*

Hole 1208A																														
SAMPLE: Core-Section, Interval (cm)	**DEPTH** (meters below sea floor: mbsf)	**GEOCHRONOLOGY**	**NANNOFOSSIL ZONE**	**NANNOFOSSIL DATUM**	**PRESERVATION**	**ABUNDANCE**	*Amaurolithus* spp.	*Amaurolithus delicatus*	*Amaurolithus primus*	*Calcidiscus leptoporus*	*Calcidiscus macintyrei*	*Calcidiscus tropicus*	*Catinaster calyculus*	*Catinaster coalitus*	*Ceratolithus cristatus*	*Ceratolithus rugosus*	*Ceratolithus telesmus*	*Coccolithus miopelagicus*	*Coccolithus pelagicus*	*Coccolithus pelagicus* (w. bar)	*Cryptococcolithus mediaperforatus*	*Cyclicargolithus floridanus*	*Discoaster asymmetricus*	*Discoaster bellus*	*Discoaster bellus-berggrenii*	*Discoaster berggrenii*	*Discoaster blackstockiae*	*Discoaster bollii*	*Discoaster braarudii*	*Discoaster brouweri*
23X-CC	208.18				P	A	•	•	•	F	F								R											F
24X-2, 121	212.51				M	A				C		F							F	R										1
24X-5, 120	217.00				P-M	A	•		•	F	•	F							F											
24X-CC	218.74				P-M	A	•	F	•	A	C								C		R									F
25X-2, 119	222.09				M	A	R	•	R	C		R							F		1									
25X-4, 90	224.50				M	A	•		•	F									C		1									
25X-CC	224.59				P-M	A	R	•	R	C	C								C				R			R				R
26X-2, 119	231.79				M	A	1		1	F	R								F											
26X-5, 82	235.37				M	A	1		1	F	R								F		F					F				•
26X-CC	235.52				P-M	A	•		•	F	F								C		R					•				F
27X-2, 120	241.40				M	A				F	F								F	1	F					F				
27X-5, 120	245.90				P-M	A				C	F	R							C		R			1		F				C
27X-CC	246.14				M	A				C	F								C		F		R			F		F		C
28X-2, 120	250.80				M	A				C	F	R							C		F			R		R				C
28X-5, 120	255.30				M	A				C	F	F							C		C			•					F	C
28X-CC	256.71				P	A				R	F								F		R			•					R	C
29X-2, 120	260.50				P-M	A				F	F	R							F		F			R						C
29X-5, 120	265.00				M	A				F	F	R							C		R			F	R					C
29X-CC	265.94				P	A				F	F		•						C		R		•	F	R					C
30X-2, 120	270.10				P-M	A					R		R	R					F					R				F		R

NAME

Discoaster pentaradiatus	*Discoaster prepentaradiatus*	*Discoaster quinqueramus*	*Discoaster surculus*	*Discoaster tamalis*	*Discoaster triradiatus*	*Discoaster variabilis*	***Emiliania huxleyi***	*Florisphaera profunda*	*Gephyrocapsa aperta*	*Gephyrocapsa caribbeanica*	*Gephyrocapsa oceanica*	*Gephyrocapsa omega* (= *G. parallela*)	*Gephyrocapsa* spp. (medium)	*Gephyrocapsa* spp. (small)	*Gephyrocapsa* spp. (large)	*Helicosphaera carteri*	*Helicosphaera inversa*	*Helicosphaera sellii*	*Helicosphaera wallichii*	*Hughesius gizoensis*	*Minylitha convalis*	*Pontosphaera discopora*	*Pontosphaera japonica*	*Pontosphaera multipora*	***Pseudoemiliania lacunosa***	*Pseudoemiliania ovata*	*Reticulofenestra asanoi*	*Reticulofenestra haqii*	*Reticulofenestra minuta*	*R. pseudoumbilicus* (5–7 μm)	*R. pseudoumbilicus* (>7 μm)	*R. pseudoumbilicus* (>7 μm) w. grill	*Reticulofenestra rotaria*	*Rhabdosphaera clavigera*	*Sphenolithus abies*	*Syracosphaera pulchra*	*Triquetrorhabdulus rugosus*	*Umbilicosphaera jafari*	*Umbilicosphaera rotula*	*Umbilicosphaera sibogae*	DIATOMS	SILICOFLAGELLATES	CALCISPHERES	**SAMPLE**
F		R	R			F																						A		?F	C				R			R	R		A	R		23X-CC
F		C				C		A								1												A	C	F	A				R		?1		F		A			24X-2, 121
F		F				F		A																				C	C	C	A				R				R		C			24X-5, 120
C		C	R		•	C																						A			A		F		F				F		A	R		24X-CC
F		C				C		F																				C	A	C	C				R				F		C			25X-2, 119
R		C			1	C																						F	A	A	C		R		F				F		C			25X-4, 90
F		C	R		•	C																						A	?C	C	A		?		F						C	R		25X-CC
F		C				C																						C	F	A	R		R		•				F		C			26X-2, 119
F		F	•			C														•								C	C	C	R		1		•				F		C			26X-5, 82
F		F	R		R	C																						A	?C	C	C		F		•				R		C			26X-CC
•		R				F																						A	A	A					•				F		C			27X-2, 120
R		R			1	C															F							C		F	F										C			27X-5, 120
R		F			R	C															F							C			C						?R		R		C			27X-CC
R		R				C		C													C							F	F	F	R				R						C	R		28X-2, 120
R	•				•	C										•				•	C							F	A	F	C				F		•		C		C			28X-5, 120
S					•	C															C										C				F		R	R			C	R		28X-CC
R	R					C															C							F		C	C						•		F		C			29X-2, 120
•	R					C														•								C	C	C	C				F		F		F		C			29X-5, 120
R	F					C																						A			A				R		•		F		C	R		29X-CC
						C																						C		C	C				R		•		R		A	R		30X-2, 120

(Continued)

TABLE 3.2. *Continued*

Hole 1208A																														
SAMPLE: Core-Section, Interval (cm)	**DEPTH** (meters below sea floor: mbsf)	**GEOCHRONOLOGY**	**NANNOFOSSIL ZONE**	**NANNOFOSSIL DATUM**	**PRESERVATION**	**ABUNDANCE**	*Amaurolithus* spp.	*Amaurolithus delicatus*	*Amaurolithus primus*	*Calcidiscus leptoporus*	*Calcidiscus macintyrei*	*Calcidiscus tropicus*	*Catinaster calyculus*	*Catinaster coalitus*	*Ceratolithus cristatus*	*Ceratolithus rugosus*	*Ceratolithus telesmus*	*Coccolithus miopelagicus*	*Coccolithus pelagicus*	*Coccolithus pelagicus* (w. bar)	*Cryptococcolithus mediaperforatus*	*Cyclicargolithus floridanus*	*Discoaster asymmetricus*	*Discoaster bellus*	*Discoaster bellus-berggrenii*	*Discoaster berggrenii*	*Discoaster blackstockiae*	*Discoaster bollii*	*Discoaster braarudii*	*Discoaster brouweri*
30X-4, 120	273.10				P-M	A				F	F	R		F					F					R						R
30X-CC	274.20				P-M	A				F	C			F					C		F			R				R	R	R
31X-2, 120	279.70				M	A				F	R	F		F					F									R		R
31X-5, 97	283.97				P-M					F	F	F						2	F											
31X-CC	285.06				P-M					C	F	R						F	C		R							R		
32X-2, 120	289.30				P-M					F	F	F						F	C									F		•
32X-5, 120	303.10				P-M					F		F						F	F		R							R		•
32X-CC	295.41				P-M					C	?R	C						F	C		F	C						R		R

* Modified from: Bown, 2005; http://wwwodp.tamu.edu/publications/198_SR/104/104.htm
Note that if a zonal marker species does not occur in the data table for Hole 1208A, you will need to lump two zones or subzones together (e.g., CN10a-b).
B = base (or first occurrence, FO), T = top (or last occurrence, LO)

NAME

SAMPLE	*Discoaster pentaradiatus*	*Discoaster prepentaradiatus*	*Discoaster quinqueramus*	*Discoaster surculus*	*Discoaster tamalis*	*Discoaster triradiatus*	*Discoaster variabilis*	***Emiliania huxleyi***	*Florisphaera profunda*	*Gephyrocapsa aperta*	*Gephyrocapsa caribbeanica*	*Gephyrocapsa oceanica*	*Gephyrocapsa omega* (= *G. parallela*)	*Gephyrocapsa* spp. (medium)	*Gephyrocapsa* spp. (small)	*Gephyrocapsa* spp. (large)	*Helicosphaera carteri*	*Helicosphaera inversa*	*Helicosphaera sellii*	*Helicosphaera wallichii*	*Hughesius gizoensis*	*Minylitha convalis*	*Pontosphaera discopora*	*Pontosphaera japonica*	*Pontosphaera multipora*
30X-4, 120							F																		
30X-CC		R					C																		
31X-2, 120							F																		
31X-5, 97																									
31X-CC																									
32X-2, 120																									
32X-5, 120																									
32X-CC									?•																

SAMPLE	***Pseudoemiliania lacunosa***	*Pseudoemiliania ovata*	*Reticulofenestra asanoi*	*Reticulofenestra haqii*	*Reticulofenestra minuta*	*R. pseudoumbilicus* (5–7μm)	*R. pseudoumbilicus* (>7μm)	*R. pseudoumbilicus* (>7μm) w. grill	*Reticulofenestra rotaria*	*Rhabdosphaera clavigera*	*Sphenolithus abies*	*Syracosphaera pulchra*	*Triquetrorhabdulus rugosus*	*Umbilicosphaera jafari*	*Umbilicosphaera rotula*	*Umbilicosphaera sibogae*	DIATOMS	SILICOFLAGELLATES	CALCISPHERES
30X-4, 120				F		C	C				F		F		R		C	R	
30X-CC				C		A	A				F		F		F		A	R	
31X-2, 120				C		C	C				F		R				C		
31X-5, 97				C		C	C						R				A		
31X-CC						A	A				R				F		C	R	
32X-2, 120				C		C	C						F		R		F		
32X-5, 120				C		C	C						F	F	F		C		
32X-CC				A		A	A				F		F	C	F		C	R	

NAME ____

- The base, or first occurrence of ***Emiliania huxleyi*** (distribution highlighted in gray) is used to define the **boundary between Zone CN14b and CN15**. *E. huxleyi* first occurs in **sample 2H-CC**, so this sample is the lowest sample in Zone CN15. Draw a horizontal line across the distribution table between sample 2H-CC and the underlying sample 3H-5, 107 cm. This line represents the boundary between Zone CN14b and CN15. The **age** of this level in Hole 1208A is **0.26 Ma**.
- The top, or last occurrence of ***Pseudoemiliania lacunosa*** (distribution highlighted in gray) is used to define the **boundary between Zone CN14a and CN14b**. *P. lacunosa* last occurs in **sample 4H-5, 60**, so this sample is the highest sample in Zone CN14a. Draw a horizontal line across the distribution table between sample 4H-5, 60 and the overlying sample 3H-CC. This line represents the boundary between Zone CN14a and CN14b. The **age** of this level in Hole 1208A is **0.46 Ma**.

2 **Interpret biozones** for the remainder of the Hole 1208A distribution table (Table 3.2) by locating the CN zonal marker species and drawing horizontal lines on Table 3.2 corresponding with the zone boundaries. Fill in the columns of Table 3.2 for **Geochronology**, **Nannofossil Zone**, and **Nannofossil Datum** as shown for the two examples described above. Geochronology is a geologic time unit and refers to the **Epoch/Age** of the sediment (e.g., Pleistocene, late Pliocene, early Pliocene, late Miocene, middle Miocene).

3 In this investigation you learned how to apply a biostratigraphic zonal scheme to a cored sedimentary sequence from the seafloor of the North Pacific Ocean. This process essentially enabled you to convert depth to age for that sedimentary sequence. What does knowing "age" now allow you to explore about the history of environmental and climate change?

____

____

____

____

____

NAME

Microfossils and Biostratigraphy

Part 3.4. Using Microfossil Datum Levels to Calculate Sedimentation Rates

Table 3.3 presents a summary of select calcareous nannofossil species datum levels observed at **Site 1208** (ODP Leg 198) on the central high of **Shatsky Rise in the Northwest Pacific** (from Bown, 2005). **FO** = first occurrence, **LO** = last occurrence. This list represents a subset of all the species shown on the distribution table (Tables 3.1 and 3.2) analyzed in Part 3.2 and Part 3.3 (i.e., most of these are the **zonal marker species**). These taxa have well established datum level ages and are therefore useful for establishing the age of the sedimentary sequence (**age model**) at Site 1208. **Note:** The age-depth data in Table 3.3 are derived from the biostratigraphic zonation that was done in Part 3.3.

1 Using the graph paper provided and the age and depth data in Table 3.3, **construct an age–depth plot** similar to the example below (Figure 3.13) from ODP Site 846 in the eastern equatorial Pacific (note that the data points represent calcareous nannofossil datum levels). Place **Age (Ma)** on the *x*-axis across the top of the plot, and **Depth (mbsf)** on the *y*-axis with zero depth at the top with increasing depth down the left-hand side of the page. Be sure to **plan out your graph based on the range of the *x* and *y* values**: when labeling your axes, ensure that the values between each tick mark are equal and that

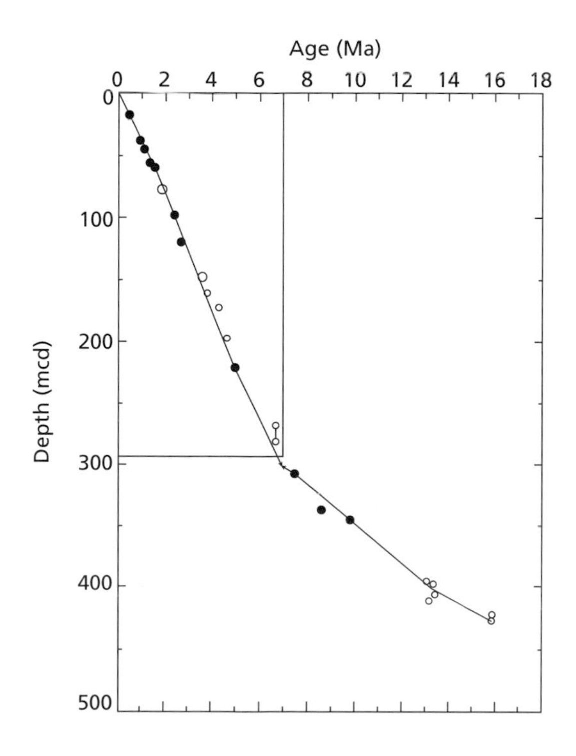

FIGURE 3.13. Example of an age–depth plot from ODP 846 in the eastern equatorial Pacific. A series of "best fit lines" illustrate changes in sediment accumulation rate. From Shipboard Scientific Party, 1992.

TABLE 3.3. Calcareous nannofossil datum levels, ages, and depths from Hole 1208A (from Bown, 2005).

Datum	Depth (mbsf)	Age (Ma)	Age data source
FO *Emiliania huxleyi*	14.24	0.26	Shipboard Scientific Party, 2002
LO *P. lacunosa*	30.30	0.46	Shipboard Scientific Party, 2002
FO *G. omega*	43.11	0.95	Shipboard Scientific Party, 2002
FO *G. caribbeanica*	87.90	1.73	Shipboard Scientific Party, 2002
LO *D. brouweri*	100.16	1.95	Shipboard Scientific Party, 2002
LO *D. pentaradiatus*	116.40	2.52	Shipboard Scientific Party, 2002
LO *D. surculis*	119.08	2.63	Shipboard Scientific Party, 2002
LO *D. tamalis*	128.70	2.78	Shipboard Scientific Party, 2002
LO Large *Reticulofenestra*	163.90	3.82	Shipboard Scientific Party, 2002
FO *D. tamalis*	166.66	4.20	Young, 1998
LO *Sphenolithus*	166.66	3.65	Shipboard Scientific Party, 2002
LO *Amaurolithus*	168.88	4.56	Shipboard Scientific Party, 2002
FO *D. asymmetricus*	168.88	4.20	Young, 1998
FO *C. cristatus*	187.90	5.07	Shipboard Scientific Party, 2002
LO *D. quinqueramus*	207.00	5.54	Shipboard Scientific Party, 2002
FO *Amaurolithus*	235.52	7.39	Shipboard Scientific Party, 2002
FO *D. quinqueramus*	250.80	8.20	Shipboard Scientific Party, 2002
FO *D. berggrenii*	250.80	8.20	Shipboard Scientific Party, 2002
LO *D. hamatus*	265.94	9.63	Shipboard Scientific Party, 2002
FO *C. calyculus*	270.10	10.10	Young, 1998
FO *D. hamatus*	274.20	10.48	Shipboard Scientific Party, 2002
FO *C. coalitus*	279.70	10.79	Shipboard Scientific Party, 2002
LO *C. miopelagicus*	285.06	10.90	Young, 1998
LO *C. premacintyrei*	295.41	12.30	Young, 1998
LO *C. floridanus*	295.41	13.19	Shipboard Scientific Party, 2002

Note: LO large *Reticulofenestra* = LO *Reticulofenestra pseudoumbilicus*; LO *Amaurolithus* = LO *Amaurolithus primus*.

your graph can accommodate all the data (i.e., approximately 14 million years of record and approximately 300 m of drilled section).

The **rate of sediment accumulation** can be calculated by dividing depth by age. For example, in the age–depth plot in Figure 3.13, the interval from 300 mcd (meters composite depth) to the surface accumulated at a rate of approximately 43 m/myr (meters per million years): 300 m/7 myr = 42.9 m/myr.

NAME

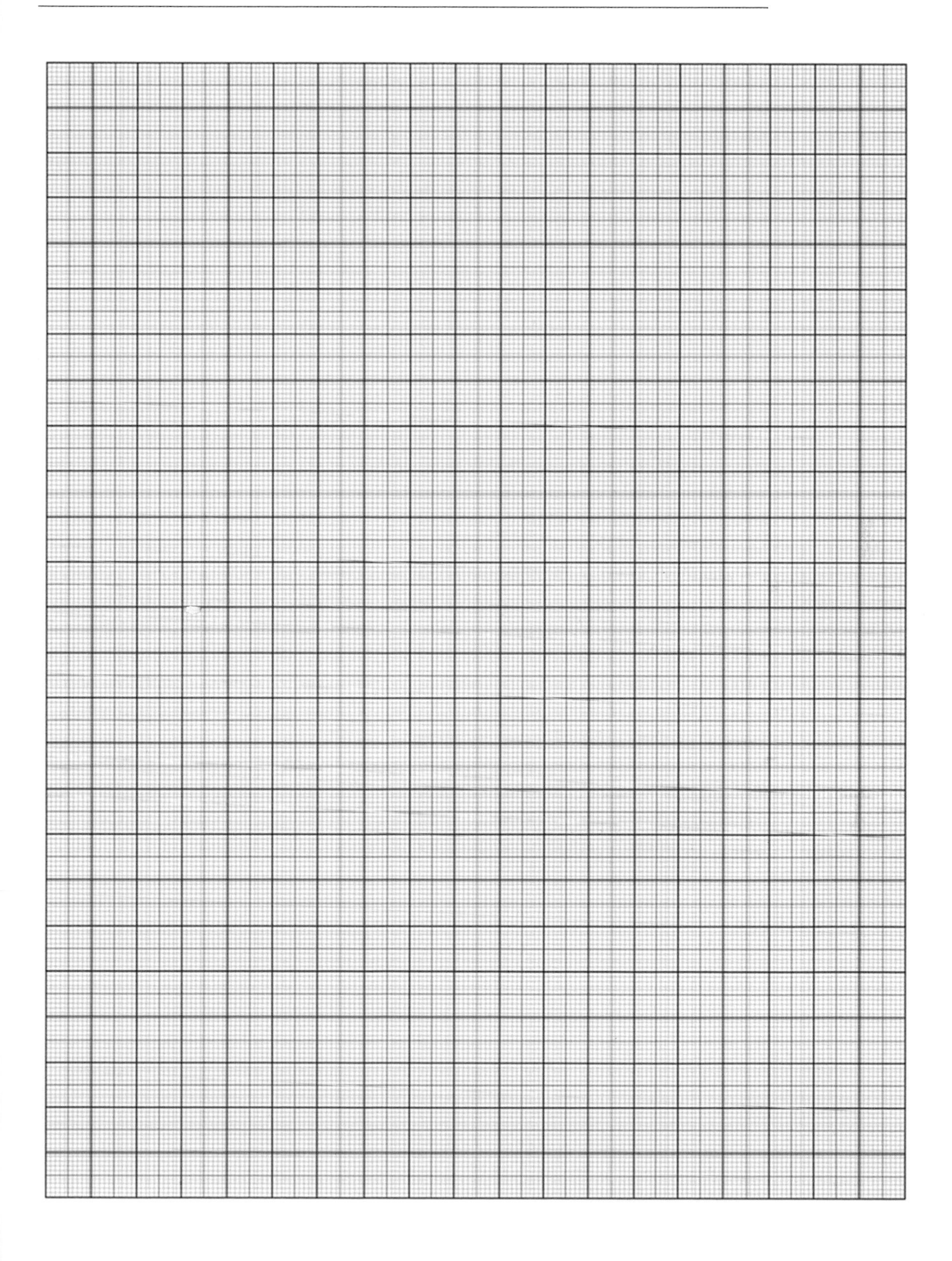

NAME ______________________________

2 Note the **change in slope** at approximately 300 mcd in Figure 3.13. What was the sedimentation rate for the interval from 300 mcd to 425 mcd? Show your work.

3 If the slope on an age–depth plot is zero (i.e., horizontal), what would this tell you about the rate of sediment accumulation? What would be the implication for the sedimentation history if a segment of the age-depth plot is horizontal?

4 Draw a **best-fit line** or **multiple best-fit lines** through the Site 1208 data that you plotted.

(a) How can your graph help you determine the **rate** at which sediment accumulated at Site 1208?

NAME ______________________________

(b) Why does the slope of the best-fit lines change?

____

____

____

____

____

____

5 (a) **Calculate the sedimentation rate** of each of your best-fit line segments (i.e., where you interpret a change in slope). Use units of meters per million years (m/myr). Show your work. **Label** each line segment with its sedimentation rate.

____

____

____

____

____

____

____

____

(b) Was the sedimentation rate constant at Site 1208 over the past 13 million years? If not, describe how the rate of sedimentation changed over time.

____

____

____

____

____

____

NAME

(c) Propose some hypotheses to explain these changes in sedimentation rate at this site through time. What data would you need to test these hypotheses?

NAME ____

Microfossils and Biostratigraphy

Part 3.5. How Reliable are Microfossil Datum Levels?

As we observed in Part 3.3, the presence of microfossils in many types of deep-sea sediments provides a **basis for age determination**. The recurrent (and testable) pattern of fossil first occurrences (FOs) and last occurrences (LOs) recognized by paleontologists studying sedimentary sequences reveals the relative sequence of **evolutionary origination and extinction of species through time**. Because of these evolutionary events, microfossil species found in marine sediments are unique to a particular interval of geologic time making them very useful for age determination.

By comparing the microfossil distribution at one site with other deep-sea sites **from the same latitudinal belt** (i.e., tropical-subtropical, temperate, or subpolar-polar) geoscientists can test whether the pattern or sequence of fossil first and last occurrences is **consistent and repeatable**. In this exercise, we will compare the nannofossil distribution found in temperate Site 1208 (ODP Leg 198) on Shatsky Rise in the Northwest Pacific (latitude, longitude: 36°7.6'N, 158°12.1'E) with the nannofossil distribution found in tropical-subtropical Site 999 (ODP Leg 165) in the Caribbean Sea (12°44.6'N, 78°44.4'W).

Table 3.3 (in Part 3.4) presents a summary of select calcareous nannofossil species datum levels observed at temperate **Site 1208** (ODP Leg 198) on **Shatsky Rise in the Northwest Pacific** (from Bown, 2005). Note: **FO** = first occurrence, **LO** = last occurrence. This list represents a subset of all the species shown on the distribution table analyzed in Part 3.2 and Part 3.3; these taxa are the most useful for constructing an age model for the sediments cored at Site 1208.

Table 3.4 presents a summary of select calcareous nannofossil species datum levels observed at tropical-subtropical **Site 999** (ODP Leg 165; Figure 3.14) in the **Caribbean Sea** (from Kameo and Bralower, 2000). Note: **B** = base, **T** = top (similar to FO = first occurrence and LO = last occurrence).

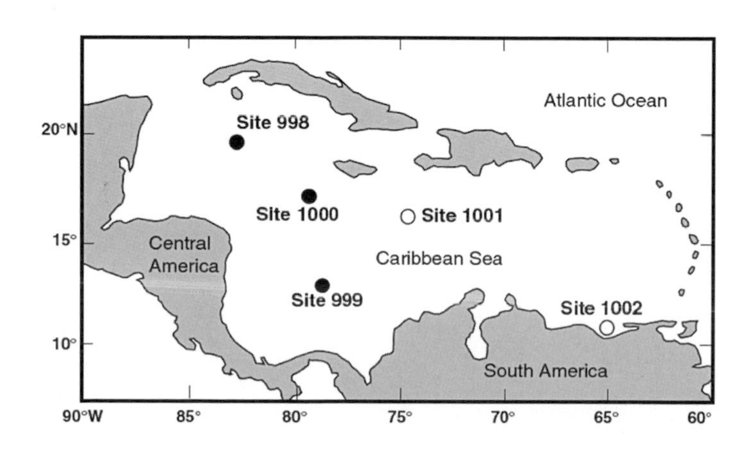

FIGURE 3.14. Sites drilled during ODP Leg 165 and location of ODP Site 999 in the Caribbean Sea. From Kameo and Bralower, 2000.

NAME ______________________

TABLE 3.4. Calcareous nannofossil datum levels, ages, and depths from Hole 999A; B = base (or first occurrence, FO), T = top (or last occurrence, LO) (from Kameo and Bralower, 2000).

Datum and bioevent	Zone (base)	Age (Ma)	Core, section, interval (cm)		Depth (mbsf)
			Upper	Lower	
			165-999A-	165-999A-	
B *Emiliania huxleyi*	CN15	0.25	2H-1, 100	2H-1, 122	8.71
T *Pseudoemiliania lacunosa*	CN14b	0.41	2H-7, 30	2H-CC	17.00
T *Reticulofenestra asanoi*		0.85	4H-4, 22	4H-4, 72	31.57
B *Gephyrocapsa parallela*	CN14a	0.95	4H-6, 100	4H-7, 30	35.50
B *Reticulofenestra asanoi*		1.16	5H-1, 72	5H-1, 100	36.96
T *Gephyrocapsa* spp. (large)		1.21	5H-4, 22	5H-4, 72	41.07
T *Helicosphaera sellii*		1.27	5H-7, 30	5H-CC	45.50
B *Gephyrocapsa* spp. (large)		1.45	6H-4, 100	6H-4, 122	51.21
B *Gephyrocapsa oceanica*		1.65	7H-1, 22	7H-1, 72	55.57
T *Calcidiscus macintyrei*		1.65	7H-1, 22	7H-1, 72	55.57
B *Gephyrocapsa caribbeanica*	CN13b	1.73	7H-1, 100	7H-1, 122	56.21
T *Discoaster brouweri*	CN13a	1.97	7H-CC	8H-1, 100	65.10
T *Discoaster pentaradiatus*	CN12d	2.38	9H-3, 72	9H-3, 122	78.07
T *Discoaster surculus*	CN12c	2.54	10H-1, 122	10H-2, 22	85.07
T *Discoaster tamalis*	CN12b	2.74	10H-5, 122	10H-6, 21	91.07
T *Reticulofenestra ampla*		2.78	10H-5, 122	10H-6, 21	91.07
T *Reticulofenestra minutula* (circ. fonn)		3.36	12H-4, 72	12H-5, 22	108.32
T *Sphenolithus* spp.		3.65	13H-1, 122	13H-2, 22	113.57
T *Reticulofenestra pseudoumbilicus*	CN12a	3.80	13H-6, 100	13H-7, 30	121.00
T *Amaurolithus primus*	CN11a	4.37	16H-5, 100	16H-6, 22	147.96
B *Ceratolithus rugosus*	CN10c	5.04	16H-CC	17H-1, 100	150.60
T *Ceratolithus acutus*	CN10c	5.04	17H-5, 100	17H-6, 22	157.46
B *Ceratolithus acutus*	CN10b	5.34	18H-1, 100	18H-2, 22	160.96
T *Triquetrorhabdulus rugosus*		5.34	18H-2, 100	18H-3, 22	162.46
T *Discoaster quinqueramus*	CN10a	5.56	18H-6, 100	18H-7, 30	168.50
T *Amaurolithus amplificus*		5.88	19H-1, 100	19H-2, 22	170.46
B *Amaurolithus amplificus*		6.50	19H-7, 22	19H-CC	178.46
B *Amaurolithus* spp.	CN9b	7.24	22X-2, 22	22X-2, 100	199.71
T paracme *Reticulofenestra pseudoumbilicus*		6.80	23X-6, 22	23X-7, 22	209.07
T *Minylitha convallis*		6.70	25X-2, 23	25X-3, 21	223.90
B *Discoaster berggrenii*	CN9a	8.35	26X-2, 22	26X-2, 100	231.61
B paracme *Reticulofenestra pseudoumbilicus*		8.85	27X-2, 24	27X-3, 24	241.59
B *Discoaster loeblichii*	CN8b	8.43	27X-5, 22	27X-5, 100	245.71

(Continued)

NAME ______________________________

TABLE 3.4. *Continued*

Datum and bioevent	Zone (base)	Age (Ma)	Core, section, interval (cm)		Depth (mbsf)
			Upper	Lower	
B *Minylitha convallis*		9.43	28X-6, 22	28X-7, 22	257.17
T *Discoaster hamatus*	CN8a	9.36	29X-1, 22	29X-1, 100	259.01
T *Catinaster calyculus*		9.36	29X-6, 22	29X-7, 22	266.87
B *Discoaster hamatus*	CN7a	10.39	29X-CC	30X-1, 22	268.11
B *Catinaster calyculus*	CN7b	10.70	30X-2, 21	30X-3, 21	270.46
T *Coccolithus miopelagicus*		10.39	30X-3, 100	30X-4, 21	272.36
B *Catinaster coalitus*	CN6	10.71	30X-5, 21	30X-5, 100	274.61
T *Discoaster kugleri*		11.50	31X-CC	32X-1, 22	287.31
B common *Discoaster kugleri*		11.74	33X-6, 100	33X-7, 30	305.60
T *Coronocyclus nitescens*		12.12	35X-2, 24	35X-3, 23	318.49
T *Cyclicargolithus floridanus*		13.19	35X-7, 20	35X-CC	325.40

In this exercise, we will **compare the sequence of calcareous nannofossil datum levels** at **Site 1208** in the Northwest Pacific with the sequence of select calcareous nannofossil datum levels observed at **Site 999** in the Caribbean Sea. There are a few minor differences in the ages of the datum levels between the two sites reflecting continued improvement in the calibration of these datum levels since the two individual data sets were published, 2005 and 2000, respectively. **Remember:** first occurrence (FO) = base (B), and last occurrence (LO) = top (T).

1 Place a **check** (√) next to the species **datum levels common to both lists** (Tables 3.3 and 3.4); place a **(X)** next to the species datum levels not used, or not recognized in the other site. Do the species datum levels occur in the same order at both localities? Explain.

____

____

____

____

____

____

____

____

NAME ______________________________

2 Provide a hypothesis or two to explain the **similarities and differences of the nannofossil assemblages at the two sites** based on the information provided in the tables. How might you test your hypotheses?

Comparison of Sediment Accumulation Rate Histories

3 On the graph paper provided **plot age vs. depth**, putting age (in millions of years, Ma) across the top (*x*-axis) and depth in the hole (meters below sea floor, mbsf) down the left side (*y*-axis). Plot the **microfossil datum levels** that are in common to both sites. **Note:** if you've already plotted the Site 1208 datum levels in Part 3.4, you could add the Site 999 datum levels to the same plot. Use solid circles (•) for the **Site 1208 datum levels** and open circles (◦) for the **Site 999 datum levels**.

(a) Calculate sedimentation rates for all best-fit line segments for both sites. Show your work.

NAME

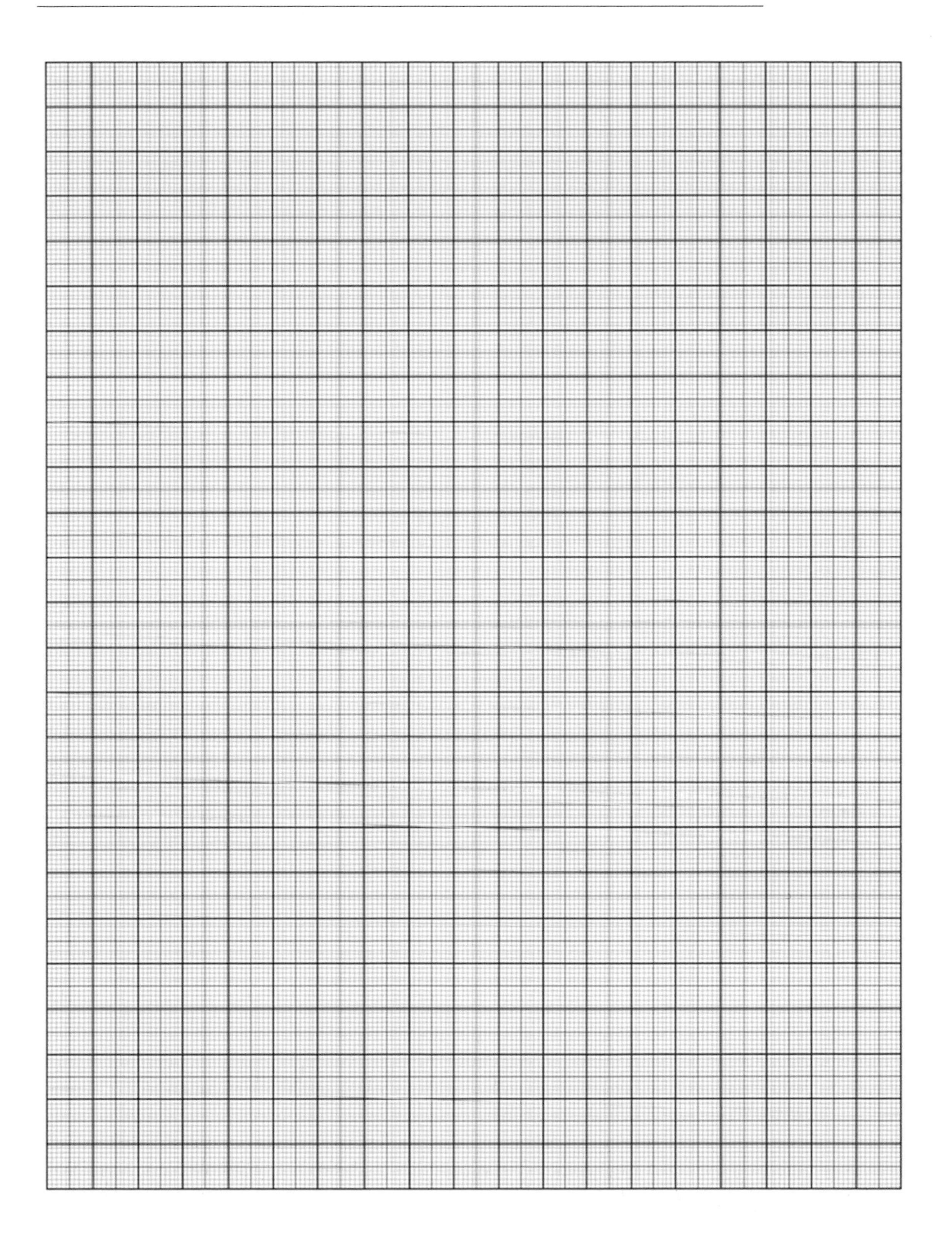

NAME

(b) Characterize the patterns of sedimentation at both sites. How are the sedimentation histories similar, and how are they different?

(c) Provide possible explanations for the similarities and/or differences.

References

Bown, P.R., 2005, Cenozoic calcareous nannofossil biostratigraphy, ODP Leg 198 Site 1208 (Shatsky Rise, Northwest Pacific Ocean). Proceedings of the Ocean Drilling Program, Scientific Results, Bralower, T.J., et al. (eds), vol. 198, College Station, TX, Ocean Drilling Program, pp. 1–44. http://www-odp.tamu.edu/publications/198_SR/104/104.htm

Gradstein, F., 1987, Report of the Second Conference on Scientific Ocean Drilling (Cosod II), European Science Foundation, Strasbourg, France, p. 109.

Kameo, K. and Bralower, T.J., 2000, Neogene calcareous nannofossil biostratigraphy of Sites 998, 999, and 1000, Caribbean Sea. In Proceedings of the Ocean Drilling Program, Scientific Results, vol. 165, Leckie, R.M., et al. (eds), College Station, TX, Ocean Drilling Program, pp. 3–17. http://www-odp.tamu.edu/publications/165_SR/chap_01/chap_01.htm

Katz, M.E., et al., 2005, Biological overprint of the geological carbon cycle. Marine Geology, 217, 323–38.

Shipboard Scientific Party, 1992, Site 846. In Initial Reports of the Ocean Drilling Program, vol. 138, Mayer, L., et al., College Station, TX, Ocean Drilling Program, pp. 256–333; doi:10.2973/odp.proc.ir.138.111.1992. http://www-odp.tamu.edu/publications/138_IR/VOLUME/CHAPTERS/ir138_11.pdf

Shipboard Scientific Party, 2002, Explanatory notes. In Proceedings ODP, Initial Reports of the Ocean Drilling Program, vol. 198, Bralower, T.J., et al., College Station, TX, Ocean Drilling Program, pp. 1–63. doi:10.2973/odp.proc.ir.198.102.2002. http://www-odp.tamu.edu/publications/198_IR/198TOC.HTM

Young, J.R., 1998, Neogene. In Calcareous Nannofossil Biostratigraphy, Bown, P.R. (ed.), Kluwer Academic Publishers, Dordrecht, The Netherlands, pp. 225–65.

Chapter 4 Paleomagnetism and Magnetostratigraphy

FIGURE 4.1. The Earth's magnetic field protects our planet from much of the deadly radiation and charged particles (**solar wind**) streaming away from the Sun, as depicted in this NASA diagram. This **magnetosphere** is compressed on the day (Sun) side of the Earth and smeared out on the night side. Times of strong solar wind interact with the magnetosphere near the north and south magnetic poles and are observable on Earth as **aurora** ("Northern" and "Southern Lights"). Figure courtesy of NASA.

SUMMARY

Reversals of our planet's magnetic field throughout geologic time have provided Earth scientists with a distinct **'barcode' record of normal and reversed polarity** that is preserved in deep-sea sediments and in oceanic crust. In **Part 4.1**, you will consider the nature of **Earth's magnetic field**

Reconstructing Earth's Climate History: Inquiry-Based Exercises for Lab and Class,
First Edition. Kristen St John, R Mark Leckie, Kate Pound, Megan Jones and Lawrence Krissek.

NAME

today and explore the nature of a **paleomagnetic record** preserved in deep-sea sediments. **Part 4.2** explores the nature of the paleomagnetic record preserved in ocean crust as positive and negative **magnetic anomalies** and relates the anomaly pattern ('barcode') observed in the ocean basins to the paleomagnetic record preserved in deep-sea sediments. In **Part 4.3**, you will see how paleomagnetism was used to test the **Seafloor Spreading Hypothesis**. In **Part 4.4**, you will learn how paleomagnetism has been used to create the **geomagnetic polarity timescale** of the past 160 million years and how paleomagnetic, radiometric, and biostratigraphic data are integrated to provide methods of age determination.

Paleomagnetism and Magnetostratigraphy

Part 4.1. Earth's Magnetic Field Today and the Paleomagnetic Record of Deep-Sea Sediments

1 Do the following to describe the Earth's magnetic field as you currently understand it:

(a) The circle below represents the Earth. **Draw a sketch** of the Earth and its magnetic field by putting in the **magnetic lines of force**; **label** the equator and the North and South poles.

NAME

(b) Do the magnetic lines of force intersect the Earth's surface everywhere at the same angle?

2 **Inclination** (or **magnetic dip**) is the angle between the orientation of a magnetic mineral grain in a rock or in sediments and the (horizontal) surface of the Earth at a particular location. Based on the similarity in behavior of the Earth's magnetic field and the magnetic lines of force generated by a bar magnet, predict the following:

(a) Where on Earth's surface would you expect magnetic minerals found in sediments and igneous rocks to have **very little or no inclination**? Why?

EARTH'S MAGNETIC FIELD

Figure 4.1 depicts the Earth's magnetosphere and Figure 4.2 depicts the behavior of iron filings in the presence of a magnetic field (you may remember doing this little experiment in elementary or middle school). Note the general similarity in the shape of the magnetic field in these two diagrams. The **Earth's magnetic field** approximates a **dipole bar magnet** and is generated within the **liquid outer core** owing to our planet's rotation about an axis defined by the North and South geographic (true) poles. The magnetic poles are close to Earth's rotational axis (i.e., the magnetic field approximates the spin axis of the Earth), but today the magnetic north pole and magnetic south pole are offset from true North and South by approximately 11°. The **Earth's magnetic field is dynamic and complex** and the magnetic poles are not stationary. In 2005, the north magnetic pole was at 82.7°N, 114.4°W, moving northwest at approximately 40 km/yr, while the south magnetic pole (2001) was at 64.7°S, 138.0°E (NOAA National Geophysical Center, http://www.ngdc.noaa.gov/geomag/, see also "Frequently Asked Questions on the NGDC site") . A compass points to the **magnetic north pole** in response to the magnetic lines of force generated by our magnetic field, similar to the bar magnet and iron filings (Figure 4.2). Likewise, **magnetic minerals** in sediments and rocks will behave like tiny compass needles that align with the Earth's magnetic field at the time they are deposited (in the case of sediments), or crystallized as molten magma cools (in the case of igneous rocks). In other words, **magnetic minerals in sediments and rocks become locked in and therefore capture and preserve the nature of the magnetic field at that particular location for that particular time in Earth history**. The study of the geologic record of Earth's magnetic field and its changes is called **Paleomagnetism**.

FIGURE 4.2. Bar magnet with iron filings showing magnetic lines of force. From http://www.askamathematician.com/?p = 4129

(b) Where on Earth's surface would you expect the magnetic minerals to have the **steepest inclination**? Why?

3 If the inclination is **into** the Earth at or near the magnetic north pole, what is the nature of the inclination at or near the south magnetic pole? Explain.

4 Drawing from your answers in Questions 2 and 3 above, and your general knowledge of plate tectonics, predict how the inclination (magnetic dip) preserved in older rocks might be used to determine **past lithospheric plate positions**?

NAME

NATURAL REMANENT MAGNETIZATION IN SEDIMENTS AND IGNEOUS ROCKS

The magnetic signal that geoscientists are interested in measuring is called **natural remanent magnetization (NRM)**. This signal is carried by magnetic minerals such as **magnetite**. The NRM is preserved in ocean crust (basalt) as it cools and crystallizes past the **Curie temperature** (approximately 580°C for the mineral magnetite), or in sediments as detrital magnetic mineral grains (eroded from another source) accumulate on the seafloor. In both situations, the magnetic minerals become aligned with the Earth's magnetic field at the time of crystallization and deposition, respectively. Not all rocks are good carriers of a magnetic signal. The **basalt** that makes up **oceanic crust** is a magnetite-bearing rock, while granite, a typical igneous rock of continental crust, has a generally poor content of magnetic minerals. Likewise, **terrigenous sediments** (e.g., sand, mud, and clay) derived from the erosion of continental rocks have a much higher concentration of detrital magnetic mineral grains than does a pure biogenic sediment, such as calcareous or siliceous ooze.

Figures 4.3(a) and (b) depict the magnetic character of **two deep-sea sediment cores**. **Site 1149** (Figure 4.3a) is located in the **northwest Pacific (approximately 31°N latitude)** near the Marianas Trench. **Site 1172** (Figure 4.3b) is located in the **southwest Pacific (approximately 44°S latitude)** near Tasmania (Figure 4.4). Both sedimentary sequences represent continuous sediment deposition over the past 6.5 million years or so. The *x*-axis shows **inclination** (magnetic dip); **positive inclination values** are into the Earth, **negative values** are out of the Earth. The *y*-axis is **depth in the core** in meters below seafloor (mbsf).

5 **Make a list of observations** about the nature of the paleomagnetic record in these two deep-sea sediment sequences. How are they similar and how are they different?

NAME

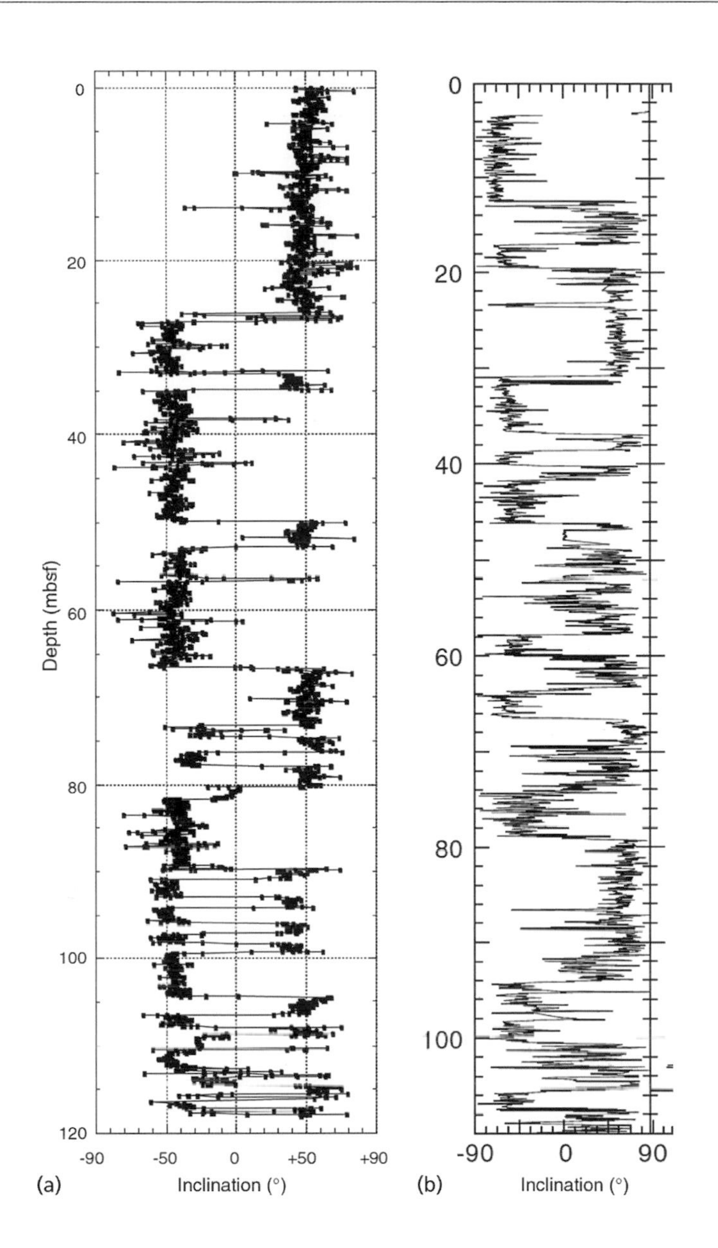

FIGURE 4.3. Two paleomagnetic records from deep-sea sediments cored in the Pacific Ocean. **Magnetic inclination** (degrees; °) is plotted against **depth in hole** (meters below seafloor; mbsf). (a) **ODP Hole 1149A** is located southeast of Japan near the north end of the Marianas Trench (31°20.52′N, 143°21.07′E). From Shipboard Scientific Party, 2000. (b) **ODP Hole 1172B** is located east of Tasmania (43°57.57′S, 149°55.70′E). From Shipboard Scientific Party, 2001a.

Similarities	Differences

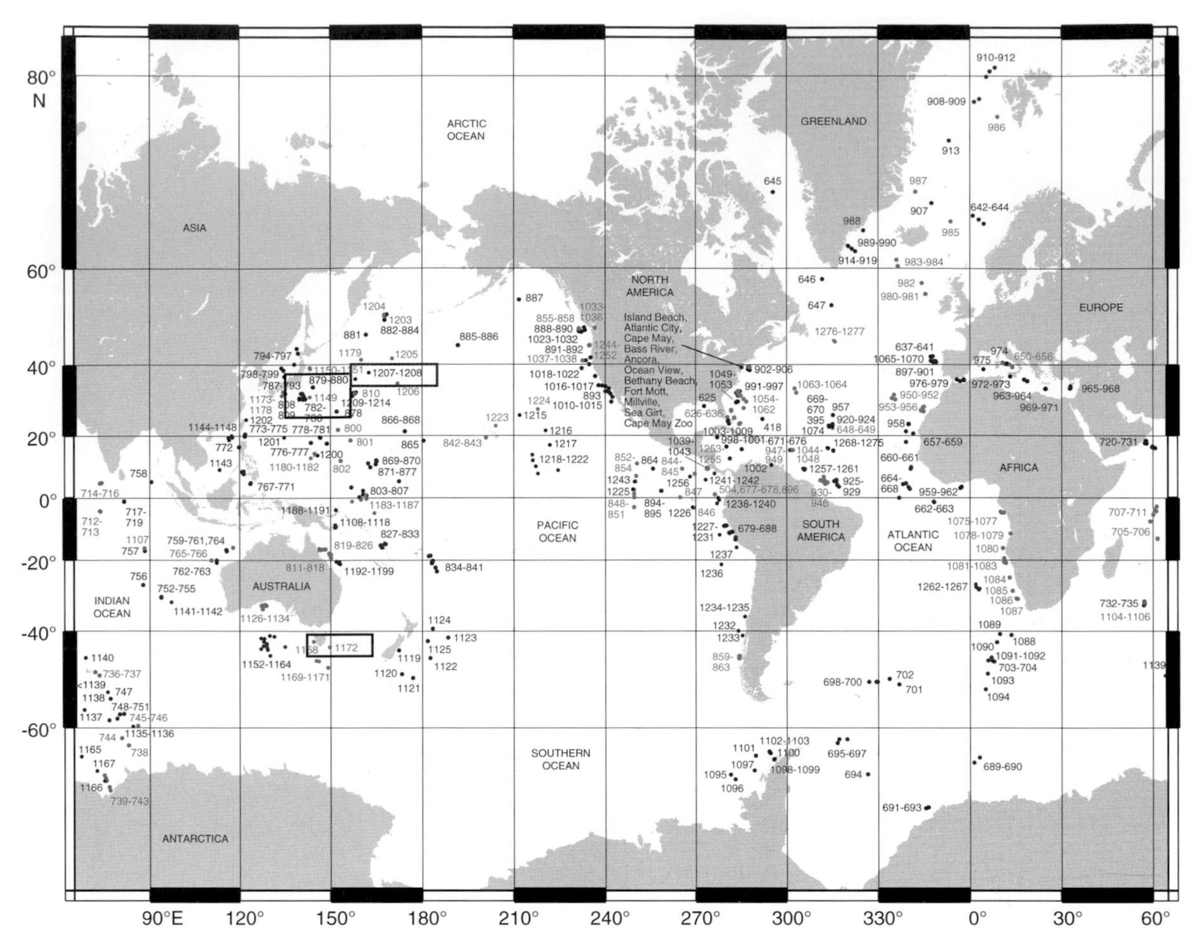

FIGURE 4.4. World map showing all Ocean Drilling Program sites (1985 to 2003; Legs 100-203). Black boxes show locations of Site 1149 (North Pacific) and Site 1172 (South Pacific). Blue box shows location of Site 1208 (Figure 4.11). Courtesy of IPDP.

6 How do your observations of the **modern sediments** (those closest to the seafloor and at the shallowest depths) at these two sites compare with the predictions that you made for Questions 2 and 3?

7 Assuming that magnetic reversals are geologically instantaneous, the sediments at approximately 25–33 mbsf at Site 1149 were deposited at the same time as **what interval** at Site 1172? (By doing this you are essentially **correlating** an interval at Site 1149 with an age equivalent interval at Site 1172, just as paleomagnetists do!)

NAME

8 (a) What is the **average inclination** of this interval at Site 1149?

(b) What is the **average inclination** of this interval at Site 1172?

(c) How does the average inclination of this interval at each site compare with the average inclination of the "modern" sediments (those at the top of these cores) at each site?

(d) Do these observations support or refute the notion that the Earth's magnetic north and south poles generally approximate the geographic (true) north and south poles over geologic time? Explain.

PALEOMAGNETIC REVERSALS (REVERSALS OF MAGNETIC POLARITY)

The Earth's magnetic field episodically reverses polarity. Prior to a **reversal of the magnetic field**, the field fluctuates and becomes weaker and there may be multiple poles for a short time. The process is **geologically rapid**, taking several thousand years for the field to completely reverse polarity and stabilize again (Figure 4.5). Today's field is referred to as "**normal polarity**". The last reversal of the magnetic field occurred approximately 780,000 years ago. Refer to http://www.ngdc.noaa.gov/geomag/ for additional information about Earth's magnetic field (see also "Frequently asked questions").

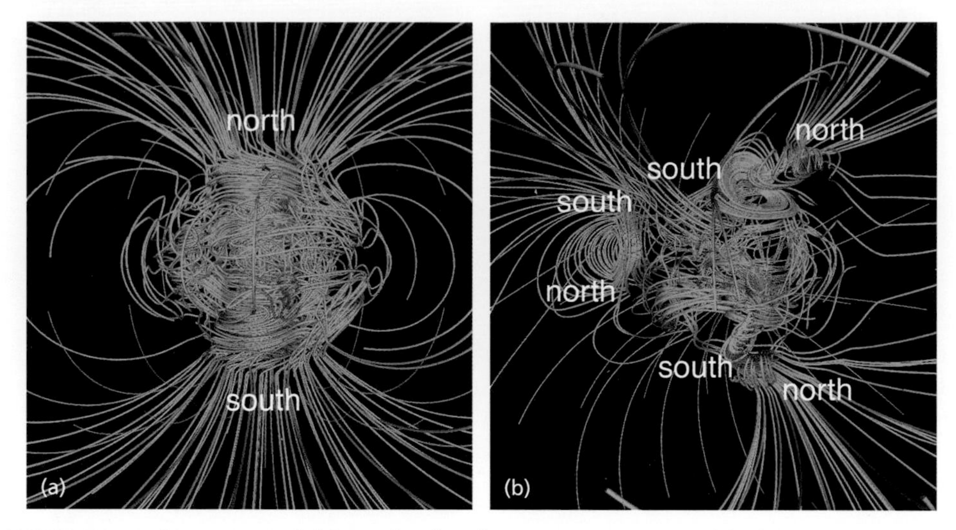

FIGURE 4.5. Earth's magnetic field simulated with a supercomputer model of the geodynamo by G.A. Glatzmaier and P.H. Roberts, University of California (es.ucsc.edu/~glatz/geodynamo). Yellow magnetic field lines are directed outward and blue magnetic field lines are directed inward. The **stable normal dipolar magnetic field** is shown in the picture on the left between reversals. The picture on the right during a reversal provides an idea of the **complicated, but rapid nature of a reversal of magnetic polarity**. Note: the field does not go away during a reversal.

PALEOMAGNETISM IN SEDIMENT CORES

Paleomagnetic records (i.e., the ancient or fossil magnetic signal preserved in sedimentary layers; Figure 4.3) are measured in deep-sea sediments cored by the Ocean Drilling Program. These data were collected using a **cryogenic magnetometer** (Figure 4.6). Sediments cored from the seafloor are passed through the magnetometer and the **inclination**, or magnetic dip, of the **magnetic minerals** is measured, as well as the magnetic **intensity**. In Figure 4.3, the data are plotted with the x-axis showing **magnetic inclination** (in degrees of inclination, or dip, from the horizontal) and the y-axis as **depth in the drill hole** (in meters seafloor, mbsf). See http://www-odp.tamu.edu/sciops/labs/pmag/ (or http://www.odplegacy.org/operations/labs/paleomagnetism/) for a detailed discussion of how shipboard paleomagnetic data are collected.

NAME

FIGURE 4.6. Cryogenic magnetometer aboard the *JOIDES Resolution* drillship. The 2G 760-R superconducting rock (cryogenic) magnetometer is used primarily for continuous measurements of magnetic properties on 1.5-m long core halves. The wide silver cylinder shields the magnetometer from the present day magnetic field. The photo shows the tray where the core half is placed and the opening into the magnetometer. Photo courtesy of IODP.

NAME

Paleomagnetism and Magnetostratigraphy
Part 4.2. Paleomagnetism in Ocean Crust

Figure 4.7 shows two transects of ship-towed **magnetometer data** across the **East Pacific Rise**, a volcanically active ridge in the eastern Pacific (Pitman and Heirtzler, 1966). Each transect displays two types of data: **magnetic intensity** (in gammas) and **bathymetry** (water depth, in kilometers) plotted on the *y*-axes, and **distance** (in km) on either side of the ocean ridge plotted on the

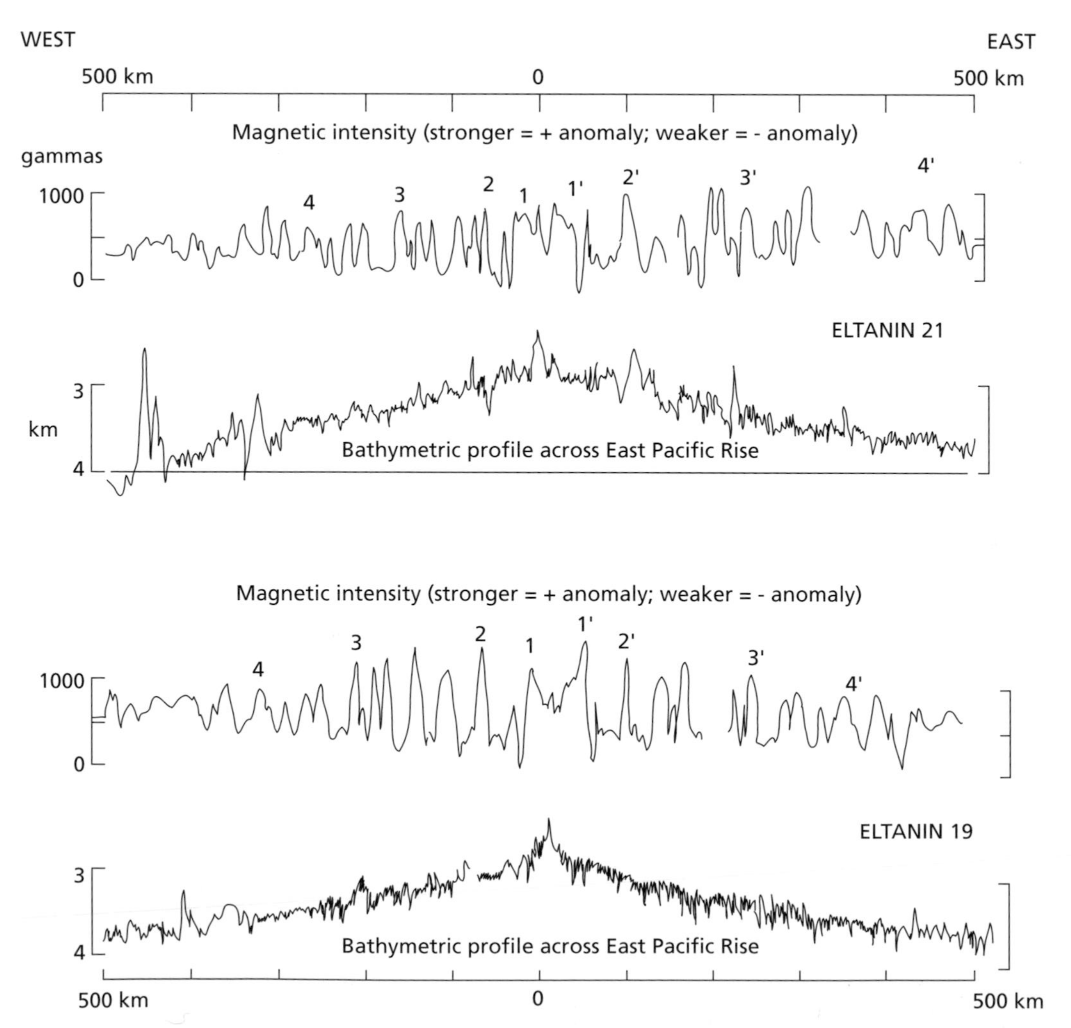

FIGURE 4.7. Two transects (numbers 21 and 19) of **magnetometer data** collected by the research vessel *Eltanin*. The East Pacific Rise is the crest (bathymetric high) in both transects. **Stronger magnetic intensity values are called positive magnetic anomalies**; several examples on either side of the ridge are number 1–4 and 1′–4′, respectively. From Pitman and Heirtzler, 1966.

NAME

---

x-axes. The **ridge crest** at 0 km is the axis of the ocean ridge. 1 and 1′, 2 and 2′, and so on correspond to distinctive measurements of the magnetic character of seafloor rocks (called "**positive magnetic anomalies**") on either side of the ridge.

1 (a) What common features do you see in the magnetic intensity data from these two transects across the East Pacific Rise (Figure 4.7)?

(b) Read the box below on seafloor magnetic anomalies. What do paleomagnetic records from such places as the East Pacific Rise (Figure 4.7) tell us about the nature of Earth's magnetic field preserved in oceanic crust?

SEAFLOOR MAGNETIC ANOMALIES

Figure 4.8 depicts a **paleomagnetic record** of ocean crust across a volcanically active **ocean ridge**. A ship towing a **magnetometer** records signals of stronger (**positive magnetic anomalies**) and weaker (**negative magnetic anomalies**) magnetic intensities. What the magnetometer measures is the magnetic character of the **oceanic crust**, consisting of the igneous rock called **basalt** (which is relatively enriched in magnetic minerals such as **magnetite**), as well as the much weaker overlying sedimentary cover and its detrital magnetic minerals. **Positive magnetic anomalies** correspond to ocean crust that has **normal polarity**, like today (today's field plus the ancient normal field result in a stronger signal); **negative magnetic anomalies** correspond to ocean crust with **reversed polarity** (today's field plus the ancient reversed field result in a weaker signal). The magnetic anomalies are mapped as **magnetic "stripes"**, which depict an alternating pattern of positive and negative magnetic anomalies preserved in the ocean crust.

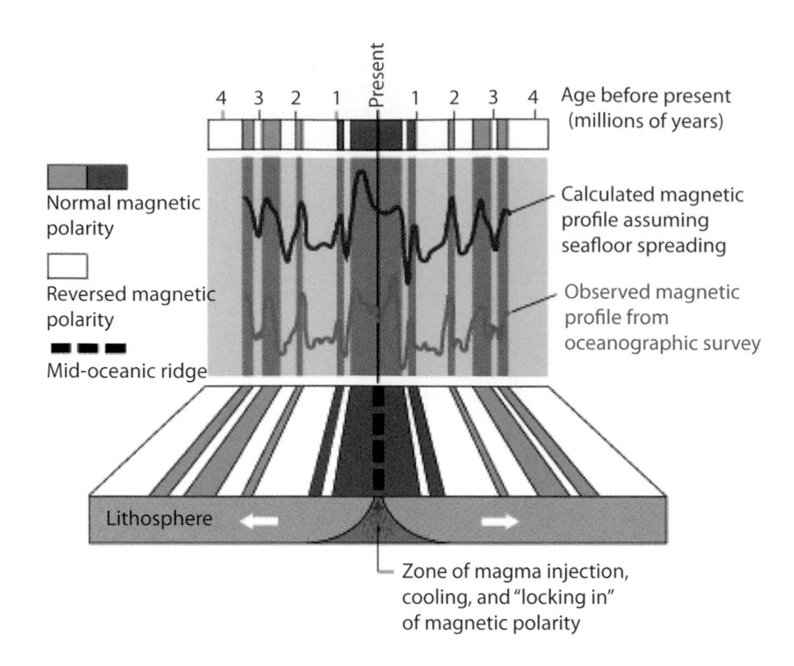

FIGURE 4.8. An observed magnetic profile (blue) for the ocean floor across the East Pacific Rise is matched quite well by a calculated profile (red) based on the Earth's magnetic reversals for the past 4 million years and an assumed constant rate of movement of ocean floor away from a hypothetical spreading center (bottom). The remarkable similarity of these two profiles provided one of the decisive arguments in support of the seafloor spreading hypothesis. From http://library.thinkquest.org/18282/lesson4.html. Reproduced from USGS.

Figure 4.9 shows an **interpretation of the pattern** of positive and negative magnetic anomalies measured by the ship-towed magnetometer on either side of East Pacific Rise and a corresponding **geomagnetic polarity time scale** (Figure 4.10). **Polarity epochs** represent times of dominantly normal or dominantly reversed polarity. This **"barcode" pattern of polarity changes** led to the development of the geomagnetic polarity time scale.

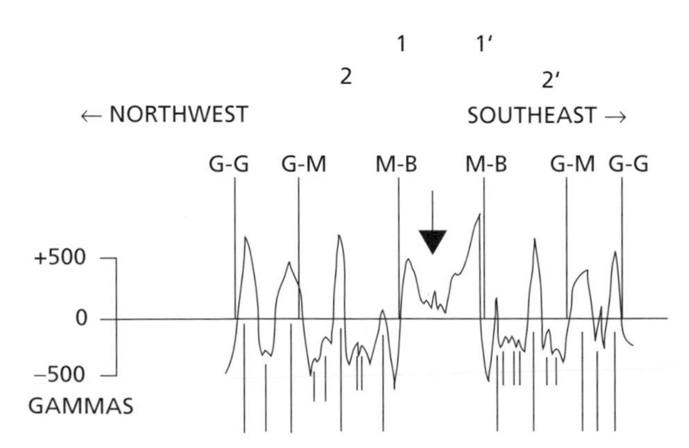

FIGURE 4.9. Magnetic profiles depicting **positive and negative magnetic anomalies** across the East Pacific Rise at latitude 52°S, longitude 118°W (notice the similarity to magnetic profiles in Figure 4.7). The heavy arrow shows the location of the crest of the East Pacific Rise (i.e., spreading center = zero-age crust). 1 and 1′ and 2 and 2′ correspond to distinctive positive anomalies. M-B, G-M, G-G = boundaries between the Bruhnes, Matuyama, Gauss, and Gilbert polarity epochs. From Cox, 1969.

NAME

THE GEOMAGNETIC POLARITY TIME SCALE

Figure 4.10 shows the first geomagnetic polarity time scale (GPTS), which was derived from continental lava flows with paleomagnetic records from numerous locations (Cox, 1969). These lava flows and the polarity reversals that they contained were **radiometrically dated**. This compilation demonstrated the potential of using magnetic reversals as a means of building a **geologic time scale**. More details on the GPTS are explored in Part 4.4.

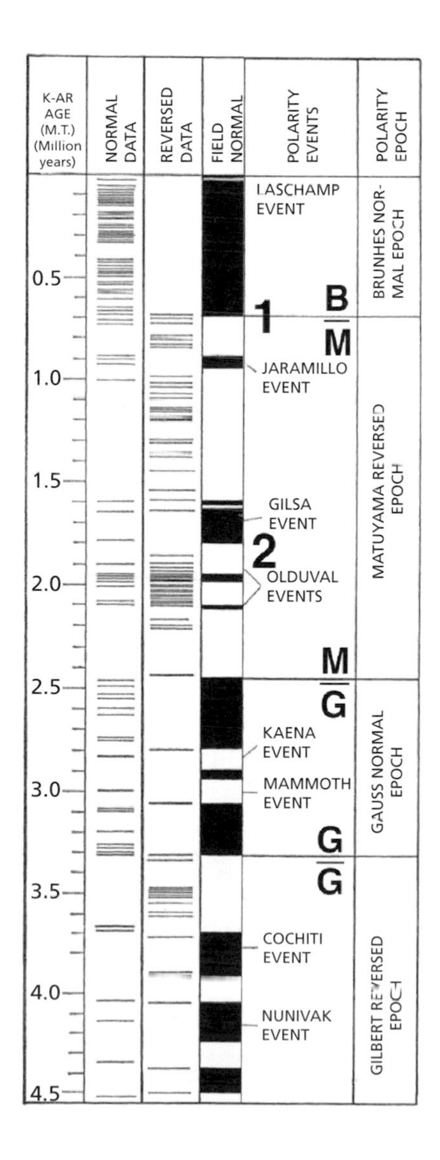

FIGURE 4.10. Time scale for geomagnetic reversals. **Normal polarity intervals** are shown by the black portions and correspond to **positive magnetic anomalies** on the seafloor; **reversed polarity intervals** are shown by the white portions and correspond to **negative polarity intervals**. Note that the **Bruhnes Epoch** is normal polarity, the **Matuyama Epoch** is dominantly reversed polarity, the **Gauss Epoch** is dominantly normal polarity and the **Gilbert Epoch** is dominantly reversed polarity. From Cox, 1969.

NAME ______________________________

2 Describe the nature of the magnetic record preserved in the ocean crust of the East Pacific Rise shown in Figure 4.9 by answering the questions below:

(a) What pattern do you see in magnetic anomaly data preserved at the crest of the ocean ridge?

____

____

____

(b) How does this pattern of magnetic anomaly data relate to distance from the crest of the ocean ridge?

____

____

____

____

(c) Provide an explanation for the observed pattern of magnetic anomalies on either side of the ocean ridge.

____

____

____

____

____

____

3 Figure 4.11 depicts the paleomagnetic record from ODP Site 1208 in the northwest Pacific Ocean (site location shown in Figure 4.4).

(a) How do the paleomagnetic data preserved in the **vertical sequence** of deep-sea sediments cored at ODP Site 1208 (Figure 4.11) compare with the **horizontal pattern** of seafloor magnetic anomaly data from the eastern Pacific shown in Figures 4.7 & 4.9? List the similarities and differences.

Similarities	Differences

NAME

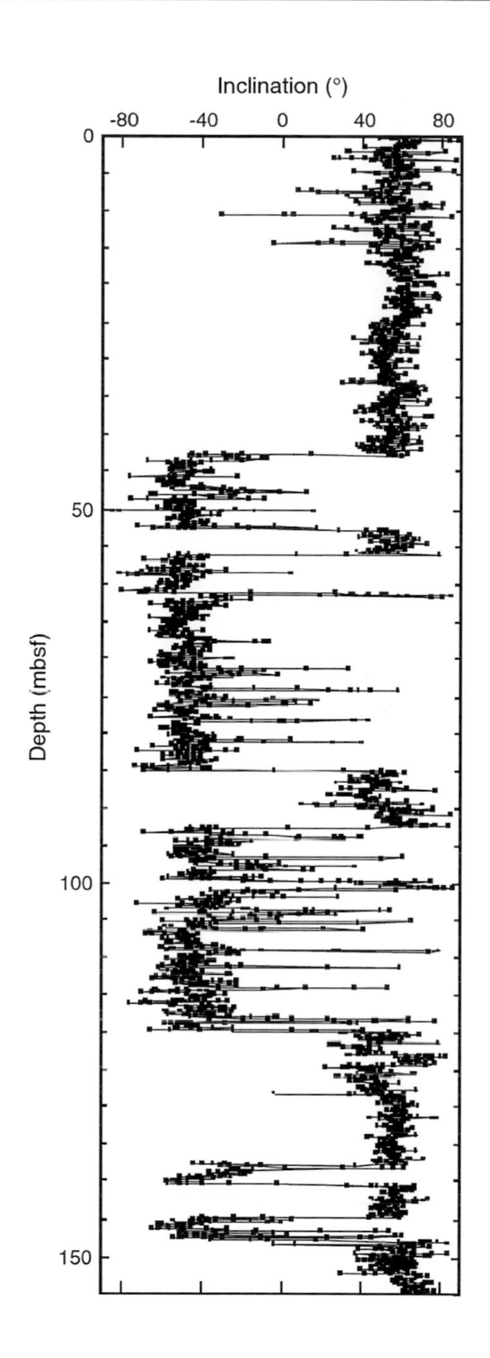

FIGURE 4.11. Paleomagnetic record preserved in the upper 155 m of Hole 1208A, located east of Japan on Shatsky Rise (36°7.6301′N, 158°12.0952′E). From Shipboard Scientific Party, 2002.

NAME ______________________________

(b) The **Bruhnes-Matuyama** and **Matuyama-Gauss** boundaries are defined in Figures 4.9 & 4.10. At what **depths** would you place these boundaries in Site 1208?

____

____

(c) What assumptions did you have to make to assign these two paleomagnetic boundaries to specific depths at Site 1208?

____

____

____

____

(d) Using the geomagnetic polarity time scale shown in Figure 4.10, what is the **age** of the sedimentary sequence at **150 mbsf**?

____

4 Reflecting on the paleomagnetic data you have worked with in this exercise, explain how paleomagnetic records preserved in sedimentary sequences and/or ocean crust can be used to construct a geologic time scale (e.g., Figure 4.10). In your response be sure to include a description of the assumptions you would need to make, as well as reference to other data that would be needed to support the construction of this time scale.

____

____

____

____

____

____

____

____

NAME

Paleomagnetism and Magnetostratigraphy

Part 4.3. Using Paleomagnetism to Test the Seafloor Spreading Hypothesis

In 1969, Deep Sea Drilling Project (DSDP) Leg 3 set out to test the **Seafloor Spreading Hypothesis**. The drillship *Glomar Challenger* cored nine sites in the South Atlantic, seven to the west of the Mid-Atlantic Ridge and two to the east of the ridge (Figure 4.12).

Table 4.1 summarizes the **age of ocean crust** ("basement") at each of the drill sites based on the paleomagnetic age, plus the **age of the sediment** over-

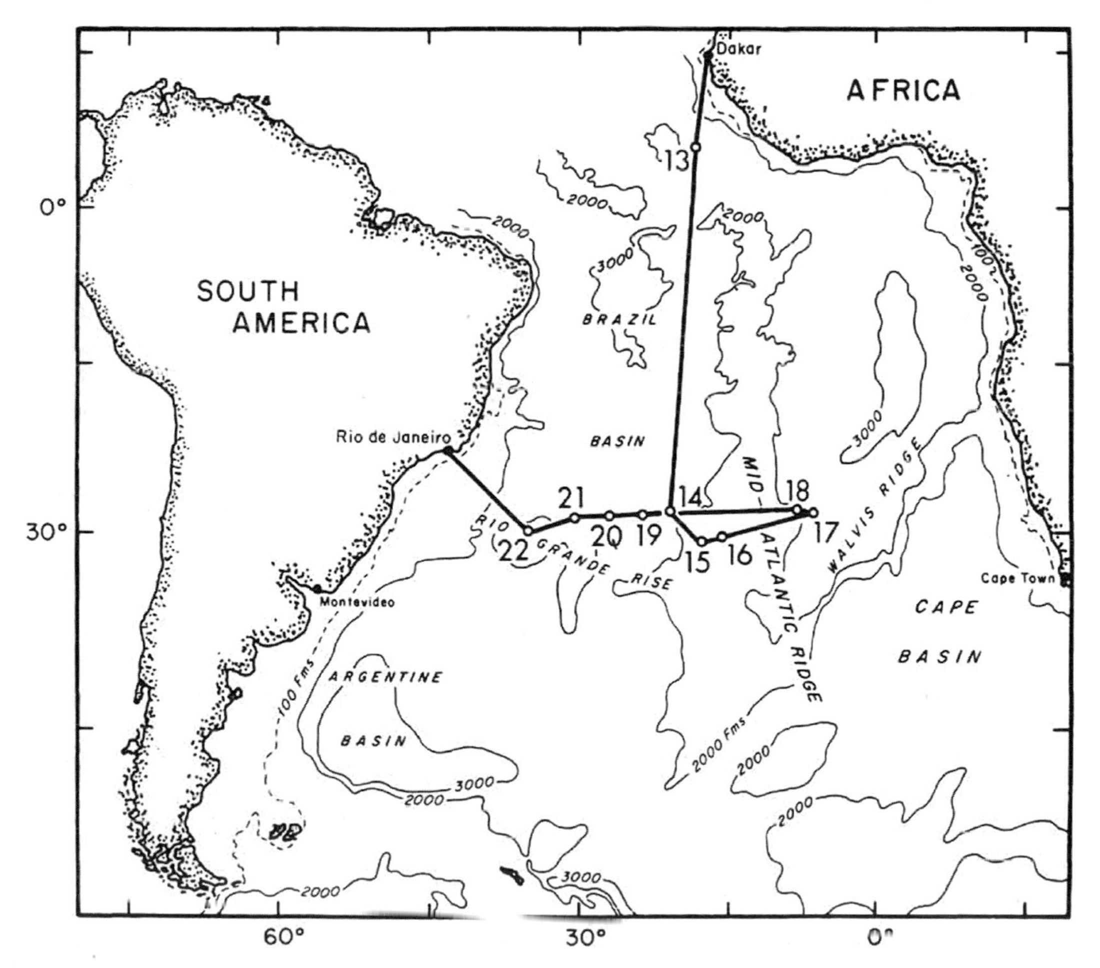

FIGURE 4.12. Map of the South Atlantic showing the cruise path and sites drilled during DSDP Leg 3. Notice that Sites 14-16 and 19-22 were drilled to the west of the Mid-Atlantic Ridge and Sites 17 and 18 were drilled to the east of the ridge. From Maxwell et al., 1970.

NAME

TABLE 4.1. Estimates for the age of ocean crust (basement) at the DSDP Leg 3 drill sites and distance of each site from the Mid-Atlantic Ridge axis. From Maxwell et al., 1970.

Site No.	Magnetic age of basement (million years)	Paleontological age sediment above basement (million years)	Distance from ridge axis (km)	
			Linear	Rotation at 62°N, 36°W
16	9	11 ± 1	191 ± 5	211 ± 20
15	21	24 ± 1	380 ± 10	422 ± 20
18		26 ± 1	506 ± 20	506 ± 20
17	34–38	33 ± 2	643 ± 20	718 ± 20
14	38–39	40 ± 1.5	727 ± 10	745 ± 10
19	53	49 ± 1	990 ± 10	1010 ± 10
20	70–72	67 ± 1	1270 ± 20	1303 ± 10
21		>76	1617 ± 20	1686 ± 10

lying ocean crust at the bottom of each drill site based on fossil evidence (from Maxwell et al., 1970). The **distance from the ridge axis** is given as a straight-line (linear) value, as well as a distance along a curve, which is based on an axis of rotation on a spherical surface.

1 What two types of data are used to determine **age** of each location on the seafloor? See Table 4.1.

____

____

2 On the graph paper provided, plot **paleomagnetic age of basement** (oceanic crust) **vs. distance** from the ridge axis using solid circles. On the same graph, plot the **paleontological age of sediment** overlying basement **vs. distance** from the ridge axis using open circles. Draw a visual **best-fit line** through the data points.

3 What is the relationship between the **age of basement** and **distance** from ridge axis?

____

____

____

4 What is the relationship between the **age of sediment** overlying basement and **distance** from ridge axis?

NAME

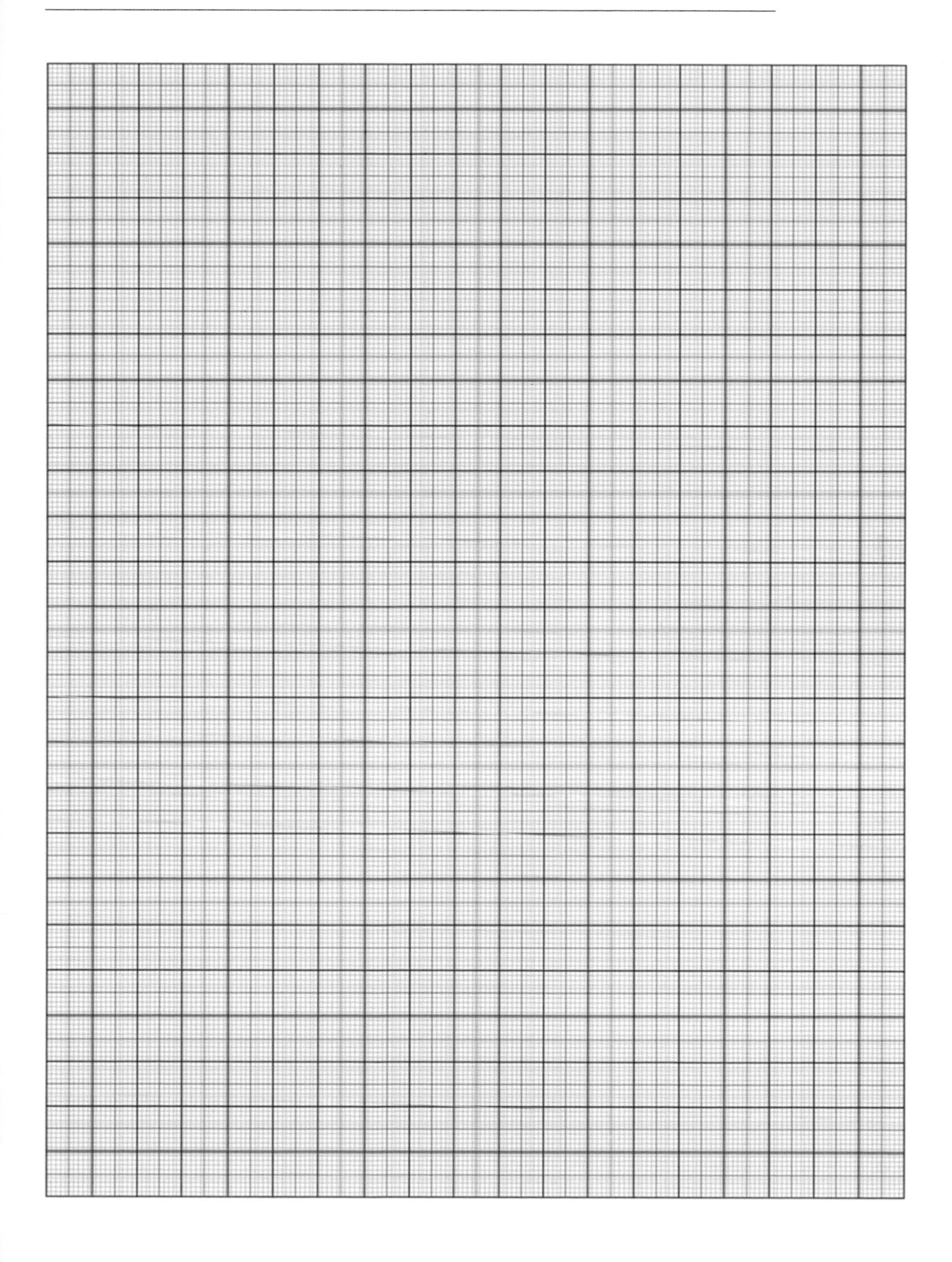

NAME ______________________________

____

____

____

The basalt that makes up oceanic crust basement **crystallized** at the ridge axis as volcanic lava cooled. This is also the time when the ocean crust acquired its magnetic character. Over time, the ocean crust at each of the drill sites has moved away from the ridge axis with slow conveyor-like motion to its present geographic location on the adjacent seafloor.

5 Hypothesize why the **ages of sediment** immediately overlying the basement (i.e., the basalt) also show increasing age from the ridge axis:

____

____

____

____

6 The **average rate of seafloor spreading** in the South Atlantic can be calculated from the data provided in Table 4.1 and in the graphs you made. In preparation for this calculation, do the following:

- In the space below sketch a **profile** across the Mid-Atlantic Ridge in the South Atlantic depicting the relationships of distance and age that you plotted.
- Label the spreading axis with its age.
- Provide units on your profile (distance in km, age in myr)
- Based on this sketch, write a formula to calculate the average spreading rate.

7 Use your graph with best-fit line through the data points (Question 2 above) to calculate seafloor spreading rates:

(a) What was the average rate of seafloor spreading on the western side of the Mid-Atlantic Ridge in **km/myr** (kilometers per million years)? Show your work.

____

NAME

(b) What was the average rate of seafloor spreading in **cm/yr** (centimeters per year)? Show your work.

8 Calculate how fast the South Atlantic Ocean is widening (i.e., opening). Show your work.

THE SEAFLOOR SPREADING HYPOTHESIS

In 1960, Professor **Harry Hess** of Princeton University advanced the hypothesis that **new ocean crust is produced at the mid-ocean ridges** such as the Mid-Atlantic Ridge and the East Pacific Rise. The youngest ocean crust should lie in the axis of each volcanically active ocean ridge and the oldest ocean crust should lie farthest from the ridge axis. According to his hypothesis, the seafloor moves laterally away from these **spreading centers** and explains why North and South America have moved away from Europe and Africa over time. This became known as the **seafloor spreading hypothesis.**

By the early 1960s, abundant paleomagnetic data from ship-towed magnetometers had demonstrated the existence of alternating positive (normal polarity) and negative (reversed polarity) **magnetic anomalies** on the seafloor (e.g., Vine and Matthews, 1963). When mapped, they reveal a pattern of **magnetic "stripes"** that form a symmetrical pattern on either side of a spreading center (i.e., the pattern on one side is a mirror image of the pattern on the other side). For example, Figure 4.13 is a seafloor map showing magnetic anomalies on the Mid-Atlantic Ridge south of Iceland.

NAME

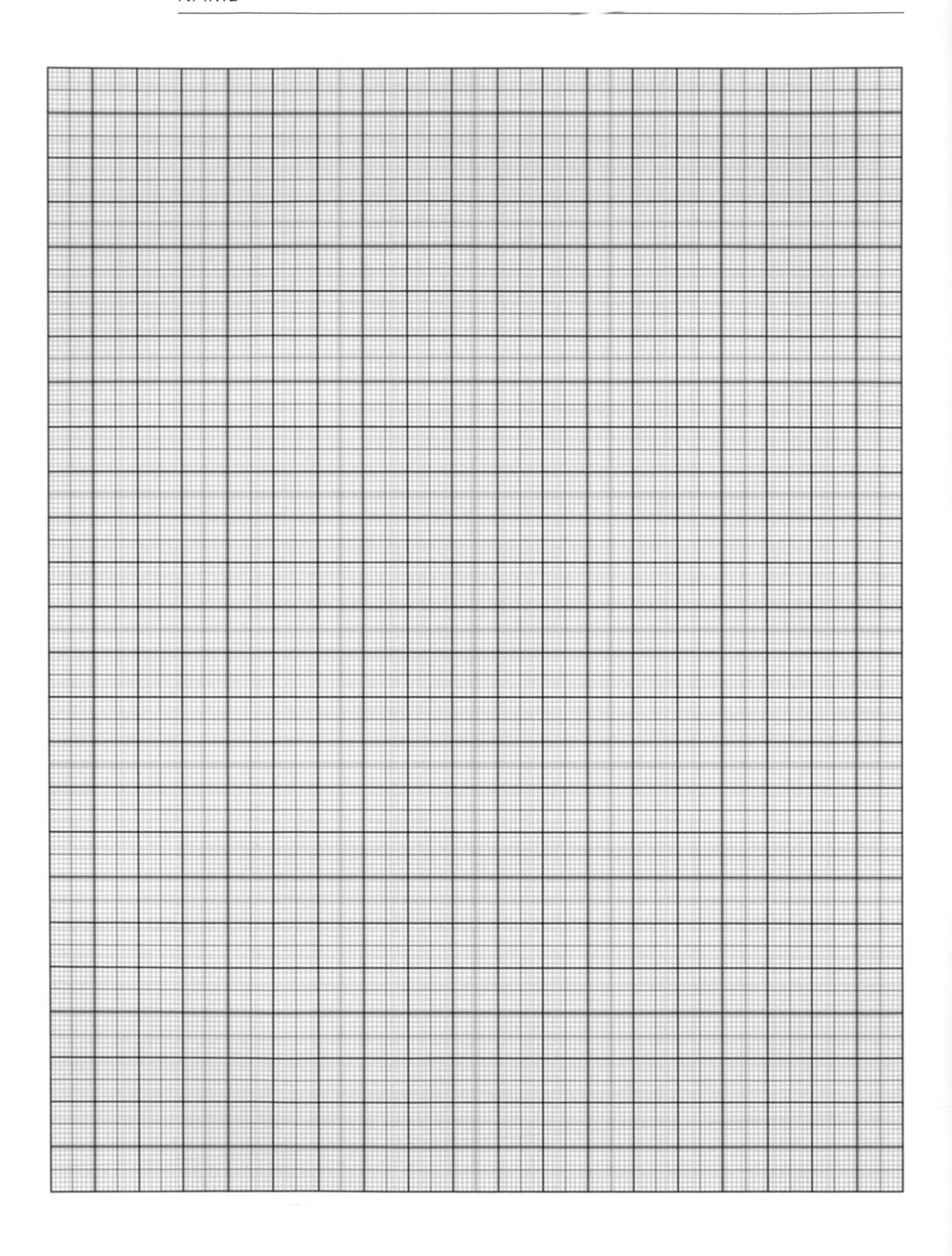

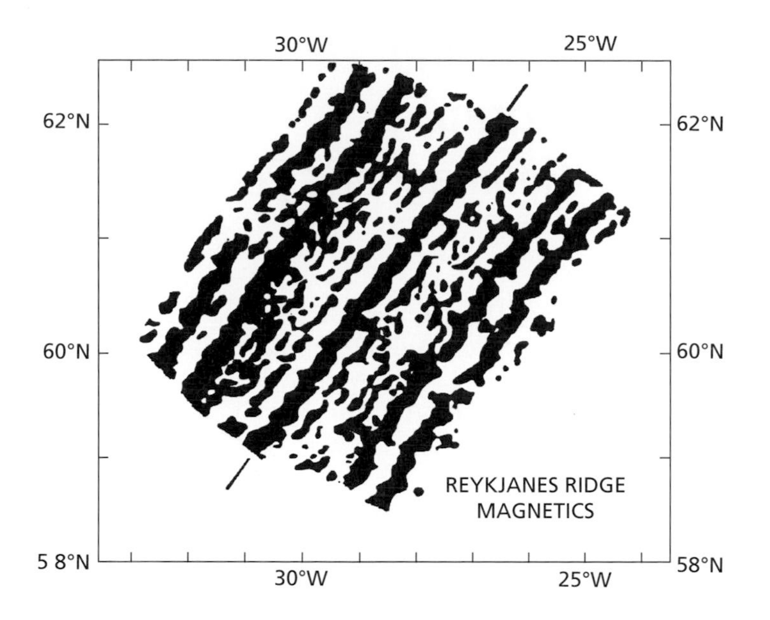

FIGURE 4.13. Map of magnetic "stripes" on the seafloor south of Iceland. The straight lines mark the ridge axis and the central positive magnetic anomaly. From Vine, 1966.

9 How could these magnetic "stripes" be used to test the hypothesis of seafloor spreading? Make a list of observations based on the patterns of magnetic stripes that would be useful for evaluating this hypothesis.

10 What additional data would you like to collect in order to test the seafloor spreading hypothesis?

NAME

Paleomagnetism and Magnetostratigraphy

Part 4.4. The Geomagnetic Polarity Time Scale

The **Geomagnetic Polarity Time Scale (GPTS)** is a composite geomagnetic polarity sequence that has been calibrated with radiometric age dates. The Late Cretaceous and Cenozoic (83 to 0 Ma) GPTS is based on an analysis of marine magnetic profiles from the Atlantic, Pacific, and Indian ocean basins

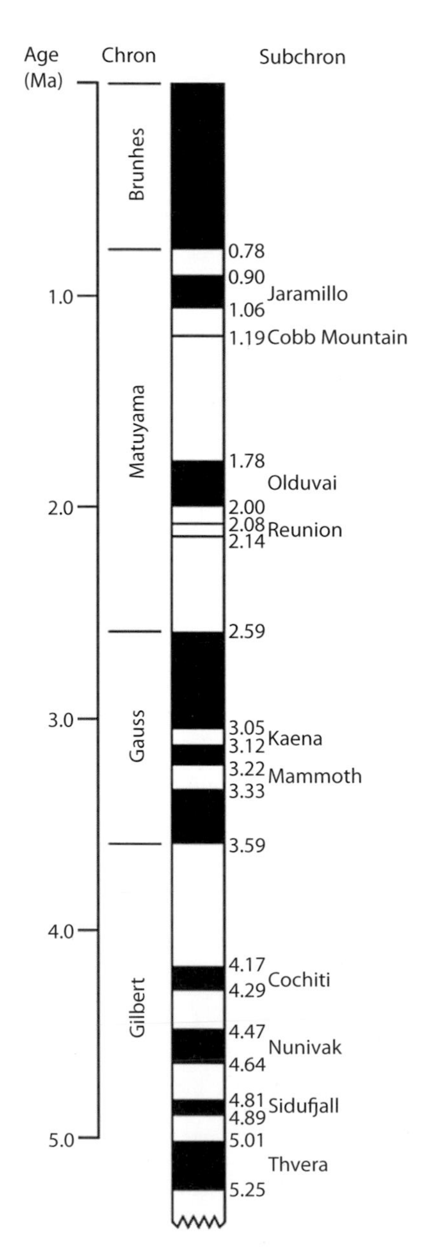

FIGURE 4.14. GPTS for the past 5 myr. Ages (in Ma) of the chron and subchron boundaries are shown as well as the names of the chrons and subchron events. From Mankinen and Wentworth, 2003.

NAME ______________________________

(Cande and Kent, 1992, 1995). The marine magnetic anomaly sequence recorded in the South Atlantic Ocean serves as the principal polarity sequence for this interval (Heirtzler et al., 1968; Cande and Kent, 1992). Radiometric dating of volcanic ash fall deposits (bentonites) in land-based and marine sections with reliable biostratigraphic and magnetostratigraphic control provides the numerical calibration for most of the anomaly pattern (paleomagnetic 'barcode'). An orbitally-tuned age provides the calibration age of the Gauss/Matuyama boundary (Figure 4.14). Nine age calibration points, plus the zero-age ridge axis, provide the age control on the Late Cretaceous-Cenozoic GPTS (Cande and Kent, 1992).

Figure 4.15 shows a **portion** of the **geomagnetic polarity time scale (GPTS)**. The GPTS is divided into **magnetic epochs**, which are themselves divided into **chrons.** Distinctive normal **polarity events** occur within the chrons. The suffix "**n**" after the anomaly number refers to normal polarity and the "**r**" refers to reversed polarity.

In the investigation that follows, you will use the GPTS to make an **interpretation** of the paleomagnetic data recovered from ODP Site 1208 in the northwest Pacific (Figure 4.4). In other words, **you will correlate the Site 1208 data to the GPTS** (Figure 4.16).

1 Interpret the paleomagnetic record of the upper 300 m of sediments cored at ODP Site 1208 (Figure 4.16) by correlating with the GPTS shown in Figure 4.15. Use the blank columns on the right of each panel of paleomagnetic inclination data to interpret **normal polarity** (color in) or **reversed polarity** (leave blank/white). As much as possible, **determine the magnetic chrons and the polarity events** and **label** these on your polarity interpretation. Assume that the top of the sequence at 0 mbsf is the top of the Brunhes Magnetic Epoch (= Chron C1n). Note that there are missing data at 225–230, 235–238, and 273–277 mbsf.

2 What **assumption(s)** must you make in order to correlate the Site 1208 paleomagnetic data with the GPTS?

____

____

____

____

3 What is the **approximate age** of the sedimentary sequence at **150** and **300 mbsf** (meters below seafloor)? What paleomagnetic **chrons** do these levels correlate with?

____

____

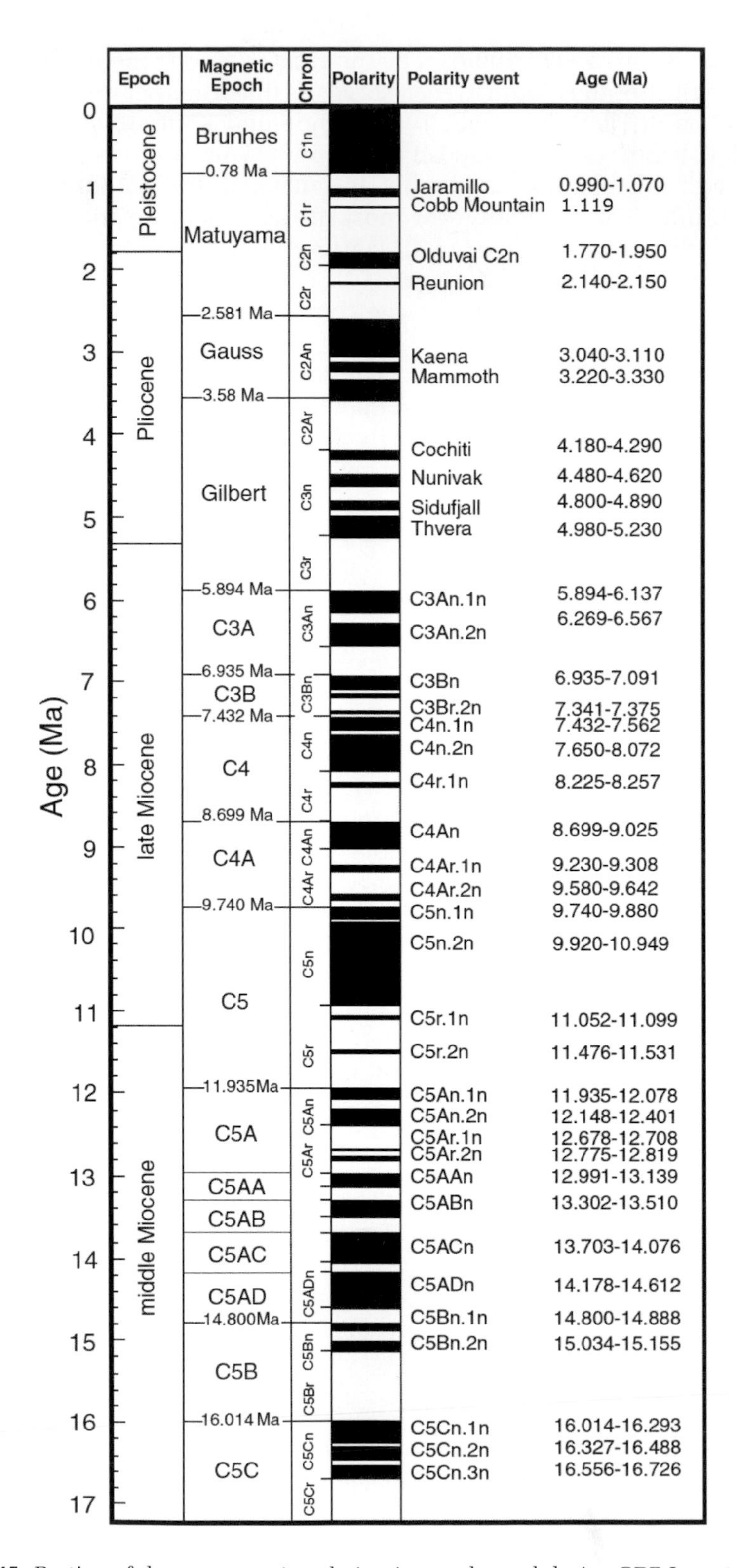

FIGURE 4.15. Portion of the geomagnetic polarity time scale used during ODP Leg 191. From Shipboard Scientific Party, 2001b.

NAME ____________________

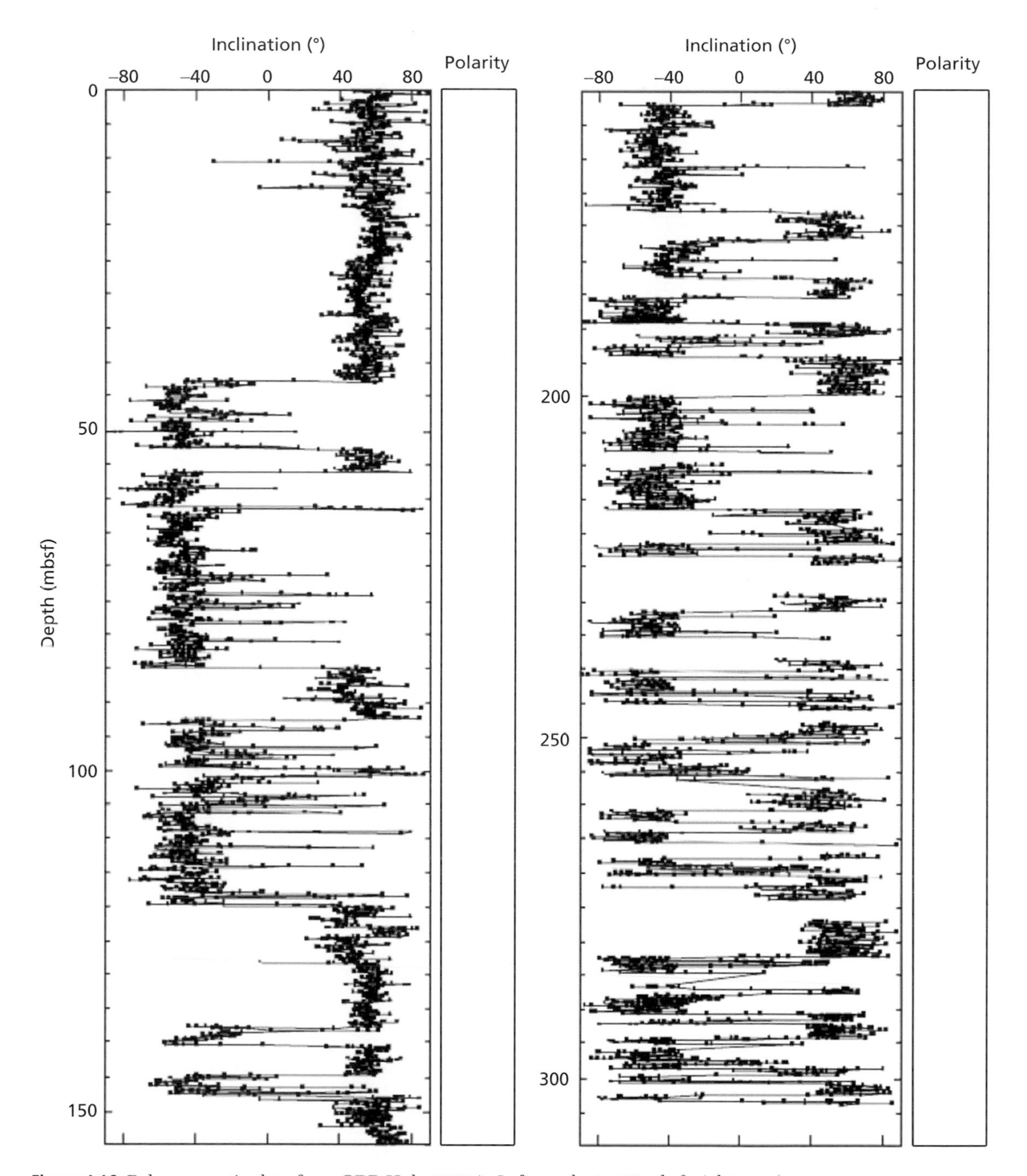

Figure 4.16. Paleomagnetic data from ODP Hole 1208A. Left graph: 0–155 mbsf; right graph: 155–305 mbsf. From Shipboard Scientific Party, 2002.

NAME ______________________________

4 As a summary of what you have learned in this chapter, list several ways that magnetic reversal data can be useful to Earth scientists.

____

____

____

____

____

____

____

Integrating Paleomagnetic and Biostratigraphic Age Control Data

THE GEOMAGNETIC POLARITY TIME SCALE (GPTS)

Figure 4.17 shows the **record of Earth's magnetic polarity changes back** approximately **160 Ma** compiled from seafloor magnetic anomalies and uplifted marine sequences. **Black = normal polarity, white = reversed polarity.** The left plot is from Lowrie (1982). The right plot is a composite record of seafloor magnetic anomalies that extends back to the top of the 'Cretaceous Long Normal', approximately 83.0 Ma, compiled by Cande and Kent (1995). This is the currently accepted **geomagnetic polarity time scale** or **GPTS** for the later part of the Cretaceous Period and Cenozoic Era (Gradstein et al., 2004). **Nine radiometric age tie points**, plus the zero-age ridge axis, are used to calibrate this part of the time scale (ages for the marine magnetic anomalies between the calibration points are interpolated by a cubic spline function; Cande and Kent, 1992; Berggren et al., 1995).

For the Late Cretaceous to present, the prominent positive magnetic anomalies (i.e., those corresponding to time intervals, or **chrons**, of normal geomagnetic polarity) have been numbered from Chron 1n in the central axis of the spreading centers to Chron 34n at the younger end of the **Cretaceous Long Normal** (also called the Cretaceous Quiet Zone; 83.0 Ma). Many of the younger chrons are divided into shorter polarity intervals, or **subchrons**. The GPTS based on marine magnetic anomalies extends back through the late Middle Jurassic (Callovian stage, approximately 162 Ma; e.g., Kent and Gradstein, 1986), while land-based polarity records go back through the Triassic (e.g., Gradstein et al., 1995). A nomenclature called the "M-sequence" is used for marine magnetic polarity chrons of the Early Cretaceous to late Middle Jurassic extending from Chron M0r at the base the Cretaceous Long Normal (Chron 34n; base of the Aptian stage in the Early Cretaceous, approximately 121 Ma) to Chron M39 in the Jurassic (Callovian stage).

NAME

5 Make a list of observations about the characteristics of the paleomagnetic record for the past 160 million years based on Figure 4.17.

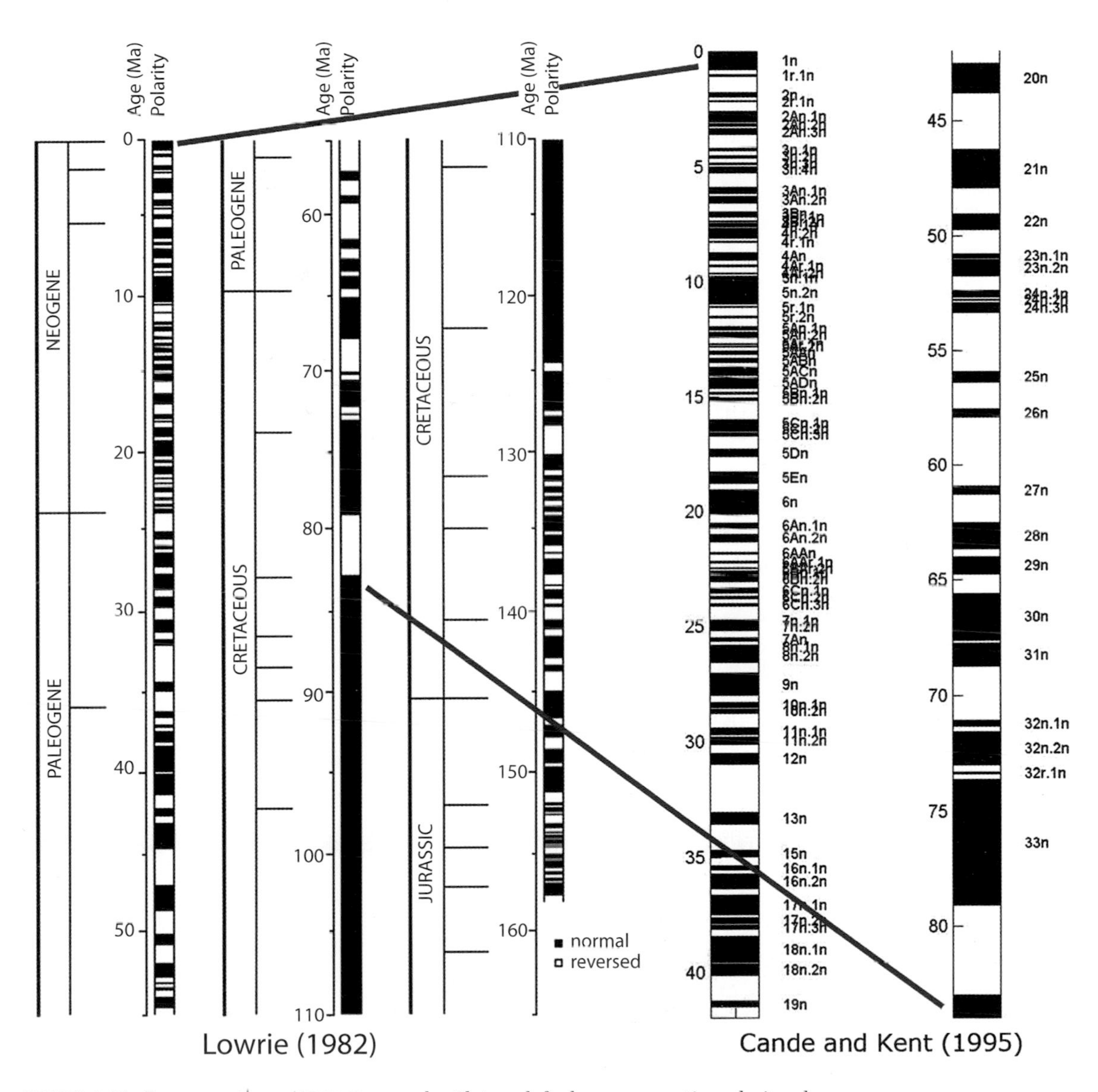

FIGURE 4.17. Geomagnetic polarity time scale. Plot on left shows magnetic polarity changes back to the Middle Jurassic (nearly 160 Ma; from Lowrie, 1982). Note the long interval in the Cretaceous Period with no magnetic reversals; this is called the "Cretaceous Long Normal" chron. The plot on the right shows the details of the marine anomaly record of the South Atlantic (from Cande and Kent, 1995). Red lines show how the South Atlantic anomaly record correlates with the longer Lowrie (1982) record.

NAME

6 Have magnetic reversals occurred in a regular or cyclic fashion (i.e., periodic), or have the reversals occurred in an irregular or random fashion (i.e., episodic)? Explain.

Since the late 1960s, the **geomagnetic polarity time scale** has been extended further back into the Cenozoic and Mesozoic Eras, refined with greater resolution (i.e., recognition of subchrons and polarity events; Figure 4.15) and **calibrated with radiometric age dates**. In addition, **biostratigraphic datum levels** (first occurrences or bases and last occurrences or tops) have been integrated (correlated) with the geomagnetic polarity time scale (Figure 4.18). In doing so, **microfossil datum levels have become primary indicators of geologic age in deep-sea sediments**. Recall from Chapter 3 that biostratigraphic datum levels define the boundaries of biostratigraphic zones, or **biozones**.

7 Drawing from your work in Chapters 3 and 4, explain how biostratigraphic datum levels can be **dated** using the geomagnetic polarity time scale. In other words, **how can ages be assigned to microfossil first occurrences (bases) and last occurrences (tops) as illustrated in Figure 4.18**?

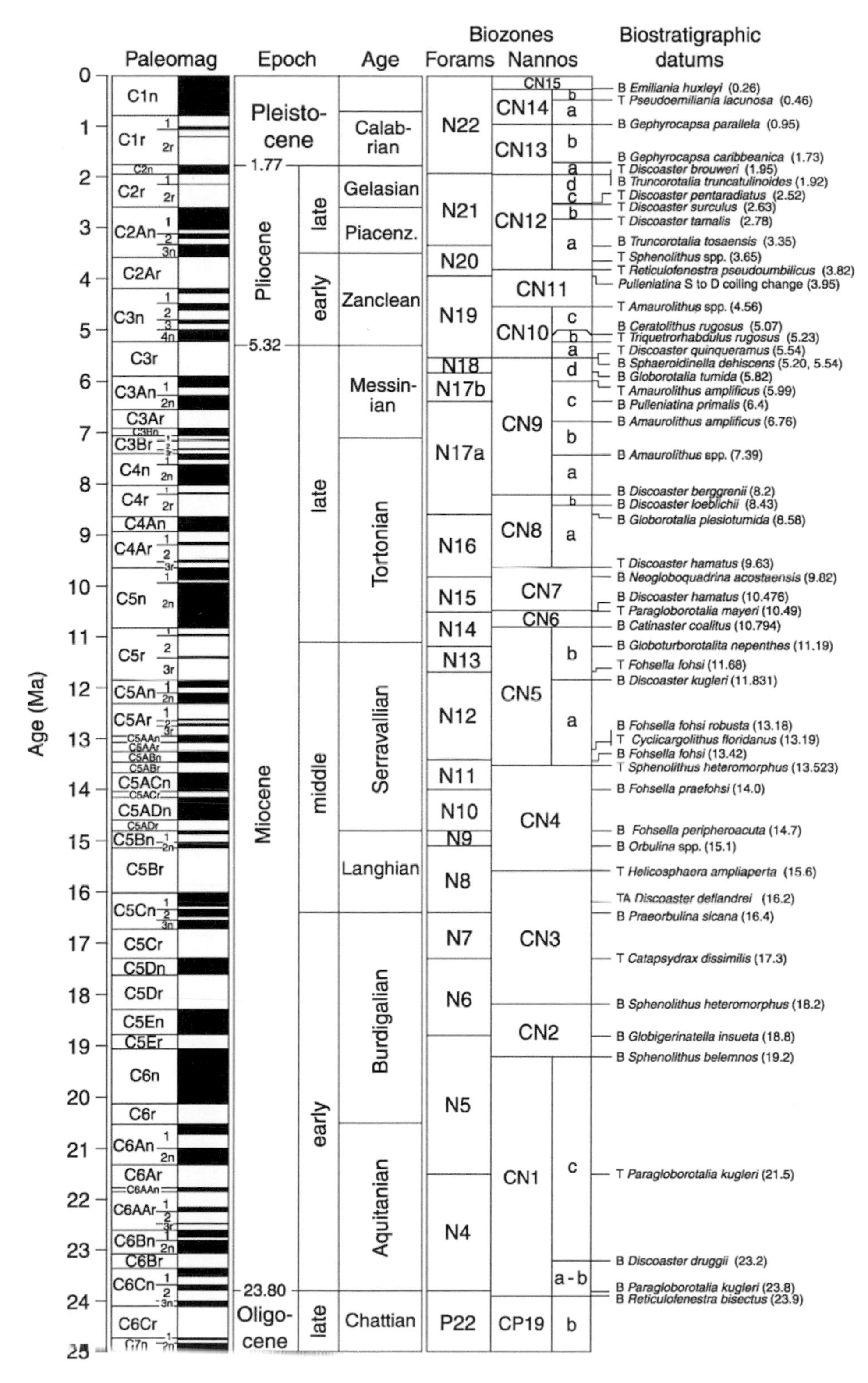

FIGURE 4.18. Portion of **geomagnetic polarity time scale** from 0 to 25 Ma, including geologic epochs and ages (see time scale inside front cover of book), and showing **integration with microfossil biozones** for planktic forams (N zones) and calcareous nannofossils (CN zones). **Microfossil datum levels are shown with ages in parentheses**. B = base or first occurrence; T = top or last occurrence. Shipboard Scientific Party, 2002, Explanatory notes.

NAME

8 Reconsider Question 2 (Part 4.4) which asked about the **assumption(s)** you needed to make in order to correlate the Site 1208 paleomagnetic data with the GPTS. How could you **test your assumption(s) using microfossils**?

NAME

References

Berggren, et al., 1995, Late Neogene chronology; new perspectives in high-resolution stratigraphy. Geological Society of America Bulletin, **107**(11), 1272–87.

Cande, S.C. and Kent, D.V., 1992, A new geomagnetic polarity time scale for the Late Cretaceous and Cenozoic: J. Geophys. Res., vol. **97**(B10), p.13, 917–13, 951.

Cande, S.C. and Kent, D.V., 1995, Revised calibration of the geomagnetic polarity timescale for the Late Cretaceous and Cenozoic. Journal of Geophysical Research, **100**(B4), 6093–5.

Cox, A., 1969, Geomagnetic reversals. Science, **163**, 239–45.

Gradstein, F.M., et al., 1995, A Triassic, Jurassic and Cretaceous time scale. In Geochronology, Time Scales and Global Stratigraphic Correlation, Berggren, W.A., et al. (eds), Special Publication of the Society of Economic Paleontologists and Mineralogists, vol. 54, pp. 95–126.

Gradstein, F.M., Ogg, J.G. and Smith, A.G. (eds), 2004, A Geologic Time Scale 2004. Cambridge University Press, Cambridge, UK, 589 pp.

Heirtzler, J.R., Dickson, G.O., Herron, E.M., Pitman III, W.W., and LePichon, X., 1968, Marine magnetic anomalies, geomagnetic field reversals, and motions of the ocean floor and continents: J. Geophys. Res., vol. 73, p. 2119–2136.

Kent, D. and Gradstein, F.M., 1986, A Jurassic to Recent geochronology. The Western North Atlantic Region: The Geology of North America, vol. M, Geological Society of America, Boulder, CO, pp. 45–50.

Lowrie, W., 1982, A revised magnetic polarity timescale for the Cretaceous and Cainozoic. Philosophical Transactions Royal Society of London, **306**, 129–36.

Mankinen, E., and Wentworth, C., 2003, Preliminary paleomagnetic results from the Coyote Creek outdoor classroom drill hole, Santa Clara Valley, California, USGS Open-File Report 03-187, 32p. http://pubs.usgs.gov/of/2003/of03-187/of03-187.pdf.

Maxwell, A.E., et al., 1970, Initial Reports of the Deep Sea Drilling Project, vol. 3, US Government Printing Office, Washington, DC. http://www.deepseadrilling.org/03/dsdp_toc.htm

Pitman, W.C., III and Heirtzler, J.R., 1966, Magnetic anomalies over the Pacific-Antarctic Ridge. Science, **154**, 1164–71.

Vine, F.J., 1966, Spreading of the ocean floor: New evidence. Science, **154**, 1405–15.

Vine, F.J., and Matthews, D.H., 1963, Magnetic anomalies over oceanic ridges: Nature, vol. **199**, p. 947–949.

Shipboard Scientific Party, 2000. Site 1149. In Proceedings ODP, Initial Reports, vol. 185, Plank, T., et al. (eds), College Station, TX, Ocean Drilling Program. http://www-odp.tamu.edu/publications/185_IR/chap_04/chap_04.htm

Shipboard Scientific Party, 2001a. Site 1172. In Proceedings ODP, Initial Reports, vol. 189, Exon, N.F., et al. (eds), College Station, TX, Ocean Drilling Program. http://www-odp.tamu.edu/publications/189_IR/chap_07/chap_07.htm

Shipboard Scientific Party, 2001b. Explanatory notes. In Proceedings ODP, Initial Reports, vol. 191, Kanazawa, T., et al. (eds), College Station, TX, Ocean Drilling Program. http://www-odp.tamu.edu/publications/191_IR/chap_02/chap_02.htm

Shipboard Scientific Party, 2002, Explanatory Notes. In Proceedings ODP, Initial Reports of the Ocean Drilling Program, vol. 198, Bralower, T.J., et al., College Station, TX, Ocean Drilling Program. pp. 1–63. doi:10.2973/odp.proc.ir.198.102.2002. http://www-odp.tamu.edu/publications/198_IR/198TOC.HTM

Shipboard Scientific Party, 2002. Site 1208. In Proceedings ODP, Initial Reports, vol. 198, Bralower, T.J., et al. (eds), College Station, TX, Ocean Drilling Program. http://www-odp.tamu.edu/publications/198_IR/chap_04/chap_04.htm

Chapter 5 CO_2 as a Climate Regulator during the Phanerozoic and Today

FIGURE 5.1. (a) Volcanic outgasing and lava; (reproduced by permission of Reuters/Bernado de Riz), (b) Oil refinery, Alaska (Reproduced by permission of AccentAlaska.com).

SUMMARY

This chapter explores how the exchange of carbon into and out of the atmosphere is a primary factor in regulating climate over time scales of years to hundreds of millions of years. In **Part 5.1**, you will make initial observations about the short-term global carbon cycle, its reservoirs, and the rates of carbon transfer from one reservoir to another. In **Part 5.2**, you will investigate how atmospheric CO_2 directly and indirectly affects temperature. This understanding is developed though quantitative analysis of the changes in radiative forcing of CO_2 and other factors (e.g., land

Reconstructing Earth's Climate History: Inquiry-Based Exercises for Lab and Class, First Edition. Kristen St John, R Mark Leckie, Kate Pound, Megan Jones and Lawrence Krissek.

NAME

surface albedo) between 1750 and 2005 based on climate models consolidated within the Intergovernmental Panel on Climate Change (IPCC) reports and by constructing qualitative logic scenarios of positive and negative feedback in the Earth's climate system. In **Part 5.3**, you will examine direct instrumental and ice core records of atmospheric CO_2 levels and identify which parts of the carbon cycle are most important in regulating climate over historical time periods. In **Part 5.4**, you will investigate the long-term global carbon cycle, atmospheric CO_2, and Phanerozoic climate history. Using proxy data and general circulation model results, you will identify times of greenhouse and icehouse world conditions and place modern climate change in a geologic context.

CO_2 as a Climate Regulator during the Phanerozoic and Today

Part 5.1. The Short-Term Global Carbon Cycle

Introduction

Examine the **carbon cycle** diagram (Figure 5.2). You can consider this to be a modern snapshot of the major **reservoirs** in which carbon is stored, the processes that transfer carbon between those reservoirs each year, and the rates at which that carbon is transferred from one reservoir to another in the Earth's system. The major carbon reservoirs in the Earth's system are the **atmosphere**, the **terrestrial biosphere**, **soil**, the **oceans** (including surface waters, deep waters, and the marine biosphere), and the **lithosphere** (including carbon in sediments and rocks, and fossil fuels).

1 Using data from Figure 5.2 compare the **relative size** of each of the five major carbon reservoirs of the Earth's system by listing the reservoirs in rank order from largest to smallest in column 2 of Table 5.1. From Figure 5.2, the "ocean" reservoir includes "dissolved organic carbon", the "deep ocean", the "surface waters", and "marine organisms". The "lithosphere" reservoir includes "marine sediment and sedimentary rocks", "sediments", "oil and gas field", and "coalfield".

2 In column 3 of Table 5.1, record your ideas about the **form(s)** of the stored carbon. For example, is it a solid, liquid, and/or gas? What type of gas? Is the carbon in living plant and animal matter? Is it in fossilized organic matter? Or is in minerals in rocks? Be as specific as possible.

3 Examine Figure 5.2 and answer the following two questions. Record your answers in Table 5.2.

NAME

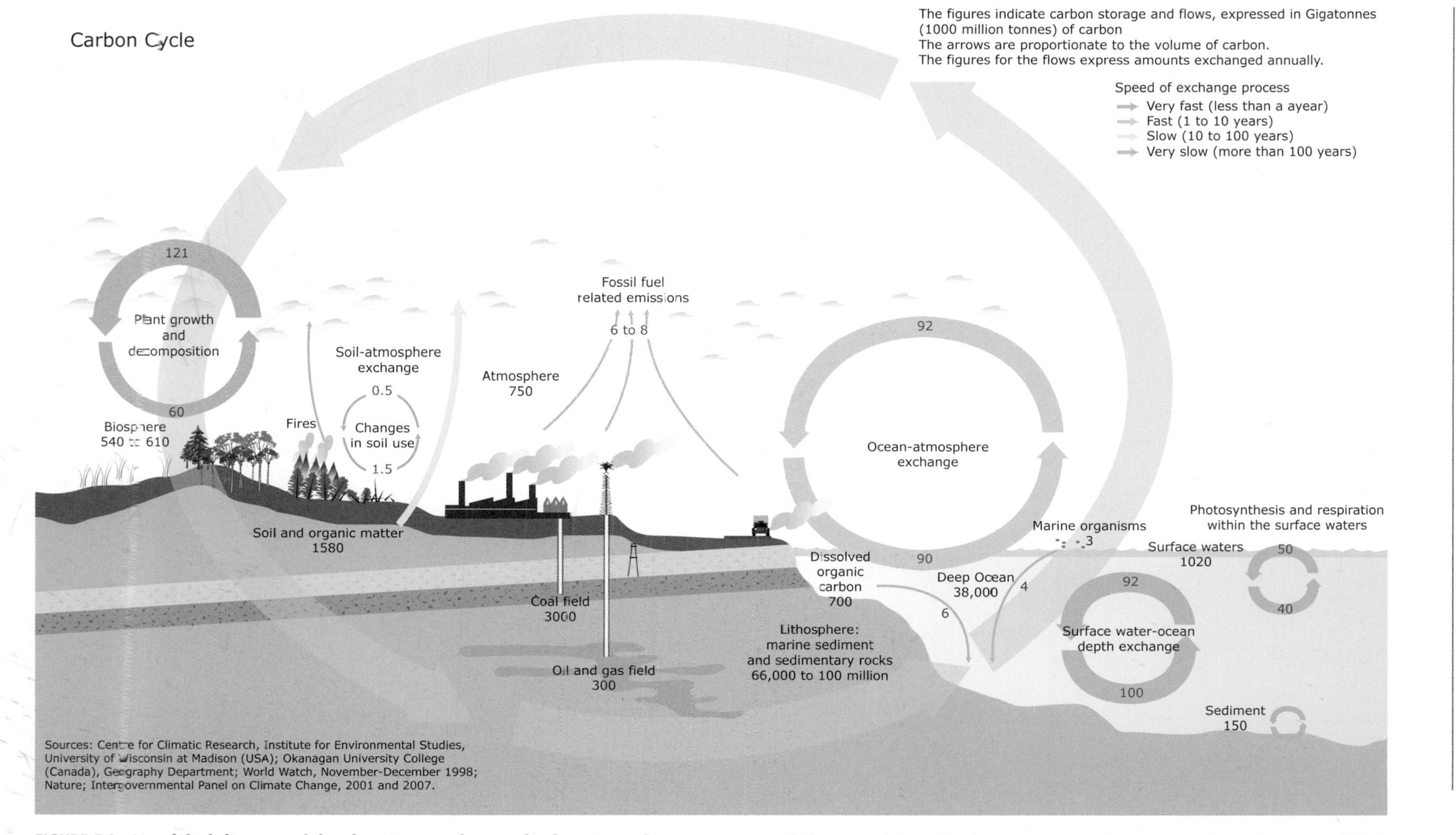

FIGURE 5.2 Simplified diagram of the short-term carbon cycle showing carbon reservoirs and the annual transfers between reservoirs, expressed in gigatonnes (1000 million tonnes) of carbon per year. The arrows are proportional to the volume of carbon transferred. Modified from UNEP/GRID-Arendal, http://maps.grida.no/go/graphic/the-carbon-cycle1, based on information from NASA http://earthobservatory.nasa.gov/Library/CarbonCycle/carbon_cycle4.html.

NAME

TABLE 5.1. Carbon Cycle Reservoirs

Column 1 Relative Size	Column 2 Reservoir Name	Column 3 Form(s) of Carbon in Reservoir
1 (largest)		
2		
3		
4		
5 (smallest)		

(a) What are the **natural processes** that transfer carbon **to the atmosphere**?
(b) What are **relative rates** [i.e., very fast (<1 yr), fast (1–10 yrs), slow (10–100 yrs), or very slow (>100 yrs)] of natural (non-anthropogenic) carbon transfer **to the atmosphere**?

Another way to consider the rate at which each carbon reservoir is affected by these processes is to calculate the average amount of time that an atom of carbon spends in that reservoir. This is known as the **residence time** for that reservoir. Under steady state conditions, the residence time is constant, because the input and output rates are equal. However, steady state conditions are not always the case in nature. **An estimate of the residence time can be calculated by dividing the total amount of carbon in that reservoir by the rate at which carbon is added or removed each year.**

4 Using data from Figure 5.2 complete Table 5.3 in order to estimate the residence times of carbon in the atmosphere, surface ocean, terrestrial biosphere, and soil owing to **natural processes only**. **Use the rates at which carbon is added to the reservoir of interest** each year for these calculations.

5 What are the two primary **anthropogenic processes** shown in Figure 5.2 that affect the carbon cycle?

NAME

TABLE 5.2. Comparison of the rates and processes of carbon transfer from land and ocean reservoirs to the atmosphere

Reservoir From → to	Process of Carbon Transfer to the Atmosphere	Relative Rate of Transfer to the Atmosphere
Terrestrial Biosphere to Atmosphere		
Soil to Atmosphere		
Ocean to Atmosphere		
Lithosphere to Atmosphere*	volcanism	n/a

* This row is completed for you because the natural process of carbon transfer from the lithosphere to the atmosphere via volcanism is so slow they are not included in the short term carbon cycle (Figure 5.2). However, note that volcanism will be an important process of the long-term carbon cycle which is introduced in Part 5.4.

6 (a) Using data in Figure 5.2, calculate the total amount of carbon transferred **to** the atmosphere each year from **anthropogenic processes**. Show your work.

(b) Do anthropogenic (see question 6a) or natural (see Table 5.3) processes transfer **more** carbon to the atmosphere each year?

(c) If the amount of carbon transferred **out** of the atmosphere does not change, what effects will anthropogenic activity have on the total amount of carbon in the atmospheric reservoir over time?

NAME ____________________

TABLE 5.3. Residence time of carbon in different reservoirs

Reservoir	Addition Process	Amount Added (Gigatonnes/yr)	Total Added (Gigatonnes/yr)	Total in Reservoir (Gigatonnes)	Residence Time (yrs)
Atmosphere	Plant Decomposition				
	Soil–atmosphere exchange				
	Ocean–atmosphere exchange				
Surface Ocean	Ocean–atmosphere exchange				
Terrestrial biosphere	Plant growth				
Soil	Addition from vegetation: calculate this as the difference between "plant growth" and "plant decomposition"				

____

____

____

____

NAME

CO_2 as a Climate Regulator during the Phanerozoic and Today

Part 5.2. CO_2 and Temperature

The Greenhouse Effect

The transfer of carbon to and from the atmosphere is a primary factor in regulating climate over time scales of years to hundreds of millions of years. This is because two of the atmospheric **greenhouse gases** with the greatest influence on Earth's surface temperature are carbon compounds: carbon dioxide (CO_2) and methane (CH_4). Other important greenhouse gases include stratospheric water vapor (H_2O), nitrous oxide (N_2O), ozone (O_3), and halocarbons (e.g., chlorofluorocarbons, CFCs).

1 Read the box below on the greenhouse effect and think about the carbon cycle. If the amount of carbon added to the atmosphere is greater than the amount of carbon removed from the atmosphere, how would you expect Earth's surface temperature to change? Explain.

THE GREENHOUSE EFFECT

Energy from the Sun is the primary driver of the Earth's climate system. Energy travels through space as electromagnetic radiation, in the form of electromagnetic waves with a wide range of wavelengths. However, the heat energy that drives the Earth's climate system is only a narrow spectrum of relatively short wavelengths of this broad **electromagnetic spectrum**; the important radiation on the Earth is predominantly **visible light**, which has a relatively short wavelength and is used by plants, algae, and other organisms to drive photosynthesis.

Because the the Earth gains heat from the Sun and has a temperature above absolute zero (–273°C or 0K), the Earth radiates heat to its surroundings. Because of the temperature of the Earth (averaging 15–16°C), the energy released by the Earth to its atmosphere (also known as the Earth's **back radiation**) has a longer wavelength than the incoming visible radiation from the Sun. This back radiation is also known as **longwave radiation**.

Some gases absorb the longwave radiation emitted by the Earth and reradiate that energy, providing additional warming of the atmosphere and Earth's surface. This **greenhouse effect** is what makes the Earth habitable; without out it, the Earth's surface temperature would be on average 31°C cooler, well below the freezing temperature of water. Clearly the greenhouse effect is essential for life as we know it on Earth, although too much of a "good" thing can also have serious repercussions!

Gases that behave in this way are known as **greenhouse gases** and they exhibit this behavior as a direct result of their composition and molecular structure. The two most abundant gases in the Earth's atmosphere, nitrogen (N_2) and oxygen (O_2), do not behave as greenhouse gases. Several less abundant gases, however, are important greenhouse gases on Earth. Table 5.4 summarizes the estimated contribution of the four primary greenhouse gases to Earth's total greennouse effect. Overlaps in the absorbing spectra of the greenhouse gases means that the exact percentage contribution of each gas to the total greenhouse effect is difficult to determine.

Direct Effect of CO_2 on Temperature: Understanding Radiative Forcing

While it may be relatively simple to understand the qualitative relationships between changing greenhouse gases levels and changing temperatures, quantifying these relationships is more challenging. Complications arise because there are both **direct** relationships and **indirect** relationships between each of the greenhouse gases and temperature. The indirect relationships are known as positive and negative Earth system feedbacks that amplify or diminish the temperature change, respectively. Determining the **temperature sensitivity** (also called "climate sensitivity") refers to quantifying how much the temperature will change for a given change in **radiative forcing** (see box below), such as an increase (or decrease) in a particular greenhouse gas.

2 Measurements of the trapped atmospheric gas in ice cores indicate that the **1750 (pre-industrial)** concentration of atmospheric CO_2 was 280 ppm. CO_2

TABLE 5.4. Estimated contribution of the four primary individual gases to Earth's total greenhouse gas effect

Gas[1]	Chemical Formula	Abundance in Earth's Atmosphere (%)	Greenhouse Effect Contribution (%)
Water Vapor	H_2O	<1–3	36–72
Carbon Dioxide	CO_2	approx. 0.038	9–26
Methane	CH_4	approx. 0.00018	4–9
Ozone	O_3	approx. 0.00006	3–7

[1]Other important greenhouse gases include nitrous oxide (N_2O) and halocarbons (e.g., chlorofluorocarbons, CFCs). Data from Kiehl and Trenberth, 1997.

NAME

RADIATIVE FORCING (RF)

Radiative forcing is a measure of the change in the **direct influence** that a factor has in altering the balance of incoming and outgoing energy in the Earth–atmosphere system. It is used to rank quantitatively the importance of climate change factors (Figure 5.3). **Radiative forcing does not take into account indirect influences on climate** (i.e., Earth system feedbacks). The unit for radiative forcing is watts per square meter (W/m^2), the same unit used to measure incoming solar radiation ("insolation") and Earth's back radiation. **Positive forcing increases surface temperature, whereas negative forcing cools it.**

The radiative forcing of any gas is determined from atmospheric radiative transfer models. It is calculated by taking the natural logarithm of the change in that gas concentration (ppm) and multiplying by a gas-dependent constant. For example the following formula is used to calculate the radiative forcing of CO_2:

$$RF = 5.35 \times \ln\left[\frac{\text{Present concentration of } CO_2}{\text{Initial concentration of } CO_2}\right] W/m^2$$

(RF discussion adapted from Forster et al., 2007; RF calculation from http://www.esrl.noaa.gov/gmd/aggi/)

concentrations have been directly measured from the atmosphere since 1959 at the NOAA Mauna Loa Observatory in Hawaii. The 2005 concentration of CO_2 was 378 ppm. The 2011 concentration of CO_2 was >392 ppm and it continues to rise.

Calculate the radiative forcing of CO_2 on climate from 1750 to 2005 using the equation in the box above. Show your work.

3 How does your answer to Question 2 compare to the radiative forcing of CO_2 from 1750 to 2005 as reported in Figure 5.3?

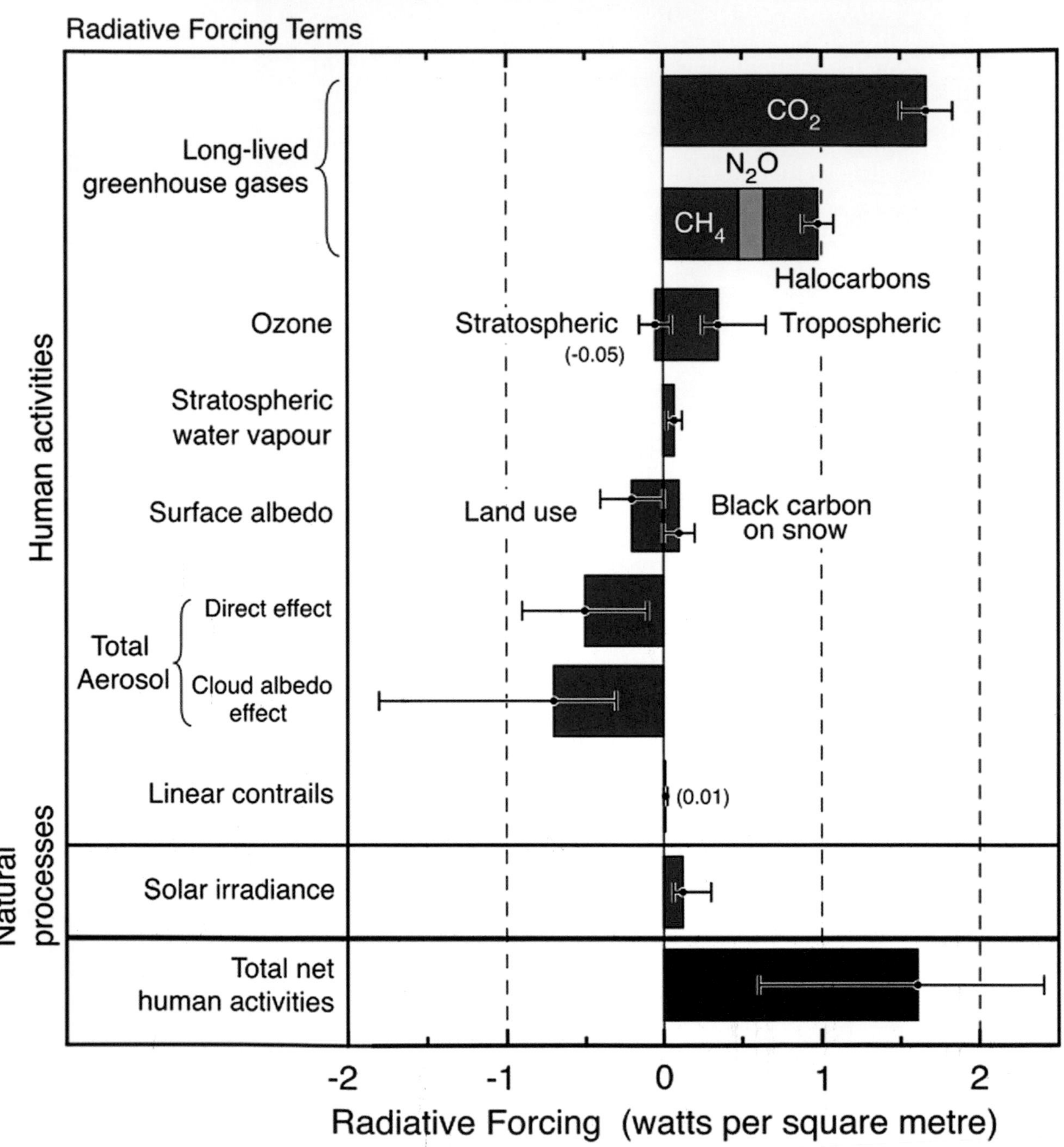

FIGURE 5.3. Anthropogenic (human) and natural radiative forcing terms that affect climate change over a 255-yr window, from 1750 to 2005. Uncertainties are represented by the thin black lines. From Forster et al., 2007.

NAME

4 Which radiative forcing term had the highest radiative forcing from 1750 to 2005, based on the data in Figure 5.3?

A NOTE ON WATER VAPOR

Recall (see box on the greenhouse effect) that water vapor (H_2O) is the largest contributor to the greenhouse effect. As water moves through the hydrologic cycle, the amount of atmospheric water vapor varies both regionally (e.g., think of the humidity in tropical rainforest compared with that in a mid-latitude desert) and temporally (e.g., think of a humid summer day compared with a dry winter day in your region). The **direct** impact of human activity on the levels of water vapor in the atmosphere is primarily linked to (a) irrigation for agriculture and (b) methane (CH_4) emissions, since CH_4 oxidizes in the stratosphere and converts convert to CO_2 and H_2O (vapor). Model estimates indicate that regional irrigation produces ≤1% change in tropospheric (i.e., lower atmospheric) water vapor content, resulting in a change of radiative forcing (RF) of +0.03 W/m^2 (Forster et al., 2007). This **direct** effect is significantly smaller than other RF factors and tropospheric water vapor is therefore **not** included in Figure 5.3. The estimated RF of stratospheric water vapor is a little larger (+0.2 W/m^2) and is therefore included in Figure 5.3. In comparison, **feedbacks** involving water vapor are quite significant; these will be addressed later in this exercise.

5 Which is larger, the radiative forcing of water vapor from human activities (box above) or the radiative forcing of atmospheric CO_2 from human activities (Figure 5.3)?

6 Based on your understanding of the carbon cycle from Part 5.1, why might the increase in CO_2 be categorized under **"human activities"** and not "natural processes" for this 255-yr time period (Figure 5.3)?

NAME ____

7 By listing CO_2 under the category of human activity (Figure 5.3), does this mean that no carbon was transferred to or from the atmosphere during this time period by **natural processes**? Explain.

____

____

____

____

A fundamental goal of climate science is to determine how sensitive global average temperature is to changes in atmospheric CO_2 concentrations. In climate models, this "temperature sensitivity" is typically quantified on the basis of a **doubling of pre-industrial level atmospheric CO_2 concentrations** and involves determining both the direct (RF) and indirect relationships between CO_2 and temperature.

8 What would the RF of CO_2 be if the atmospheric level of CO_2 rose to twice its pre-industrial level? Show your work. (Hint: refer back to question 2, Part 5.2.)

____

____

____

____

9 Read the box on Land Surface Albedo. What **direct changes in land use** could cause land surface albedo to **increase**? If possible, provide examples of changing land use in your own region or state that might affect the reflectivity of the local land surface.

____

____

____

____

____

____

____

NAME

LAND SURFACE ALBEDO: A DIFFERENT STORY OF RADIATIVE FORCING

While CO_2 has a high radiative forcing and is therefore an important factor in directly regulating atmospheric temperature, it is not the only factor. For example, land surface albedo also affects atmospheric temperature. **Albedo** is a measure of the reflectivity of a surface. Lighter surfaces (e.g., fresh snow and ice) are more reflective (i.e., have a higher albedo) than darker surfaces (e.g., road pavement, dark soil, forests, lakes, and the ocean). The more reflective the surface, the less incoming solar radiation is absorbed by that surface.

Model estimates indicate a decrease in the radiative forcing of the **land surface albedo** between 1750 and 2005 (Figure 5.3). In other words, the land surface was absorbing less (and reflecting more) incoming solar radiation in 2005 than in 1750 (i.e., the land surface albedo increased). It is important to understand **that changes in snow and ice cover caused by** the **Earth system feedback were not factored in this calculation**, since such changes would be **indirect** and radiative forcing is a measure of only direct effects on energy transfer in the Earth–atmosphere system.

Indirect Effects of CO_2 on Temperature: Earth System Feedbacks

In general, as radiative forcing increases, the Earth's average temperature also increases. However, the relationship between a change in radiative forcing and the corresponding change in temperature is not simple or straightforward. It is calculated within large scale computer models of the Earth's atmosphere, called **general circulation models (GCMs)**. In general, GCM results indicate that the direct effect of doubling CO_2 would produce an increase in radiative forcing of 3.7 W/m^2 (refer back to Question 8) and an increase in global average temperature of approximately 1.25°C (Rahmstorf, 2008; Ruddiman, 2008).

This is not the only way that changing CO_2 levels affect temperature. There are also **indirect effects** on other parts of the Earth system, which in turn either augment (**positive feedback**) or diminish (**negative feedback**) the direct effect that CO_2 has on temperature. These **earth system feedbacks** include: (a) water vapor, (b) surface albedo, and (c) clouds. To determine qualitatively how these Earth system feedbacks affect temperature, consider Table 5.5.

TABLE 5.5. Earth System Feedback Factors

Feedback	Key Physical Relationships/Conditions
Water Vapor	1 A warmer atmosphere can store more water vapor than a colder atmosphere. [This is based on the Clausius–Clapeyron relationship between temperature and saturation vapor pressure]. 2 A warmer atmosphere will evaporate more water from the ocean, which increases the amount of water vapor in the atmosphere. 3 Water vapor is a greenhouse gas.

NAME ______________________________

TABLE 5.5. *Continued*

Feedback	Key Physical Relationships/Conditions
Surface Albedo	4 Albedo is a measure of the reflectivity of a surface. 5 Lighter surfaces (e.g., fresh snow and ice) are more reflective (have a higher albedo) than darker surfaces (e.g., road pavement, dark soil, forests, lakes, ocean). 6 The more reflective the surface, the less incoming solar radiation is absorbed by that surface.
Clouds	7 Clouds vary in space (altitude, latitude, longitude) and time. Cloud cover, height, albedo, water content, and particle size all affect cloud feedbacks. 8 A warmer atmosphere evaporates more water from the ocean, thereby supplying more moisture to the atmosphere. A warmer atmosphere also can store more water vapor than a cold atmosphere, thereby reducing condensation. Both of these factors affect the abundance, type, and size of clouds. 9 Cloud thickness increases with temperature. 10 The upper surface of some clouds reflects incoming solar radiation; these have a high albedo. Other clouds are less reflective and therefore have only a moderate albedo. 11 The lower surface of some clouds re-radiates long wavelength radiation from the Earth. 12 Higher clouds are colder and therefore radiate less long wavelength energy back to space [i.e., Stefan–Boltzmann relationship].

10 With the information from Table 5.5 in mind, complete the blanks in each of the following six **feedback scenarios** and justify the selections you make. In each case, indicate how each of the feedback factors will change (will it **increase** or will it **decrease**?) and how the resulting temperature will change (will it **increase** or will it **decrease**?). For each scenario indicate whether it outlines a positive (P) or negative (N) feedback.

(a) Atmospheric CO_2 increases → temperature increases → water vapor __________ → temperature __________. P or N?

Justification:

____

____

____

____

(b) Atmospheric CO_2 increases → temperature increase → snow and ice __________ → surface albedo __________ → temperature __________. P or N?

NAME ____

Justification:

____

____

____

____

____

____

(c) Atmospheric CO_2 increases → temperature increases → clouds __________ → temperature __________. P or N?

Justification:

____

____

____

____

____

____

(d) Atmospheric CO_2 decreases → temperature decreases → water vapor __________ → temperature __________. P or N?

Justification:

____

____

____

____

____

(e) Atmospheric CO_2 decreases → temperature decreases → snow and ice __________ → surface albedo __________ → temperature __________. P or N?

NAME ____

Justification:

____

____

____

____

____

____

(f) Atmospheric CO_2 decreases → temperaturedecreases → clouds ________ → temperature __________. P or N?

Justification:

____

____

____

____

____

____

11 Based on your responses to the six scenarios in Question 10, which of these feedbacks introduces the most uncertainty in predicting its ultimate effect on Earth's climate? Why might that be?

____

____

____

____

____

____

____

NAME

Combined Effect (Direct and Indirect) of CO_2 on Temperature

In the same way that you have identified uncertainties in your qualitative predictions, each GCM identifies uncertainties and has its own way of dealing with those uncertainties in its calculations. As a result, different GCMs often predict different temperature changes for the same change in the concentration of atmospheric CO_2.

From the range of GCM results, the "average", or mid-range, model projections for a doubling of CO_2 are:

- an **increase of approximately 2.5°C** caused by the water vapor feedback,
- an **increase of approximately 0.6°C** caused by the surface albedo feedback (including snow and ice albedo changes), and
- a **decrease of approximately 1.85°C** caused by the cloud feedback (Ruddiman, 2008).

12 Calculate the projected combined effect of these three climate system feedbacks on temperature, given a doubling of CO_2. Show your work.

13 As stated earlier, the direct radiative effect of doubling the pre-industrial level of CO_2 is 3.7 W/m^2, resulting in an increase of global average temperature of 1.25°C (Rahmstorf 2008; Ruddiman 2008). Calculate the projected **total warming** caused by this direct radiative effect plus the warming from the climate system feedbacks (from questions 12) when CO_2 is doubled?

NAME

CO_2 as a Climate Regulator during the Phanerozoic and Today

Part 5.3. Recent Changes in CO_2

Consider the atmospheric CO_2 data from air samples collected at Mauna Loa Observatory for the period 1958–2010 (Figure 5.4).

1 Is this carbon dioxide record a **direct measure** of carbon dioxide concentrations in our atmosphere or is it a **proxy (indirect indicator)** of carbon dioxide concentrations?

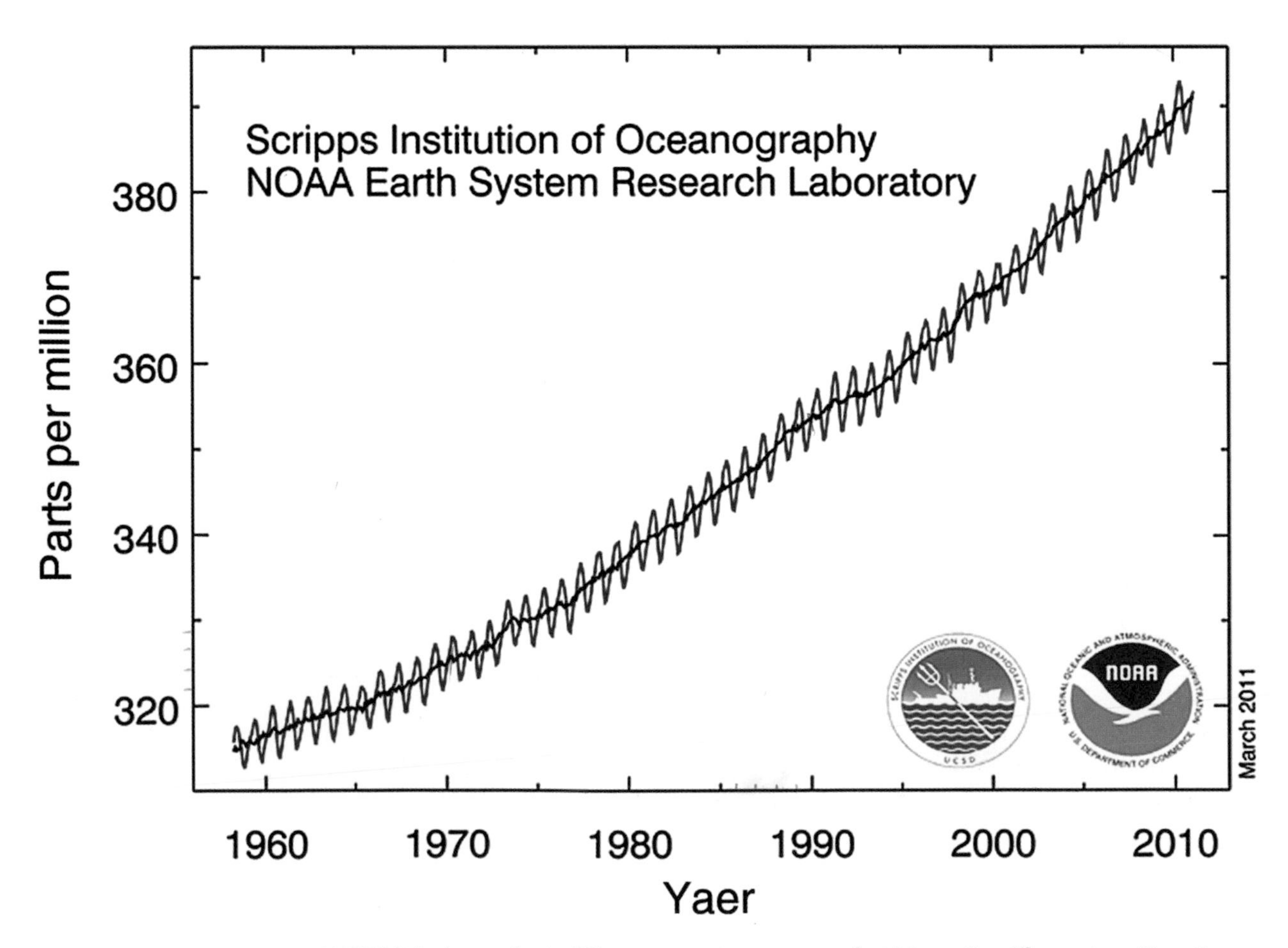

FIGURE 5.4. Atmospheric CO_2 concentrations measured at Mauna Loa Observatory, Hawaii. From: Dr. Pieter Tans, NOAA/ESRL, www.esrl.noaa.gov/gmd/ccgg/trends/.

NAME ____________________

2 What do you think the **solid black curve** represents?

____

____

3 What do you think the **red saw-toothed curve** represents?

____

____

4 How many peaks (or "teeth") are present in a 5-year interval on the saw-toothed curve? Based on your observation, what is the length of time between adjacent peaks on the saw-toothed curve?

____

____

5 Speculate on the **process** controlling the **saw-toothed** nature of the red curve.

____

____

____

6 What **reservoirs** from the global carbon cycle (Figure 5.2) and what **form(s) of carbon** (organic or inorganic) might be involved in producing the saw-toothed pattern plotted in Figure 5.4?

____

____

____

7 (a) Use Figure 5.4 to determine the atmospheric CO_2 concentration for the year you were born?

Year _____ CO_2 concentration _____ (ppm)

(b) Go to the NOAA web link listed in the caption for Figure 5.4. Based on the data provided on the NOAA web site, determine the modern CO_2 concentration.

NAME ______________________________

Today's Date ______________ CO_2 concentration __________(ppm)

8 From Part 5.2, recall that climate models often focus on predicting when the atmospheric CO_2 concentration will have doubled and what the resulting temperatures will be at that time. The baseline for these models is the 1750 (or pre-industrial) level of 280 ppm CO_2. Are the CO_2 levels today double the pre-industrial concentration?

____

9 Based on the rate of atmospheric CO_2 increase over the past decade (which you can calculate from the data in Figure 5.4), predict when CO_2 level will double from the pre-industrial 280 ppm base level. Show your work.

____

____

____

____

____

____

____

____

____

Now consider the atmospheric CO_2 data for the last 20,000 years, measured from bubbles of trapped gas in ice cores and directly from the atmosphere (Figure 5.5).

10 The last glacial maximum occurred approximately 20,000 years ago. At this time continental ice sheets were widespread, covering much of the landscape north of 40°N (e.g., southern Ohio). What was the atmospheric CO_2 concentration during the last glacial maximum (Figure 5.5)?

____

11 Describe the trends of CO_2 over the last 20,000 years (Figure 5.5). In particular, identify times when the slope of the CO_2 trend changed noticeably.

____

____

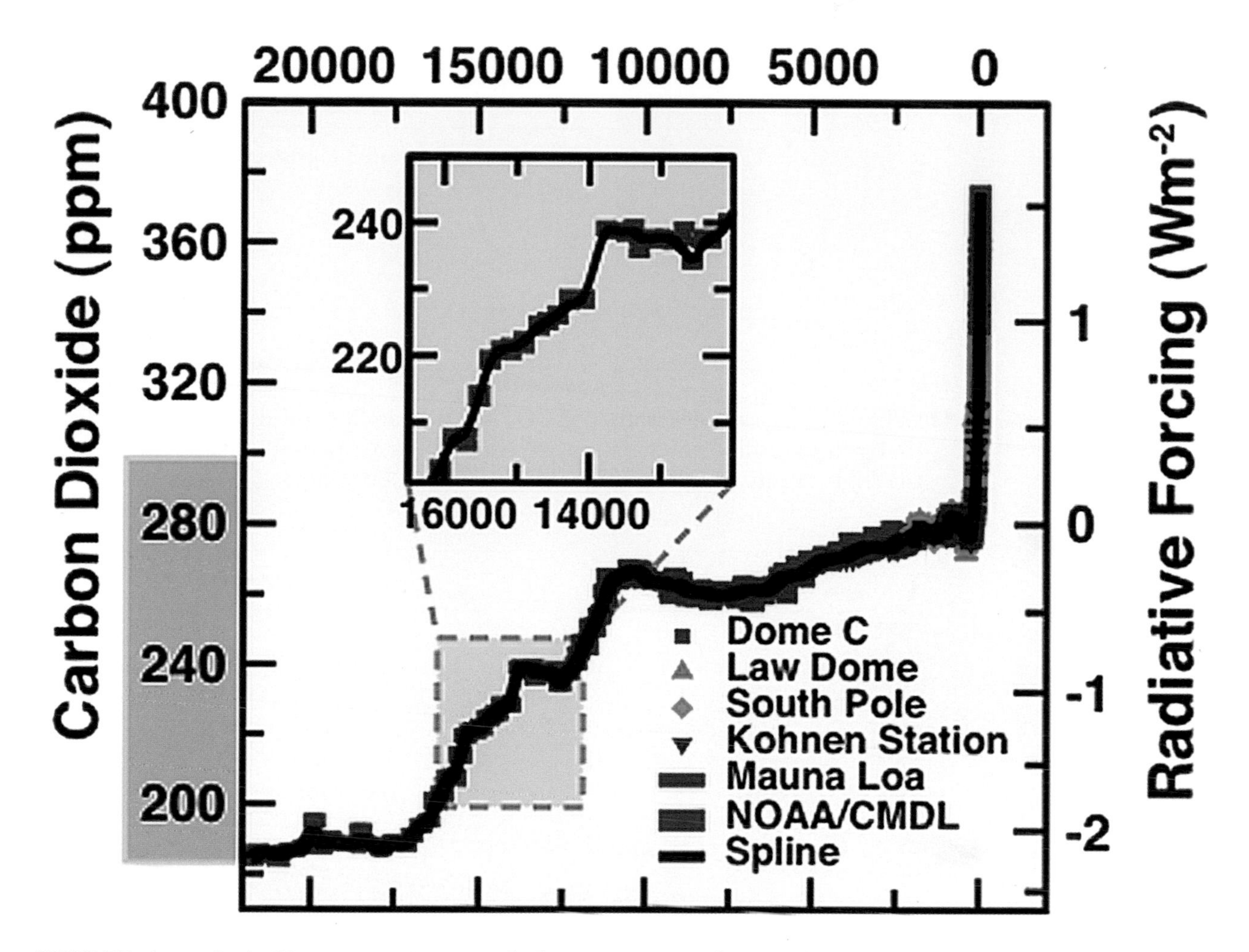

FIGURE 5.5. Atmospheric CO_2 concentrations over the last 20,000 years. The area of gray shading indicates pre-industrial natural range of CO_2. Data are from ice cores (colored symbols are from different studies), atmospheric samples (colored bars), and merged (spline function; black). The corresponding radiative forcing relative to pre-industrial (1780 baseline) levels are shown on the right axis of the large panel. From Joos and Spahni, 2008. **Note that approximately 260 years ago (1750) the CO_2 level was 280 ppm.**

NAME

12 Calculate the **rate of change of CO_2** (ppm/year) for the different trend lines that you identified in Question 11. Show your work. Which time period has the highest rate of CO_2 change in the last 20,000 years?

13 Speculate on the **processes** of the carbon cycle that controlled the CO_2 trends over the last 20,000 years.

NAME ______________________________

CO_2 as a Climate Regulator during the Phanerozoic and Today

Part 5.4. Long-Term Global Carbon Cycle, CO_2, and Phanerozoic Climate History

Long-Term Carbon Cycle

As you learned in Part 5.1, carbon reservoirs include the lithosphere, the atmosphere, the ocean, and the biosphere. Carbon is transferred to and from the atmosphere over short time scales (Figure 5.2) and also over long time scales for which **plate tectonics** and weathering becomes important climatic influences (Figure 5.6).

1 There are over **66,000,000 gigatonnes of carbon** stored in the lithosphere (Figure 5.2). What are the long-term processes (Figure 5.6) that transfer this geologically stored carbon to the atmosphere?

____

____

____

____

2 The rate at which each process transfers carbon from one reservoir to another can change over geologic (i.e., "long") time scales. Use Figure 5.6 to explain how changing rates of seafloor spreading would affect the rate of CO_2 transfer to the atmosphere.

____

____

____

____

____

____

____

____

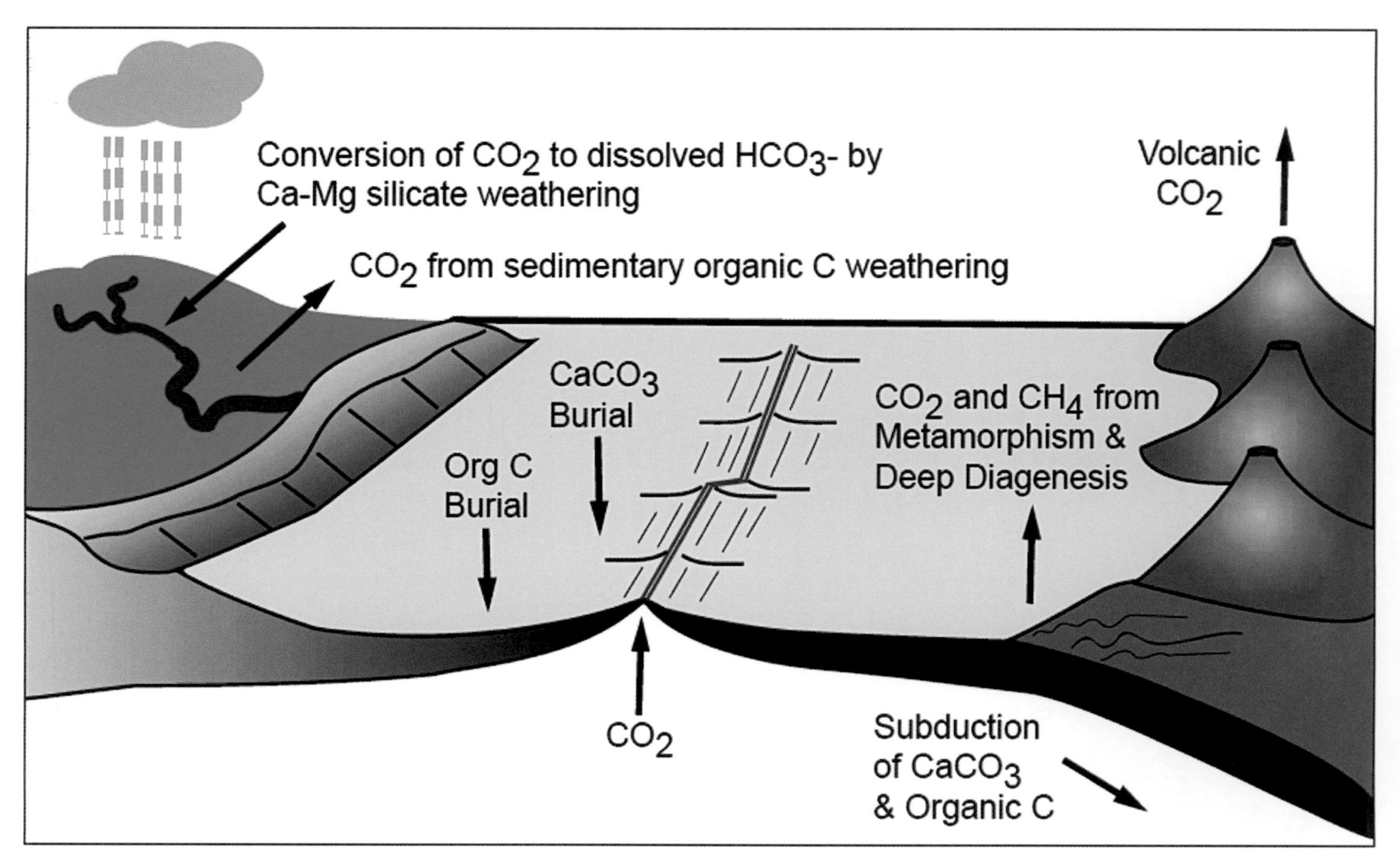

FIGURE 5.6. Simplified diagram of the **long-term carbon cycle**. From Berner, 1999. **Note that CO_2 released at oceanic ridges is considered to be from volcanic sources.** Originally drawn by Steve Petsch, University of Massachusetts, Amherst.

NAME

3 Examine the age distribution map of the seafloor (Figure 5.7). How could the geographic distribution of seafloor ages help us determine if seafloor spreading rates have changed in the past?

4 The natural transfer of carbon from the lithosphere to the atmosphere typically operates on slow (i.e., tectonic) timescales; with significant changes taking place over hundreds of thousands to millions of years (although there are important exceptions; see Chapter 9). Think back to Part 5.1, Figure 5.2; how have humans accelerated the transfer of geologically stored carbon to the atmosphere?

CO_2 and Phanerozoic Climate History

The importance of CO_2 in the Earth's recent climate history was demonstrated in Parts 5.1–5.3. Here, you will examine proxy data and model results which reconstruct the CO_2 concentrations of the Earth's atmosphere for the Phanerozoic Eon. The Phanerozoic encompasses the last 540 million years of the Earth's history (see geologic time scale inside front cover of this book). The goal of this section is to evaluate the importance of CO_2 as a regulator of long-term climate change.

The paleo-atmospheric CO_2 proxy data (Figure 5.8) come largely from geochemical and paleontological sources (Royer et al., 2004, and references therein). These include the stable carbon-isotope records ($\delta^{13}C$) of minerals of fossil soils (paleosols) and of phytoplankton, the stable boron isotope record ($\delta^{11}B$) from planktonic foraminifera, and the changes in characteristics of fossil plant leaves (stomatal density). The GEOCARB III model also supplies data on the long-term CO_2 record. This model is based on estimates of the rate of transfer of carbon by processes in the long-term carbon cycle (Figure 5.6).

5 For what time period would you be most confident of the CO_2 paleo atmospheric proxy data (Figure 5.8) and why?

NAME

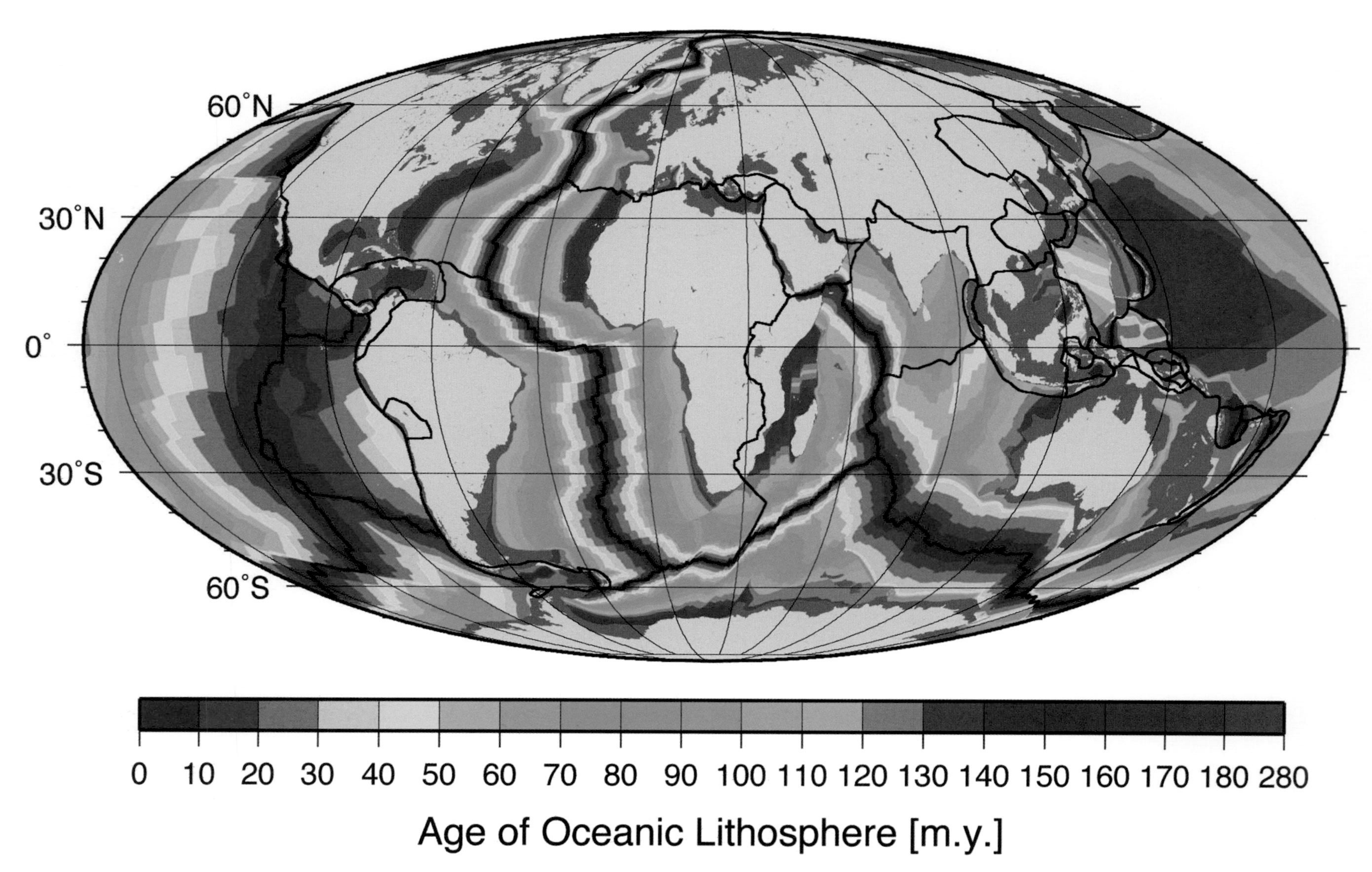

FIGURE 5.7. Seafloor age in millions of years. Age determined from the paleomagnetic reversal patterns preserved in ocean crust basalts (see Chapter 3) Map courtesy of NOAA (http://www.ngdc.noaa.gov/mgg/ocean_age/data/2008/image/age_oceanic_lith.jpg).

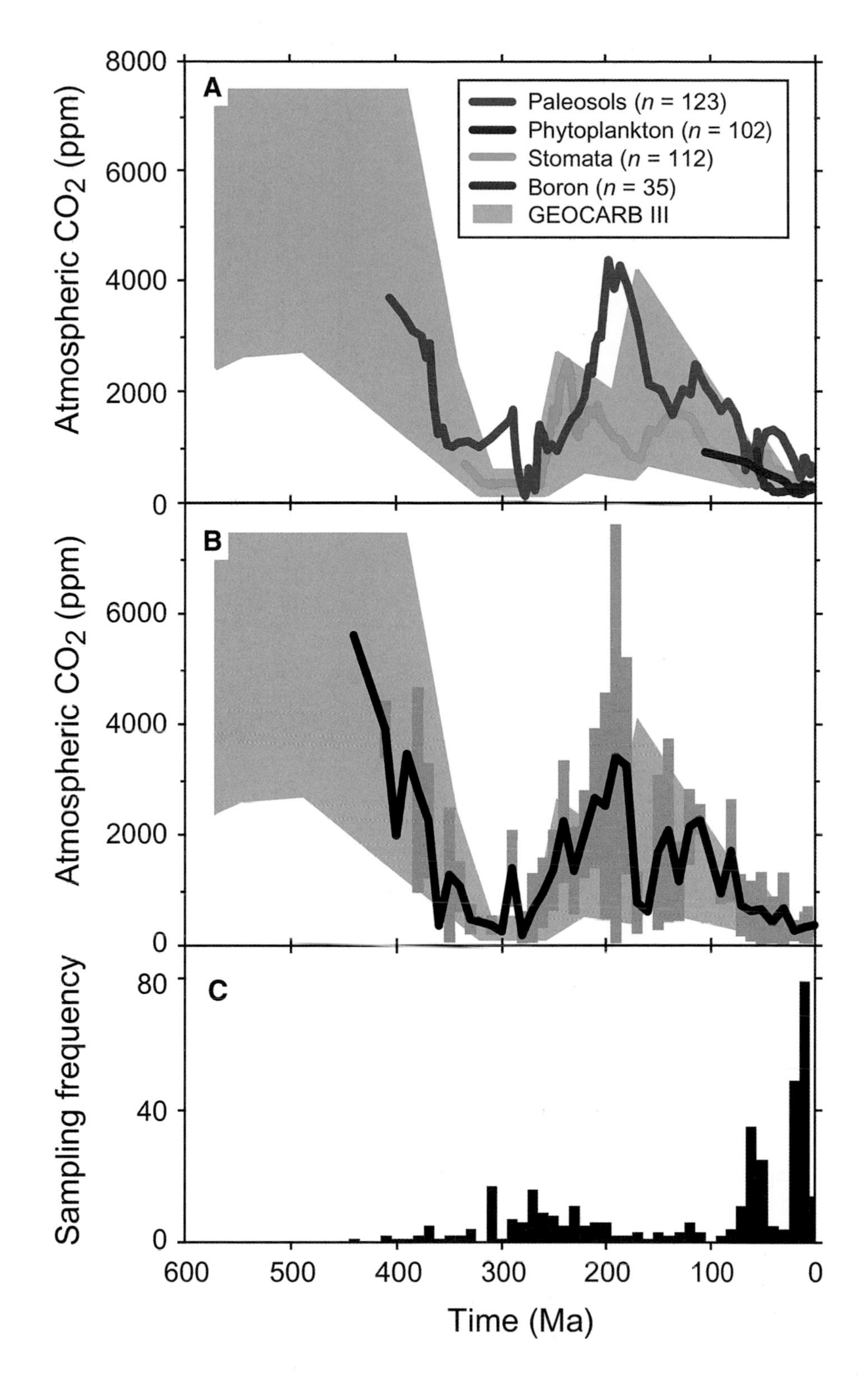

FIGURE 5.8. (A) Paleo-atmospheric CO_2 proxy and model data for the Phanerozoic Eon; salmon colored area plot shows envelope of modeled CO_2 values, including uncertainty; four colored lines represent proxy data; (B) Average value of the combined CO_2 proxy data (black line). Standard deviation of the combined data is shown in gray and represents the uncertainty of the data; (C) Frequency distribution of the CO_2 proxy data. From Royer et al., 2004.

NAME

6 How do the records from the different independent paleo-atmospheric CO_2 proxies compare with the results from the paleo-atmospheric CO_2 geochemical model (GEOCARB III; Figure 5.8)?

7 In Part 5.3 (Question 7) you looked up the modern concentration of atmospheric CO_2. Draw a horizontal line on Figure 5.8(A) indicating this modern value. Estimate the percentage of the Phanerozoic time (i.e., the last 540 million years) that had a higher CO_2 level than today.

Direct evidence for Phanerozoic climate change comes from the rock record. The temporal and spatial distributions of glacial erosion features (e.g., striations on bedrock) and glacial deposits (e.g., end moraines; see Ch. 14 for photos and maps of recent glacial deposits) can be used to constrain the timing and geographic extent of glaciations of the past (Figure 5.9).

8 Compare the Phanerozoic records of paleo-atmospheric CO_2 and glaciation. Is there any correlation between the two records? Describe.

9 (a) According to the data in Figure 5.9, what is the maximum CO_2 level that can support the existence of large ice sheets – those extending past the Arctic and Antarctic circles (66°e paleolatitude)?

(b) Refer back to the sampling frequency and standard deviation data in Figure 5.8. Do you want to modify your answer to question 9a, based on this uncertainty data? How so?

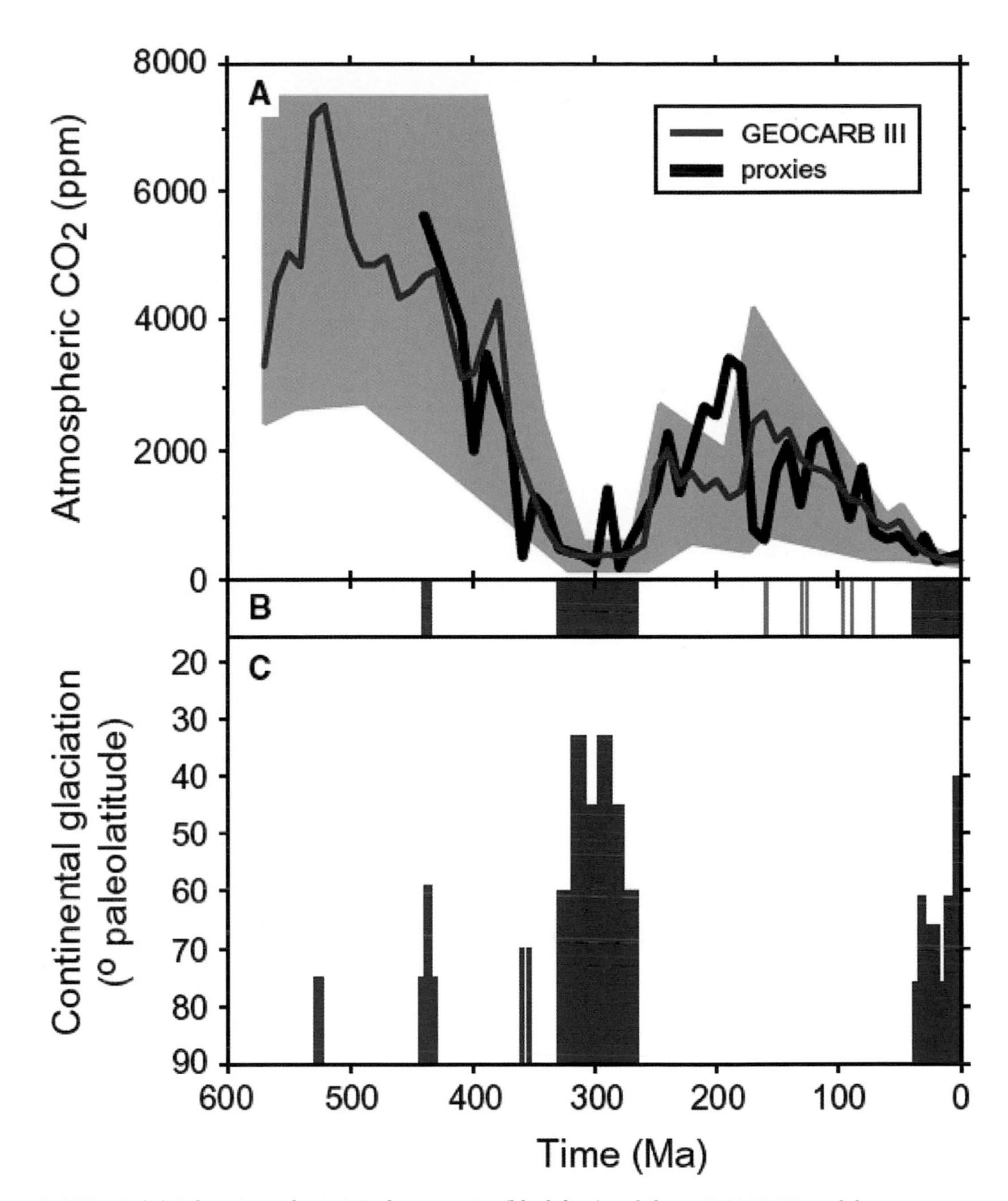

FIGURE 5.9. (A) Paleo-atmospheric CO_2 from proxies (black line) and the GEOCARB III model (red line); model uncertainty is represented by the orange shading, (B) time periods of glaciation (dark blue) or cool climates (light blue), and (C) the latitudinal distribution of glaciation. Modified from Royer et al., 2004.

NAME ______________________________

GREENHOUSE AND ICEHOUSE WORLDS

Through approximately the last 2 billion years, Earth's climate has fluctuated – at timescales of tens of millions of years to approximately 100 million years – between relatively cold times, with low atmospheric CO_2 levels, long-lived large ice sheets, and low sea levels, and relatively warm times, with high atmospheric CO_2 levels, no long-lived large ice sheets, and high sea levels. The warm episodes are described as **greenhouse worlds**, whereas the cold episodes are described as **icehouse worlds**.

The occurrence of greenhouse worlds and icehouse worlds in the Earth's history appears to have been triggered by one or more of several factors: changes in the latitudinal positions of large continents; opening or closing of oceanic gateways, which affect ocean circulation patterns and the movement of heat around the planet; and **changes in the global carbon cycle**, such as by the uplift and chemical weathering of mountain ranges (see Chapter 10 on the Oi1 event), by changes in the rates of seafloor spreading or other volcanic activity, or by changes in biological cycling of carbon.

10 Read the box above. Was the Cretaceous Period (145.5 Ma to 65.5 ma) a greenhouse world or an icehouse world?

____

11 Do we currently live during a greenhouse world or an icehouse world?

____

12 How can you reconcile your answer to Question 11, with the documented global warming that is occurring today (e.g., Part 5.3)?

____

____

____

____

13 Examine Figure 5.10. What is the range of model-projected atmospheric CO_2 concentrations by the year 2100?

____

14 Compare the data in Figure 5.10 to that in Figure 5.9. In the past, has the Earth ever experienced the atmospheric CO_2 levels that are predicted to be reached in the next 100 years?

____

NAME

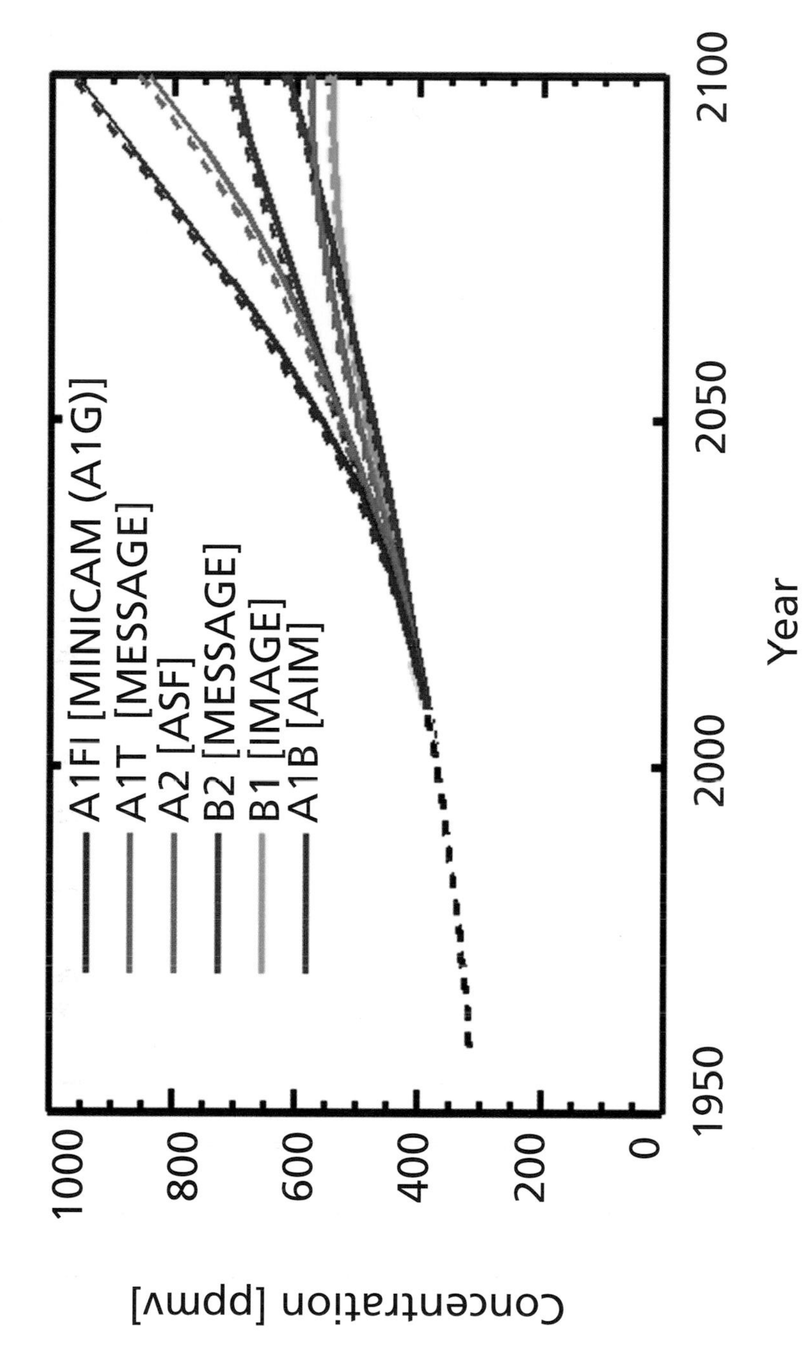

FIGURE 5.10. Atmospheric CO_2 concentrations (parts per million vapor) observed at Mauna Loa (black dashed line, for 1958–2008), and six projections of atmospheric CO2 concentration (colored lines, for 2008–2100) from different scientific research groups. The colored solid vs. colored dashed lines represent two different carbon cycle models used in each scenario. Graph courtesy of the IPCC.

NAME

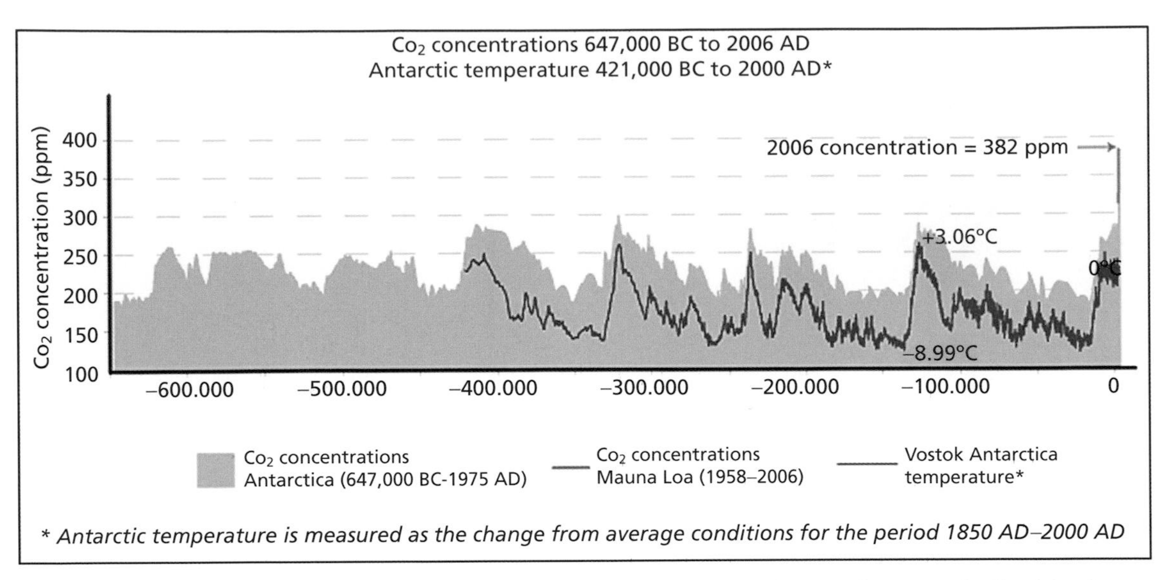

FIGURE 5.11. Inferred temperature (red line) and measured atmospheric CO_2 (yellow) for the past 649,000 years from Antarctic ice cores. The orange vertical bar represents the increase in atmospheric CO_2 over the past two centuries, measured from gases in ice cores and directly from the atmosphere at Mauna Loa. Graph courtesy of the EPA.

15 The data from the field of anthropology indicate that our species (*Homo sapiens*) evolved approximately 200,000 years ago. Examine Figure 5.11. In the past 200,000 years, have humans ever experienced the atmospheric CO_2 levels that are predicted to be reached in the next 100 years?

16 Based on your answer to Question 15 and what you know about the relationship between CO_2 and temperature, how would you compare the global average temperature that humans are likely to face in the next 100 years to the global average temperatures faced by humans during the past 200,000 years? Explain your answer.

17 What effects do you think these projected temperatures are likely to have on human populations and societies in the next 100 years? (These may include both direct effects and indirect effects.)

NAME

NAME

References

Berner, R.A., 1999, A new look at the long-term carbon cycle. GSA Today, **9**, 2–6.

Forster, P., et al., 2007, Changes in atmospheric constituents and in radiative forcing. In Climate Change 2007: The Physical Science Basis. Contribution of Working Group I to the Fourth Assessment Report of the Intergovernmental Panel on Climate Change, Solomon, S., et al. (eds), Cambridge University Press, Cambridge, UK.

Joos, F. and Spahni, R., 2008. Rates of change in natural and anthropogenic radiative forcing over the past 20,000 years. Proceedings of the National Academy of Sciences of the USA, **105** (5), 1425–30, http://www.pnas.org/content/105/5/1425.full.

Kiehl, J. T. and Trenberth K.E., 1997, Earth's annual global mean energy budget. Bulletin of the American Meteorological Society, **78** (2), 197–208.

Rahmstorf, S., 2008, Anthropogenic climate change: revisiting the facts. In Global Warming: Looking Beyond Kyoto, Zedillo, E. (ed), Brookings Institution Press, Washington DC, and Yale Center for the Study of Globalization, Newhaven, Connecticut, pp. 34–53. http://www.pik-potsdam.de/~stefan/Publications/Book_chapters/Rahmstorf_Zedillo_2008.pdf.

Royer, D.L., et al., 2004, CO_2 as a primary driver of Phanerozoic climate. GSA Today, **14**, 4–10.

Ruddiman, W. F., 2008, Earth's Climate Past and Future, W.H. Freeman and Co., New York, 388 pp.

Chapter 6 The Benthic Foraminiferal Oxygen Isotope Record of Cenozoic Climate Change

FIGURE 6.1. Hydrologic cycle reservoirs. (a) Iceland Vatnajökull ice cap meltwater and runoff into North Atlantic Ocean, (b) evaporative haze over Lake Superior, and (c) rain and fog at Höfn, Iceland. Photos courtesy of Megan Jones.

SUMMARY

This chapter explores one of the most widely accepted and scientifically cited lines of evidence for global climate change, the 65 Ma-long composite stable oxygen isotope record derived from benthic foraminifera. In **Part**

Reconstructing Earth's Climate History: Inquiry-Based Exercises for Lab and Class,
First Edition. Kristen St John, R Mark Leckie, Kate Pound, Megan Jones and Lawrence Krissek.

NAME

6.1, you will familiarize yourself with some of the graphic elements of the oxygen isotope record and make some initial observations. In **Part 6.2**, a "primer" on stable isotope geochemistry, you will unravel the implications of the physical states of water and the movement of water through the hydrologic cycle in the distribution of the two most common stable isotopes of oxygen. In **Part 6.3**, you will examine the biogeochemical connections that allow scientists to extract such a detailed record of long-term oceanic, continental, and atmospheric change. In **Part 6.4**, you will integrate and apply your knowledge from Parts 6.1–6.3 by making a detailed examination of the Cenozoic benthic foraminiferal $\delta^{18}O$ record by **identifying** patterns and changes in trends and by **interpreting** what the patterns, trends and changes indicate, thereby recognizing some of the most significant climatic events that have occurred over the past 65 Ma.

The Benthic Foraminiferal Oxygen Isotope Record of Cenozoic Climate Change

Part 6.1. Introduction

One of the most valuable proxies of past climate change is the **benthic foraminiferal stable oxygen isotope record** (Figure 6.2). Fundamentals of chemistry, biology, geology, and physics are all important in understanding the nature of this type of data and how it is useful as a paleoclimate proxy. We will explore it in detail in this chapter and revisit it again in Chapters 7, 10, and 11 as it applies to particular climate patterns and events.

Start by becoming familiar with the data shown in Figure 6.2.

1 What are the variable and its units for the **x-axis**?

2 What are the variable and its units for the ***y*-axis**?

3 There are thousands of individual data points on this graph. Using a colored pencil, draw a line on the graph representing the mean values of the data as they change over time. Note this is not a straight line, but a curved line representing your best visual estimate of the running mean.

4 Make a list of your initial observations about changes displayed in the data (Figure 6.2).

NAME

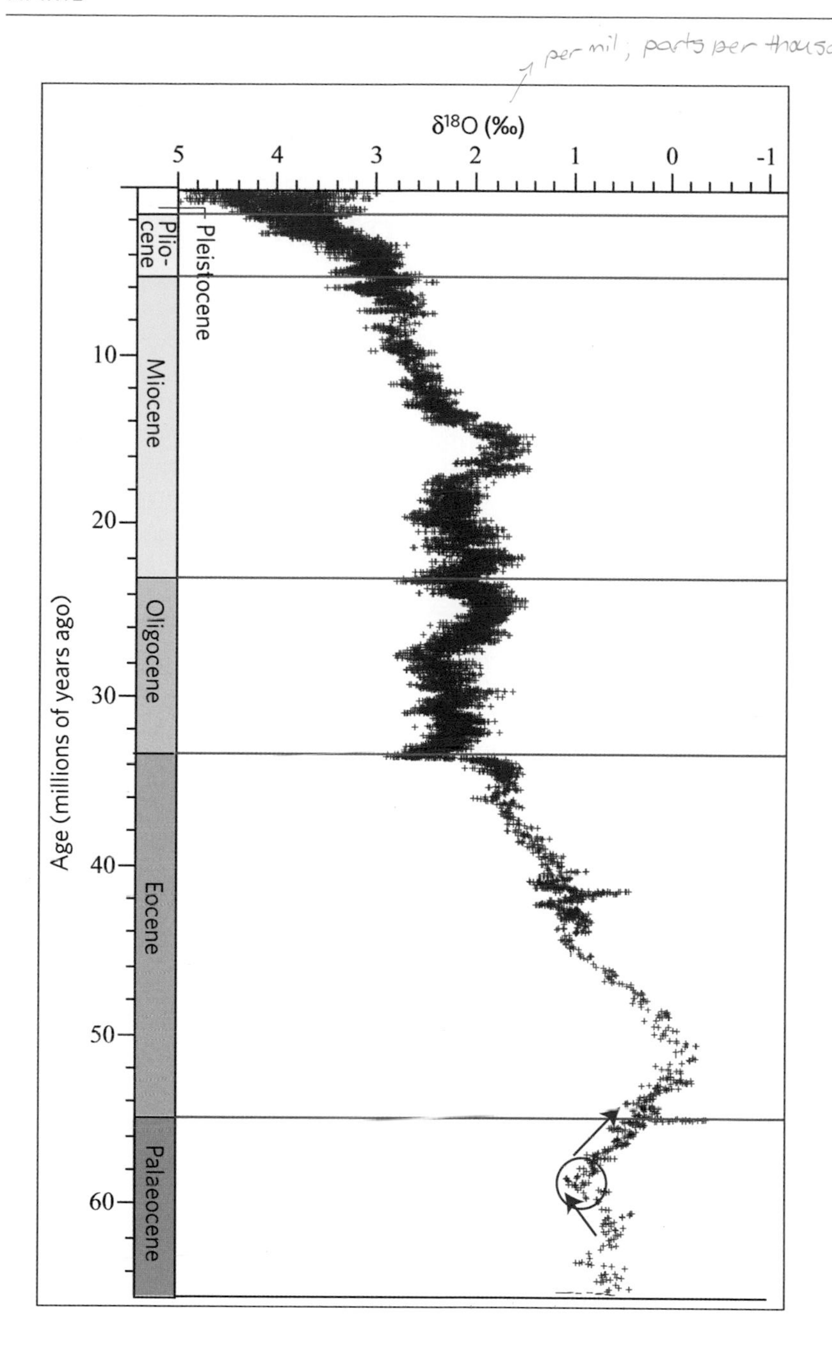

FIGURE 6.2. 65 million year **composite record** of marine benthic foraminiferal stable oxygen isotope values ($\delta^{18}O$) from Deep Sea Drilling Project (DSDP) and Ocean Drilling Program (ODP) sediment cores (modified from Zachos et al., 2008, which is an update of Zachos et al., 2001). Note that the blue arrows and the red circle on this diagram are annotations added specifically for work you will do in Part 6.4, and can be ignored for your work in Part 6.1.

NAME

The Benthic Foraminiferal Oxygen Isotope Record of Cenozoic Climate Change
Part 6.2. Stable Isotope Geochemistry

Atomic Nuclei

All atoms contain a nucleus (protons plus neutrons) surrounded by an electron cloud. The groups of blue and white spheres (Figure 6.3) represent generalizations of the **atomic nuclei of two isotopes** of the element oxygen.

1 How are the two nuclei similar and how are they different?

Similarities	Differences

2 Go to the link in the caption for Figure 6.3 to rotate (view all sides) of the atomic nuclei via animation.

(a) What do the blue spheres represent?

FIGURE 6.3. Generalization of the atomic nuclei of two different oxygen isotopes. From: http://earthobservatory.nasa.gov/Features/Paleoclimatology_OxygenBalance/

NAME

(b) What do the white spheres represent?

Neutrons

3 Based on your prior knowledge, what do we call nuclei that all have the same number of protons?

4 Based on your prior knowledge, what do we call nuclei that have the same number of protons, but have different numbers of neutrons?

5 Write a new caption for Figure 6.3. Use your answers from Questions 1–5 to help you construct an accurate, descriptive, and concise caption in the space below.

Physical Properties and Processes

6 How would water (H_2O) containing ^{16}O be similar to water that contains ^{18}O and how would it be different?

Similarities	Differences

NAME

HYDROLOGIC CYCLE AND ISOTOPIC FRACTIONATION

Water molecules change physical state (liquid, gas, solid) and move from reservoir to reservoir (e.g., ocean, atmosphere, glacial ice, Figure 6.1) over long distances via the **hydrologic cycle** (Figure 6.4).

The differences in mass between water molecules that contain ^{18}O and water molecules that contain ^{16}O are large enough for physical and biological processes to **fractionate,** or partially separate water containing ^{18}O from water containing ^{16}O. Processes that move water between reservoirs via the hydrologic cycle will result in **"isotopic fractionation"**. Such fractionation has been demonstrated by measuring the relative abundance of ^{18}O and ^{16}O in modern water samples from different reservoirs of the hydrologic cycle at different locations.

You will have the opportunity to work through examples of this fractionation effect in Questions 7–9.

7 Read the box above on isotopic fractionation.

(a) Which would tend to **evaporate** preferentially from a glass of water: a water molecule containing ^{16}O ($H_2{}^{16}O$) or a water molecule containing ^{18}O ($H_2{}^{18}O$)?

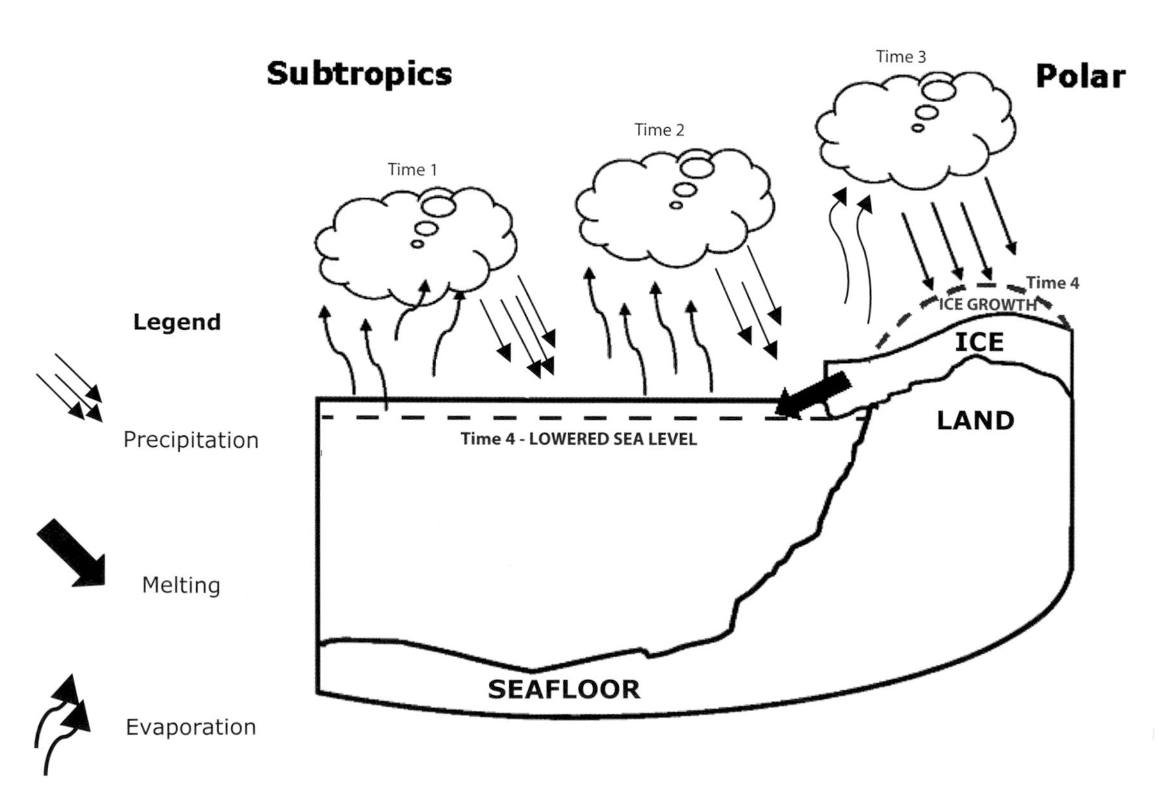

FIGURE 6.4. Schematic diagram showing subtropical to polar water movement between ocean, atmosphere, and continental ice sheet reservoirs under present day conditions (heavy black line) and under conditions of lowered sea level and increased ice sheet volume (dashed blue lines). Diagram drawn by Megan Jones.

NAME ______________________________

(b) Why?

____

____

____

____

____

____

Refer to Figure 6.4 as you complete Questions 8–10.

8 Select the correct oxygen isotopes for the multi-part scenario described below by **circling** them. (**Note**: annotating Figure 6.4 may help you reason through the scenarios and determine your answers.)

(a) When **water evaporates** from the ocean in the **tropics or subtropics**, the water vapor is **enriched** in (has more) ^{16}O ^{18}O relative to the original water. When the resulting **precipitation** falls, the remaining water vapor is **enriched** in ^{16}O ^{18}O.

(b) Water vapor is transferred from the tropics to the poles through the hydrologic cycle. When this water vapor condenses and **precipitates as snow on land in the polar regions** and later recrystallizes as ice, this ^{16}O ^{18}O enriched water becomes "locked up" in the ice. Because of this transfer, the **oceans** become relatively **depleted** in ^{16}O ^{18}O (i.e., have less ^{16}O ^{18}O than they did before evaporation removed the water vapor) and therefore are relatively **enriched** in ^{16}O ^{18}O.

(c) If the **precipitation falls as rain**, and is **NOT** "locked up" in ice, but **is returned to the ocean**, then the ocean does **NOT become enriched** in ^{16}O ^{18}O.

(d) Hence, during times of continental ice sheet formation (**glacial periods**) the oceans are **relatively enriched** in ^{16}O ^{18}O and during warmer times with reduced or no ice sheet formation (**inter-glacial periods**), the oceans are **relatively enriched** in ^{16}O ^{18}O.

9 Demonstrate your understanding of the isotopic fractionation effect on oxygen isotopes in the hydrologic cycle by filling in the blanks below with **"increase"** or **"decrease"** as appropriate.

(a) As the global volume of **ice sheets increases**, the relative abundance of ^{16}O in the ocean will decrease and the relative abundance of ^{18}O in the ocean will increase

(b) As the global volume of **ice sheets decreases**, the relative abundance of ^{16}O in the ocean will increase and the relative abundance of ^{18}O in the ocean will decrease.

NAME

The Benthic Foraminiferal Oxygen Isotope Record of Cenozoic Climate Change
Part 6.3. A Biogeochemical Proxy

Application of Important Marine Microfossils

One of the main types of microscopic life in the ocean that has a mineralized shell of **calcium carbonate ($CaCO_3$)** are the **foraminifera** (Chapters 2 and 3).

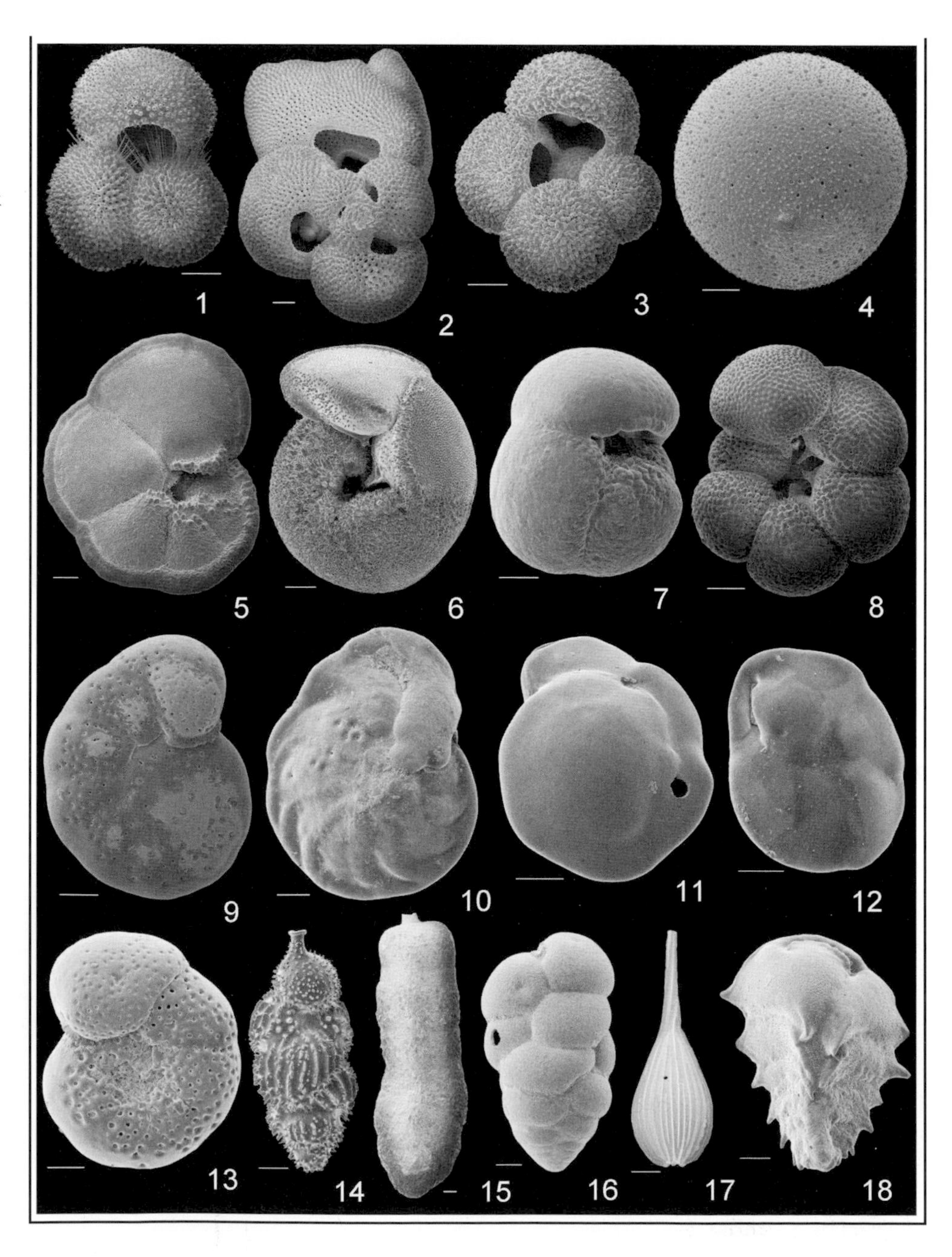

FIGURE 6.5. Representative mix of microfossil shells of benthic and planktic foraminifera. Scale bar = 100 μm (=0.1 mm). Scanning electron microscope images by Mark Leckie.

NAME ____________________

1 Microfossil shells of some representative and important foraminiferal species are shown in Figure 6.5. The top two rows show **planktic** species and the bottom two rows show **benthic** species. Planktic foraminiferal species live in the near surface ocean, (i.e., upper 200 m of water column). In contrast, benthic foraminiferal species live on the seafloor, either just above (**epifaunal**) or within (**infaunal**) the sediments.

Make a list of your observations about foraminifera based on the species shown in Figure 6.5.

____

____

____

A BIOGEOCHEMICAL PALEOCLIMATE PROXY

Both benthic and planktic foraminifera (Figure 6.5) construct (precipitate from seawater) their $CaCO_3$ shells either in isotopic equilibrium or in a consistent isotopic disequilibria with the seawater in which they live (Figure 6.6). Such disequilibria "vital effect" maybe due to the influence of respiratory CO_2, but appear unrelated to environmental conditions such as depth or temperature (Faure, 1986), and can be corrected for. In the case of data in Figure 6.2, the genus-specific vital effects were corrected for by adjusting the $\delta^{18}O$ values by +0.64‰ for *Cibicidoides* and +0.4‰ for *Nuttalides* (Barker et al., 1999).

The isotopic composition of seawater (see Part 6.2) is an important factor affecting the isotopic composition of the shells of the foraminifera, but so is the temperature of the seawater in which they live. This is because the isotopic fractionation is temperature dependent. Thus, the seawater temperature creates an additional fractionation effect that influences the isotopic composition of the shell. As a result of this **temperature fractionation** effect, in isotopically identical waters, calcite ($CaCO_3$) shells precipitated in cooler water temperatures are enriched with ^{18}O relative to ^{16}O, whereas calcite precipitated in warmer water temperatures is depleted of ^{18}O relative to ^{16}O. **Therefore, both the isotopic composition of seawater and the temperature of seawater affect the isotopic composition of the foraminiferal shells.**

Based on these principles, the foraminifera (or other marine organisms that form calcareous shells) **serve as proxies for the chemical and isotopic composition of the waters in which they live**. Their shells can accumulate on the seafloor after death (Figure 6.7), over time becoming part of the sedimentary record. Thus, we can use the isotopic signature in their shells to reconstruct the oceanographic and climatic history through time.

Read the box above on biogeochemical paleoclimate proxies and examine Figures 6.2, and 6.5–6.7 to answer Questions 2–5.

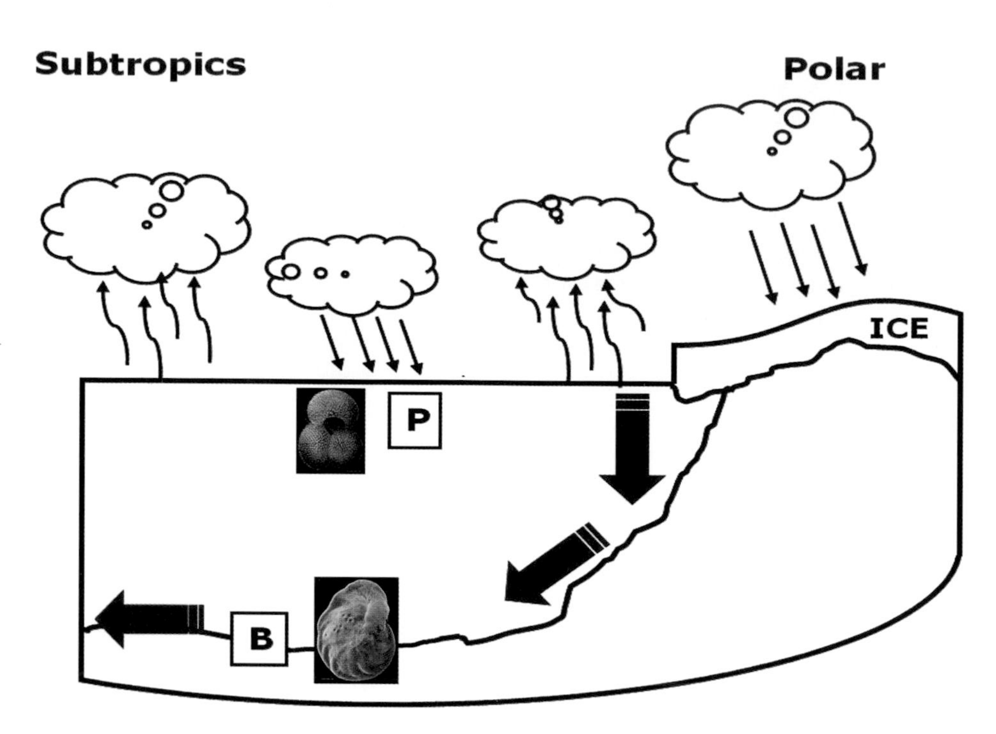

FIGURE 6.6. Schematic diagram from subtropic to polar latitudes showing where benthic (B) and planktic (P) foraminifers live and the location of deep and bottom water formation under today's sea level and ice volume conditions. **Heavy arrows** indicate locations of deep and bottom water formation and flow. The fossil planktic foraminifera shown is the species *Globigerinoides ruber*. The fossil benthic foraminifera shown is the species *Planulina wuellerstorfi*. Diagram drawn by Megan Jones, microfossil images from Mark Leckie.

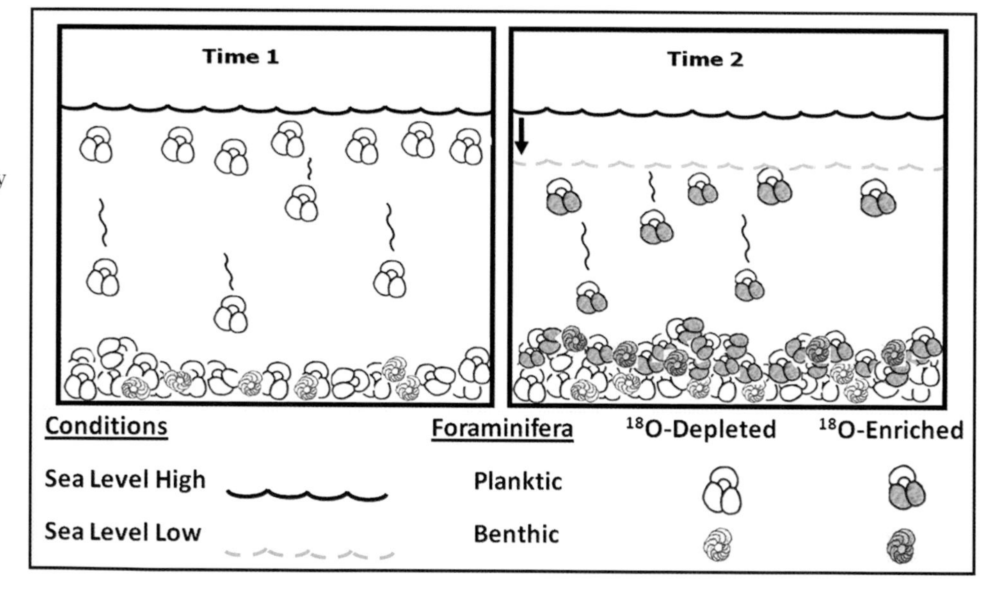

FIGURE 6.7. Schematic diagram illustrating the accumulation of benthic and planktic foraminifera shells on the seafloor. Note that the size of the foraminifera shells is highly exaggerated in comparison to the water depth in these diagrams. Diagram drawn by Megan Jones.

2 Foraminifera use Ca^{2+} and CO_3^{2-} ions dissolved in seawater to make their $CaCO_3$ shells. Think about the composition of the microfossil shells of foraminifera. How might benthic foraminifera be related to the stable isotope record you examined in Figure 6.2?

NAME

3 How would the stable oxygen isotope composition of foraminifera shells change if the **local temperature** of the seawater in which they live became colder?

4 How would the stable oxygen isotope composition of foraminifera shells change if there was an increase in **global ice volume**?

5 An increase in global ice volume and a decrease in seawater temperature are both indications of **climatic cooling**. Write a statement predicting how climatic cooling would affect stable oxygen isotopes of marine biogenic calcite.

MEASURING STABLE ISOTOPES

The relative abundance of stable oxygen isotopes in foraminifera shells is measured using a **mass spectrometer**. Typically, the calcite ($CaCO_3$) is dissolved in acid to liberate CO_2 gas. The CO_2 gas is ionized in a vacuum and sent through a strong magnetic field in a curved tube. The trajectories of the ions bend (i.e., are deflected) depending on their masses; lighter ions are deflected more than heavier ions. By putting sensors at different

NAME

positions, or by changing the strength of the magnetic field, the relative abundance of ions with different masses (i.e., different isotopes) can be measured. The typical way of recording the measured isotopic mass differences is in a ratio format.

The measured ratios of **$^{18}O/^{16}O$** in a sample are compared to the isotopic ratios in a known standard, as shown in the equation below which calculates the difference between the ratios in the sample and the standard. The use of a common standard in all geochemical labs around the world helps ensure consistency among the data produced from different mass spectrometers (Wright, 1999).

$$\delta^{18}O \text{ (in ‰ or per mil)} = \left(\frac{(^{18}O/^{16}O)_{sample} - (^{18}O/^{16}O)_{standard}}{(^{18}O/^{16}O)_{standard}} \right) \times 1000$$

These "delta" (δ) values are reported in units of **parts per thousand** (‰), or "**per mil**" (in contrast to units of parts per hundred, or percent).

A positive or increasing value per mil (or ‰) indicates that more ^{18}O than ^{16}O is present in the marine microfossil sample than is present in the standard. This is commonly referred to as being a "**high or increase in**" $\delta^{18}O$ value and is interpreted as reflecting a cooler climate. In contrast, a negative or decreasing value per mil (or ‰) indicates less ^{18}O than ^{16}O is present in the marine microfossil sample than is present in the standard. This is commonly referred to as being a "**low or decrease in**" $\delta^{18}O$ value and is interpreted as reflecting a warmer climate.

Read about how scientists measure stable oxygen isotopes in the box (above) and complete Questions 6–8:

6 What would the **$\delta^{18}O$ value** be if the ratio of $^{18}O/^{16}O$ in a sample is the same as the ratio of $^{18}O/^{16}O$ in the standard?

zero

7 Would the $\delta^{18}O$ value be a **positive** or **negative number** if the ratio of $^{18}O/^{16}O$ in a sample is **greater** than the ratio of $^{18}O/^{16}O$ in the standard?

8 Would the $\delta^{18}O$ value be considered **high** or **low** if the ratio of $^{18}O/^{16}O$ in a sample is **less** than the ratio of $^{18}O/^{16}O$ in the standard?

NAME ________________________________

The Benthic Foraminiferal Oxygen Isotope Record of Cenozoic Climate Change

Part 6.4. Patterns, Trends and Implications for Cenozoic Climate

With background knowledge gained from Parts 6.1–6.3, you will now focus on identifying patterns and the trends in the Cenozoic benthic foraminiferal oxygen isotope record (Figure 6.2).

1 Working independently, examine and annotate the oxygen isotope record (Figure 6.2) as follows: **Starting from the bottom** of the figure, work your way up and mark any trends in the data with arrows (see example in Figure 6.2: blue arrows = trend); mark the position of any changes in trends with a circle (see example in Figure 6.2: red circle = change in trend).

2 In Table 6.1 (**Column 1**), record the **observations** of the trends and changes in trends in the $\delta^{18}O$ record that you recognized as you completed question 1. Leave Column 2 blank for now! Always, **start from the bottom and work your way up epoch by epoch**. Use descriptive terminology (e.g., **high, low, constant, abrupt increase, gradual decrease, etc.**) and realize that an epoch may be marked by more than one trend or change in trend.

3 (a) Share/compare your written **observations** for each epoch with the person next to you or those at your lab group.
(b) How similar or different are your **observations** for each epoch compared with those of your classmates? Explain with an example.

____

____

____

____

4 Working independently and referring as needed to Parts 6.2 and 6.3, **interpret** your observations (from Column 1 of Table 6.1) of the $\delta^{18}O$ record (Figure 6.2). Record your interpretations on the table below. As you did for Question 2, start at the bottom of Table 6.1, read your observations for one epoch, then write your interpretations for that epoch in **Column 2**. Continue working your way up the table epoch by epoch. Use descriptive terminology (e.g., **abrupt warming, gradual increase in ice volume, etc.**).

5 (a) Share/compare your written **interpretations** for each epoch with the person next to you or those at your lab group.

NAME

TABLE 6.1. Observations and Interpretations of the Cenozoic Benthic Foraminiferal Stable Oxygen Isotope Record.

Epoch	Column 1 Observations of the Cenozoic Benthic Foraminiferal $\delta^{18}O$ record	Column 2 Interpretations of the Cenozoic Benthic Foraminiferal $\delta^{18}O$ record
Pleistocene		
Pliocene		
Miocene		
Oligocene		
Eocene		
Paleocene		

NAME

(b) How similar or different are your **interpretations** for each epoch compared with those of your classmates? Explain with an example.

6 Recall (Part 6.3) that changes in global ice volume, as well as changes in local seawater temperature, influence benthic foraminifera stable oxygen isotope ($\delta^{18}O$) values. However, if no continental ice sheets were present, the variation in $\delta^{18}O$ would solely reflect changes in temperature.

(a) Would we know from the $\delta^{18}O$ data **alone** whether continental ice sheets were present in the past? Explain.

(b) What other sources of data might you use to determine whether continental ice sheets were present in the past?

(c) Look back at Chapter 5, Figure 5.9 and the related discussion on Greenhouse and Icehouse times. Based on this information, when did Cenozoic ice sheets first form?

(d) What does this imply about the Cenozoic changes in $\delta^{18}O$ (Figure 6.2) prior to this time? What were the early Cenozoic $\delta^{18}O$ changes driven by?

NAME ____________________

7 The epoch boundaries on the geologic timescale were largely defined over 200 years ago based on the fossil record of relatively large fossils (e.g., clams, snails) deposited in shallow marine sediments that were subsequently exposed on land. How do the changes in **benthic foraminiferal $\delta^{18}O$** compare to the already defined epoch boundaries?

____

____

____

____

8 Summarize your thoughts about, and understanding of, the isotopic evidence that indicates how the Earth's climate has changed over the past 65 Ma.

____

____

____

____

____

____

____

NAME

References

Barker, P., Barrett, P., Cooper, A. and Huybrechts, P. 1999, Antarctic glacial history from numerical models and continental margin sediments, Palaeogeography, Palaeoclimatology, Palaeoecology, **150**, pp. 247–267.

Faure, G., 1986, Principles of Isotope Geology, Wiley, New York, **589**, p. ISBN 0–471–86412–9.

Wright, J.D., 1999, Global climate change in marine stable isotope records. In Quaternary Geochronology: Applications in Quaternary Geology and Paleoseismology. Noller, J.S., et al. (eds), U.S. Nuclear Regulatory Commission, NUREG/CR 5562, 2–671–682.

Zachos, J., et al., 2001, Trends, rhythms, and aberrations in global climate change 65 Ma to present. Science, **292**, 686–93.

Zachos, J., et al., 2008. An early Cenozoic perspective on greenhouse warming and carbon-cycle dymanics. Nature, **451**, 279–83.

Chapter 7 Scientific Drilling in the Arctic Ocean: A Lesson on the Nature of Science

FIGURE 7.1. Integrated Ocean Drilling Program (IODP) Arctic Coring Expedition at the North Pole in 2004. Photo courtesy of MARUM; http://www.marum.de/Alexius_Wuelbers.html.

SUMMARY

This set of activities investigates the **scientific motivations, logistical challenges, and recent history of Arctic Ocean scientific drilling**. In **Part 7.1**, IPCC **climate models** are introduced as a complimentary method of studying climate change which possesses its own benefits and challenges. In **Part 7.1**, you will address the question "Why drill there?" from a modeling

Reconstructing Earth's Climate History: Inquiry-Based Exercises for Lab and Class, First Edition. Kristen St John, R Mark Leckie, Kate Pound, Megan Jones and Lawrence Krissek.

perspective. In **Part 7.2**, you will explore the technological and logistical question of "How to recover deep cores from the Arctic seafloor?" In **Part 7.3**, you will evaluate the spatial and temporal records of past Arctic seafloor coring expeditions and build a scientific rationale for drilling deep into the Arctic Ocean seafloor to study regional and global climate change. In **Part 7.4**, you will identify some of the key findings of the Arctic Coring Expedition (ACEX), as well as the unexpected challenges encountered in this core.

Scientific Drilling in the Arctic Ocean: A Lesson on the Nature of Science

Part 7.1. Climate Models and Regional Climate Change

Introduction

One of the ways scientists work to understand how the Earth's complex climate system works is to reconstruct Earth's climate history based on proxies of temperature, precipitation, and so on. Another approach is to develop physical **models** (see box below) of the climate system that can be refined and manipulated to explore different "what if" scenarios. Such model scenarios can project forward or backward in time, depending on the goals of the study. Common to all models is the need to simulate as closely as possible the real conditions and interactions of the natural world.

GLOBAL CIRCULATION MODELS

Global circulation models (also called general circulation models, **GCMs**) are used to study global climate dynamics and change. These GCMs are computer models based on the current scientific understanding of the physics, chemistry, biology, oceanography and geology of the different components of Earth's complex climate system and their interactions. The model output depends on several factors. These include, but are not limited to: (1) the specific input of boundary conditions, (e.g., the historical record of temperature and CO_2, the position of continents, the location direction of ocean currents, the topography of the land), (2) the understanding of scientific theory (e.g., radiation, atmospheric, and oceanic circulation), (3) software programming issues, including how the model links (or "couples") different components of the Earth's climate system (e.g., interactions between the ocean and the atmosphere), (4) the model's resolution, which depends on the three-dimensional grid size of the many

(Continued)

NAME

cells that the Earth's atmosphere and ocean is broken into by the model, and (5) the strategies by which the model handles numerical problems with the poles (i.e., the distance between longitude grid spacing goes to zero at poles). Future climate projections also must make assumptions about human factors that can influence climate, such as the rate of greenhouse gas emissions (CO_2, CH_4), the rate of population growth, the rate of economic growth, and the introduction of new and more efficient technologies.

Figures 7.2 and 7.3 (below) show temperature projections based on GCM simulations, reported in the Intergovernmental Panel on Climate Change (IPCC) 2007 Report (http://www.ipcc.ch/).

1 Examine Figure 7.2 which shows projected **changes** in global average temperature (not the global average temperature itself). Notice that model outputs for scenarios B1, A1B, A2 are shown both in the Figure 7.2 graph and in the bars to the right of the graph. Model outputs for A1T, B2, and A1F1 are only shown on the bars to the right of the graph. What is the range of the projected global **average temperature increases** for the year 2100 based on the IPCC 2007 models?

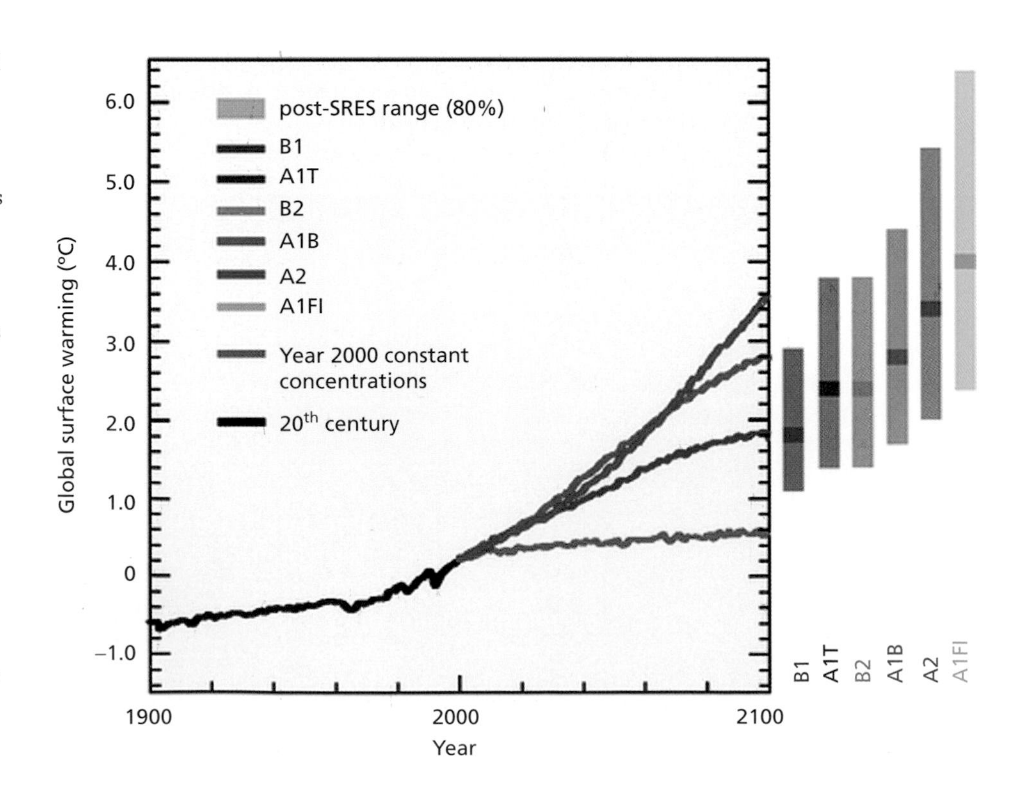

FIGURE 7.2. Projected global averages of Earth's surface warming based on global circulation models. Scenarios A2, A1B, and B1 take into account emissions of short-lived greenhouse gases and aerosols. The pink line projects temperature increases if atmospheric concentrations of greenhouse gases and aerosols are held constant at year 2000 values. The bars on the right of the figure indicate the best estimate (solid line within each bar) and the likely range assessed for six scenarios at the end of the 21st century. All temperature increases are relative to the period 1980–1999. Graph courtesy of IPCC.

NAME

2 Based on the general introductory information provided (or more detailed readings supplied by your instructor), what are some of the sources of uncertainty in these global temperature models?

3 In your opinion, how does the model uncertainty affect (a) interpretation of the model results, and (b) potential action (e.g., societal, political, economic) based on the model results?

4 Based on the information in the preceding box about GCMs, what are some ways that model uncertainty could be reduced?

5 (a) Examine Figure 7.3, which shows the projected mean annual surface temperature changes at the end of this century. How does the predicted warming of continents compare to the predicted warming of the surface ocean?

NAME

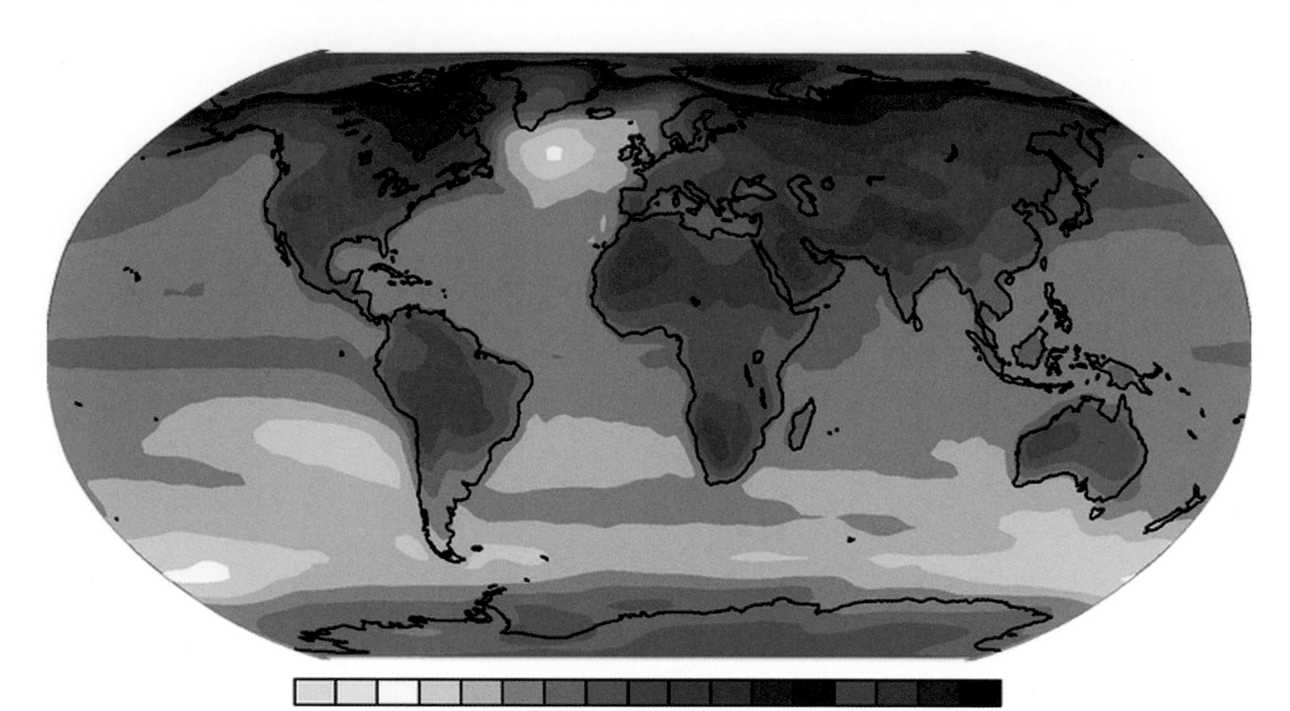

FIGURE 7.3. Projected mean annual surface temperature changes for the late 21st century (2090–2099) based on the A1B SRES general circulation computer model scenario. (Note: A1B SRES model results are also represented by the green line on the graph in Figure 7.2). Temperatures are relative to the period 1980–1999. Map courtesy of IPCC.

(b) What could account for the difference in the predicted warming of continents compared with that of the surface ocean?

6 (a) In what geographic region is the warming predicted to be the greatest?

NAME ____

(b) Compare results from model scenario A1B in Figures 7.2 and 7.3 to determine how much warmer this **region** (from Question 6a) is predicted to be relative to the predicted **global** average warming by the year 2100.

(c) List your ideas for why the Arctic might experience the greatest warming.

7 Do these model results provide any scientific motivation to investigate the history of climate change in the Arctic? Explain.

NAME ______________________________

Scientific Drilling in the Arctic Ocean: A Lesson on the Nature of Science

Part 7.2. Arctic Drilling Challenges and Solutions

1 Make a list of what you think might be unique challenges to drilling into the seafloor in the Arctic Ocean to recover sediment/rock cores.

____

____

____

____

2 How might these challenges be overcome?

____

____

____

____

____

3 Watch a 5 minute movie on the technical approach to drilling in the Arctic by going to: http://recordings.wun.ac.uk/conf/nwo/oceandrilling2006 then selecting the movie "Drilling the Arctic". How was the challenge of staying on-site (or "station") long enough to drill a deep core solved?

____

____

____

____

____

____

____

NAME

Scientific Drilling in the Arctic Ocean: A Lesson on the Nature of Science

Part 7.3. Need for Scientific Drilling

More than 700 piston cores have been collected in the Arctic Ocean, 13 of which are included on Figure 7.4 and Table 7.1 below. These selected sites are representative of the general geographic distribution of Arctic core sites, as well as the sub-seafloor depths and ages of the sediment cores recovered.

TABLE 7.1. Selected Arctic Core Sites (approximately 1988–2004)[1]

Site-Core Identification	Geographic Location	Latitude & Longitude	Water Depth (m)	Max Coring Depth Below Seafloor (m)	Max Age of Sediment Core
NP26-5	Mendeleev Ridge	78.58°N 178.09°W	1435	2.10	approximately 650,000 yrs; marine isotope stage 16
PI-88-AR-P5	Northwind Ridge, Canada Basin	74.37°N 157.53°W	1089	4.76	approximately 1 million yrs; just below the Jaramillo polarity event
PI-88-AR-P7	Northwind Ridge, Canada Basin	74.38°N 157.23°W	3513	5.36	approximately 500,000 yrs; marine isotope stage 13
96/12-1pc	Lomonosov Ridge	87.06°N 144.46°E	1003	7.22	approximately 900,000 yrs; marine isotope stage 22
PS2185-6	Lomonosov Ridge	87°32N 144°55E	1052	7.68	5.1 million years; Gilbert Chron
PS2189-5	Lomonosov Ridge	88.48°N 144.1°W	1001	10.35	mid-Pleistocene[2]
PS2167-1	Gakkel Ridge	86.57°N 59.1°W	4434	6.40	
PS2175-5	Amundsen Basin	87.40°N 104.05°W	4313	16.92	
PS2190-1	Amundsen Basin/ North Pole	90°N	4275	4.27	
PS2159-6	Nansen Basin	83.60°N 30.17°W	4010	2.18	
PS2200-5	Morris Jessup Rise	85.19°N 14°E	1073	7.70	
PS2213-6	Yermak Plateau	80.28°N 8.3°W	853	13.09	
PS2180-2	Makarov Basin	87.39°N 56.58°W	3991	12.96	mid-Pleistocene[2]
PS2138-1	Near Svalbard	81.32°N 30.35°E	995	6.29	approximately 150,000 yrs; marine isotope stage 6
PL-380	Alpha Ridge	84.37°N 128.28°W	2401	3.45	mid-Pleistocene[2] (with possible older clasts at base)

[1] Data in Table 7.1 from: Clark et al., 2000; Fütterer, 1992; Jakobsson et al., 2000, 2001; Matthiessen et al., 2001; Polyak et al., 2004; Poore, et al., 1994; Spielhagen et al., 1997.
[2] Maximum age was difficult to constrain due to limited biostratigraphic datums. An estimate of approximately 1 million years is a reasonable assumption for the maximum age represented by these cores.

NAME

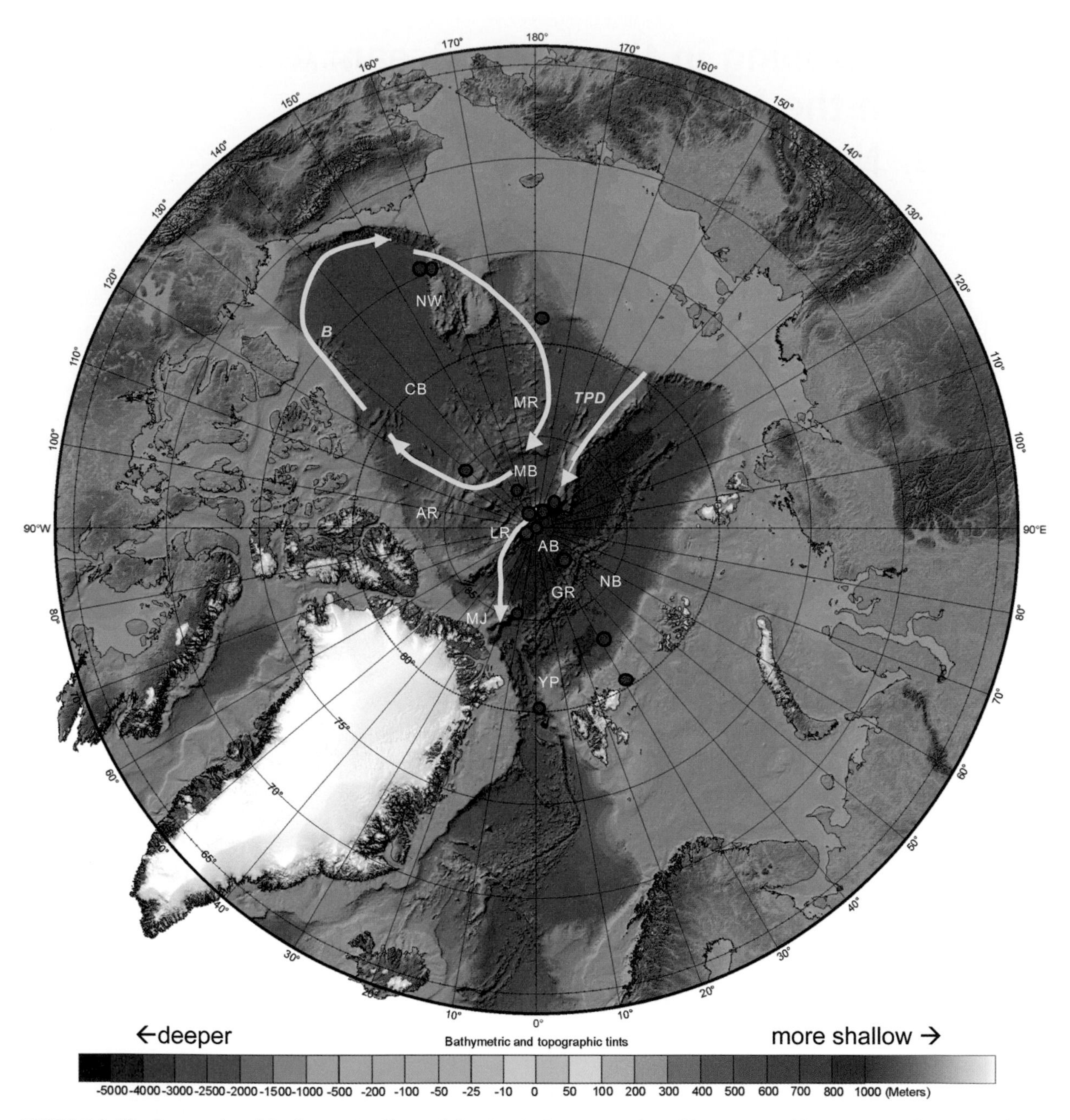

FIGURE 7.4. The International Bathymetric Chart of the Arctic Ocean, produced by a team of investigators from Canada, Denmark, Germany, Iceland, Norway, Russia, Sweden, and the USA (see Jakobsson et al., 2008). This is a physiographic map of the Arctic Ocean. Bathymetric features (e.g., ridges and basins) of the Arctic seafloor are shown and water depths are indicated by shading (blue deeper, brown more shallow). AB = Amundsen Basin, AR = Alpha Ridge, CB = Canada Basin, GR = Gakkel Ridge, LR = Lomonosov Ridge, MB = Makarov Basin, MJ = Morris Jessup Rise, MR = Mendeleev Ridge, NB = Nansen Basin, NW = Northwind Ridge, YP = Yermak Plateau. Yellow arrows indicate the location of modern major surface currents of the Beaufort Gyre (BG) and transpolar drift (TPD). Map modified from: http://geology.com/world/arctic-ocean-map.shtml. Core locations (Table 7.1) are indicated by red circles.

NAME

1 From the data shown in Table 7.1 and Figure 7.4, **describe the distribution of the coring locations** in terms of geography, water depth, and age, by answering (a)–(c) below.
(a) Are cores mostly on bathymetric highs, mostly in basins, or distributed between these two settings?

(b) What is the maximum sub-seafloor depth reached by these cores?

(c) What is the maximum age of the sediment in these cores?

2 Use data in Table 7.1 to **calculate** the **percentage of Cenozoic time** that the sediment records of these Arctic cores represents. Note the Cenozoic Era includes the last 65.5 million years. Show your work.

3 Based on what you have learned in this exercise so far, do you think that the Cenozoic geologic history (including climate history) of the Arctic Ocean was well understood from the sediments recovered by piston cores? Explain.

INTEGRATED OCEAN DRILLING PROGRAM ARCTIC CORING EXPEDITION (ACEX)

In 2004 a team of scientists and technicians drilled five holes within 16 km of each other along the crest of the Lomonosov Ridge (Figure 7.5) in the central Arctic Ocean. The holes were distributed between 87°N and 88°N in water depths between 1100 m and 1200 m and were all in international

NAME

waters. Some of the pieces of information that helped the science team select these particular locations were seismic reflection profiles (see Figure 7.5 inset; like "sonograms" of the subsurface seafloor) from seismic survey cruises that suggested the sedimentary sequence along this part of the ridge was very thick. It was estimated that there were 450 m of sediments overlying the harder basement rock in this area.

It was predicted that this 450 m sedimentary sequence represented about **56 million years** of history. This estimate is based on the previously known sedimentation rate on the Lomonosov Ridge of 7.22 m/0.9 myr = 8 m/myr (calculated from Table 7.1) and assumes a constant sedimentation rate for the entire 450 m thickness of sediment inferred from the seismic profile.

4 Read the information about the IODP Arctic Coring Expedition in the box above. Answer (a) and (b) below.

(a) Calculate the percentage of Cenozoic time that the inferred sedimentary sequence on the Lomonosov Ridge would represent. Show your work.

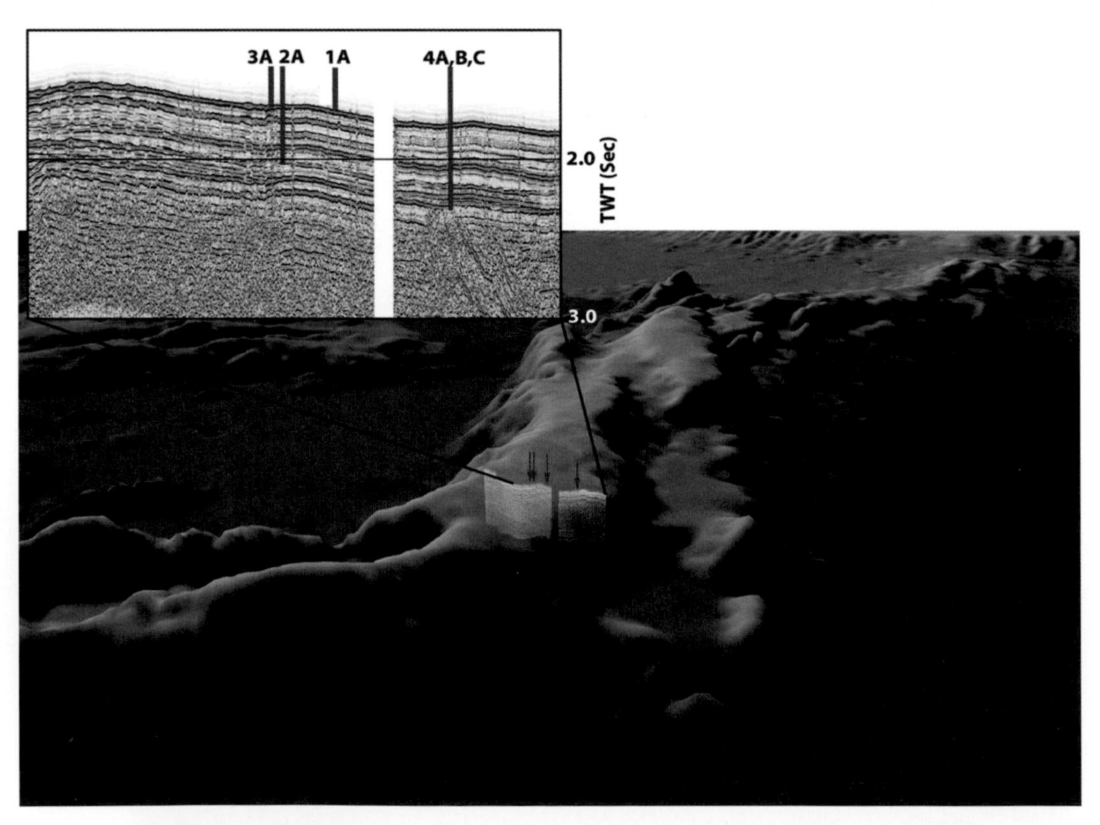

FIGURE 7.5. **Map and cross sections showing the location of the holes cored on IODP Expedition 302 on the** Lomonosov Ridge. Reflection seismic cross-sections of Lomonosov Ridge (inset figure). From Jakobsson et al., 2006.

NAME ______________________________

(b) Compare this to your answer to Question 2. Would this be a significant improvement in Cenozoic sediment recovery from the Arctic? Explain.

5 Propose three scientific questions that the drilling and recovery of a continuous sedimentary sequence from today back to 56 million years ago from the Arctic could potentially answer. For your reference, a chart which includes 65-myr composite benthic oxygen and carbon isotope records and reference to the major tectonic, climatic, and biotic events of the Cenozoic, is shown on the following page (Figure 7.6; Zachos et al., 2001). Alternatively, you may wish to examine the updated marine oxygen-isotope record for the Cenozoic in Chapter 6 (Figure 6.2; Zachos et al., 2008).

NAME

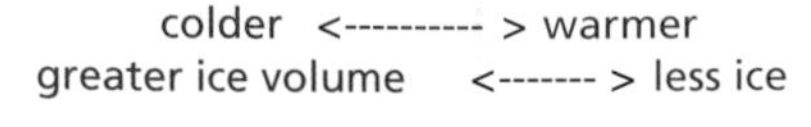

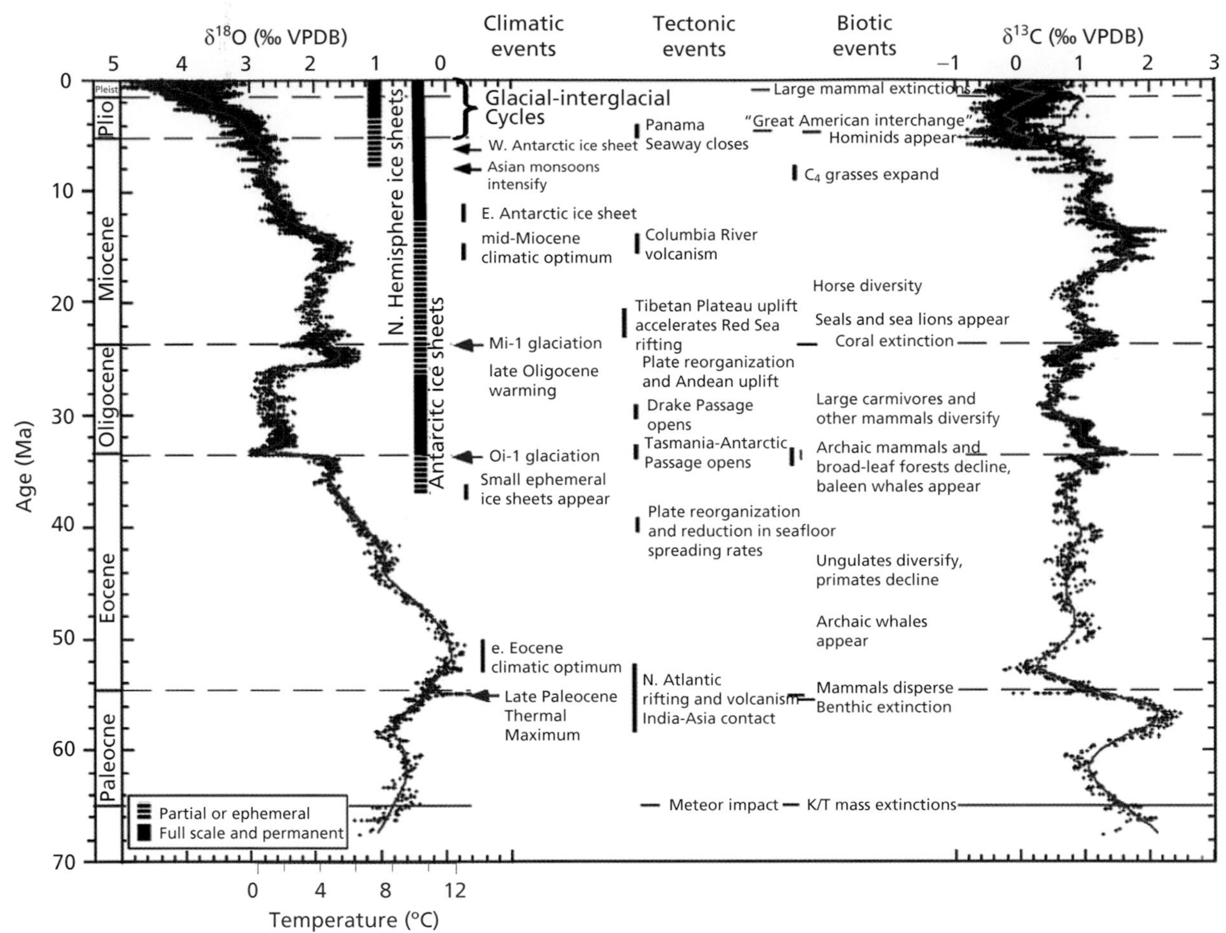

FIGURE 7.6. 65 million year composite record of benthic foraminifer stable isotope values ($\delta^{18}O$ and $\delta^{13}C$) from Deep Sea Drilling Project (DSDP) and Ocean Drilling Program (ODP) sediment cores, and a summary of major climatic, tectonic, and biotic events during the Cenozoic. Modified from Zachos et al., 2001.

NAME

Scientific Drilling in the Arctic Ocean: A Lesson on the Nature of Science

Part 7.4. Results of the Arctic Drilling Expedition

The Arctic Coring Expedition (ACEX; IODP Expedition 302) was considered "transformative" in the more than 40 year history of scientific ocean drilling, because of both the logistical challenges it overcame to recover a long sedimentary sequence from the central Arctic Ocean and the scientific results that had an impact on our perceptions of Cenozoic climate. Some of these scientific results are embedded in other chapters of this book (see Chapter 8, Figure 8.4; Chapter 9, Figures 9.20 & 9.21; Chapter 14, Figures 14.6 & 14.7). In this investigation, you will read a synthesis paper on the Arctic Coring Expedition to identify some of the key findings.

1 Below is the abstract from a synthesis paper summarizing the results of the Arctic Coring Expedition. **Read** this abstract. Use your library resources to find and read the full peer-reviewed article that this abstract came from: **Backman and Moran, 2009**. Expanding the Cenozoic paleoceanographic record in the Central Arctic Ocean: IODP Expedition 302 Synthesis, *Central European Journal of Geosciences*, v. 1 (2), p. 157–175; DOI: 10.2478/v10085-009-0015-6.

ABSTRACT FROM BACKMAN AND MORAN (2009)

The Arctic Coring Expedition (ACEX) proved to be one of the most transformational missions in almost 40 years of scientific ocean drilling. ACEX recovered the first Cenozoic sedimentary sequence from the Arctic Ocean and extended earlier piston core records from approximately 1.5 Ma back to approximately 56 Ma. The results have had a major impact in paleoceanography even though the recovered sediments represent only 29% of Cenozoic time. The missing time intervals were primarily the result of two unexpected hiatuses. This important Cenozoic paleoceanographic record was reconstructed from a total of 339 m of sediments. The wide range of analyses conducted on the recovered material, along with studies that integrated regional tectonics and geophysical data, produced surprising results including high Arctic Ocean surface water temperatures and a hydrologically active climate during the Paleocene Eocene Thermal Maximum (PETM), the occurrence of a fresher water Arctic in the Eocene, ice-rafted debris as old as middle Eocene, a middle Eocene environment rife with organic carbon, and ventilation of the Arctic Ocean to the North Atlantic through the Fram Strait near the early-middle Miocene boundary. Taken together, these results have transformed our view of the Cenozoic Arctic Ocean and its role in the Earth climate system.

NAME ______________________________

2 Based on your reading of Backman and Moran (2009), **list** at least three key results from the Arctic Coring Expedition. Briefly **explain** how each of these results has an impact on our understanding of Cenozoic climate. You may also wish to refer back to Figure 7.6, which is a 2001 summary of Cenozoic climatic, tectonic, and biotic events (Zachos et al., 2001).

____

____

____

____

____

____

____

____

____

____

3 The reconstructed paleogeographic setting of the Arctic Ocean in the early Cenozoic (Figure 7.7) shares some similarities with the geography of the Arctic Ocean today (Figure 7.4), but also includes some important differences. List two similarities and two differences between the paleogeographic settings of the Arctic in the Early Cenozoic compared to today.

Similarities	Differences

4 The ACEX age-depth model is shown in Figure 7.8. Lithologic units are indicated by the column on the right with numbers (e.g., 1/2 = unit 1 subunit 2). What reasons are there for including more than one type of **age control data** (e.g., paleomagnetic stratigraphy, dinocyst biostratigraphy) to construct this model?

____

NAME

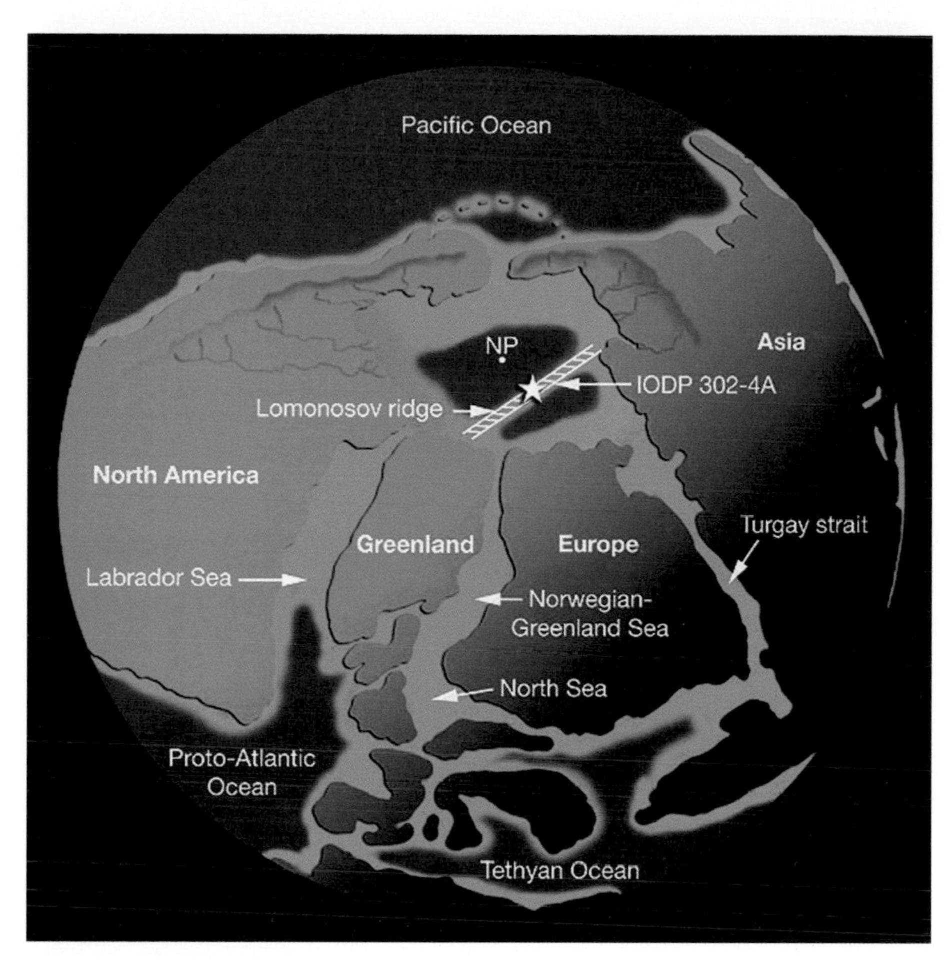

FIGURE 7.7. Paleogeographic reconstruction of the early Cenozoic **Arctic Basin** and Northern Hemisphere. Note that the Labrador Sea and Norwegian-Greenland Sea had just begun to open and Greenland was a large continental fragment between North America and northwest Europe. Location of Integrated Ocean Drilling Program (IODP) Expedition 302, Hole 4A, on the **Lomonosov Ridge**, a fragment of continental crust that rifted from the Eurasian continental margin and is now situated in the central Arctic Basin. NP = North Pole. From Sluijs et al., 2006.

5 Calculate the sedimentation rates for (a) lithologic unit 1/3 and (b) lithologic unit 1/4, in meters per million years (m/myr). Show your work below and label the age model (Figure 7.8) with these two sedimentation rates.

(a)____

NAME ______________________________

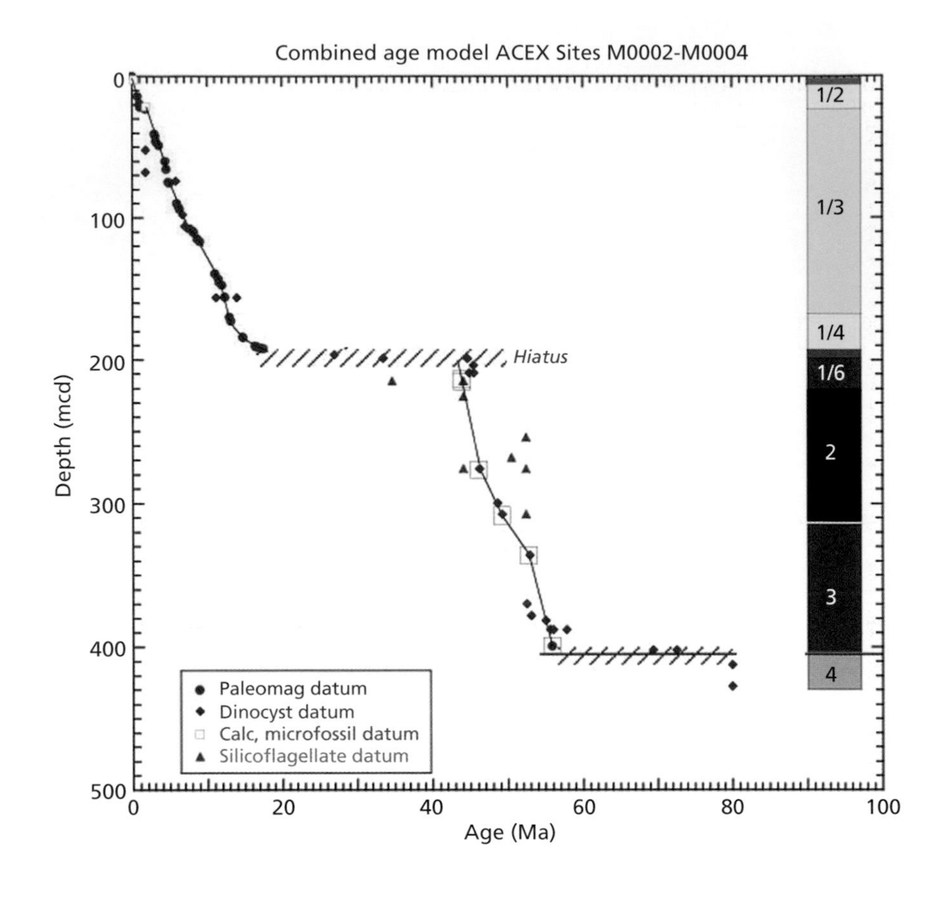

FIGURE 7.8. ACEX age-depth model. Depth is in meters composite depth (mcd); age is in millions of years (Ma; from Backman et al., 2006). For a more detailed ACEX age-depth model see Figure 4 in Backman et al. 2006.

____

____

(b)______________________________

____

6 A major hiatus occurs within lithologic unit 1/5 (Figure 7.8); this hiatus was unexpected based on the pre-drilling seismic data (Figure7.5).

(a) Based on the data in Figure 7.7, how much time does the hiatus in unit 1/5 represent?

____

____

(b) How does this hiatus impact the efforts to reconstruct Cenozoic paleoclimates from the ACEX sedimentary record?

____

NAME

(c) In the future, how might we obtain a sediment record that spans this "missing" timeframe?

NAME

References

Backman, J. and Moran, K., 2009, Expanding the Cenozoic paleoceanographic record in the Central Arctic Ocean: IODP Expedition 302 Synthesis. Central European Journal of Geosciences, **1** (2), 157–75; doi: 10.2478/v10085-009-0015-6.

Backman, J., et al., and Expedition 302 Scientists, 2006, Proceedings of the Integrated Ocean Drilling Program, vol. 302, College Station, TX, Ocean Drilling Program, doi.10.2204/iodp.proc.302.104.2006.

Backman, J., et al., 2008, Age model and core seismic integration for the Cenozoic ACEX sediments from the Lomonosov Ridge. Paleoceanography, **23**, doi:10.1029/2007PA001476.

Clark, D.L., et al., 2000, Orphan Arctic Ocean metasediment clasts: Local derivation of from Alpha Ridge or pre-2.6 Ma ice rafting? Geology, **28**, 1143–6.

Fütterer, D.K., 1992, ARCTIC'91: The expedition ARK-VIII/3 of R/V Polarstern in 1991. Alfred Wegener Institute for Polar and Marine Research, Bremerhaven, Germany. Report on Polar Research, vol. 107, 267 pp.

Jakobsson, M., et al., 2000, Manganese and color cycles in Arctic Ocean sediments contrain Pleistocene chronology. Geology, **28**, 23–6.

Jakobsson, M., et al., 2001, Pleistocene stratigraphy and paleoenvironmental variations from Lomonosov Ridge sediments, central Arctic Ocean. Global Planetary Change, **31**, 1–22.

Jakobsson, M., Flodén, T., and the Expedition 302 Scientists, 2006. Expedition 302 geophysics: integrating past data with new results," in Backman, J., Moran, K., McInroy, D.B., et al., *Proceedings of the Integrated Ocean Drilling Program* 302 (Edinburgh: Integrated Ocean Drilling Program Management International, Inc., 2006). doi:10.2204/iodp.proc.302.102.2006.

Jakobsson, M., et al., 2008. An improved bathymetric portrayal of the Arctic Ocean: Implications for ocean modeling and geological, geophysical and oceanographic analyses. Geophysical Research Letters, **35**, L07602, doi:10.1029/2008GL033520

Moran, K., et al., 2006. The Cenozoic palaeoenvironment of the Arctic Ocean. Nature, **441**, 601–6.

Matthiessen, J., et al., 2001. Late Quaternary dinoflagellate cyst stratigraphy at the Eurasian continental margin, Arctic Ocean: indications for Atlantic water inflow in the past 150,000 years. Global and Planetary Change, **31**, 65–86.

Polyak, L., et al., 2004, Contrasting glacial/interglacial regimes in the western Arctic Ocean as exemplified by a sedimentary record from the Mendeleev Ridge. Palaeogeography, Palaeoclimatology, Palaeoecology, **203**, 73–93.

Poore, R.Z., et al., 1994, Quaternary stratigraphy and paleoceanography of the Canada Basin, Western Arctic Ocean US. Geological Survey Bulletin, vol. **2080**, 34 pp.

Sluijs, A., et al., and the Expedition 302 Scientists, 2006, Subtropical Arctic Ocean temperatures during the Palaeocene/Eocene thermal maximum. Nature, **441**, 610–13.

Spielhagen, R.F., et al., 1997, Arctic Ocean evidence for late Quaternary initiation of northern Eurasian ice sheet. Geology, **25** (9), 783–6.

Zachos, J., et al., 2001, Trends, rhythms, and aberrations in global climate 65 Ma to present. Science, **292**, 686–93.

Zachos, J., et al., 2008. An early Cenozoic perspective on greenhouse warming and carbon-cycle dymanics. Nature, **451**, 279–83.

Chapter 8 Climate Cycles

FIGURE 8.1. The Earth and Sun. Image from NASA; http://www.nasa.gov/centers/goddard/news/topstory/2008/solar_variability.html.

SUMMARY

This exercise explores **cyclic climate change** from the geologic record and the explanation of that change using astronomical theory. In **Part 8.1**, you will examine a variety of records displaying cyclic climate change, calculate the periodicities of these records, and reflect on sources and implications of scientific uncertainty. In **Part 8.2**, you will reflect on your knowledge of seasonality. Then you will be introduced to the long-term orbital variations of eccentricity, obliquity, and precession and connect these orbital drivers to the periodicities in the climate proxy records from Part 8.1. In **Part 8.3**, you will dissect the CO_2 record of the last 400 kyr to characterize greenhouse gas levels during past glacial–interglacial cycles and today. You will identify a distinct break in the cyclicity and develop hypotheses to explain this change in climate.

Reconstructing Earth's Climate History: Inquiry-Based Exercises for Lab and Class, First Edition. Kristen St John, R Mark Leckie, Kate Pound, Megan Jones and Lawrence Krissek.

NAME

Climate Cycles
Part 8.1. Patterns and Periodicities

Introduction

This exercise examines paleoclimate records from a variety of archives, locations, proxies, and geologic epochs. However, they all have something in common – the data display **cyclicity** or the repetition of some distinct pattern. Cyclicity in the time dimension (vs. space dimension) is termed **periodicity**. Your instructor will assign one or more of these records to you to examine and ultimately determine the periodicity of the observed cycles. As an introduction to periodicity, examine the hypothetical data below (Figure 8.2). These data show three cycles (high-low-high) in some variable over approximately 30 million years. Thus, the cycles have a periodicity of approximately 10 million years. Note periodicity (time/cycle) is the inverse of **frequency** (number of cycles/time). Furthermore, when determining periodicity in a record it is important to identify the number of full cycles, rather than simply counting the number of peaks (or troughs) in the curve. The importance of this distinction is apparent in the example below as there are four peaks (highs), yet only three full cycles.

The records you will work with contain real data from natural climate archives and therefore will not be as "clean" as the simple sine curve in this example. Nevertheless, one or more cycles are present in each record you will work with.

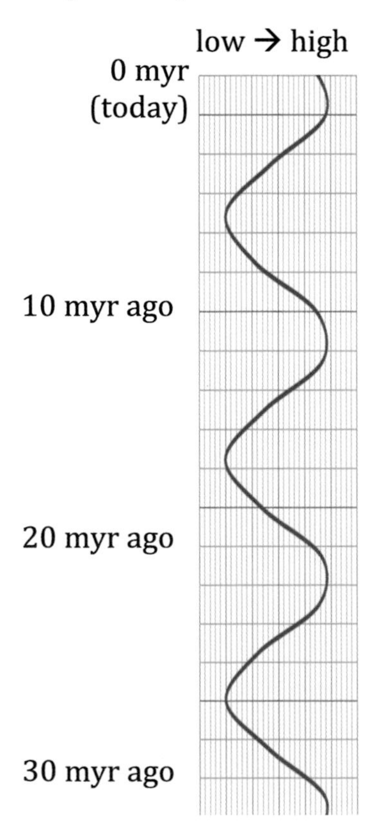

FIGURE 8.2. Hypothetical data displaying cyclicity.

NAME ______________________________

To do:

Answer the questions for **your record**. Next, add information about your record to **Table 8.1**. Then be prepared to present your record to the class and explain how you determined the periodicity and any other observations or questions you have about your record.

Record 1: Shatsky Rise, North Pacific Ocean (Figure 8.3)

1 (a) What evidence is there of cyclicity in the digital core image?

____

____

____

(b) The **reflectance** (Figure 8.3) of visible light from the sediment cores was measured using a spectrophotometer. This quantitative measurement provides a high-resolution stratigraphic record of color variations for visible wavelengths (400–700 nm). Notice that light colored sediments have high reflectance values, whereas dark colored sediments have low reflectance values. What evidence is there of cyclicity in the reflectance data?

____

____

____

(c) Approximately how many meters of sediment were deposited within a typical cycle (Figure 8.3)?

____

2 If the age at 62 m below seafloor (mbsf) in this core (Figure 8.3) is determined to be 1.21 million years (Ma) and the age at 85 mbsf is 1.77 Ma, and if we assume a constant sedimentation rate between these two age control points, what is the likely periodicity of the cyclicity represented in this core? Show your work.

____

____

____

____

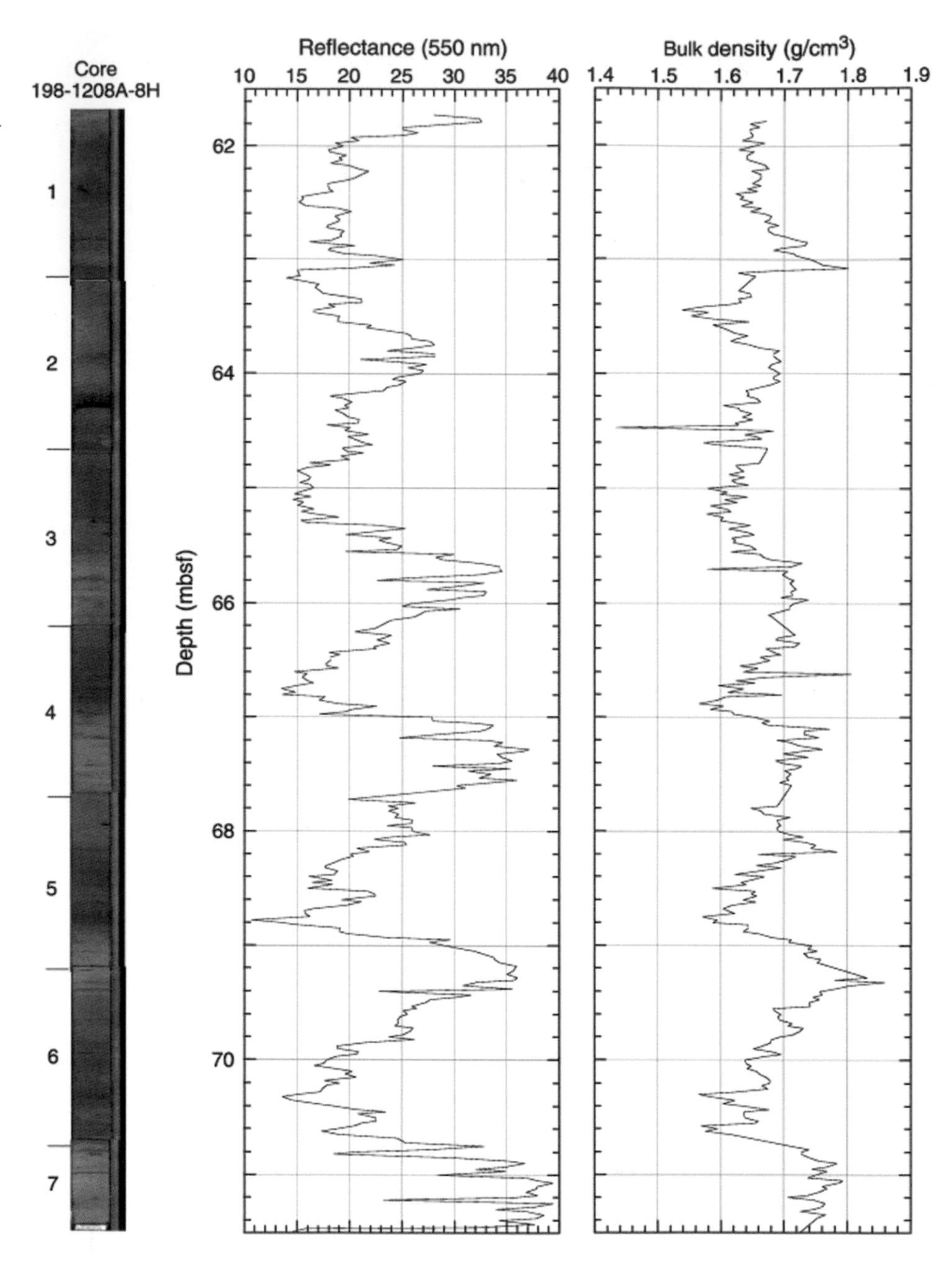

FIGURE 8.3. Composite digital photograph, color reflectance, and bulk density for Core 198-1208A-8H. This core is composed of nannofossil ooze and nannofossil clay of Pleistocene age from Shatsky Rise, NW Pacific Ocean. Ages in this core were determined using magnetostratigraphy. From Shipboard Scientific Party, 2002.

NAME

3 What additional information would you like to know to reduce the uncertainties in your periodicity estimate above? In other words, how could you test this hypothesis of periodicity?

Record 2: Lomonosov Ridge, central Arctic Ocean (Figure 8.4)

4 (a) What evidence is there of cyclicity in the data (Figure 8.4)? (b) Approximately how many meters of sediment were deposited within a typical cycle?

(a)

(b)

5 Based on the age data, the sedimentation rate for this core was determined to be approximately 25 m/myr (assuming a constant sedimentation rate). What is the likely periodicity of the cyclicity represented in this core (Figure 8.4)? Show your work.

6 What additional information would you like to know to reduce the uncertainties in your periodicity estimate above? In other words, how could you test this hypothesis of periodicity?

NAME

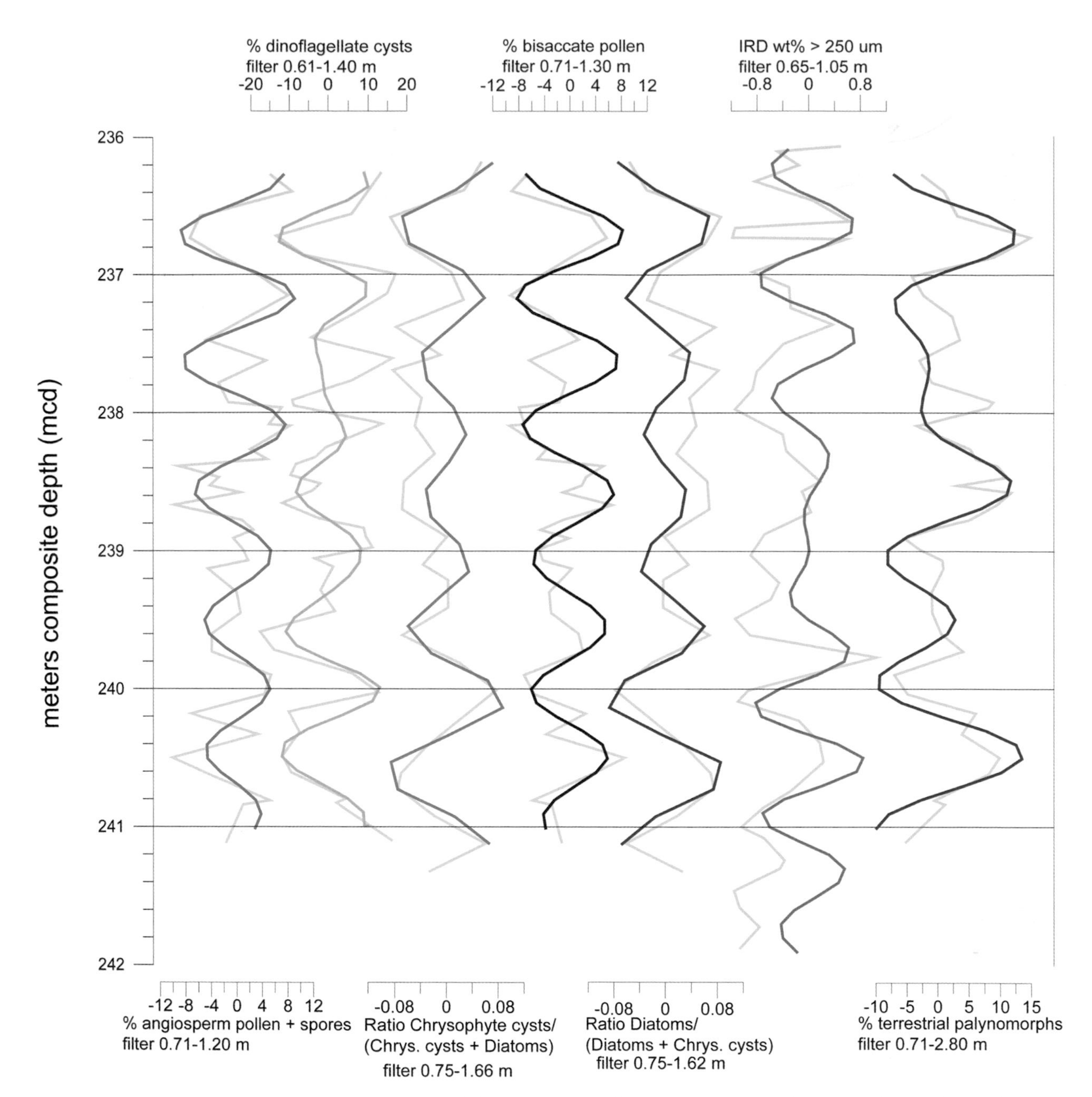

FIGURE 8.4. Records of the relative abundances of various "palynomorph" microfossils (i.e., angiosperm pollen and spores, dinoflagellate cysts, bisaccate pollen, terrestrial palynomorphs) and the weight-percent of ice-rafted debris (IRD, i.e., sediment transported from land by icebergs and/or sea ice) in samples from middle Eocene age sediments in Core 302-2A-55X from the Lomonosov Ridge, central Arctic Ocean. Ages in this core were determined using biostratigraphy (dinocyst and silcoflagellate index fossils). Both the "raw" data (gray lines), and the filtered data (colored lines) are included. Data are filtered as part of the signal processing in spectral analysis. From Sangiorgi et al., 2008.

NAME ____

FIGURE 8.5. Mudstone and limestone sediments in a Miocene age outcrop of the Cascante sedimentary section in NE Spain. Age is determined using magnetostratigraphy (the magnetic time scale). Cycle numbers are listed from the base to the top of the outcrop (see also corresponding cycle numbers in Figure 8.6). From Abels et al., 2009.

Record 3: Cascante outcrop, NE Spain (Figures 8.5 and 8.6)

7 In Figure 8.6: (a) What evidence is there of cyclicity in the reflectance data? (b) In the lithology column? (c) Approximately how many meters of sediment were deposited within a typical cycle?

(a)____

____

(b)____

____

(c)____

8 Based on the magnetostratigraphy of this outcrop (Figure 8.6), the age at the top of Chron C4Ar.2n is 9580 ka and the age at top of Chron C5n.2n is 9920 ka (ka = thousands of years ago). Unlike core depths, the stratigraphic position

FIGURE 8.6. Data and interpretation of the Cascante sedimentary section (Figure 8.5). Gray shading in the magnetic polarity record represents uncertainty intervals between samples with certain polarity. Gray in the lithology column represent color differences (white = limestone, light grey = green-yellow mudstone, medium grey = red-orange mudstone, and dark grey = red-brown mudstone). Cycles of dark mudstone and paler limey-mudstone/limestone are numbered on the left of the column. Locations of limestone samples collected for thin sections are also indicated. The graph shows smoothed color reflectance data. Modified from Abels et al., 2009.

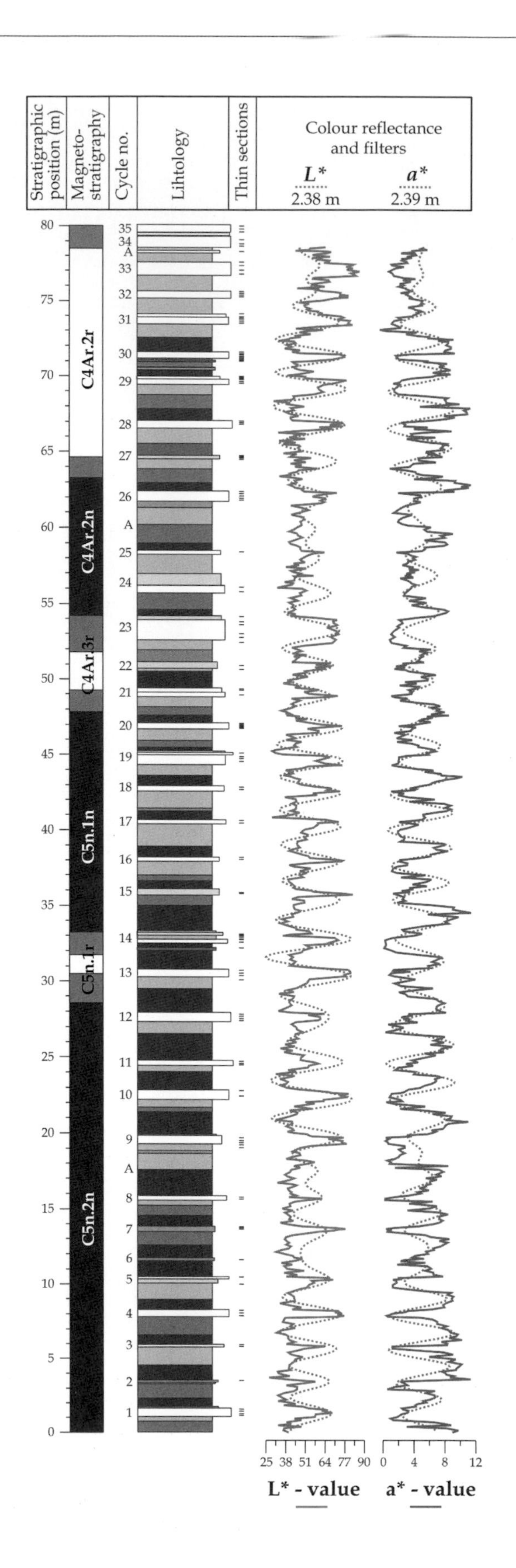

NAME ________________________________

(i.e., vertical scale) in this outcrop is numbered from the bottom up because field data collection started at the base of the outcrop and the scientists worked their way up-section (and up the mountainside, see (Figure 8.5)). Assuming a constant sedimentation rate, what is the likely periodicity of the cyclicity represented in this outcrop of ancient lake sediments? Show your work.

9 What additional information would you like to know to reduce the uncertainties in your periodicity estimate above? In other words, how could you test this hypothesis of periodicity?

Record 4: Global Ocean (Figures 8.7–8.8)

10 Describe any evidence of cyclicity in this record (Figure 8.8).

11 Notice that Figure 8.8 is a composite record of benthic oxygen isotope data obtained from 57 marine sites (Figure 8.7). What evidence is there that the cycles in these data are globally synchronous? Explain.

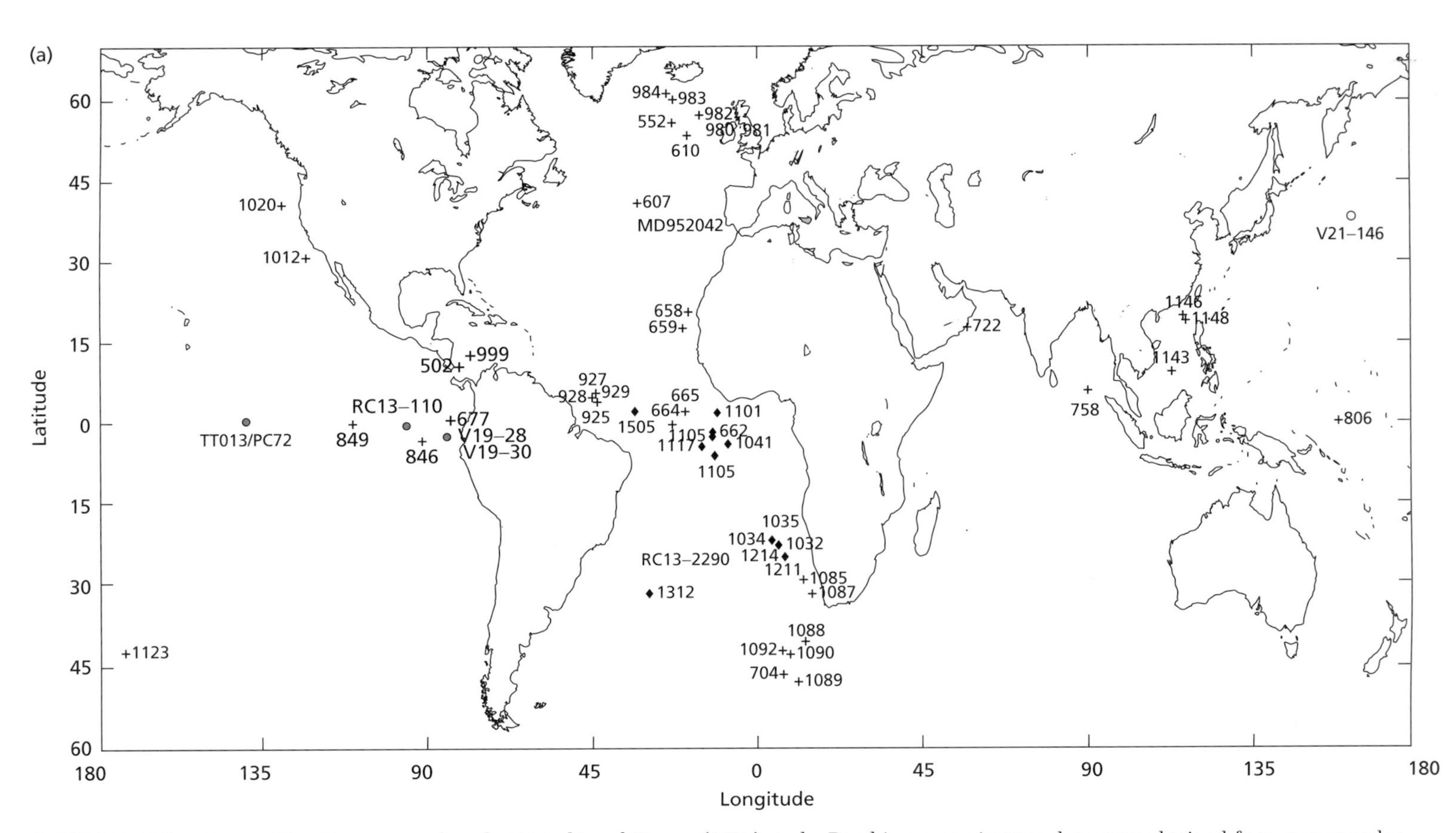

FIGURE 8.7. (a) Locations of the 57 cores used in the Lisiecki and Raymo (2005) study. Benthic oxygen isotope data were obtained from core samples from the Deep Sea Drilling Program (DSDP) and Ocean Drilling Program (ODP) sites (crosses), GeoB sites (diamonds), and others (circles).

NAME

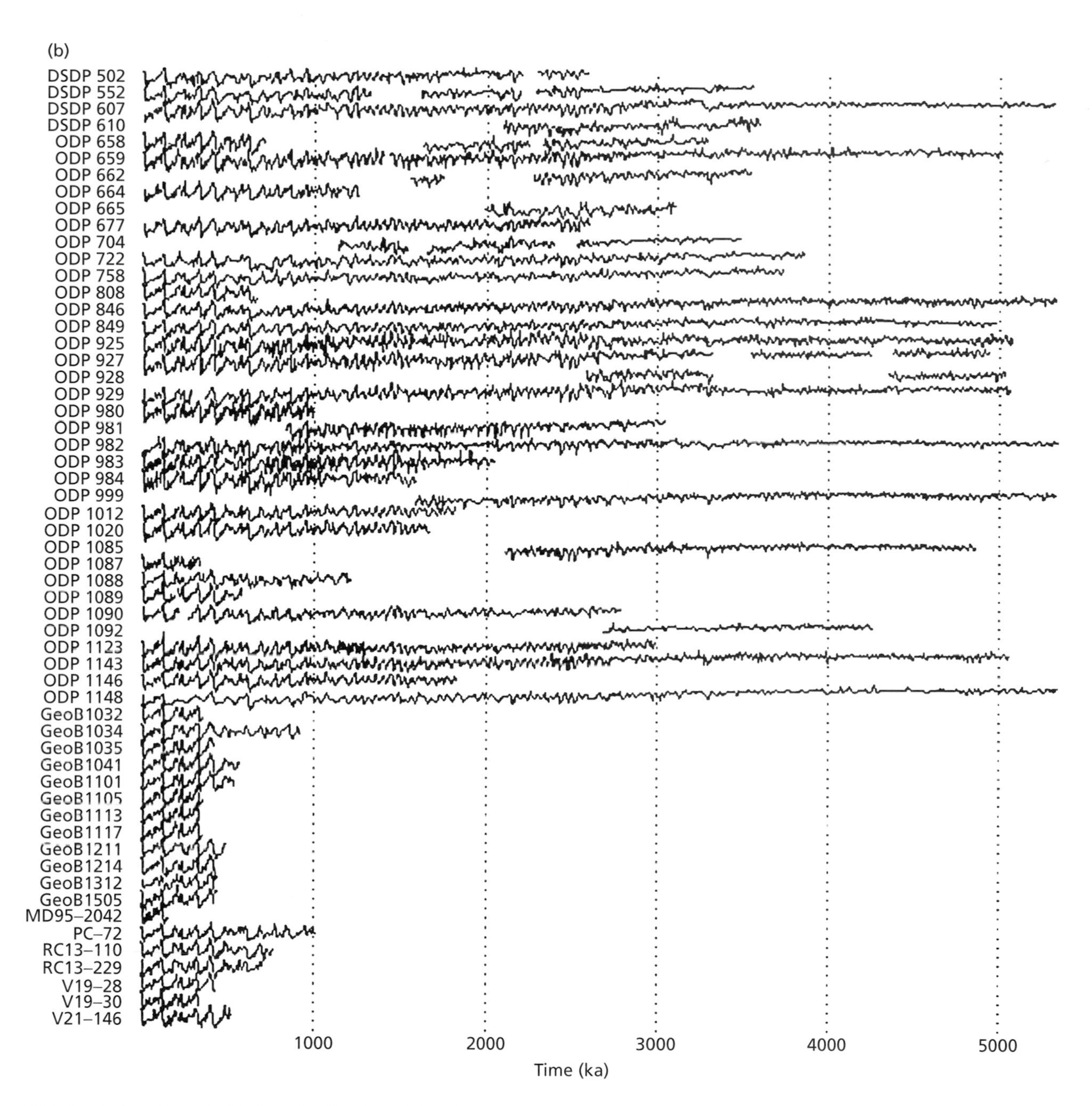

FIGURE 8.7. (*Continued*) (b) Graphically aligned benthic oxygen isotope data from 57 sites, offset vertically, on the page so that the variability within each can be seen. **These data are compiled to make Figure 8.8.**

NAME

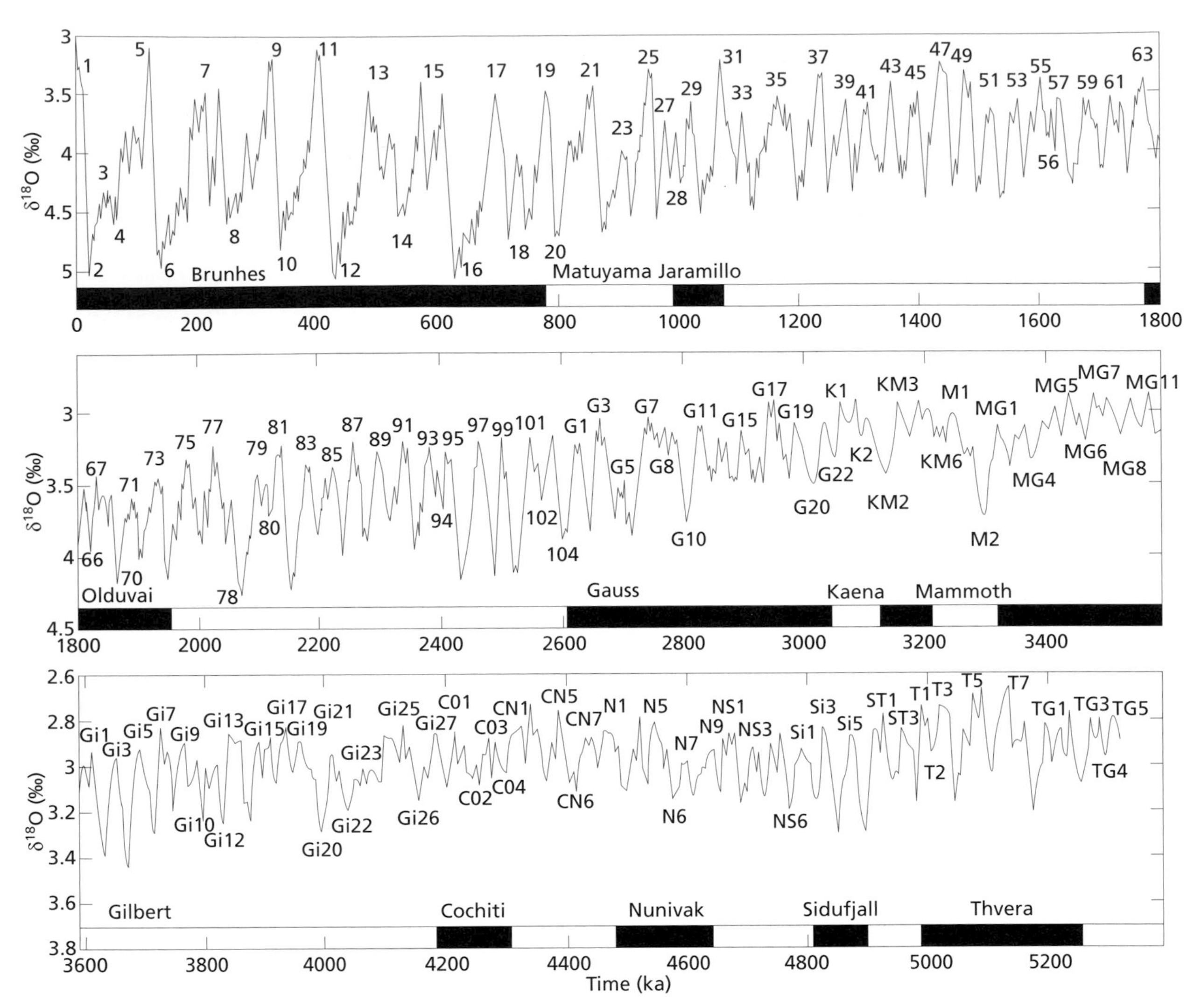

FIGURE 8.8. "Stacked" benthic oxygen isotope record constructed by graphically correlating and combining the 57 globally distributed benthic $\delta^{18}O$ records shown in **Figure 8.7**. Ages are determined using magnetostratigraphy. The magnetic polarity record is shown at the base of each panel in the figure. The numbering of peaks and troughs on this curve represents marine isotope stage numbers with odd numbers representing warmer (interglacial or interstadial) times and even numbers representing colder (glacial or stadial) times. Note ka = thousand years ago. From Lisiecki and Raymo, 2005.

NAME

12 What is the periodicity of the cycles (Figure 8.8) between 0 and 1000 ka (1 Ma)? Show your work.

13 What is the periodicity of the cycles (Figure 8.8) between 1000 and 2600 ka (1–2.6 Ma)? Show your work.

Record 5: Vostok, Antarctica (Figure 8.9)

14 Describe any evidence of cyclicity in this record (Figure 8.9).

15 What are the likely periodicities of the cycles represented in this ice core (Figure 8.9)? Show your work.

NAME ______________________________

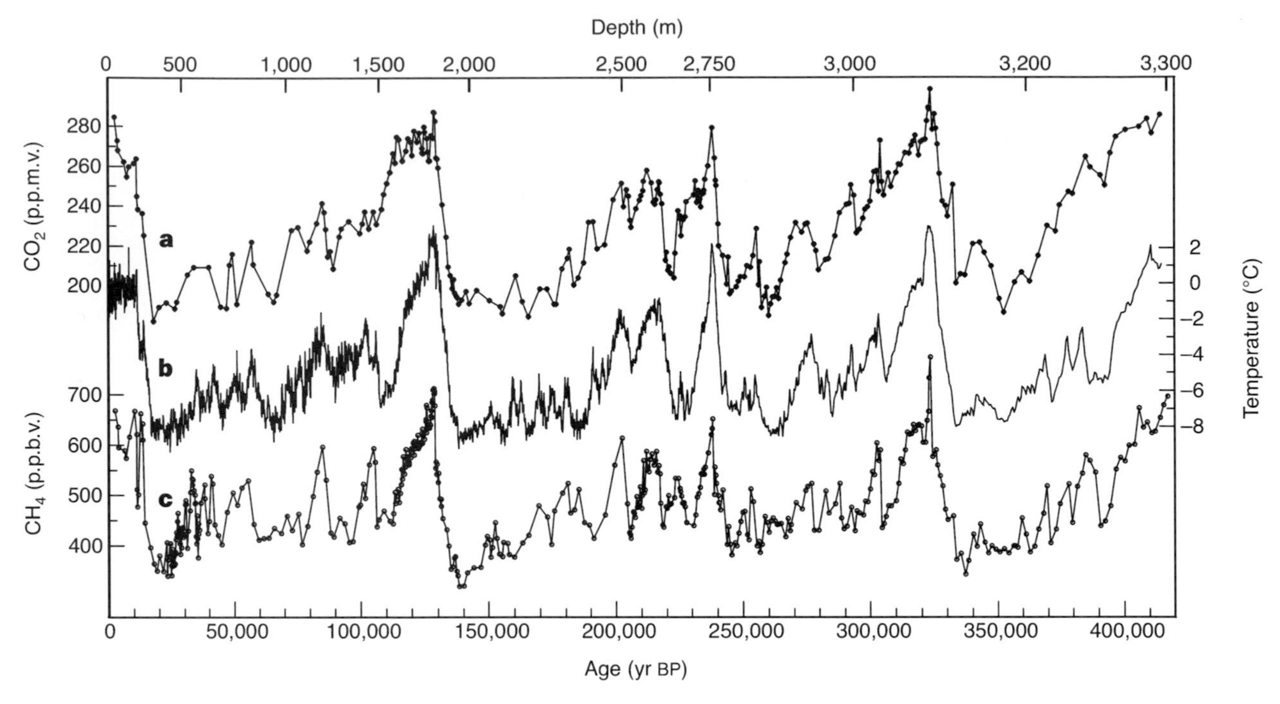

FIGURE 8.9. (a) Carbon dioxide (CO_2) and (c) methane (CH_4) concentrations measured in air bubbles trapped in the Holocene to Pleistocene-age Vostok ice core, Antarctica. Ancient atmospheric air temperature (b) is calculated from the oxygen isotope data measured from the ice. Ages are based on an ice-flow model and an ice accumulation model. From Petit et al., 1999.

____

____

____

____

____

____

16 What additional information would you like to know to reduce the uncertainties in your periodicity estimate above? In other words, how could you test this hypothesis of periodicity?

____

____

____

____

NAME

Record 6: Xifeng and Luochuan, central China (Figure 8.10)

17 Is the magnetic susceptibility higher in the loess or in the soils?

18 Describe any evidence for cyclicity in this record (Figure 8.10).

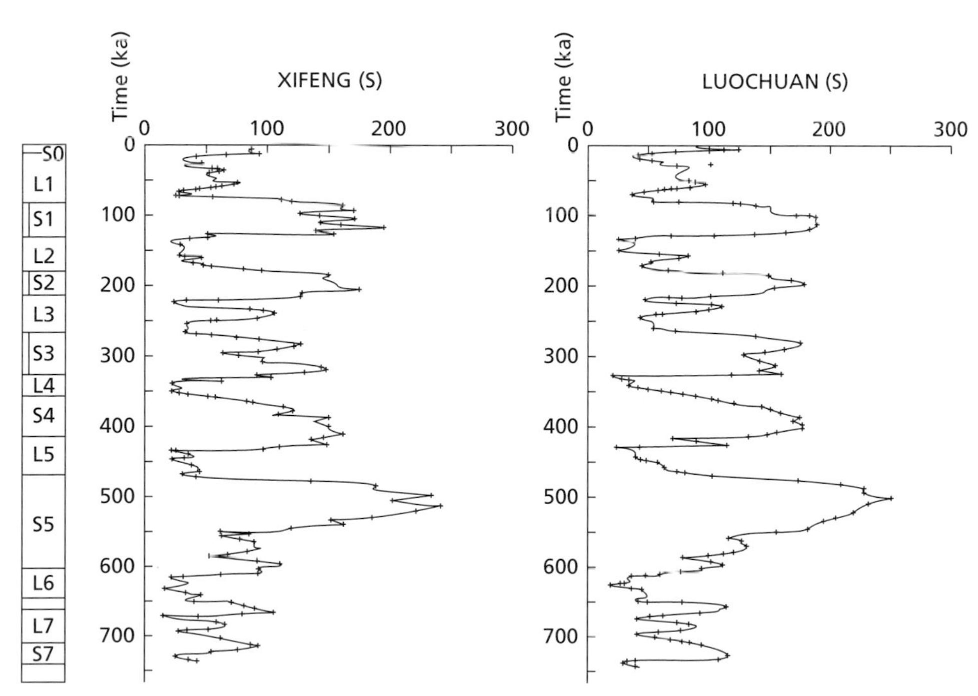

FIGURE 8.10. Magnetic susceptibility of loess (wind blown dust) outcrops and interbedded soils at two locations (Xifeng and Luochuan) in central China. The stratigraphic column on the left shows different **loess** (L1, L2 . . .) and **soil** (S1, S2 . . .) units. The time scale was developed based on a model that assumes a constant deposition rate of ferromagnetic minerals. This model was also compared with the established marine oxygen isotope time scale (but was not astronomically tuned). Two thermoluminescence dates were used to constrain the ages of the upper units. Note ka = thousand years ago. From Kukla et al., 1988.

NAME ______________________________

19 What is the likely periodicity of the cyclicity represented in this sedimentary sequence (Figure 8.10)? Show your work.

____

____

____

____

20 What additional information would you like to know to reduce the uncertainties in your periodicity estimate above? In other words, how could you test this hypothesis of periodicity?

____

____

____

____

____

____

Record 7: North Atlantic Ocean (Figure 8.11)

21 Describe evidence of cyclicity in the diatom record (Figure 8.11b).

____

____

____

____

____

22 Based on the diatom abundances (Figure 8.11b), what is the likely periodicity of the cyclicity represented in this core? Show your work.

____

____

____

NAME

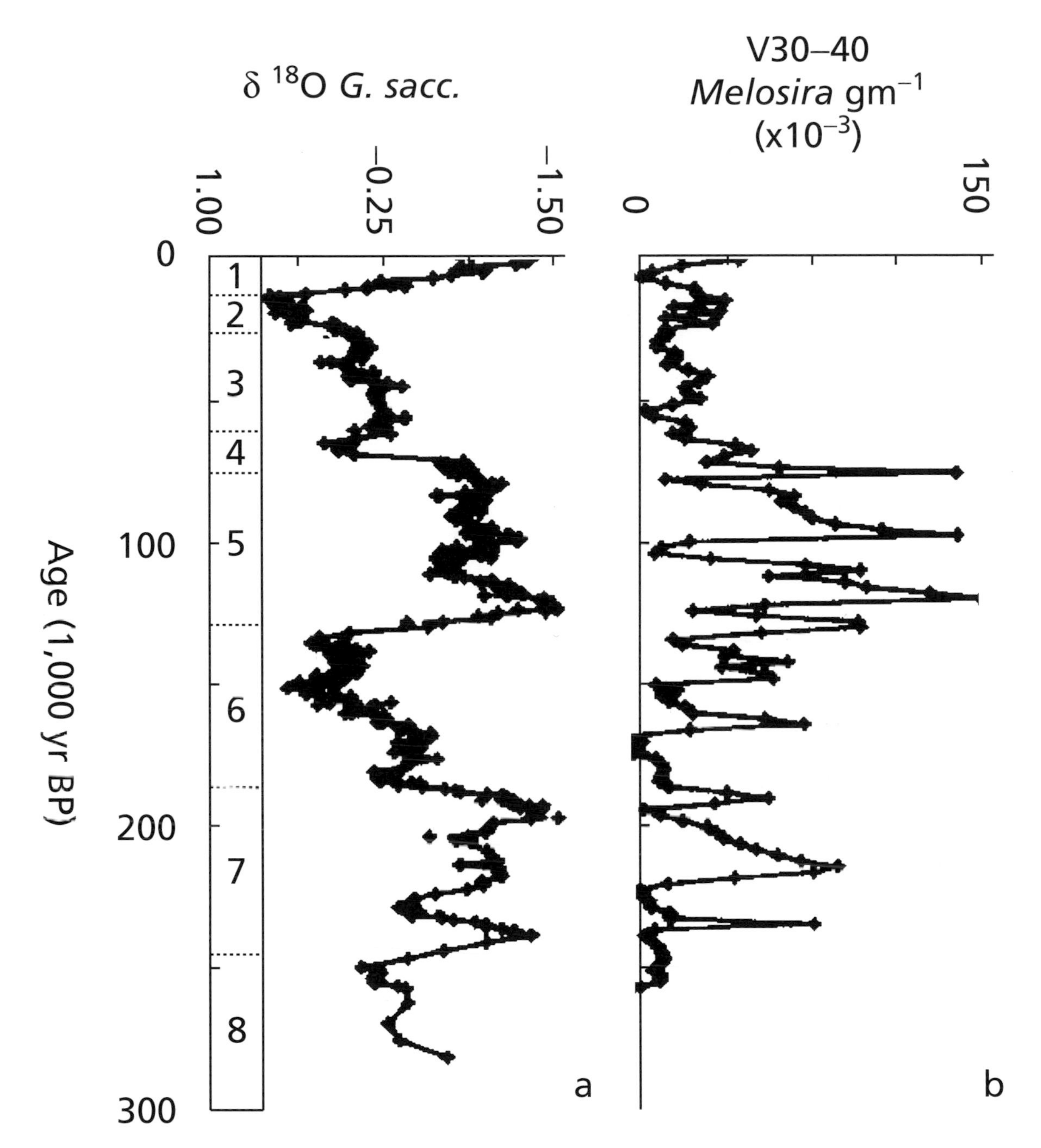

FIGURE 8.11. Abundance of freshwater African diatom *Melosira* in sediment core V30-40 from the North Atlantic (b), plotted alongside the oxygen isotope record (a) derived from a planktic foraminifera species *Globigerinoides sacculifer* in the same core. Ages were determined based on correlation of this oxygen isotope record with the established marine oxygen isotope (SPECMAP) time scale. Marine oxygen isotope stages (1-8) are shown to the left. From Pokras and Mix, 1987

NAME

23 What additional information would you like to know to reduce the uncertainties in your periodicity estimate above? In other words, how could you test this hypothesis of periodicity?

Synthesis: Refer to Table 8.1 and the figures for each of the seven records.

24 Are cycles restricted to a certain type of archive, location, proxy, or geologic epoch?

25 In particular, compare the stacked deep-sea oxygen isotope record (Record 4) with the Vostok ice core record (Record 5) for the interval 0–400 ka. What similarities do you see?

NAME

TABLE 8.1. Data summary

Record (Key Figures)	What Type of Archive is It?	Where is the Record From?	What type of Data (Proxy) is It?	What Geologic Time Span Does This Record Cover?	What is the Periodicity That You Determined?	[This column to be completed in Part 8.2] What Orbital Cycle Best Matches the Climate Cycle?
1 (Figure 8.3)						
2 (Figure 8.4)						
3 (Figures 8.5 & 8.6)						
4 (Figures 8.7 & 8.8)						
5 (Figure 8.9)						
6 (Figure 8.10)						
7 (Figure 8.11)						

NAME ____________________

26 What periodicities are "common", in other words, occur in two or more of the seven records (Table 8.1)?

____
____
____
____
____
____

27 What factors introduce uncertainty in the periodicity calculations?

____
____
____
____
____

28 How much uncertainty do you think is acceptable? Explain.

____
____
____
____

29 Speculate on what might have caused the cyclicity observed in this diverse collection of paleoclimate archives. List your ideas:

____
____
____
____

NAME ______________________________

30 The records for which you estimated periodicities were fairly straightforward. Speculate about how the periodicities of a more complex record, such as one of the wiggle plots in Figure 8.12, could be determined. List your ideas:

____

____

____

____

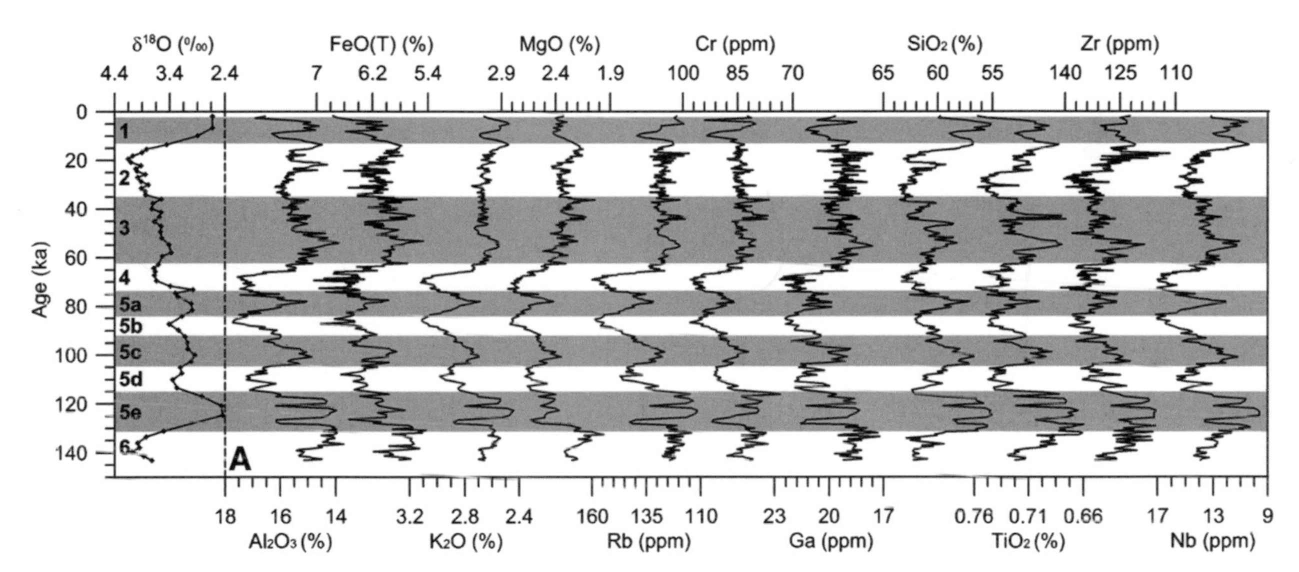

FIGURE 8.12. Temporal variations of elements enriched in clay- and silt-sized minerals at ODP Site 1145, in the northern South China Sea. From Sun et al., 2008.

NAME

Climate Cycles
Part 8.2. Orbital Metronome

Annual Cyclicity

1 Are the Earth's seasons a type of cyclic climate change? Why or why not?

2 Why does the Earth have seasons? Explain as completely as possible and include a sketch.

NAME

Long-Term Changes in Earth's Geometry in Space Relative to the Sun

It may be surprising to learn that the shape of Earth's orbit, the angle of Earth's tilt, and the direction of Earth's axis (with respect to fixed stars) is not constant over long time scales. The variations of Earth's geometry in space are known from contributions in the scientific fields of astrophysics and astronomy. These changes in the Earth's orbit around the Sun (Figure 8.1) occur in a cyclic or rhythmic way and therefore have particular periodicities. The three major **orbital cycles** are **eccentricity**, **obliquity**, and **precession**. These cyclic changes in Earth's orbital geometry largely result from gravitational pull from large planets in our solar system, especially Jupiter, and from the Sun and moon. Each orbital cycle is briefly discussed below. See Ruddiman (2001) for a more thorough description of orbital cycles.

Eccentricity describes the degree of deviation from a perfect circle (Figure 8.13 left); the greater the eccentricity, the greater the elliptical deviation from a circle. A perfect circle has an eccentricity of 0 and a flattened circle (i.e., a straight line) has an eccentricity of 1. Earth's orbit around the Sun has an eccentricity that ranges from 0.005 to 0.0607 with a mean of 0.028. The eccentricity today is 0.0167. The eccentricity of Earth's orbit varies at a range of periodicities (95 kyr to 136 kyr) with an average of approximately 100 kyr (100,000 years). There is also a long eccentricity cycle with a periodicity of approximately 413 kyr and an even longer cycle with a period of 2.1 myr.

3 (a) How would the Earth–Sun **distance** change in a highly eccentric orbit (Figure 8.13 left), compared to a more circular orbit (Figure 8.13 right)?

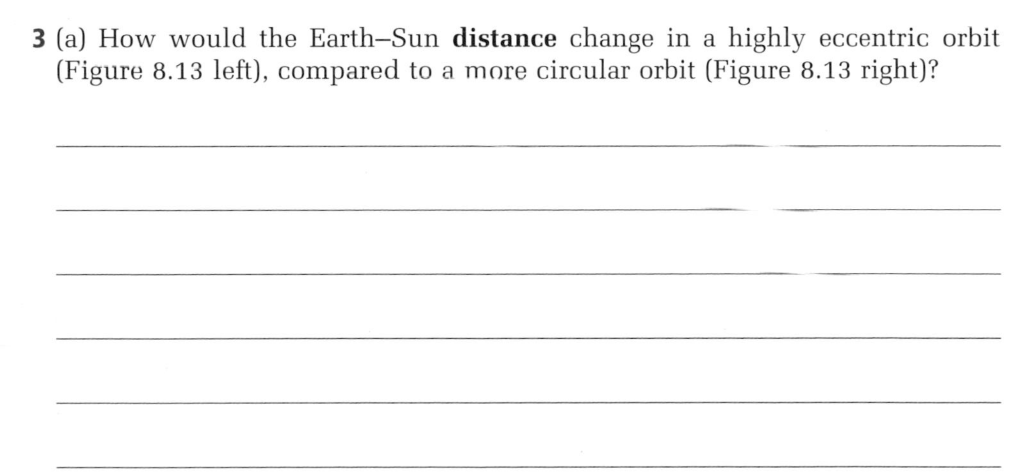

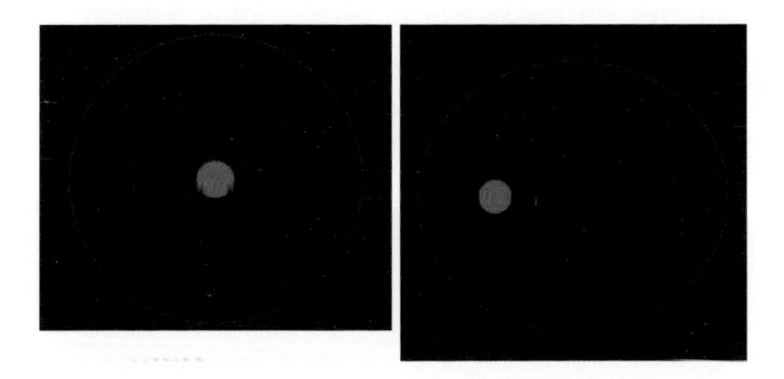

FIGURE 8.13. (left) Zero eccentricity and (right) 0.5 eccentricity, which greatly exaggerates Earth's actual maximum eccentricity. From http://en.wikipedia.org/wiki/Milankovitch_cycles

NAME

(b) How would this affect the **intensity** of solar radiation striking the Earth's atmosphere?

4 The Earth completes one revolution around the Sun each year regardless of the shape of the orbital path; however changing the shape of the orbital path (i.e., changes in eccentricity) will have some affect on seasonality. Which orbital configuration would result in the greatest **seasonal contrast** (i.e., summer vs. winter temperatures): a highly eccentric orbit or a more circular orbit? Why?

Obliquity describes the Earth's axial tilt with respect to the plane of the Earth's orbit around the Sun. Today this tilt is 23.5° and is estimated to vary from 22.1° to 24.5° through time (Figure 8.14). The periodicity of this variation is approximately 41 kyr.

5 Predict how a **smaller tilt angle** would affect polar climate in the summer and in the winter.

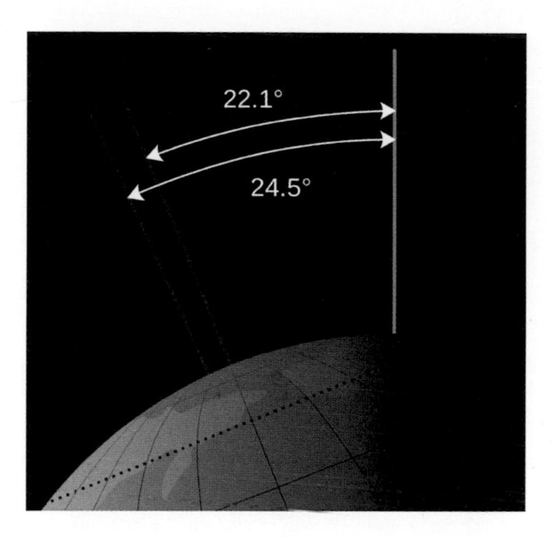

FIGURE 8.14. Schematic diagram of different tilt angles. From http://en.wikipedia.org/wiki/Milankovitch_cycles

6 Would high latitudes or low latitudes be more affected by changes in obliquity? Why?

Precession of the equinoxes has a cyclicity of 19 kyr to 26 kyr. This cycle exists primarily because the spinning Earth wobbles like a toy top, which affects the direction of the Earth's rotational axis (Figure 8.15), and with it, the position of solstices and equinoxes around the orbital plane and the time of year that the Earth is closest to the Sun (owing to our slightly elliptical orbit around the Sun). Therefore the distance from the Earth to the Sun has varied with time for each of the seasons. Today, we are closest to the Sun (**perihelion**) on January 3 (Northern Hemisphere winter), and farthest from the Sun (**aphelion**) on July 4 (Northern Hemisphere summer). About 11,000 years ago, the Northern Hemisphere was at its closest distance to the Sun during the summer months.

NAME ______________________________

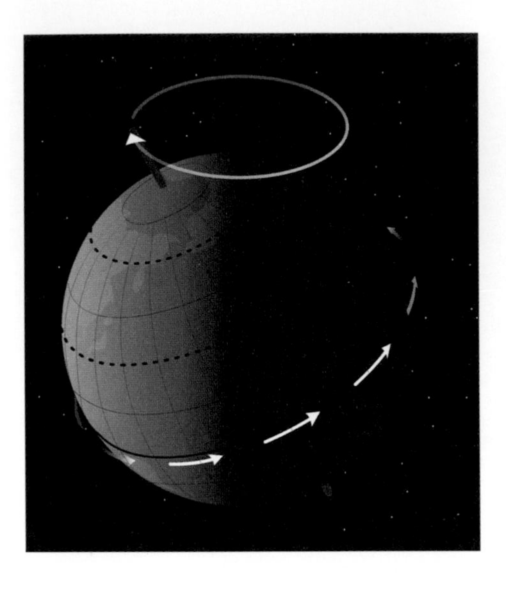

FIGURE 8.15. Schematic diagram showing the changing direction of Earth's spin axis. From http://en.wikipedia.org/wiki/Milankovitch_cycles

7 How would the intensity of solar radiation striking the Northern Hemisphere in **January** been different 11,000 years ago compared to today?

____

____

____

____

____

____

8 How would the intensity of solar radiation striking the Northern Hemisphere in **July** been different 11,000 years ago compared with today?

____

____

____

____

____

____

NAME ______________________________

9 Based on your answers to questions 7 and 8, compare the **seasonal contrast** (i.e., summer vs. winter temperatures) of 11,000 years ago to the seasonal contrast of today.

____

____

____

____

____

____

Orbital Control of Earth's Climate Cycles

The natural variations in the Earth's orbit influence the amount of **in**coming **sol**ar radia**tion** (**insolation**) received and therefore affect Earth's climate in a cyclic way. Such orbitally driven climate cycles are called **Milankovitch cycles**, after the Serbian scientist who first made this connection between variations in the Earth's orbit and changes in the Earth's climate in the early 20th century.

10 **Return to Table 8.1**. In the last column labeled **orbital cycle** write down which orbital variation (eccentricity, obliquity, or precession) has a periodicity that best matches the periodicity in that climate record. Note that some climate records exhibit more than one periodicity, so you should list each corresponding orbital variation.

11 It may be relatively easy to see how changes in insolation could directly cause changes in temperature and indirectly cause changes ice volume (expanding ice during cold times and melting ice during warm times), but what about some of the other proxy records of cyclic climate change? Speculate about how cyclic orbitally driven changes in insolation might have an indirect effect on vegetation (Record 2), lake levels (Record 3), and greenhouse gas levels (Record 5). This is not an easy question ☺ (scientists are working on this question themselves), but have a go at it!

____

____

____

____

____

____

NAME

NAME

Climate Cycles
Part 8.3. A Break in the Pattern

Cyclic changes are part of the Earth's climate history. In the Pleistocene, climate cycles produced relatively cooler **glacial** intervals and relatively warmer **interglacial** intervals. As shown in Part 8.1, changes in temperature, ice volume, precipitation, and greenhouse gases all followed a similar pacing. How do modern climate changes compare with this pattern? To put this question into context, examine Figure 8.16 and answer the questions that follow.

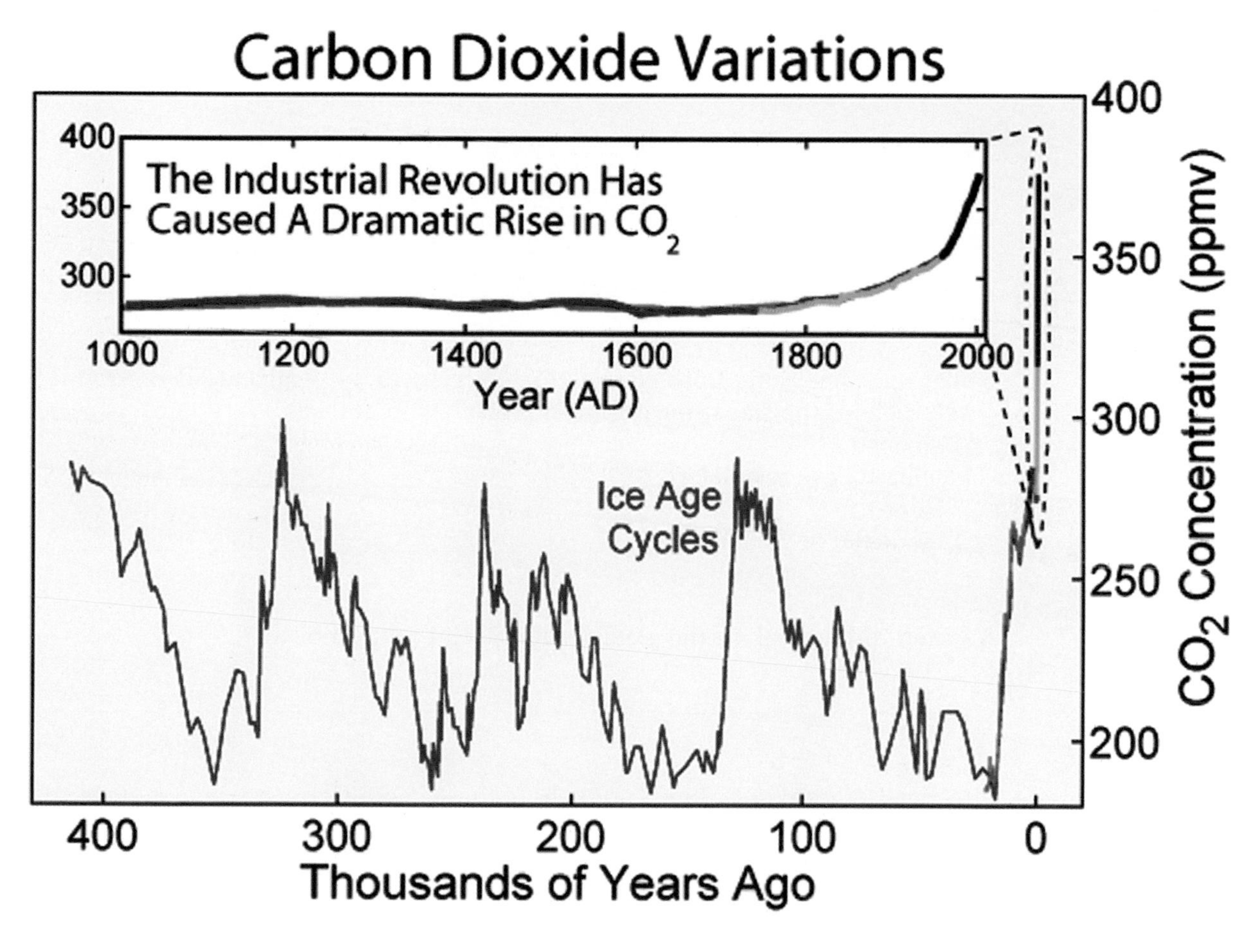

FIGURE 8.16. Variations in the **concentration of carbon dioxide (CO_2) in the atmosphere during the last 400 thousand years.** Data sources include: (blue) Vostok ice core, Antarctica (Fischer et al., 1999); (green) EPICA ice core, Antarctica (Monnin et al., 2004); (red) Law Dome ice core, Antarctica (Etheridge et al., 1998); (cyan) Siple Dome ice core, Antarctica (Neftel et al., 1994); (black) Mauna Loa Observatory, Hawaii (Keeling and Whorf, 2004). Throughout most of the record, the largest changes can be related to **glacial–interglacial cycles**. Although the glacial cycles are most directly caused by changes in the Earth's orbit (i.e., **Milankovitch cycles**), these changes also influence the carbon cycle, which in turn feeds back into the glacial system. This figure was originally prepared by Robert A. Rohde from publicly available data and is incorporated into the Global Warming Art project (http://en.wikipedia.org/wiki/File:Carbon_Dioxide_400kyr.png).

NAME ______________________________

1 **Glacials** are times of extensive ice and relatively cold temperatures. The **last glacial maximum** occurred approximately 20,000 yrs ago. **Interglacials** are times of minimal ice extent and relatively warm temperatures.

Use your understanding of the relationship between CO_2 and temperature to label each of the glacial maxima with a "**G**" and each of the interglacials with an "**I**" on Figure 8.16.

2 What was the typical atmospheric CO_2 value during glacial maxima of the last 400,000 years?

____

3 What was the typical atmospheric CO_2 value during the interglacials (glacial minima) of the last 400,000 years?

____

4 What was the atmospheric CO_2 value in the year 2000?

____

5 Are the transitions from glacials to interglacials typically rapid or gradual? What about the transitions from interglacials to glacials?

glacial to interglacial: ______________________________

interglacial to glacial: ______________________________

6 When did a break in the glacial–interglacial pattern occur?

____

7 What scientific questions and/or hypotheses about climate change does this data raise? Make a list your questions and/or hypotheses:

____

____

____

____

____

____

NAME

References

Abels, H.A., et al., 2009, Shallow lacustrine carbonate microfacies document orbitally paced lake-level history in the Miocene Teruel Basin (North-East Spain). Sedimentology, **56**, 399–419, doi: 10.1111/j.1365-3091.2008.00976.x

Etheridge, D.M., Steele, L.P., Langenfelds, R.L., Francey, R.J., Barnola, J.M., and Morgan, V.I., 1998, Historical CO_2 records from the Law Dome DE08, DE08-2, and DSS ice cores, *in* Trends: A Compendium of Data on Global Change, Carbon Dioxide Information Analysis Center, Oak Ridge National Laboratory, U.S. Department of Energy, Oak Ridge, Tenn., U.S.A, http://cdiac.ornl.gov/trends/co2/lawdome.html

Fischer, H., Wahlen, M., Smith, J., Mastroianni, D., and Deck, B., 1999, Ice core records of Atmospheric CO_2 around the last three glacial terminations: Science, v. **283**, p. 1712–1714.

Keeling, C.D., and Whorf, T.P., 2004, Atmospheric CO_2 records from sites in the SIO air sampling network, *in* Trends: A Compendium of Data on Global Change, Carbon Dioxide Information Analysis Center, Oak Ridge National Laboratory, U.S. Department of Energy, Oak Ridge, Tenn., U.S.A., http://cdiac.ornl.gov/trends/co2/sio-keel-flask/sio-keel-flaskmlo_c.html

Kukla, G., et al., 1988, Pleistocene climates in china dated by magnetic susceptibility. Geology, **16**, 811–14.

Lisiecki, L.E. and Raymo, M.E., 2005, A Pliocene-Pleistocene stack of 57 globally distributed benthic $\delta^{18}O$ records. Paleoceanography, **20**, PA1003, doi:10.1029/2004PA001071.

Monnin, E., Steig, E.J., Siegenthaler, U., Kawamura, K., Schwander, J., Stauffer, B., Stocker, T.F., Morse, D.L., Barnola, J.M., Bellier, B., Raynaud, D., and Fischer, H., 2004, Evidence for substantial accumulation rate variability in Antarctica during the Holocene, through synchronization of CO_2 in the Taylor Dome, Dome C and DML ice cores: *Earth and Planetary Science Letters*, v. **224**, p. 45–54. doi:10.1016/j.epsl.2004.05.007

Neftel, A., Friedli, H., Moor, E., Lötscher, H., Oeschger, H., Siegenthaler, U., and Stauffer, B., 1994. Historical CO_2 record from the Siple Station ice core, *in* Trends: A Compendium of Data on Global Change, Carbon Dioxide Information Analysis Center, Oak Ridge National Laboratory, U.S. Department of Energy, Oak Ridge, Tenn., U.S.A., http://cdiac.ornl.gov/trends/co2/siple.html

Petit, J.R., et al., 1999, Climate and atmospheric history of the past 420,000 years from the Vostok Ice core, Antarctica. Nature, **399**, 429–36.

Pokras, E.M. and Mix, A.C., 1987, Earth's procession cycle and Quaternary climate change in tropical Africa. Nature, **326**, 486–7.

Sangiorgi, F., et al., 2008, Cyclicity in the middle Eocene central Arctic Ocean sediment record: Orbital forcing and environmental response. Paleoceanography, **23**, PA1S08, doi:10.1029/2007PA001487.

Shipboard Scientific Party, 2002, Site 1208, In Proceedings of the Ocean Drilling Program, Initial Reports, vol. 198, Bralower, T.J., et al., (eds), College Station, TX, Ocean Drilling Program, pp. 1–93. doi:10.2973/odp.proc.ir.198.104.2002.

Sun, Y, et al., 2008, Processes controlling the geochemical composition of the South China Sea sediments during the last climatic cycle. Chemical Geology, **257**, 240–6.

Chapter 9 The Paleocene–Eocene Thermal Maximum (PETM) Event

FIGURE 9.1. Paleocene–Eocene boundary clay marking the PETM in a core collected during ODP Leg 208, Walvis Ridge, southeast Atlantic. Left: A full length of core, approximately 9.7 m in total length, which has been split lengthwise. This is the "archive half" laid out on the core description table aboard the drillship *JOIDES Resolution*; each "Section" is 1.5-m long. Section 1 is to the right in this view with "up" closest to the viewer. There are seven Sections, plus a core-catcher at the far left (approximately 30-cm long). Right: Close-up of the sharp lithologic contact in Section 6 between the light-colored uppermost Paleocene calcareous ooze and the reddish-colored clay marking the base of the Eocene Epoch and the onset of the PETM; note the holes in the core where two 2-cm^3 samples were removed from the "working half" of the core. Photos from Jim Zachos (UC-Santa Cruz).

SUMMARY

The Paleocene–Eocene Thermal Maximum (PETM) is one of the best examples of a transient climate state; it is marked by a brief and intense interval of global warming and a massive perturbation of the global carbon cycle (Röhl et al., 2007). In **Part 9.1**, you will consider how carbon isotopes are used as a proxy for climate change in the context of the global carbon cycle. To gain a sense of the scope of the PETM, in **Part 9.2** you will evaluate evidence for the impact and possible cause(s) of the event from multiple sites around the globe, including the deep-sea, continental shelf, and continental deposits. A synthesis of these observations will help you construct initial hypotheses about the cause of the PETM. In **Part 9.3**, you will read an article that introduces methane hydrates and

Reconstructing Earth's Climate History: Inquiry-Based Exercises for Lab and Class, First Edition. Kristen St John, R Mark Leckie, Kate Pound, Megan Jones and Lawrence Krissek.

NAME

assess their possible effect on oceanic and atmospheric carbon reservoirs and the role they may have played in the PETM. In **Part 9.4**, you will consider different types of data, which will allow you to evaluate various rates, to get a sense of how fast the PETM was initiated and how fast the ocean–climate system recovered from a major perturbation of the carbon cycle. In **Part 9.5**, you will apply your new understanding of the PETM and pose the question: how does it compare with the rate of global warming today?

The Paleocene–Eocene Thermal Maximum (PETM) Event

Part 9.1. The Cenozoic $\delta^{13}C$ Record and an Important Discovery

The Cenozoic $\delta^{13}C$ Record

In Chapters 1–8 you were introduced to and began using the tools, techniques, and thought processes common to scientists who study Earth's geologic and climate history. Let's now extend that practice by examining, in more detail, one of the most marked **climatic events** of the past 67 million years, the Paleocene–Eocene Thermal Maximum, PETM.

Examine Figure 9.2 showing the change in $\delta^{13}C$ over the past 67 myr. The measurement of stable carbon isotopes, and the calculation and units of $\delta^{13}C$ are very similar to the measurement of stable oxygen isotopes and the calculation and units of $\delta^{18}O$, which you learned about in Chapter 6. Apply your knowledge about stable isotopes here:

Circle one of the answers, right or left.

1 (a) In which direction does the relative proportion of ^{12}**C increase** (Figure 9.2)? **Right** **Left**

(b) In which direction does the relative proportion ^{12}**C decrease** (Figure 9.2)? **Right** **Left**

As you might expect, the element carbon has several isotopes just as oxygen does, three of which are very important in studying the Earth's climate history: **carbon-12** (6 protons and 6 neutrons), **carbon-13** (6 protons and 7 neutrons), and **carbon-14** (6 protons and 8 neutrons; an isotope that may be familiar to you). Both ^{12}**C** and ^{13}**C** are stable isotopes, whereas ^{14}**C** is an unstable (i.e., radioactive) isotope and is used primarily for age determinations over the period of the last 50–60 kyr. Scientists use the ratio of $^{13}C/^{12}C$ as a **proxy of the carbon cycle** in ways similar to the way that $^{18}O/^{16}O$ ratios from the shells of planktic and benthic foraminifera are used as a **proxy** both for sea surface and deep-water temperature and ice volume estimates (refer back to Chapter 6 on oxygen isotopes).

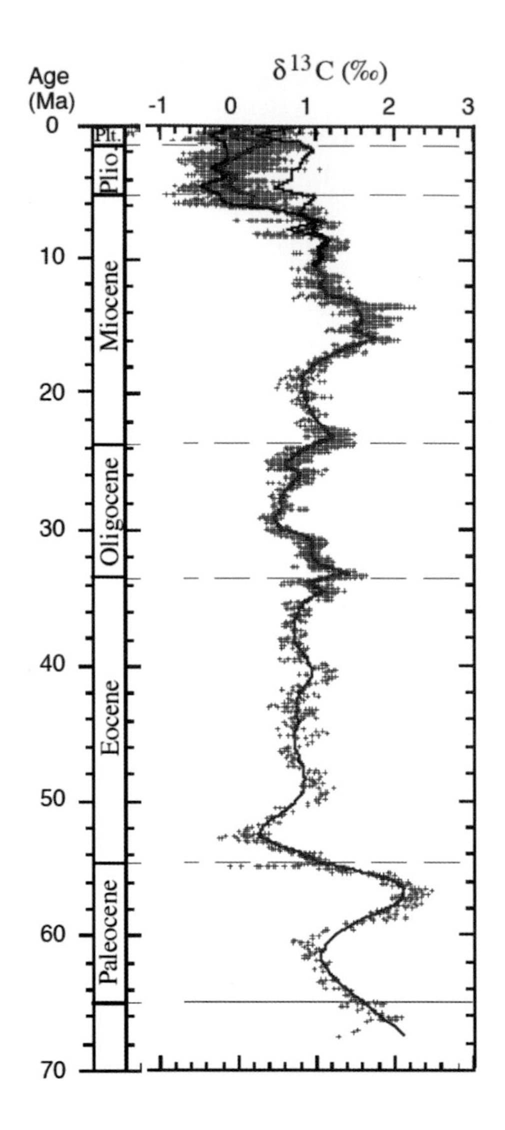

FIGURE 9.2. 67-million year **composite record of benthic foraminiferal carbon isotopes** ($\delta^{13}C$) from the Deep Sea Drilling Project (DSDP) and Ocean Drilling Program (ODP) sediment cores. Analyses were made on the calcium carbonate shells of these deep-sea microfossils. Modified from Zachos et al., 2001.

TABLE 9.1. Average reservoir $\delta^{13}C$ values (from Ruddiman, 2008).

Reservoir	Type of Carbon	$\delta^{13}C$ (‰)	Reservoir	Type of Carbon	$\delta^{13}C$ (‰)
Atmosphere	Inorganic CO_2	−7	Surface ocean	Organic C	−22
Trees	Organic C	−25	Surface ocean	Inorganic CO_2	+1
Grasses	Organic C	−13	Deep ocean	Organic C	−22
Methane Deposits	Organic C	−60	Deep ocean	Inorganic CO_2	0

Read the box on Understanding Stable Carbon Isotopes and review Table 9.1 before answering Questions 2–7.

2 Carefully examine Table 9.1 which outlines characteristic $\delta^{13}C$ values for some important carbon reservoirs. List **at least two observations** from Table 9.1 that

NAME

UNDERSTANDING STABLE CARBON ISOTOPES

The most abundant form of carbon (approximately 99%) found in nature is ^{12}**C**, whereas ^{13}**C** and the other carbon isotopes are found in very small amounts. The carbon isotopic ratios preserved in both **inorganic carbon** (e.g., $CaCO_3$ shells) and **organic carbon** (e.g., organic molecules preserved in sediments) are valuable proxies in paleoclimate research because of the importance of the **carbon cycle** in controlling global climate.

Because ^{13}**C** atoms are approximately 8% more massive than ^{12}**C** atoms, chemical reactions involving carbon have a **fractionation** affect. The mechanism of this fractionation affect is different from the fractionation of oxygen isotopes, which results from phase changes in water or is a temperature-dependent fractionation. In carbon isotopes, the mass difference between ^{12}**C** and ^{13}**C** results in ^{12}**C** being more weakly bonded to the oxygen in CO_2, and thus more chemically reactive than is ^{13}**C**. Thus, the lighter isotope ^{12}**C** is more readily incorporated into the **organic matter** (e.g., carbohydrates/sugars, lipids/fats, and proteins) created during **photosynthesis**. Recall from Chapter 3 the generalized reaction describing the process of photosynthesis:

$$\underset{\text{carbon dioxide}}{6CO_2} + \underset{\text{water}}{6H_2O} + \underset{\text{nitrates, phosphates, trace elements \& vitamins}}{\text{inorganic nutrients}} + \text{solar energy} \rightarrow \underset{\text{glucose (simple sugar)}}{C_6H_{12}O_6} + \underset{\text{free oxygen}}{6O_2}$$

Photosynthetic autotrophs **synthesize organic carbon** (carbohydrates, proteins, and fats) **from inorganic** CO_2. Therefore, the **organic matter** synthesized by terrestrial vegetation and marine phytoplankton (both are photosynthetic autotrophs) tends to be **depleted with respect to** ^{13}**C** compared to the inorganic carbon reservoir from which they extracted the CO_2 (e.g., atmosphere and oceans, respectively).

In contrast, relatively little fractionation occurs between carbon isotopes when the **inorganic calcium carbonate** ($CaCO_3$) **shells** of coccolithophorids, and benthic and planktic foraminifera are precipitated out of seawater. Thus, their shells record the $^{13}C/^{12}C$ **ratio** of the water from which they are precipitated.

As in oxygen isotopes, the measured ratios of $^{13}C/^{12}C$ in a sample are compared to a commonly agreed upon standard. These "delta" (δ) values are reported in units of parts per thousand (‰), or "**per mil**" (in contrast to units of parts per hundred, or percent). The following equation summarizes how stable carbon isotope values are calculated:

$$\delta^{13}C \text{ (in ‰ or per mil)} = \frac{(^{13}C/^{13}C)_{sample} - (^{13}C/^{12}C)_{standard}}{(^{13}C/^{12}C)_{standard}} \times 1000$$

When the $^{13}C/^{12}C$ ratio measured in calcium carbonate shells has a higher ^{13}**C** content than is present in the standard, this trend is referred to as **an increase** in $\delta^{13}C$, while a $^{13}C/^{12}C$ ratio with higher ^{12}**C** than is present in the standard, this trend is referred to as **a decrease** in $\delta^{13}C$.

NAME ______________________________

you think are significant or might be useful in understanding the record of carbon transfer through time.

____

____

3 When **benthic** and **planktic** foraminifers precipitate calcium carbonate ($CaCO_3$) **shells**, would you expect their $\delta^{13}C$ values to be **very similar to** or **somewhat different** from one another?

____

Why?

____

____

____

____

4 Complete in the sentence below for the conditions (a) to (d) based on your understanding of stable carbon isotopes (see box above) and the information in Table 9.1.

The δ^{13}**C record** of **foraminiferal shells** will (**circle one below**) if there is:

(a) burial of more organic carbon in sediments/rocks.	**increase**	**decrease**
(b) a reduction in the amount of organic carbon buried in sediments/rocks.	**increase**	**decrease**
(c) erosion of buried organic carbon and carbon oxidation forming CO_2.	**increase**	**decrease**
(d) greater emissions of volcanic CO_2 ($\delta^{13}C = -6$ to $-4‰$)	**increase**	**decrease**

5 On Figure 9.2, write the letter "B" to **identify times** during the Cenozoic when **organic carbon burial** had probably **increased**.

6 What **conditions** (e.g., changes in sea level, or productivity) might favor **organic carbon burial** in marine sediments? Recall earlier considerations of sources and sinks of carbon from Chapter 5.

____

____

NAME

7 On Figure 9.2, write the letter "E" to **identify times** during the Cenozoic when **organic matter burial** had probably **decreased**, or **previously buried organic matter was eroded, oxidized**, and returned to the atmosphere and ocean reservoirs as CO_2.

8 What **conditions** (e.g., changes in sea level, or productivity) might favor a **decrease in organic matter burial** in marine sediments, or the erosion of buried organic matter in marine sediments? Recall earlier considerations of sources and sinks of carbon from Chapter 5.

9 (a) **List below all the positive or negative $\delta^{13}C$ excursions you observe in Figure 9.2.** Speculate about possible climate significance of each.

Timing	Excursion positive or negative)	Climatic Significance (e.g., warm/ cold; high/low sea level)

(b) Which of these $\delta^{13}C$ excursions is the **largest**?

NAME

An Important Discovery

10 Carefully examine Figure 9.3 and **read the figure caption**. Review previous information as needed and complete Table 9.2 by **listing at least three observations from the data presented** and **your interpretations of what they might represent.**

TABLE 9.2. Observations and Interpretations from Figure 9.3, changes in $\delta^{18}O$ and $\delta^{13}C$.

Observations	Interpretations
1.	1.
2.	2.
3.	3.

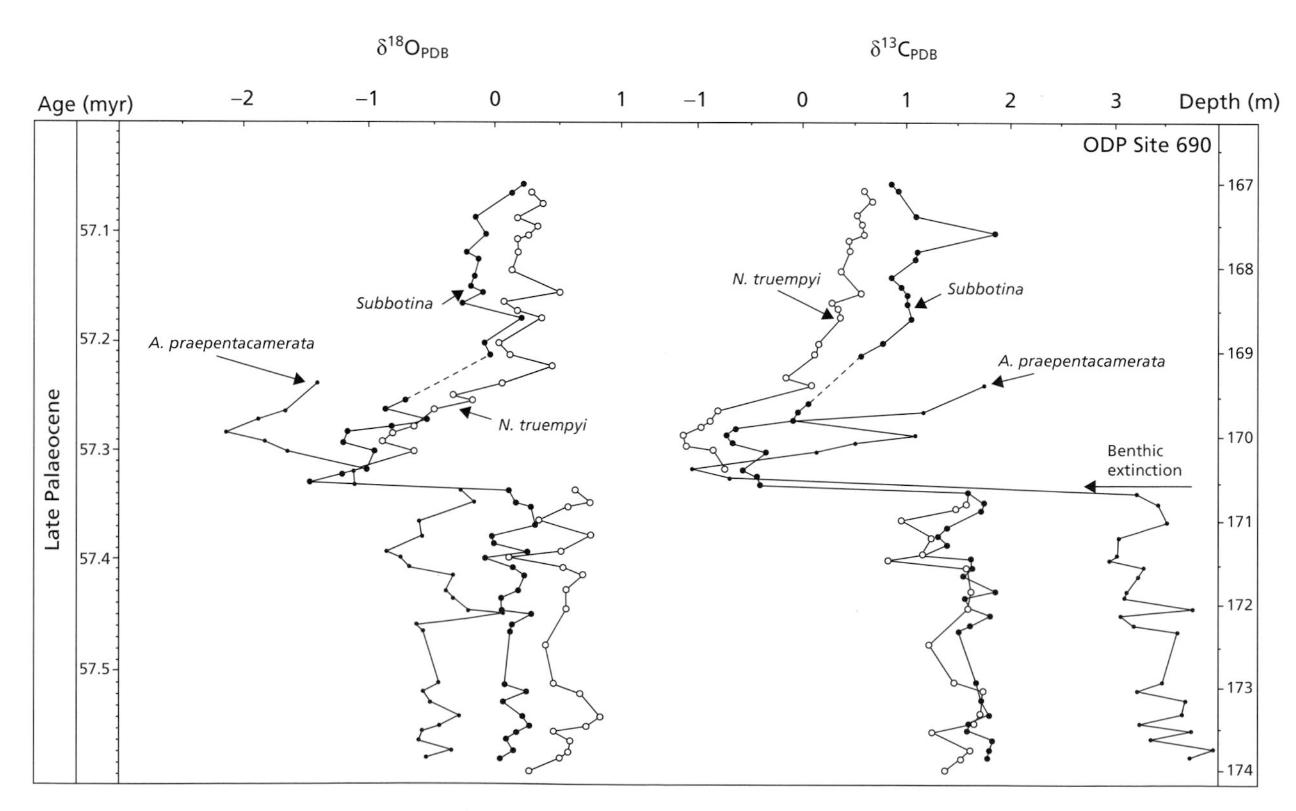

FIGURE 9.3. Changes in $\delta^{18}O$ and $\delta^{13}C$ values of **planktic foraminifera** *Acarinina praepentacamerata* (**mixed layer-dweller**) and *Subbotina* sp. (**thermocline-dweller**) and the **benthic foraminifer** *Nuttalides truempyi* (from ODP Hole 690B, Maud Rise, Antarctica) over a period of approximately 600,000 years (57.6–57.0 Ma). From Kennett and Stott, 1991.

NAME ____________________

11 (a) What part of the geologic time scale and age range are represented in Figure 9.3 (left side).

____

____

(b) Now go to the geologic time scale on the inside front cover of this book and compare and contrast the time scale in Figure 9.3 with the more recently updated time scale. Describe your findings below.

____

____

____

____

____

____

DYNAMIC SCIENCE

The data contained in Figure 9.3 above were the first to show this significant **negative excursion in both $\delta^{13}C$ and $\delta^{18}O$ values** of benthic and planktic foram shells, coincident with a **benthic foraminiferal extinction event**. This was a **major discovery** that led us, through numerous additional studies, to our present understanding of abrupt climate change in the Earth's ancient past and its potential implications for understanding future climate change. The importance of this discovery and the subsequent work cannot be emphasized enough.

The paleoclimatic event that is the focus of this chapter was initially called the **Late Paleocene Thermal Maximum (LPTM)**, but in subsequent years, the International Commission on Stratigraphy has concluded that **this event marks the Paleocene/Eocene boundary**, so the event is now called the **Paleocene-Eocene Thermal Maximum (PETM)**. In addition, the age calibration of the event (55.8 Ma according to the 2009 time scale) has improved since the Kennett and Stott paper was published in 1991 (Figure 9.3). Throughout the remainder of this chapter the PETM is described as occurring at approximately 55 Ma based on the publications used, but individual examples from the literature all refer to the same PETM event.

12 Read the box above on the dynamic nature of science and, in a sentence or two, describe the **important discovery** revealed by the data that you examined in Figure 9.3. Speculate on the **significance** of this important discovery.

____

NAME

NAME

The Paleocene–Eocene Thermal Maximum (PETM) Event

Part 9.2. Global Consequences of the PETM

In Part 9.1, you described changes in the marine carbon and oxygen stable isotope records that occurred at or near the time of the Paleocene/Eocene boundary. These data record a climatic event called the **Paleocene–Eocene Thermal Maximum (PETM)**. In this part, we will consider a variety of data from a number of locations around the world (Figure 9.4). Groups of students will each assess data for one or more sites and report back to the class. The

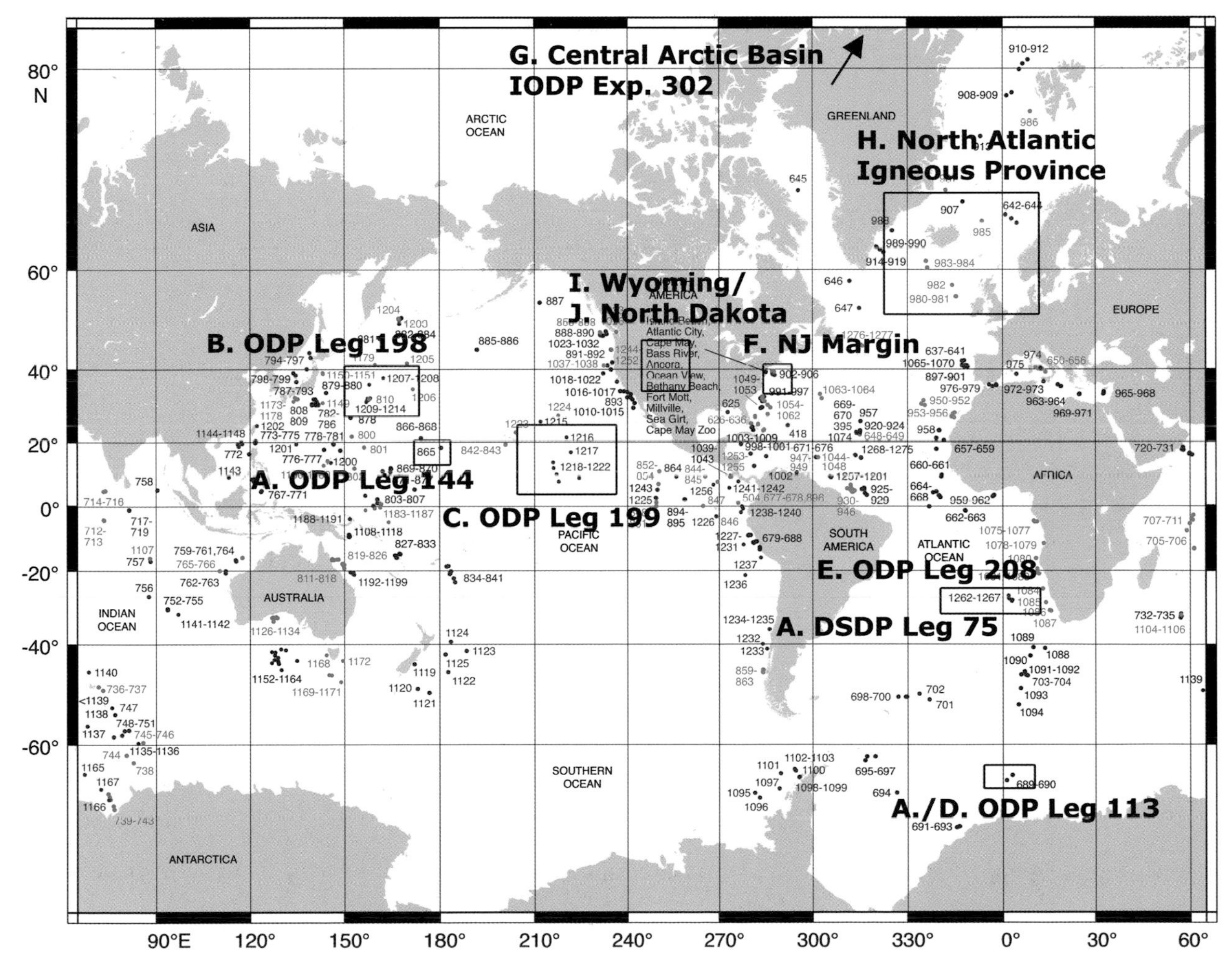

FIGURE 9.4. Map of all the Ocean Drilling Program (ODP) sites (1985–2003). Note: many more new sites have been added since 2003 with the Integrated Ocean Drilling Program (IODP). Courtesy of IODP.

NAME ______________________________

table at the end of this section will be used to compile the key findings from each site and build a **spatial (geographic) assessment of the impact of the PETM.** The goal is to make environmental connections across the globe and to hypothesize about a possible cause, or causes, for the PETM.

Data from a number of ODP and Deep Sea Drilling Project (DSDP) legs and land-based sections are presented below.

A. Benthic Foram Carbon and Oxygen Stable Isotope Records from Multiple Deep-Sea Sites

1 Describe the nature of the PETM as recorded by stable isotopes in Figure 9.5. Comment on the rate of change (rapid or gradual) through the record, magnitude of isotope excursions (positive or negative?), and similarities or differences between the $\delta^{13}C$ and $\delta^{18}O$ trends. Note that older is to the right in these plots and you will want to **frame your discussion in a temporal sequence of events** (i.e., from older to younger).

Magnitude of $\delta^{13}C$ excursion:

____

Magnitude of $\delta^{18}O$ excursion:

____

Similarities between sites:

____

____

____

____

Differences between sites:

____

____

____

____

2 Recall that paleotemperatures can be inferred from marine stable oxygen isotopes. Are Antarctic deep waters colder, warmer, or the same as South Atlantic and equatorial Pacific deep waters (see Figure 9.5)? Does this surprise you?

____

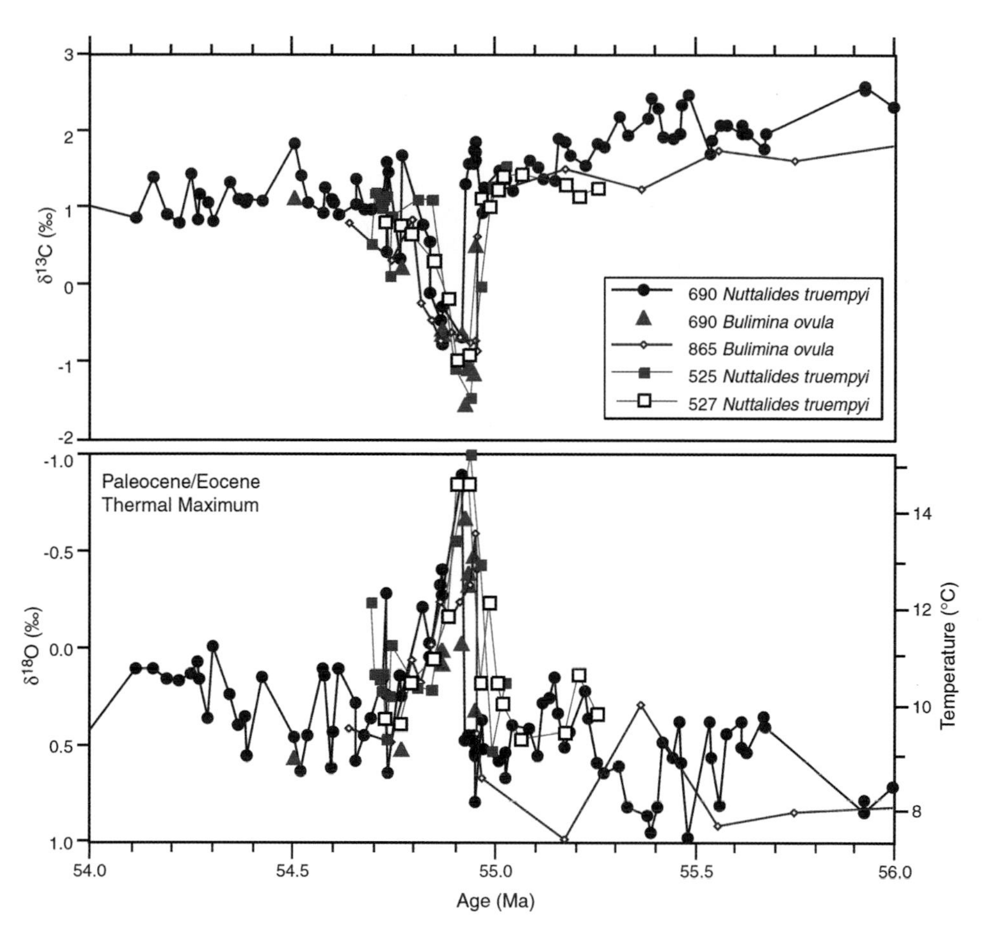

FIGURE 9.5. Benthic foraminiferal oxygen (lower panel) and carbon (upper panel) **isotope records** across the Paleocene–Eocene boundary from DSDP Sites 525 and 527 (DSDP Leg 75) on the Walvis Ridge in the southeast Atlantic, ODP Site 690 (ODP Leg 113) near the Weddell Sea of Antarctica, and ODP Site 865 (ODP Leg 44) on Allison Guyot, Mid-Pacific Mountains in the equatorial Pacific. Two different species were analyzed. Note the paleotemperature scale on the right side of the $\delta^{18}O$ plot. Multiple lines of evidence suggest that the Paleocene–early Eocene was an **ice-free world** (i.e., no ice sheets; a Greenhouse World, see Ch. 5) and **oxygen isotope values primarily represent changes in temperature only**, not changes in ice volume. Age (Ma; millions of years) increases from left to right. **The PETM occurred at approximately 55 Ma.** From Zachos et al., 2001.

3 How much did the deep-sea warm during the PETM?

NAME ______________________________

B1. Record of the PETM Event at the ODP Leg 198 Sites on Shatsky Rise in the Paleo-Subtropical Pacific

4 Describe what you observe about the sediments in the digital core photographs of the PETM at Shatsky Rise (Figures 9.6 and 9.7). Make a list of **observations**, as well as **questions** that you may have about the event as it is recorded in this region.

Observations:

____

____

____

____

Questions:

____

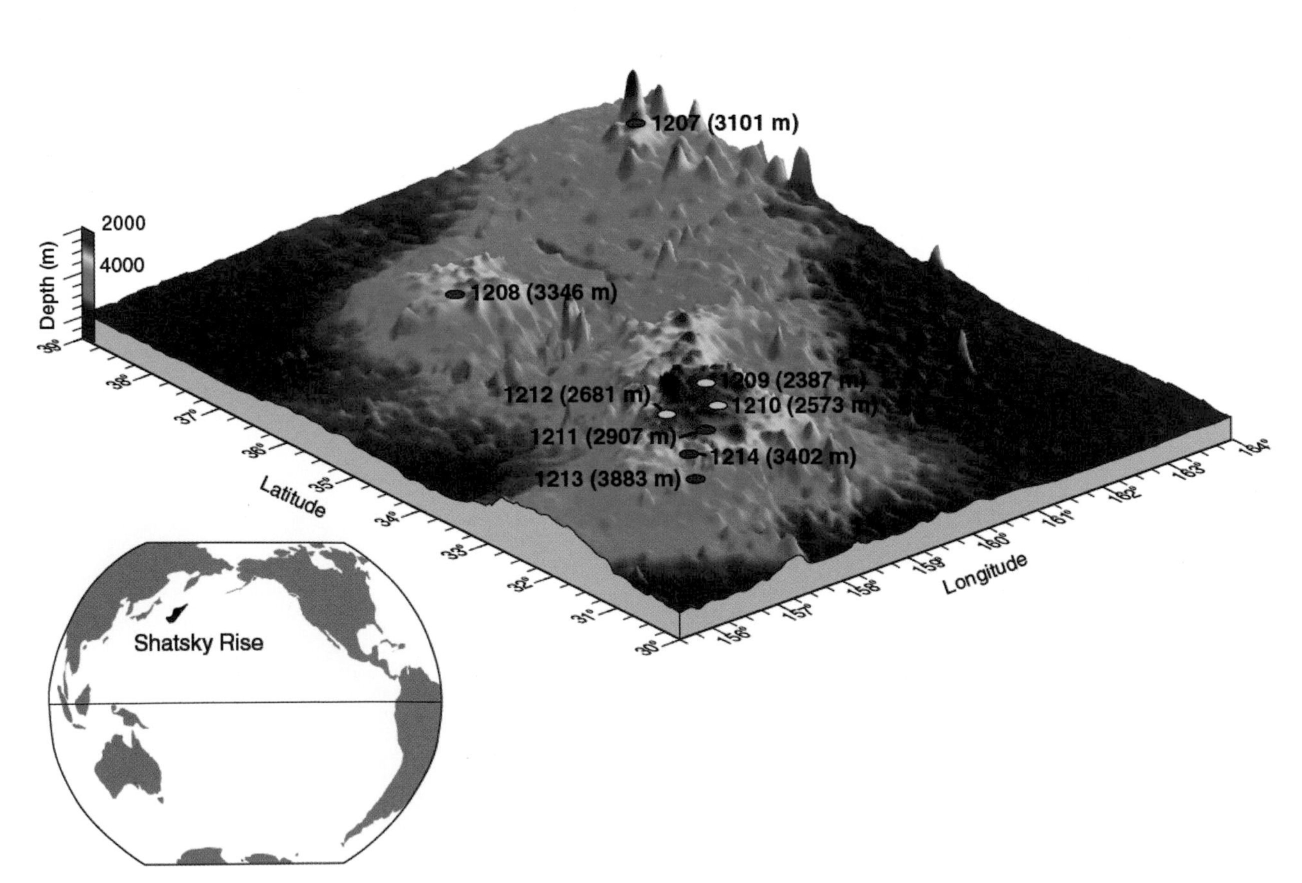

FIGURE 9.6. Location of sites drilled on Shatsky Rise during ODP Leg 198. Note the depth transect of sites on the southern Shatsky Rise. From Bralower, et al., 2002.

NAME

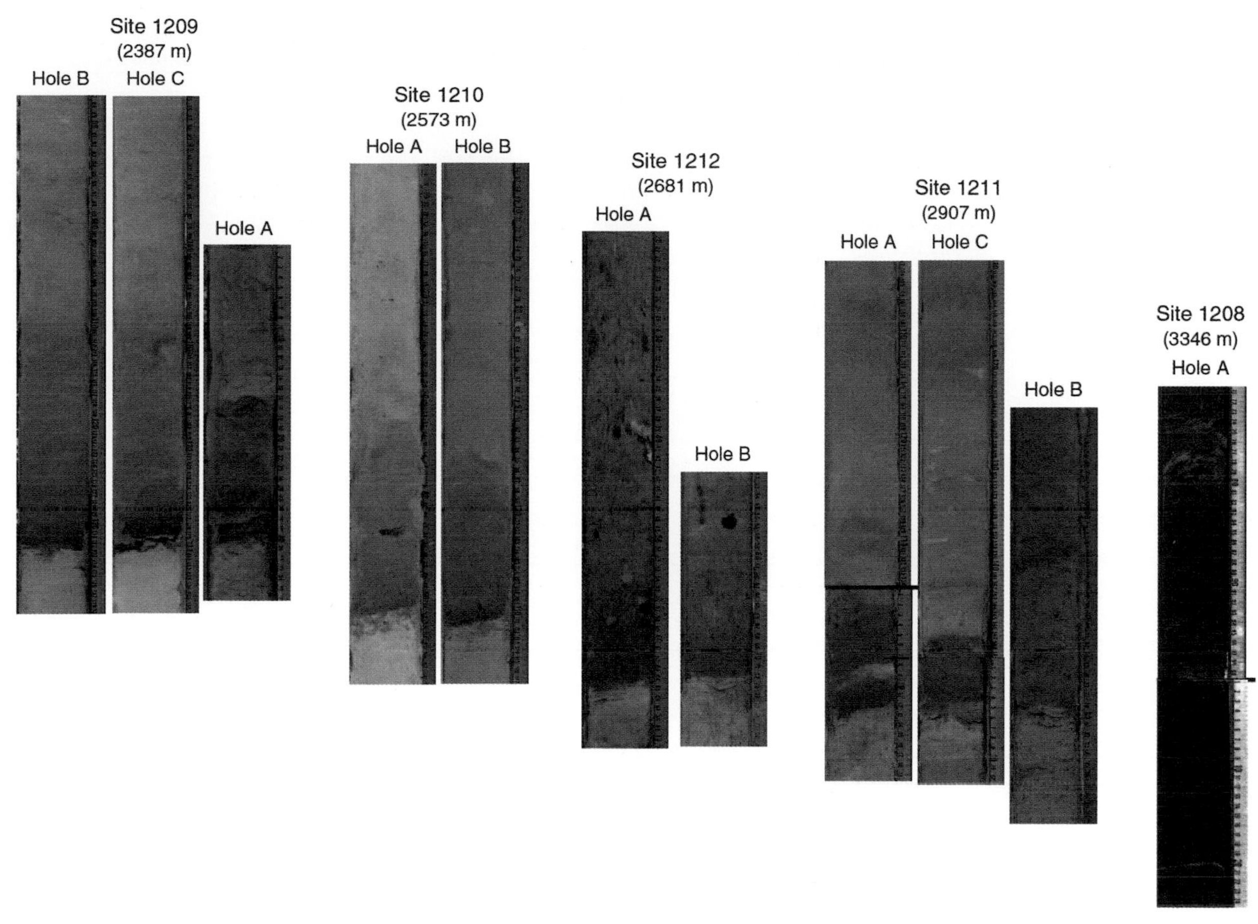

FIGURE 9.7. PETM sections on Shatsky Rise. **Sites are arranged by increasing water depth from left to right**. Note that more than one hole was cored at all sites except Site 1208. **Lighter colored sediments are calcareous** (dominated by calcareous nannofossils), while **darker colored sediments are carbonate-poor and dominated by clay minerals**. Nannofossil biostratigraphy and physical properties measurements allow correlation of Site 1208 (farthest on the right) on the central Shatsky Rise with Sites 1209–1212 on the southern Shatsky Rise. From Bralower, et al., 2002.

5 Does the onset (start) of the event appear to be sharp/abrupt or gradual? What is your evidence?

NAME

6 Does the recovery of the event appear to be sharp/abrupt or gradual? What is your evidence?

7 What is different about Site 1208 relative to the other sites? Can you suggest any reasons why the record of the PETM at Site 1208 may differ from the PETM interval on the southern Shatsky Rise? (Note: see Figure 9.6 for a map of these locations).

B2. Stable Isotope Evidence of the PETM at Site 1209 on Shatsky Rise (ODP Leg 198)

8 What is the average magnitude of the $\delta^{13}C$ excursion (Figure 9.8)? Is the excursion positive or negative?

9 What is the average magnitude of the $\delta^{18}O$ excursion of surface-dwelling planktic forams? Is the excursion positive or negative?

NAME

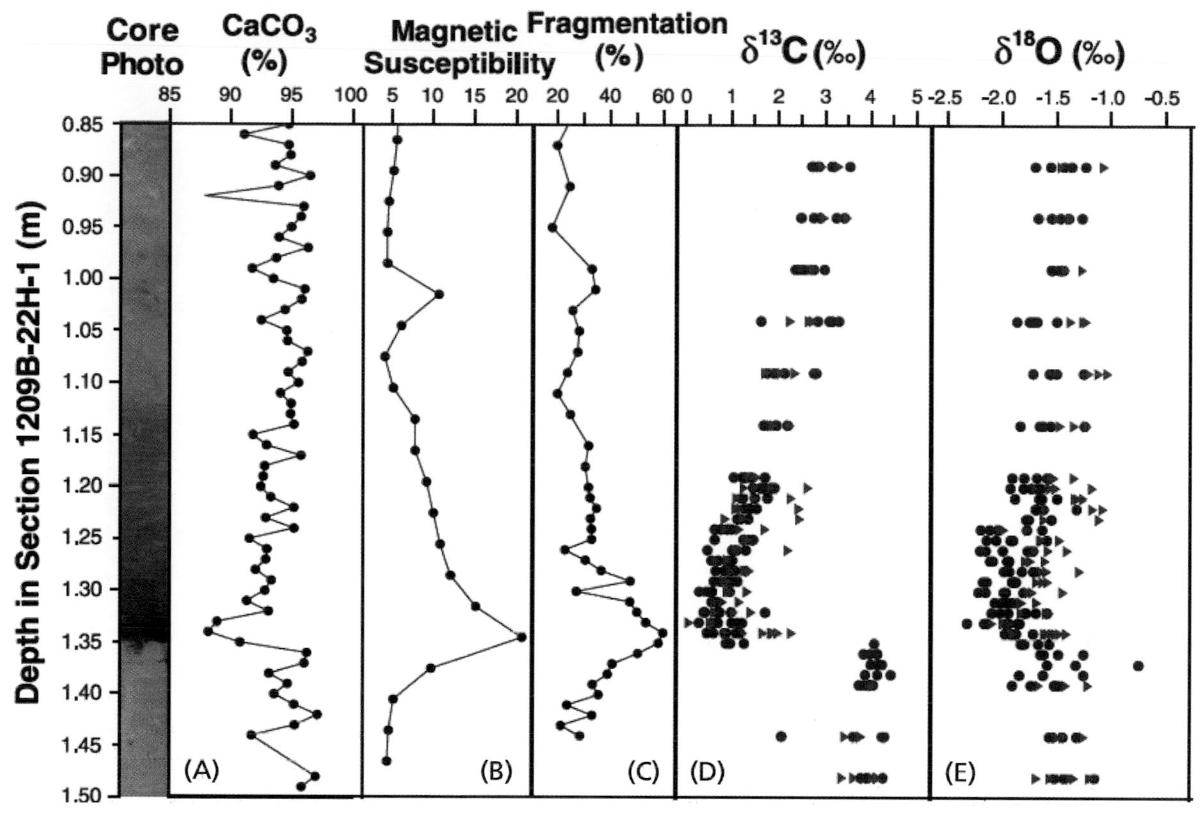

FIGURE 9.8. Hole 1209B was drilled in a water depth of 2387 m on the southern high of Shatsky Rise. The PETM occurs at 1.35 m (135 cm) in Section 1209B-22H-1. (A) %$CaCO_3$ of bulk sediment; (B) Magnetic susceptibility; (C) %fragmentation of planktic forams; (D) single-specimen $\delta^{13}C$, and (E) single-specimen $\delta^{18}O$ of **surface-dwelling species** *Morozovella velascoensis* (blue circles) and *Acarinina soldadoensis* (red triangles). Note: many studies utilize multiple specimens in each isotopic analysis; this study measured **individual foram shells** for each analysis (i.e., each data point is one specimen). From Zachos et al., 2003.

---

10 If a 1°C increase in temperature produces a 0.25‰ decrease in $\delta^{18}O$, approximately how much did sea surface temperatures change at this location (subtropics) during the PETM? Show your work.

NAME

DEEP-SEA CARBONATE AND THE CALCITE (OR CARBONATE) COMPENSATION DEPTH (CCD)

A change in carbonate content is a visual and striking aspect of many deep-sea records of the **PETM** (this chapter) and **Oi1** (Chapter 10) events. Carbonate content (percent carbonate, $\%CaCO_3$) in marine sediments is influenced by: (1) **production** of carbonate by calcareous plankton in the surface ocean, (2) **dilution** of carbonate input by mud and other terrigenous sediments coming in from the weathering of land masses or volcanic islands, and (3) **dissolution** of carbonate by corrosive deep waters. The depth in the deep ocean marking the balance between carbonate rain (i.e., microscopic shells of calcareous plankton) from above and dissolution at depth is called the **calcite (carbonate) compensation** depth, or **CCD**. In general, no carbonate accumulates below the depth of the CCD. The CCD is actually a dynamic surface that varies through space and time. The **depth of the CCD** is influenced by a number of factors, most important of which is **carbonate production in the overlying surface waters** (e.g., the CCD is deeper under areas with a high influx of pelagic carbonate such as at the equator), and **deep water age** (i.e., deep waters originate as surface waters where they acquire their characteristics, such as temperature, salinity, nutrients, and dissolved gases, including oxygen and carbon dioxide, before sinking to become deep water masses). As deep waters move through the ocean basins and get older (i.e., time away from the surface increases), they acquire nutrients and dissolved CO_2 from the **respiration** of deep-sea organisms and by the **decomposition of organic matter** raining down from the surface ocean. This accumulating **CO_2 lowers the pH and makes the deep waters more corrosive to carbonate**. The level of the CCD in the modern ocean is deeper in the Atlantic (approximately 5000 m) because deep waters form today in the Greenland-Norwegian Sea and in the Weddell Sea of Antarctica (i.e., at the north and south ends of the Atlantic); in other words, the **deep waters of the Atlantic today are "young" and CO_2-poor**. By contrast, the level of the CCD in the Pacific, particularly the North Pacific, is much shallower (approximately 4000–4500 m) because the deep waters have been away from the surface for a long time: 1000 years or more. **Deep waters of the present-day Pacific are "old", CO_2-rich, and corrosive to carbonate.** A **massive influx of CO_2** to the ocean may be another way to affect carbonate chemistry adversely in the deep sea, for example, from extensive submarine or sub-aerial volcanism, or catastrophic release of methane (CH_4) from continental margin sedimentary deposits or other sources.

Read the explanation for changing CCD levels in the box above and answer the following questions.

11 Propose a hypothesis to account for the observed change in carbonate content at ODP Site 1209 (Figure 9.8).

NAME

12 Speculate (hypothesize) about why magnetic susceptibility (terrigenous content) increased during the PETM at Site 1209 (Figure 9.8).

13 Speculate about why planktic foram fragmentation (breakage) increased during the PETM at Site 1209.

B3. Record of Planktic and Benthic Foram Changes Associated with the PETM Event at ODP Sites 1209 and 1210, Shatsky Rise.

14 Examine Figure 9.9. These data demonstrate that there were changes in the maximum sizes of both planktic foraminiferal species and benthic foraminiferal species. Summarize the trends for both planktics and benthics. Were the changes abrupt or gradual?

Planktic forams:

NAME

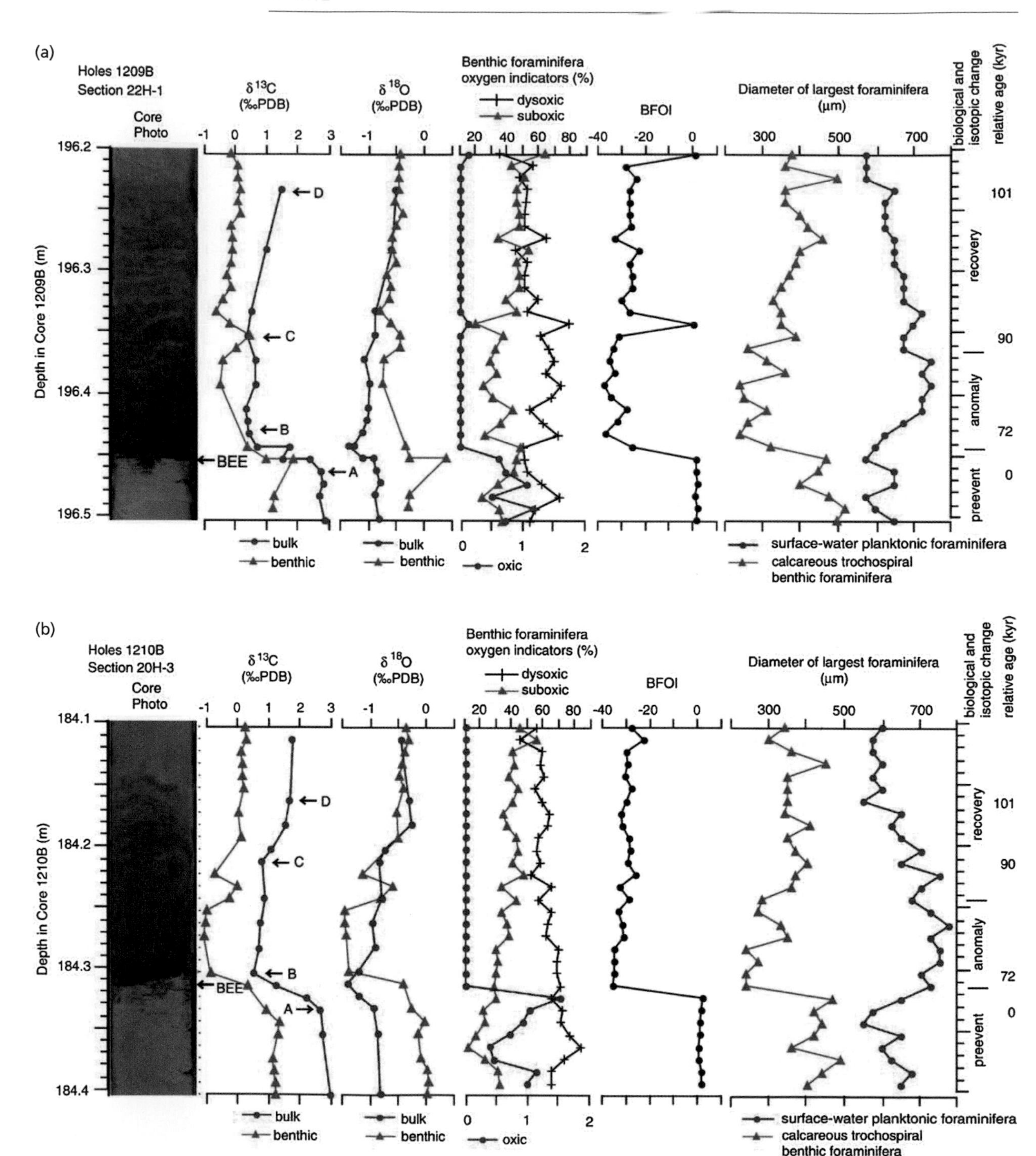

FIGURE 9.9. Stratigraphic **changes in environmental and biotic proxies** through the Paleocene–Eocene boundary interval and the PETM at **two sites on Shatsky Rise** (Holes 1209B and 1210B). Data include (1) stable carbon and oxygen isotope ratios of bulk sediment and benthic forams, (2) percentages of benthic foram oxygen indicators: oxic (species adapted to well-oxygenated conditions; red circles), suboxic (blue triangles), and dysoxic (species adapted to very low dissolved oxygen conditions; black crosses), (3) the **benthic foram oxygen index (BFOI)** is based on quantitative changes in benthic foram species abundances; the BFOI is a reliable **proxy for dissolved oxygen content** in the waters where the forams live, and (4) the maximum diameter of the largest specimens among calcareous deep-sea coiled benthic forams (there are other types of wall composition and chamber arrangements in benthic forams other than trochospiral coiling) and surface-dwelling planktic forams (see Figure 6.5). The P–E boundary interval is divided into three intervals: pre-event, anomaly (onset of the PETM), and recovery. **BEE = benthic extinction event**, which marks the Paleocene–Eocene boundary and the base/onset of the PETM. Key carbon isotope events are labeled A–D. From Kaiho et al., 2006.

NAME

Benthic forams:

15 (a) Are any of the changes in foram size (Figure 9.9) synchronous with changes in the environment based on proxy data, such as stable isotopes, %$CaCO_3$ (see Figure 9.8), or benthic oxygenation? If so, speculate on how the **environmental changes** may have influenced foram sizes.

(b) What types of benthic forams were most affected by the PETM according to the data presented in Figure 9.8?

16 Based on the benthic foraminiferal data presented here, speculate about the possible cause(s) of the **benthic extinction event (BEE)** in the deep-sea at the onset of the PETM.

NAME

C. Record of the PETM Event at Sites 1220 and 1221 in the Paleo-Tropical Central Pacific (ODP Leg 199).

17 How does the sediment color of the PETM interval in the Leg 199 sites (Figures 9.11 & 9.12) differ from the Leg 198 sites (Figure 9.7)?

18 The extinction of deep-sea benthic foraminiferal species at the onset of the PETM has been called the **benthic extinction event, or BEE** (e.g., Kennett and Stott, 1991; Thomas and Shackleton, 1996; Thomas, 2003). Based on the species shown here (Figure 9.12), what percentage of Paleocene species became extinct (i.e., do not range up into Eocene sediments)?

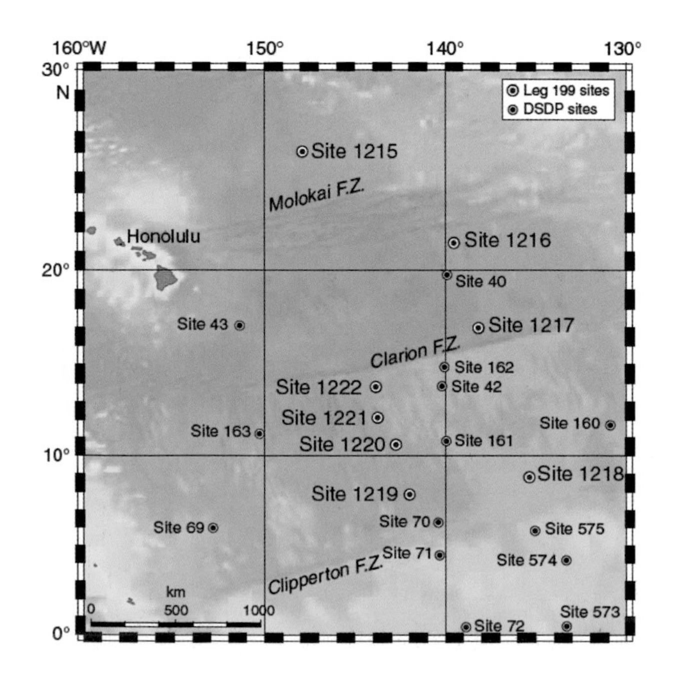

FIGURE 9.10. Map of the central tropical Pacific showing the ODP Leg 199 drill sites (larger font, red circles) and Deep Sea Drilling Project (DSDP) drill sites (smaller font, black circles) on the regional bathymetry. F.Z. = fracture zones in the ocean crust underlying the sediments. From Lyle et al., 2002.

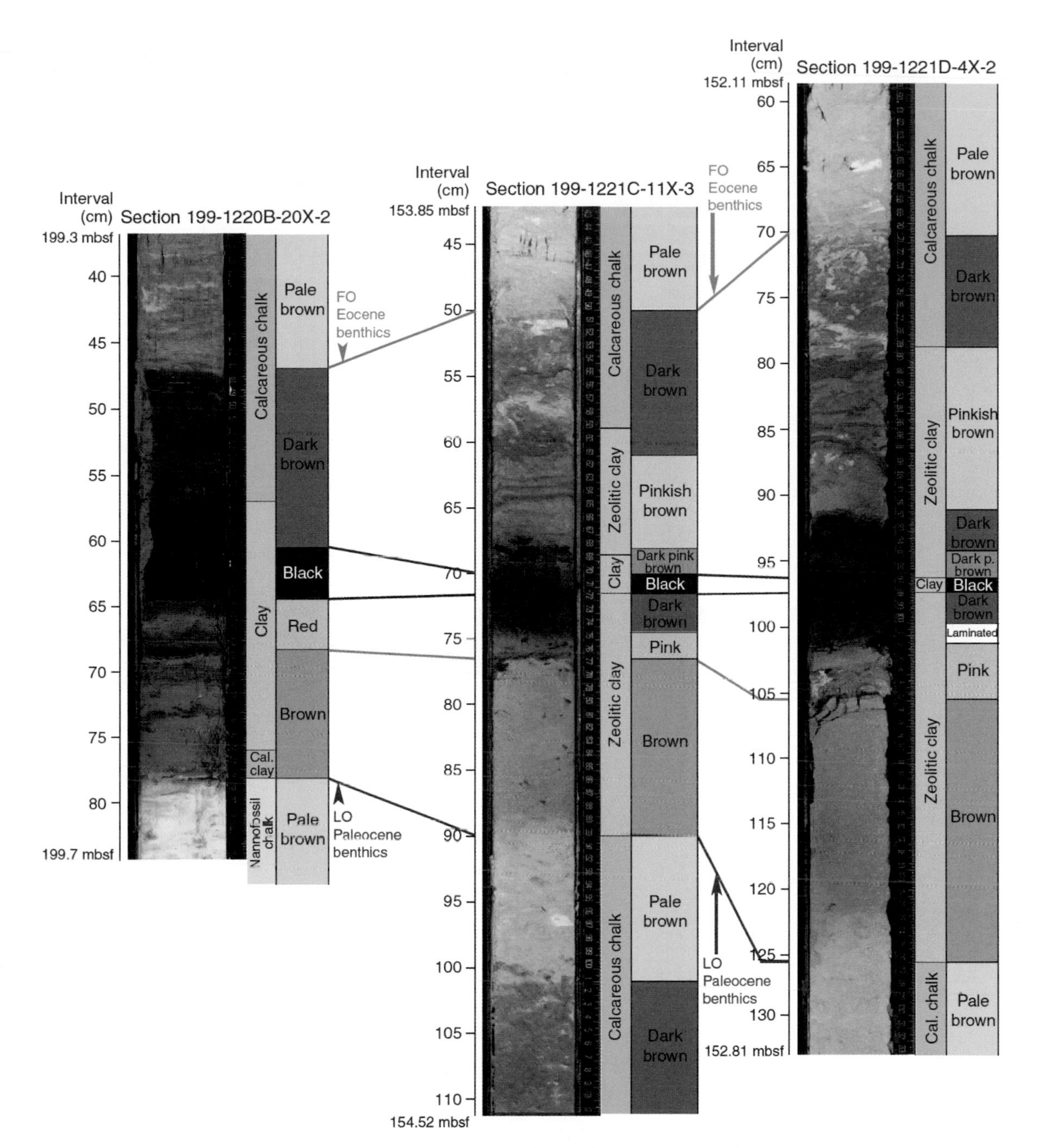

FIGURE 9.11. Digital images of the Paleocene–Eocene boundary sediments recovered at Sites 1220 (Hole 1220B) and 1221 (Holes 1221C and D). The **multicolored clay-rich interval** marks the Paleocene–Eocene boundary, with calcareous chalks above and below. The black layer in the center is a manganese oxide-rich interval. This manganese oxide-rich unit and the zeolite clays on either side indicate **severe carbonate dissolution** and **deposition of hydrothermal materials** derived from a nearby spreading center during Paleocene–Eocene boundary time. The **last occurrence (LO) of Paleocene benthic foraminiferal species** is recorded at the base of the brown zeolitic clay (see Figure 9.12). Calcareous microfossils, including nannofossils and planktic and benthic forams, are absent or poorly preserved in the multicolored clay layers where calcium carbonate levels are very low or reach 0%. From Lyle et al., 2002.

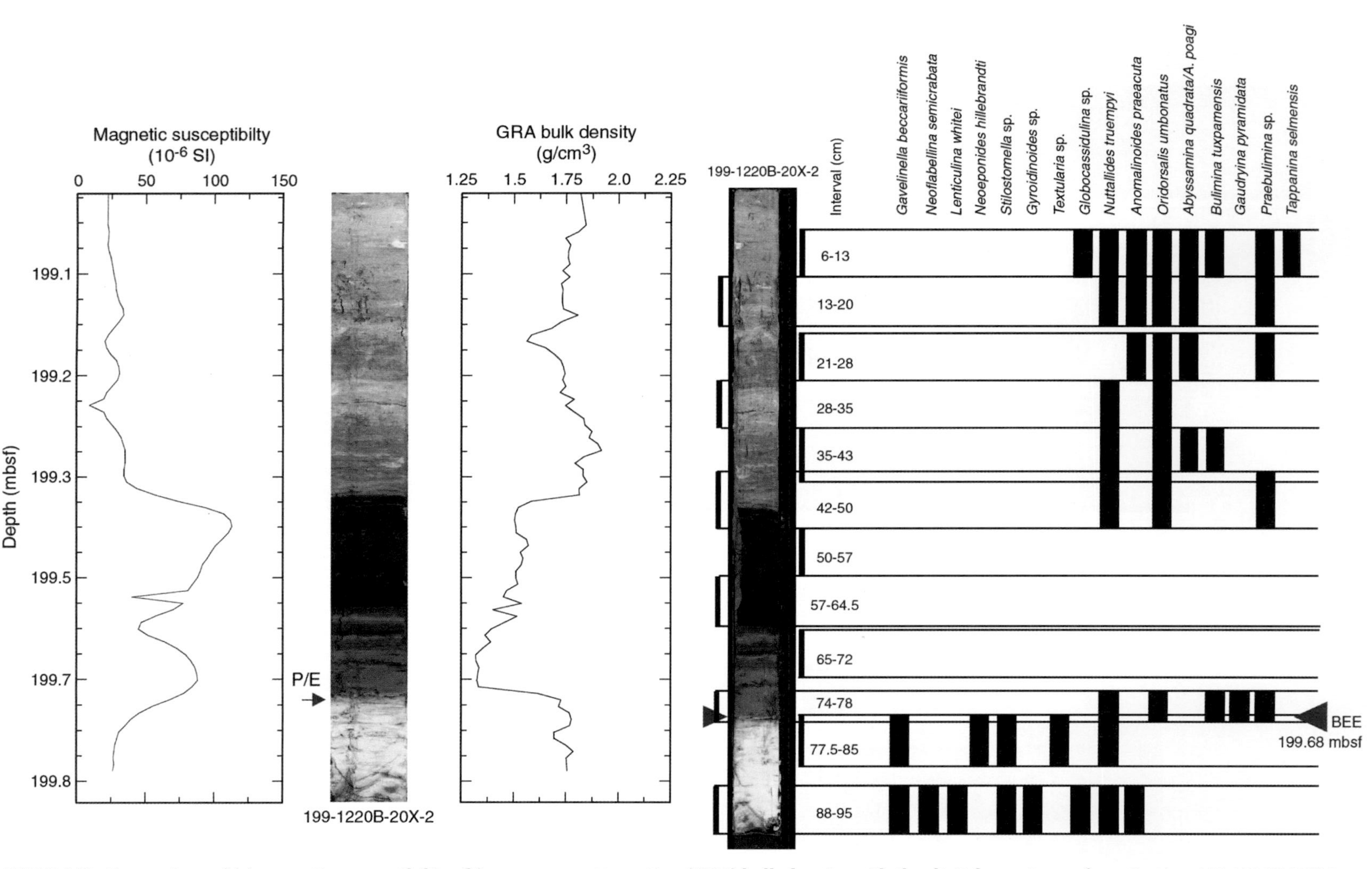

FIGURE 9.12. Comparison of (a) magnetic susceptibility, (b) gamma ray attenuation (GRA) bulk density with the digital core image from Section 199-1220B-20X-2, and (c) the distribution of benthic foraminiferal species in this core section. The magnetic susceptibility and GRA bulk density indicate different physical properties of the **carbonate-poor multicolored clay interval** compared with the calcareous chalks above and below. **The red arrow marks the Paleocene–Eocene boundary and the benthic extinction event (BEE) at the base of the PETM.** From Lyle et al., 2002.

NAME

19 Speculate about **possible causes of the BEE** based on data presented in Figures 9.9 & 9.12. Make a list.

D. Record of the PETM Event at Site 690 in the Southern High Latitudes Near Antarctica (ODP Leg 113).

20 Examine the **high-resolution oxygen isotope data** (Figure 9.13). Do surface-dwelling planktic forams, thermocline-dwelling planktic forams, and benthic forams record a negative isotope excursion at the same time? Describe the sequence of changes observed in the data.

21 Is the **surface** carbon isotope excursion instantaneous with the **surface** oxygen isotope excursion?

22 Is the **benthic** carbon isotope excursion instantaneous with the **benthic** oxygen isotope excursion?

23 If a 1°C increase in temperature produces a 0.25‰ decrease in $\delta^{18}O$, approximately how much did **sea surface temperatures** change at this location (southern high latitudes) during the PETM? Show your work.

NAME

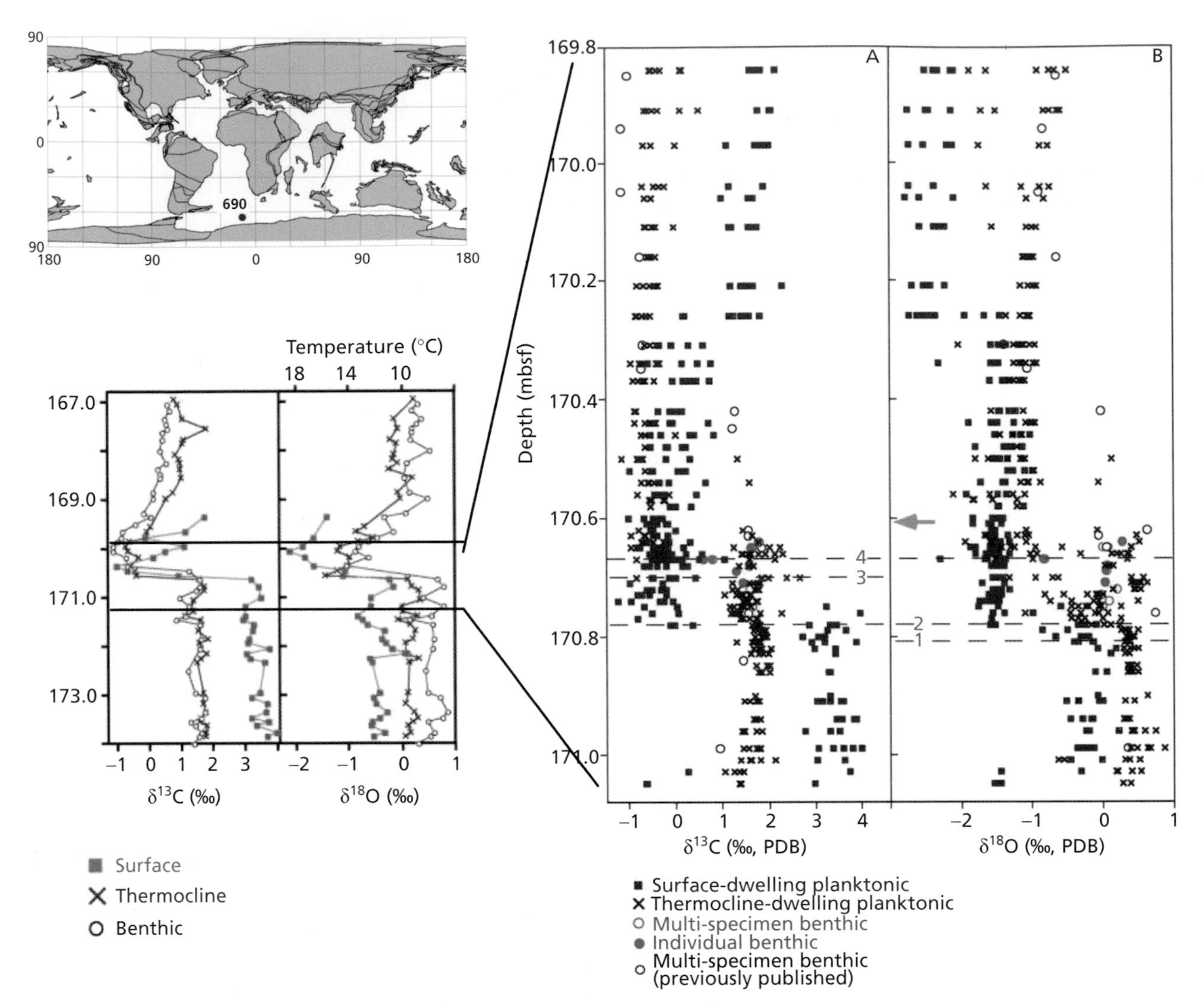

FIGURE 9.13. PETM stable isotope data from **ODP Site 690** near the Weddell Sea Antarctica (refer to map in upper left). Lower left panel shows low-resolution planktic (surface and thermocline) and benthic foram carbon and oxygen isotope data from Kennett and Stott (1991; Figure 9.3). Right panel shows **high-resolution single specimen isotope data**. The depth scale on the *y*-axis of both plots is in mbsf (meters below seafloor). Note (as in Figure 9.8): many studies utilize multiple specimens in each isotopic analysis; this study measured **individual foram shells** for each analysis (i.e., each data point is one specimen). Benthic foram data (*Nuttalides truempyi*) in green filled circles and open circles, thermocline planktic foram data (genus *Subbotina*) in blue "x", and surface-dwelling planktic foram data (genus *Acarinina*) in red filled squares. Gray dashed lines indicate important changes (1–4) during the onset of the PETM as interpreted by Thomas et al. 2002.

24 How much did **deep-sea temperatures** change at this location? Show your work.

NAME

25 Speculate about a possible cause (or causes) of the **difference in timing** between the **environmental changes** in the surface and deep ocean, based on the planktic and benthic isotopic data at Site 690.

E. Record of the PETM Event at the ODP Leg 208 Sites on Walvis Ridge in the Southeast Atlantic.

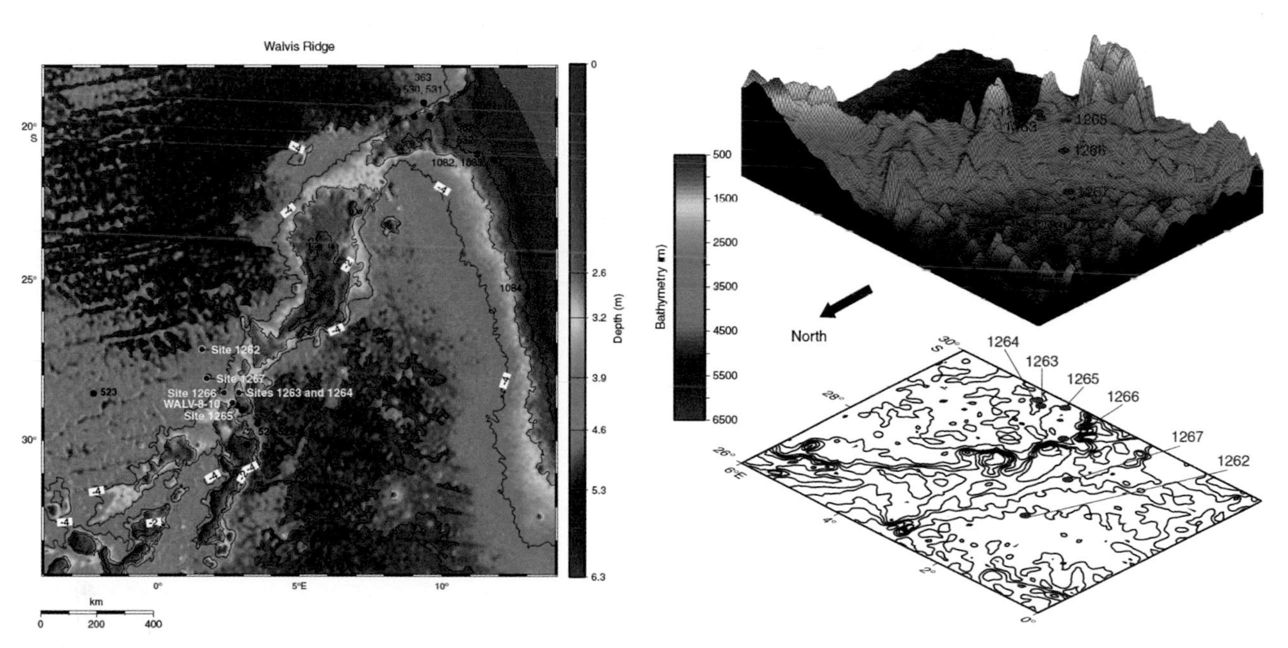

FIGURE 9.14. ODP Leg 208 drill sites on Walvis Ridge. Left: Bathymetric map with **Leg 208 sites** labeled with white circles with green centers, other DSDP and ODP sites labeled with black circles. Right: three-dimensional map showing Leg 208 sites as a depth transect down the northwest flank of Walvis Ridge. From Zachos et al., 2004.

NAME

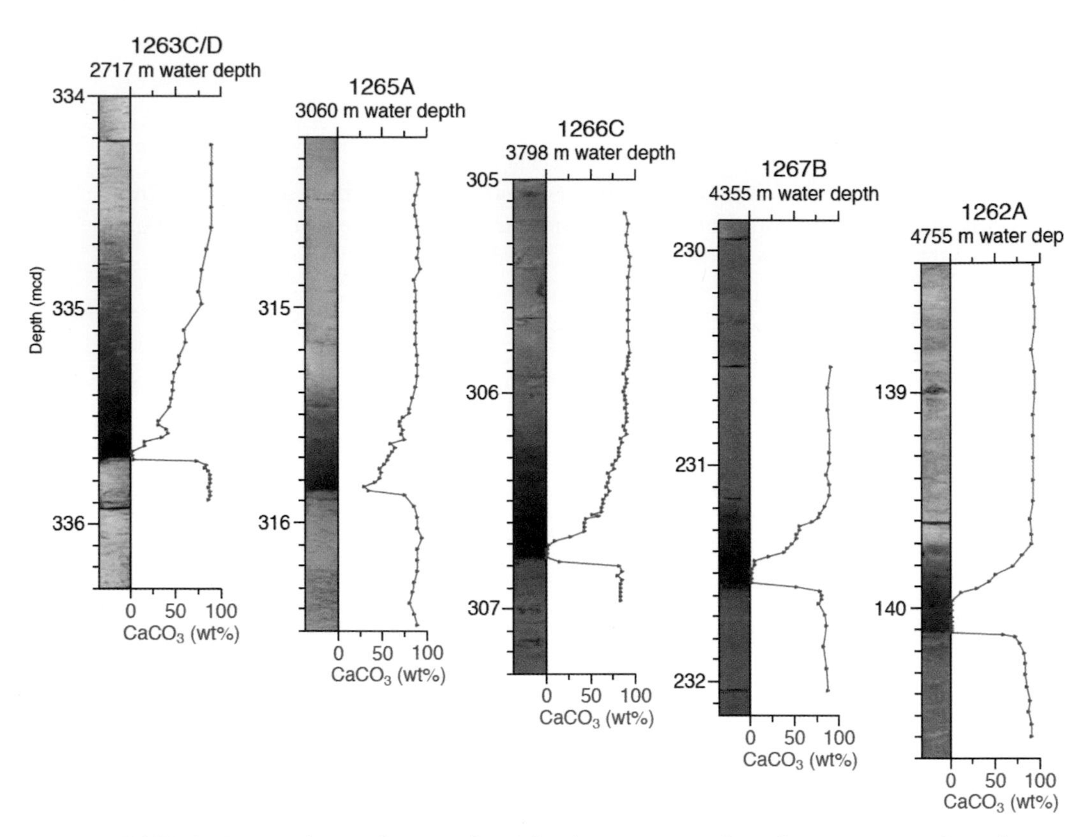

FIGURE 9.15. Digital core photos and **weight %$CaCO_3$** across the Paleocene–Eocene boundary interval at the Leg 208 drill sites on Walvis Ridge in the southeast Atlantic. The sites represent a **depth transect** down the Walvis Ridge from shallowest (left) to deepest (right). Depth is shown as meters composite depth (mcd). Also refer back to Figure 9.1. From Zachos et al., 2005.

26 Make a list of similarities between the five Walvis Ridge sites (Figure 9.14) based on the digital core photographs and carbonate content shown in Figure 9.15.

27 List any differences you observe between the five sites.

NAME

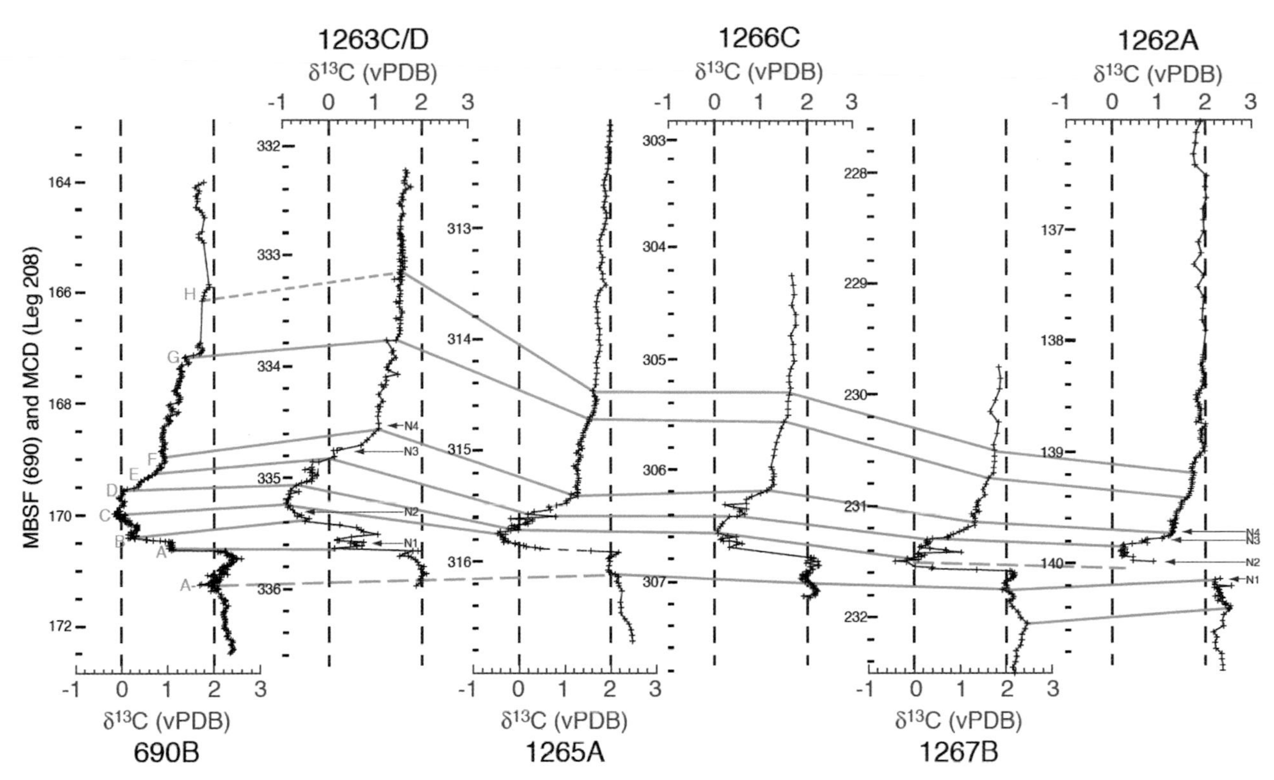

FIGURE 9.16. Bulk sediment **carbon isotope records** across Walvis Ridge from shallow (second from left) to deep (right) plotted against depth (mcd = meters composite depth); correlated to Antarctic Site 690 (far left; mbsf = meters below seafloor). Red arrows on two of the sections correspond to **calcareous nannofossil events** (first or last occurrences). Green lines show **lines of correlation** between Site 690 and the Walvis Ridge sites. From Zachos et al., 2005.

28 How are the six deep-sea sections shown in Figure 9.16 correlated to one another?

29 At which of the sites (i.e., shallowest site or deepest site) is the $\delta^{13}C$ excursion the weakest? And at which is it the strongest?

Weakest:

Strongest:

30 Were all sites in the depth transect adversely affected by carbonate dissolution? Explain.

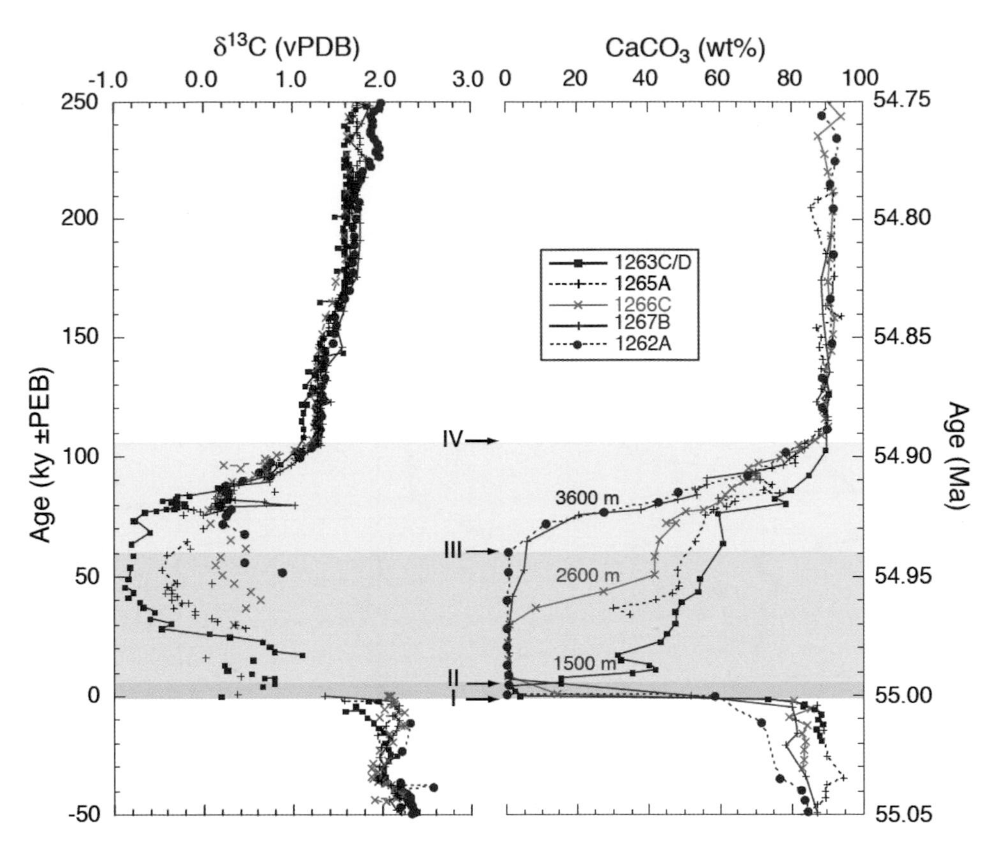

FIGURE 9.17. Composite diagram showing bulk sediment $\delta^{13}C$ and weight %$CaCO_3$ plotted against age (i.e., five Walvis Ridge sites are plotted, each with a different color and symbol). Note the different the water depths for each of the sites. Depth was converted to age using an age model for Site 690 that is correlated to Walvis Ridge sites using the $\delta^{13}C$ record of each site and the calcareous nannofossil events (datum levels; Figure 2.13). Key events in the evolution of South Atlantic carbonate chemistry are denoted by roman numerals. From Zachos et al., 2005.

31 Which site began to recover the quickest from carbonate dissolution? Which site was affected by carbonate dissolution the longest? What is your evidence?

32 Sketch a depth profile across the Walvis Ridge., arranging the Leg 208 sites down the flank. Label each of the sites. On your sketch illustrate how the Walvis Ridge sites were affected differently by carbonate dissolution (e.g., the degree of dissolution, timing of dissolution).

NAME

33 If the record of carbonate dissolution along the Walvis Ridge is best explained by the rise and fall in the level of the carbonate compensation depth (CCD), by how much did the CCD migrate vertically in the water column during the PETM interval? Use your sketch (Question 32) to document what may have happened to the five Walvis Ridge sites during the PETM. Express your answer in meters.

F. Record of Environmental Changes Associated with the PETM Event on the Paleo-Continental Shelf of Eastern North America.

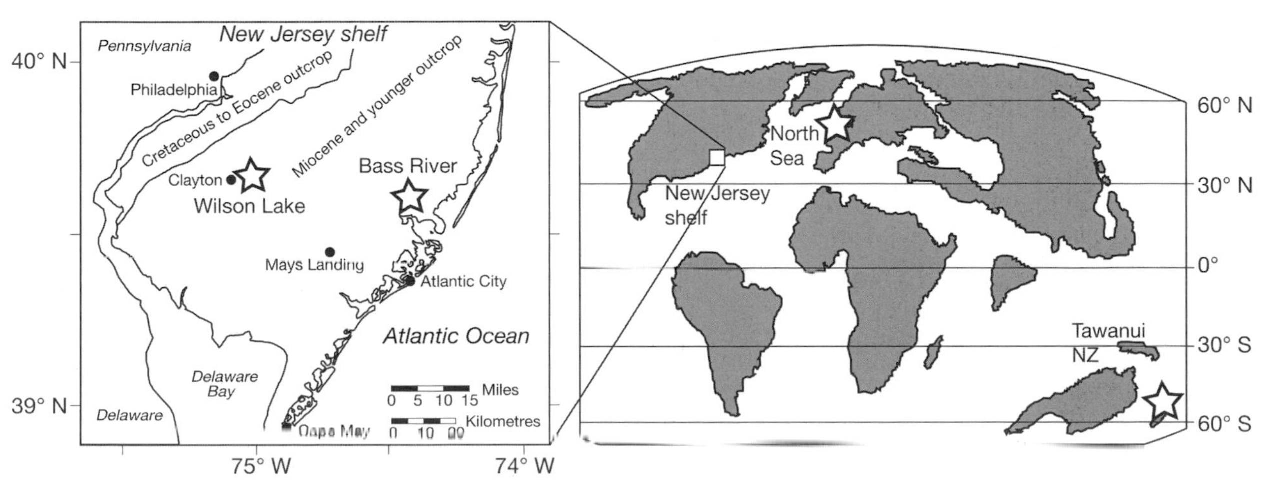

FIGURE 9.18. Location of **two drill sites in New Jersey** (Wilson Lake and Bass River), which were situated on a **continental shelf environment during the Paleocene–Eocene owing to a higher global sea level at that time**. From Sluijs et al., 2007.

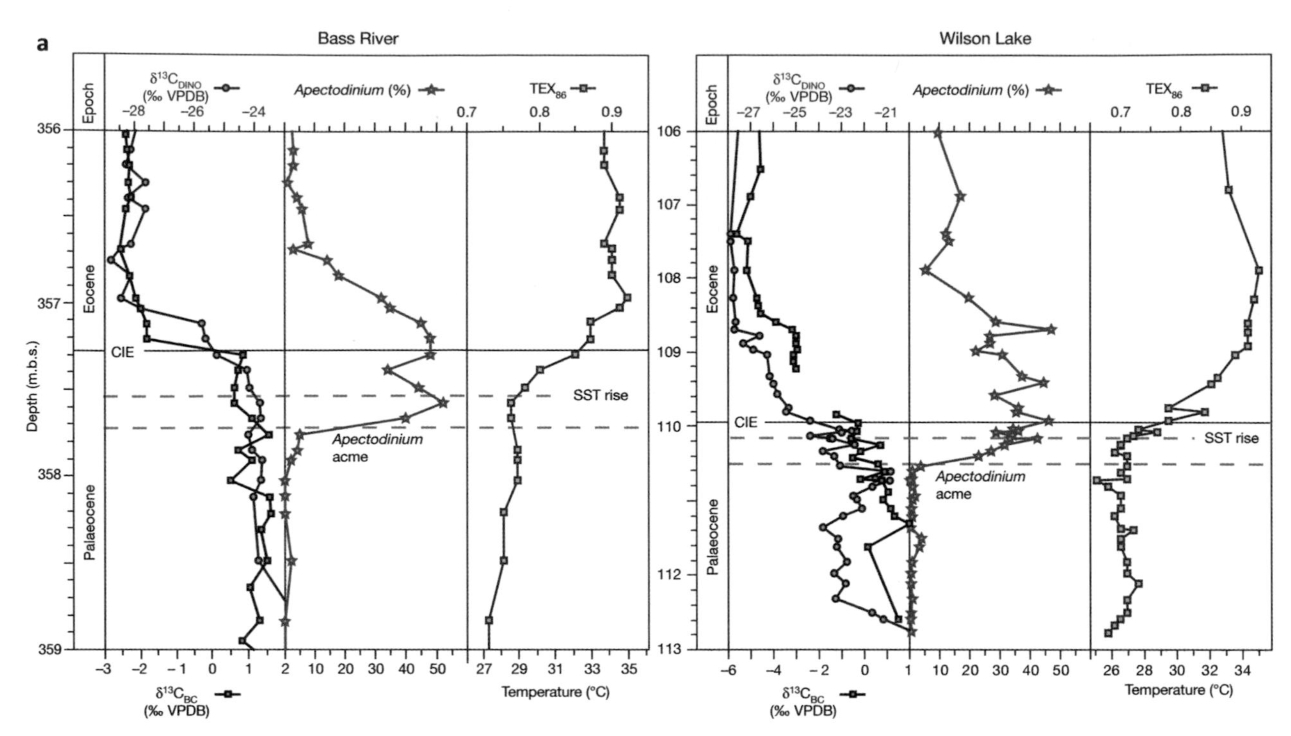

FIGURE 9.19a. High resolution details across the onset of the PETM carbon isotope excursion (CIE) at Bass River (left panel) and Wilson Lake (right panel), two drill sites in New Jersey. For each site, the following data are shown: (1) bulk sediment carbonate $\delta^{13}C$, (2) relative abundance of the subtropical dinoflagellate (organic-walled marine plankton) *Apectodinium*, a subtropical species (acme = peak in abundance), and (3) TEX_{86} paleotemperature proxy data, which is based in the abundance of a particular type of organic biomarker (crenarchaeotal membrane lipids). SST rise = onset in the rise of sea surface temperatures based on the TEX_{86} paleotemperature proxy. From Sluijs et al., 2007.

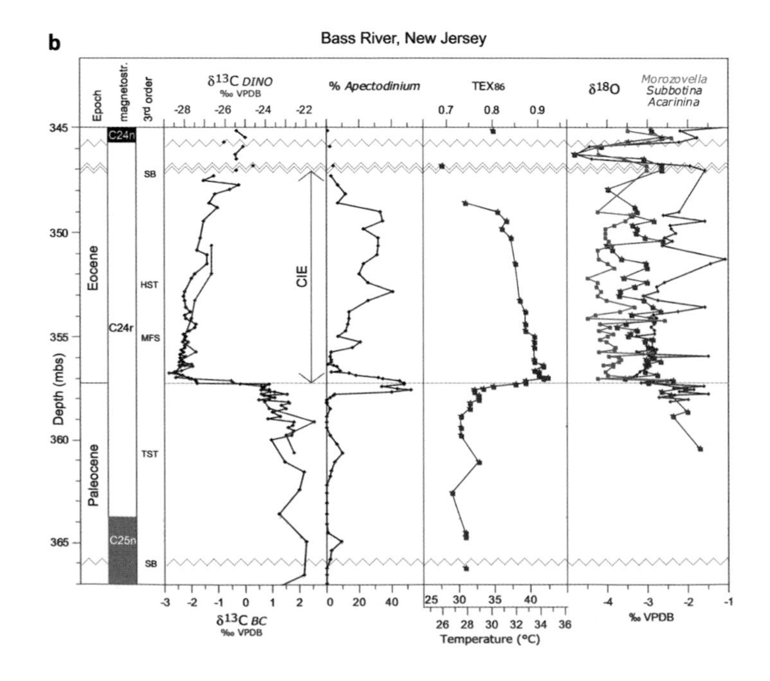

FIGURE 9.19b. High-resolution records of the **onset and recovery** of the PETM interval at Bass River, New Jersey. **This Bass River record also shows the planktic foraminiferal stable isotope record through the PETM.** *Morozovella* and *Acarinina* are mixed layer planktic foraminiferal genera; *Subbotina* is a deeper thermocline genus. From Sluijs et al., 2007.

NAME ______________________________

34 Which came first, a rise in sea surface temperatures (SSTs) or the negative carbon isotope excursion (CIE; Figure 9.19)?

35 What do the results of the above study reveal about the **sequence of events leading up to the PETM**? Cite specific evidence presented in Figure 9.19a. Hint: consider lead–lag relationships in the data ($\delta^{13}C$, subtropical dinoflagellate *Apectodinium* and TEX_{86} paleotemperature).

36 If a 1°C increase in temperature produces a 0.25‰ decrease in $\delta^{18}O$, **how much did sea surface temperatures (SSTs) rise** on the New Jersey shelf during the PETM (Figure 9.19b)? Show your work.

37 How does the **magnitude of SST increase compare with other sites** around the globe (e.g., compare Figure 9.19 TEX_{86} data with Figure 9.13 and Question 23)?

38 Propose a couple of ideas (hypotheses) about what may have caused the PETM:

NAME

G. PETM in the Arctic Ocean, Lomonosov Ridge, IODP Expedition 302

39 At what depth in Hole 302-4A (Figures 9.20 and 9.21) would you place the onset of the PETM? Explain your reason(s) for placing the boundary where you do.

40 What is the magnitude of the carbon isotope excursion (CIE) at this site (Figure 9.21)?

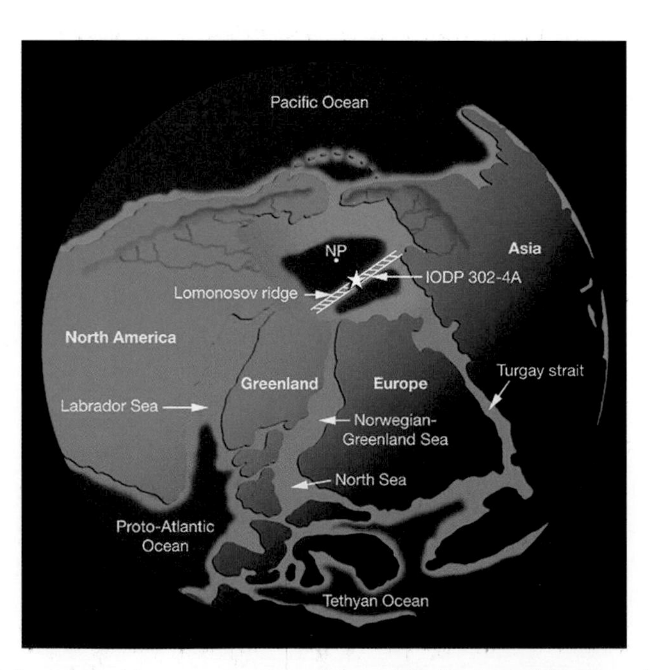

FIGURE 9.20. Paleogeographic reconstruction of the **Arctic Basin** and Northern Hemisphere at the time of the PETM, approximately 55 Ma. Note that the Labrador Sea and Norwegian-Greenland Sea had just begun to open and Greenland was a large continental fragment between North America and northwest Europe. Location of the Integrated Ocean Drilling Program (IODP) Expedition 302, Hole 4A, on the **Lomonosov Ridge**, a fragment of continental crust that rifted from the Eurasian continental margin and is now situated in the central Arctic Basin. NP = North Pole. From Sluijs et al., 2006.

NAME

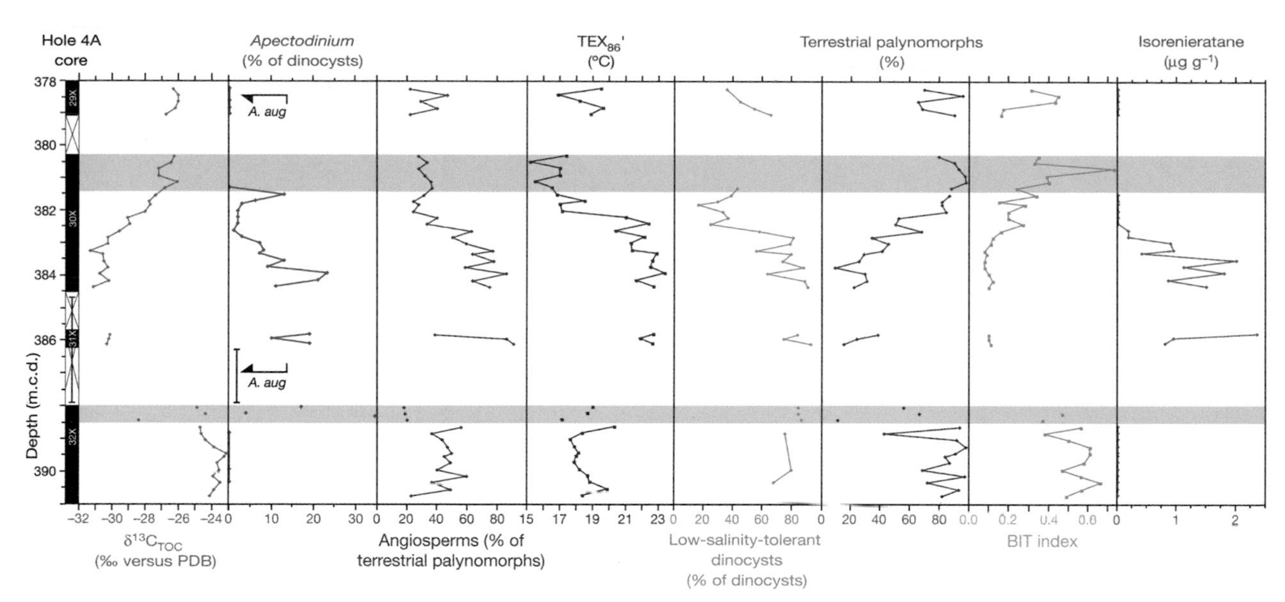

FIGURE 9.21. Core recovery in IODP Hole 302-4A, and geochemical and palynological (organic walled microfossils) data across the **Paleocene–Eocene boundary interval**, which is barren of calcareous and siliceous microfossils. Data columns from left to right: (1) **carbon isotopic values** of total organic carbon (TOC), (2) percentage of the **subtropical dinoflagellate species *Apectodinium*** (dinoflagellates are marine, organic walled plankton preserved in the sediments as "dinocysts"), (3) percentage of **angiosperm/flowering plant pollen** (palynomorphs are organic walled microfossils, in this case pollen and spores from land plants), (4) **TEX_{86}, a paleotemperature proxy** based on the abundance of a particular type of organic biomarker (from crenarchaeotal membrane lipids), probably biased towards summer paleotemperatures, (5) percentage of **low salinity tolerant dinoflagellates species**, (6) percentage of **terrestrial organic walled microfossils** (i.e., pollen and spores) compared with total organic walled microfossils (including the dinocysts), (7) **BIT index** = branched and isoprenoid tetraether, a measure of the amount of river-derived terrestrial organic matter relative to marine organic matter, and (8) **isorenieratane abundance**, which is derived from the brown strain of photosynthetic green sulfur bacteria, a marker of anoxic and sulfidic conditions in the photic zone. From Sluijs et al., 2006.

41 Describe what happened during the PETM at this location (Figures 9.20 and 9.21).

42 The mean annual sea surface temperature (SST) of the Arctic Ocean today is 0°C. What were SSTs like during the PETM based on the TEX_{86} paleotemperature proxy?

NAME

43 In addition to the warm temperatures during the time of the PETM, what other data in Figure 9.21 indicate that the Arctic was much warmer than today?

H. PETM Associated with the North Atlantic Igneous Province and the Opening of the Northern North Atlantic

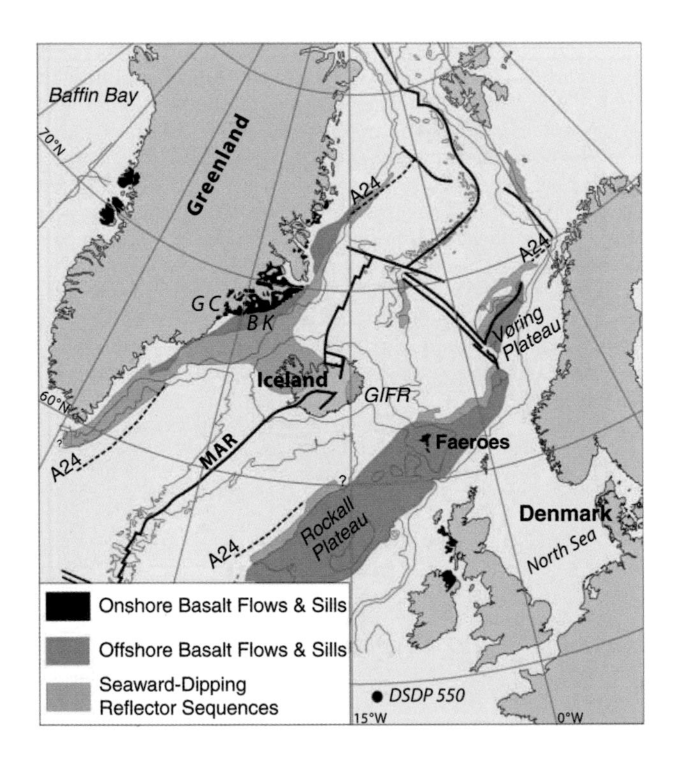

FIGURE 9.22. Map of the present-day northern North Atlantic (Greenland-Norwegian Sea) showing the distribution of onshore and offshore volcanic rocks (basalt flows and intrusive sills) of the **North Atlantic Igneous Province** (NAIP). DSDP Site 550 (Leg 80) contains Danish Ash-17 closely overlying the PETM. A24 is seafloor magnetic anomaly 24r (see Figure 9.23). MAR = Mid-Atlantic Ridge. From Storey et al., 2007.

44 What is the Mid-Atlantic Ridge?

45 How old is the ocean crust at the crest (i.e., axis) of the Mid-Atlantic Ridge?

NAME

46 Describe the relationship between the Mid-Atlantic Ridge and the offshore and onshore basalt flows and intrusions (i.e., sills; Figures 9.22 and 9.23)

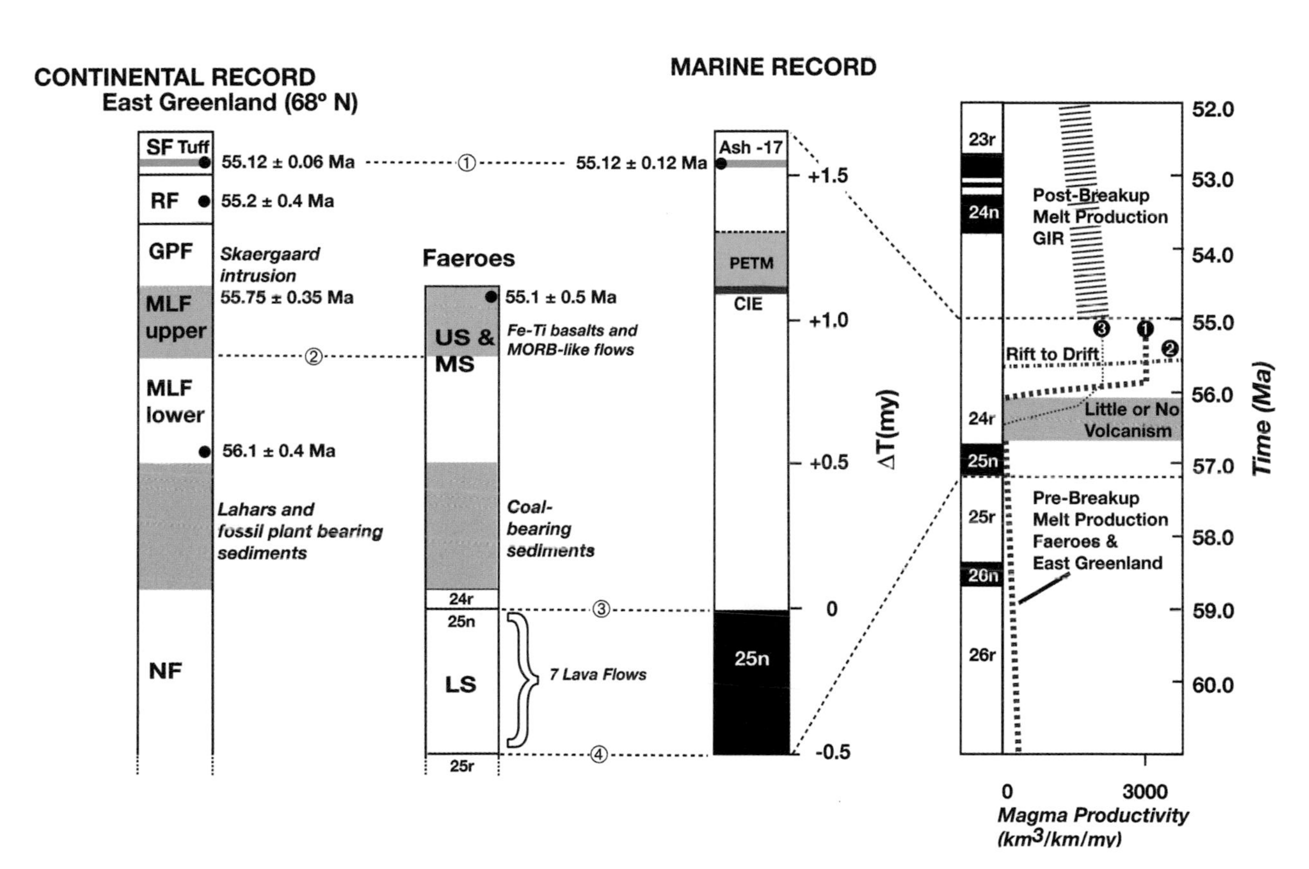

FIGURE 9.23. Correlation of **continental records** of well-dated coal-bearing terrestrial deposits, lava flows (56.1 ± 0.4 Ma; 55.75 ± 0.35 Ma; 55.1 ± 0.5 Ma), and a volcanic tuff (55.2 ± 0.4 Ma) on East Greenland and the Faeroes Islands, with a **marine record** of the PETM interval at DSDP Site 550 containing a well-dated volcanic ash (Danish Ash-17, 55.12 ± 0.12 Ma) and good **paleomagnetic polarity** control (e.g., presence of Chron 25n and 25r). The Faeroes Island flood basalts also contain a paleomagnetic polarity record that can be correlated with the marine core. Dashed lines represent **points of correlation (tie-points)** between the sections. Importantly, the lava flows on Greenland and the Faeroe Islands date to the time of the PETM as the Greenland-Norwegian basin was in a **transition from continental rifting to seafloor spreading**, the so-called transition from **"rift to drift"**. The continental rift basin that formed between Greenland and northwest Europe contained coal-bearing sediments prior to continental breakup. Curves 1–3 represent estimates of magma production as the Greenland-Norwegian basin changed from a rift basin to a new ocean basin (i.e., **future Greenland-Norwegian Sea**) very close to the time of the PETM. From Storey et al., 2007.

NAME ______________________________

47 How can **volcanism** affect the **global carbon cycle** (refer back to Chapter 5, Parts 5.1 and 5.4 for the short-term and long-term carbon cycle, respectively)?

48 Examine Figure 9.23. Describe the **timing of volcanism** in the northern North Atlantic region and compare it with the **timing of the PETM**.

49 Speculate about **possible implications of this volcanic activity** (lava flows at the surface and intrusions into the basin sediments) close to the time of the PETM?

I. Floral Changes in Wyoming Associated with the PETM

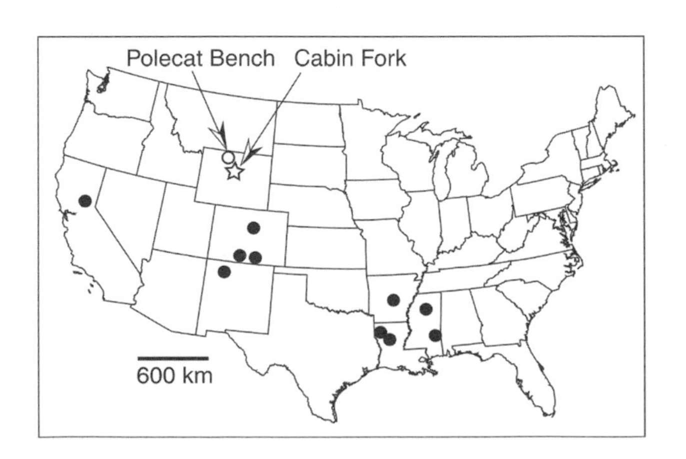

FIGURE 9.24. The open circle shows the locations of Polecat Bench and the open star shows the location of Cabin Fork **PETM sections in Wyoming**. Solid circles show localities of Paleocene–Eocene age with plant types that are restricted to the PETM in northern Wyoming. From Wing et al., 2005.

NAME

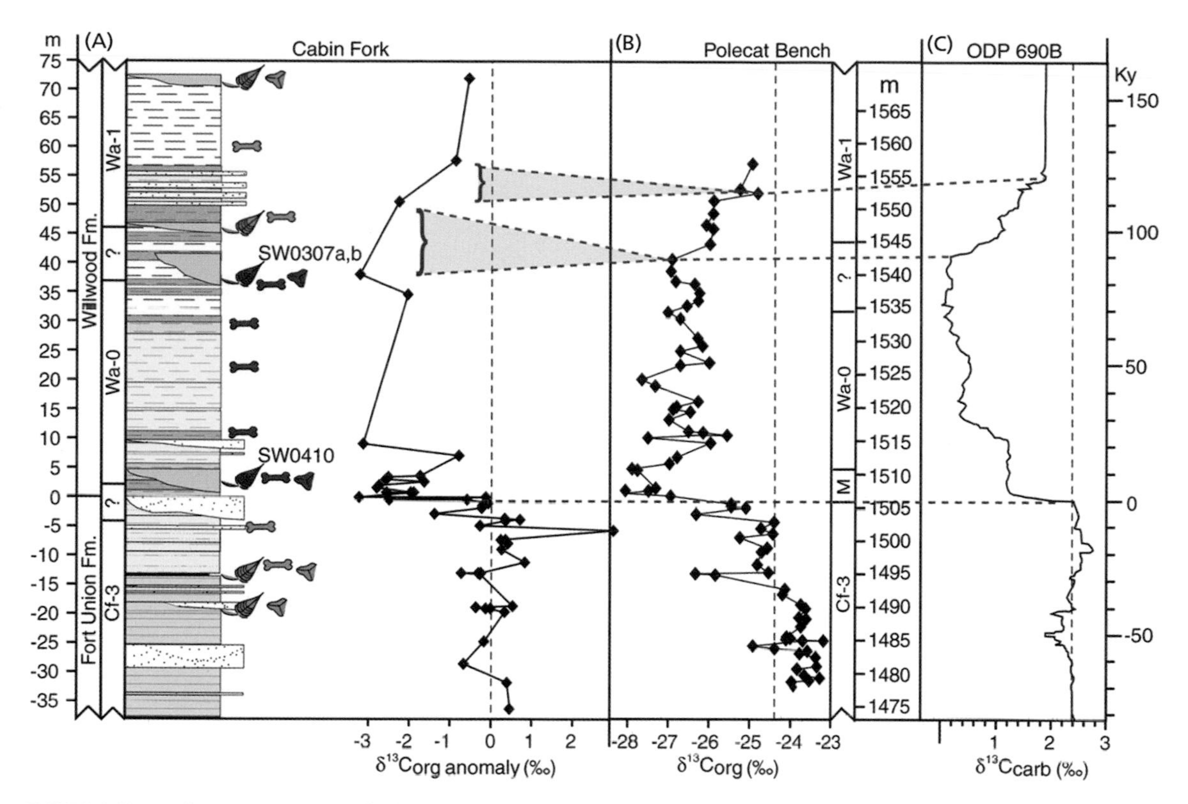

FIGURE 9.25. Carbon isotope records from Cabin Fork (A) and Polecat Bench (B) **continental sections in northern Wyoming**, correlated with a carbon isotope record from ODP Hole 690B (C), which is a **deep-sea section** from the southern high latitudes adjacent to Antarctica. **Vertebrate fossil zones** (Cf-3, M, Wa-0, Wa-1) are shown for the Cabin Fork and Polecat Bench sections. Lithologic summary of the Cabin Fork section (A) shows **alternating fluvial (river) sandstones and channel deposits, and floodplain mudstones and paleosols (ancient soils).** Blue symbols (bones, leaf fossils, pollen grains) = Paleocene age fossils; red symbols = PETM age fossils; green symbols = post-PETM Eocene age fossils. Dashed orange lines indicate lines of correlation/tie points between sections. Dashed vertical lines indicate mean δ^{13}C values for the latest Paleocene (i.e., pre-PETM). From Wing et al., 2005.

50 Which vertebrate fossil zone(s) correlate with the PETM interval (Figure 9.25)?

51 According to the study by Wing et al. (2005), PETM floras (assemblages of plant fossils and pollen grains) differ from the Paleocene floras below and the Eocene assemblages above. Specifically, the PETM floras contain a mixture of native and **migrant species**. Based on the flora data presented on the map (Figure 9.24), where did the migrant plants probably come from? What can you infer about these plant migrations?

NAME

52 What is the magnitude of the negative carbon isotope excursion at the two continental sections compared with deep-sea marine sections?

53 Speculate about the significance of discovering a negative carbon isotope signature in both marine and continental deposits.

J. Clay Mineral Changes in North Dakota Associated with the PETM

54 **Kaolinite** is a clay mineral that is formed under conditions of **intense chemical weathering**, typically associated with tropical warmth and abundant rainfall. What might the clay mineral record at Farmers Butte tell us about climate changes in North Dakota during the late Paleocene and early Eocene (Figures 9.26 and 9.27)?

NAME

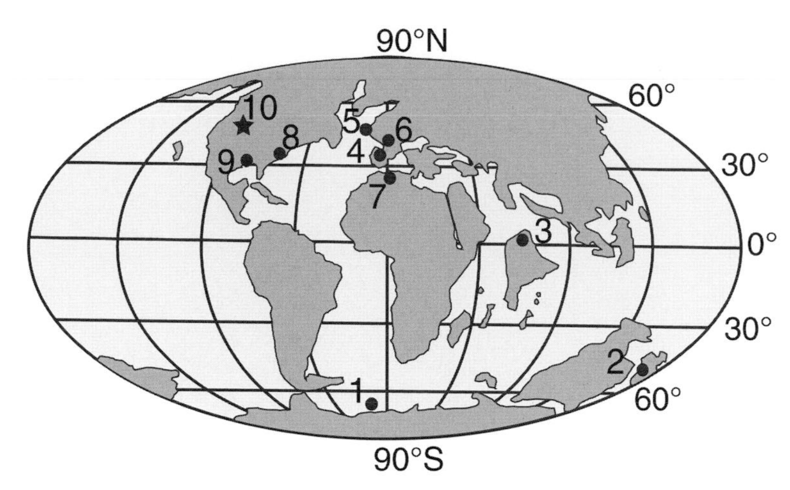

FIGURE 9.26. Paleogeographic map showing the distribution of land masses at the time of the PETM. Numbers correspond to known **kaolinite deposits** (a type of clay formed by intense chemical weathering of igneous rocks like granite) associated with the Paleocene–Eocene boundary. The site labeled number 10 with the star shows the Williston Basin of North Dakota. From Clechenko et al., 2007.

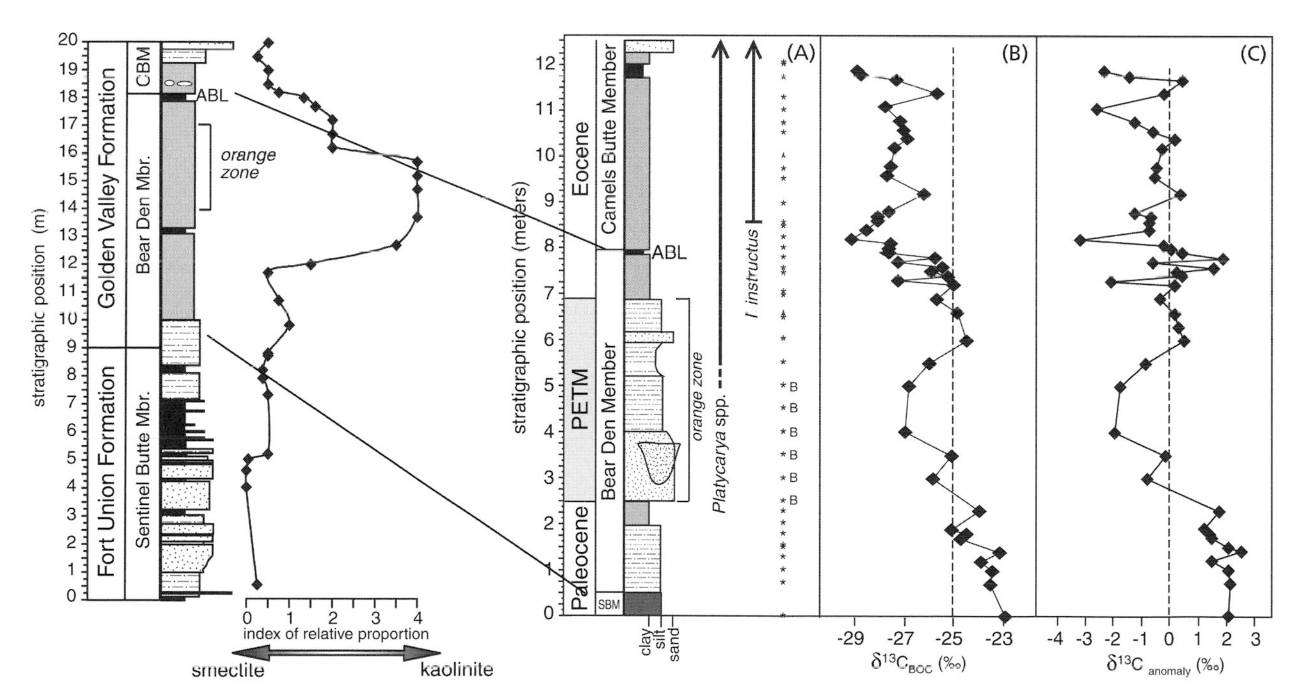

FIGURE 9.27. Left: Clay mineralogy of the upper Fort Union Formation and Golden Valley Formation plotted against lithostratigraphy and graphic summary of the Farmers Butte section. Note the **large increase in the clay mineral kaolinite relative to smectite in the "orange zone"** of the Bear Den Member. Smectite is derived from the weathering of volcanic rocks. Right: (A) Pollen biostratigraphy and (B) and (C) bulk sediment carbon isotope record from the Paleocene–Eocene boundary interval. Note the symbol "B" indicates barren of pollen grains. The stratigraphic interval labeled the "orange zone" correlates with the PETM based on age control from pollen and with the negative carbon isotope excursion which correlates with other continental and deep-sea marine sections. ABL = Alamo Bluff lignite; CBM = Camels Butte Member; SBM = Sentinel Butte Member. From Clechenko et al., 2007.

NAME ______________________________

55 Speculate why the PETM interval corresponds to the **orange-colored sediments** (i.e., stratigraphic interval labeled the "orange zone", part of the Bear Den Member of the Golden Valley Formation; Figure 9.27)?

56 What is the **climatic significance of the clay mineral changes** recorded at Farmers Butte in North Dakota relative to other terrestrial PETM sections around the world (see maps, Figures 9.24 & 9.26).

Synthesis and Discussion

57 Use Table 9.3 (below) or 9.4 (oversize format, available as an online supplement) to compile and summarize the **key observational data** from **each** of the deep-sea and land-based records presented above (records A to J). Also use Table 9.3 (or 9.4) to list any questions that you have about what you have seen in the data.

58 It is clear that the PETM is associated with a major change in the carbon cycle. What was the average magnitude of the **carbon isotope excursion (CIE)** around the globe? What were the maximum and minimum values?

NAME ______________________________

TABLE 9.3. Synthesis of observations for PETM.

Key Figures	Location (Figure 9.4)	Key Observations	Your Questions
9.5	**A.** 4 ODP sites (525, 527, 690, 865) **Atlantic–Antarctic–Pacific**		
9.7	**B1.** ODP Leg 198 (Sites 1208–1212) **Subtropical Pacific**		
9.8	**B2.** ODP Leg 198 (Site 1209) **Subtropical Pacific**		
9.9	**B3.** ODP Leg 198 (Sites 1209–1210) **Subtropical Pacific**		
9.11 9.12	**C.** ODP Leg 199 (Sites 1220–1221) **Tropical Pacific**		
9.13	**D.** ODP Leg 113 (Site 690) **Weddell Sea, Antarctica**		

NAME

TABLE 9.3. *Continued*

Key Figures	Location (Figure 9.4)	Key Observations	Your Questions
9.15 9.16 9.17	**E.** ODP Leg 208 (Sites 1262–1267) **Southeast Atlantic**		
9.19	**F.** Wilson Lake & Bass River cores **New Jersey**		
9.20 9.21	**G. Arctic Ocean** Lomonosov Ridge		
9.22 9.23	**H. Greenland- Norwegian Sea**		
9.24 9.25	**I.** Cabin Fork and Polecat Bench sections **Wyoming**		
9.26 9.27	**J.** Ft. Union & Golden Valley Formations **North Dakota**		

NAME

59 What was the average magnitude of **sea surface temperature** change around the globe? What were the maximum and minimum values?

60 What was the average magnitude of **deep water temperature** change around the globe? What were the maximum and minimum values?

61 Are the **oxygen and carbon isotope excursions** perfectly synchronous, or is there evidence that one leads or lags the other? Explain.

62 Use the global synthesis data from Tables 9.3 (or from the online supplemental Table 9.4) to **propose a general sequence of events of the PETM**.

63 Hypothesize some potential **causes for the PETM**. What sort of information would you need to test these hypotheses?

NAME

NAME ______________________________

The Paleocene–Eocene Thermal Maximum (PETM) Event

Part 9.3. Bad Gas: Is Methane to Blame?

Use your library resources to find and read the 2-page article: Schiermeier, Q., 2003. Gas Leak! *Nature*, v. 423, p. 681–682, and answer the following questions.

1 How does this brief summary article influence your synthesis of the PETM developed in Part 9.2?

____

____

____

____

____

2 Based on the article, which scenario is the better supported by the available data: (1) did a massive release of greenhouse gasses cause the observed global warming, or (2) did global warming trigger the massive release of greenhouse gasses? Explain your answer.

____

____

____

____

____

____

____

____

3 Use the space below to **revise your general sequence of events** (Part 9.2, Question 62) accordingly.

____

NAME

4 What **additional questions** does this summary raise in your understanding of what may have caused the PETM?

NAME

The Paleocene–Eocene Thermal Maximum (PETM) Event

Part 9.4. How fast? How long?

The first estimates of the duration of the PETM event, as well as the rate at which the carbon and climate perturbation occurred, were based on the numerical ages of known paleomagnetic reversals and **the assumption** that sediment accumulated at a constant rate between two **geochronological** tie-points. For example, in the original paper that defined the event (Kennett and Stott, 1991; see also Figure 9.3 in Part 9.1), the **age model** for the PETM was based on two magnetic reversal events from ODP Hole 690B:

- 154.62 mbsf: bottom of magnetic reversal 24n dated at 56.14 Ma
- 185.20 mbsf: top of magnetic reversal 25n dated at 58.64 Ma.

1 Using the above paleomagnetic information from Hole 690B, what is the average sedimentation rate of these sediments in meters per million years (m/myr)? Show your work.

2 What is the average sedimentation rate in centimeters per thousand years (cm/kyr)? Show your work.

3 Kennett and Stott's isotope sample from 170.45 mbsf in Hole 690B has pre-excursion $\delta^{13}C$ values, while their sample from 170.35 mbsf records the onset of the **carbon isotope excursion (CIE)**.

NAME ______________________________

(a) Write a formula and then calculate **how quickly the CIE occurred** based on Kennett and Stott's original data in millions of years (myr) and thousands of years (kyr). Show your work.

Formula:______________________________

Onset of CIE in millions of years (myr):

____

____

____

____

____

____

Onset of CIE in thousands of years (kyr):

____

____

____

____

____

____

(b) Does this value represent a minimum or maximum estimate? Explain.

____

____

____

____

4 Assuming that southern high latitude sea surface temperatures and global deep-water temperatures rose by at least 6°C during the PETM, convert this to a **rate of temperature increase** in degrees Celsius per thousand years (°C/kyr). Show your work.

____

____

NAME

5 How might the PETM be relevant to our understanding of **climate change** today?

Challenge: Exploring the Onset and Duration of the PETM

The ODP Site 690 off Antarctica (Figure 9.28) contains a special record of the **PETM interval** because it is **expanded** and probably **complete** (i.e., there is no evidence of erosion or serious dissolution at the onset and, therefore, no hiatus in the sediment record through the event).

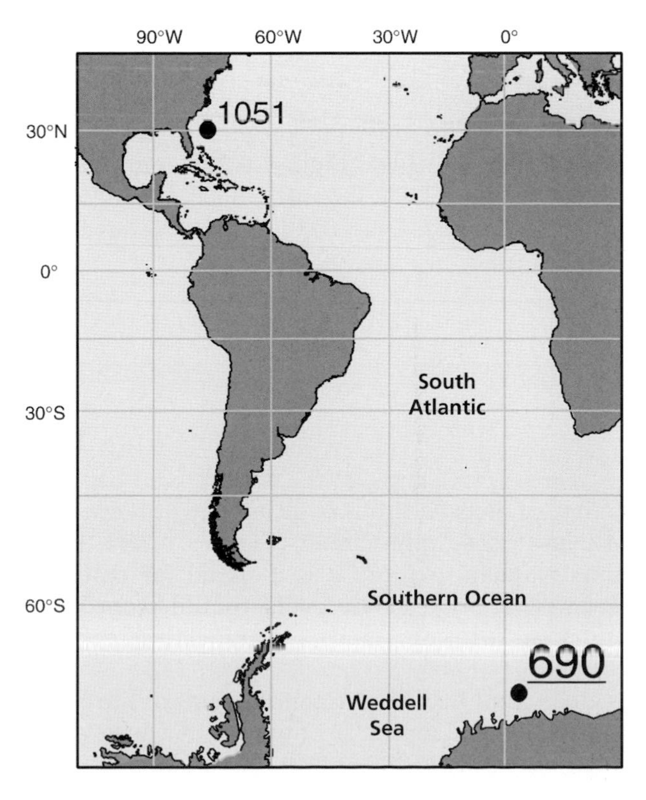

FIGURE 9.28. Location of ODP Site 690 near the Weddell Sea in the Southern Ocean, and ODP Site 1051 on Blake Nose in the western North Atlantic. From Röhl et al., 2000.

NAME

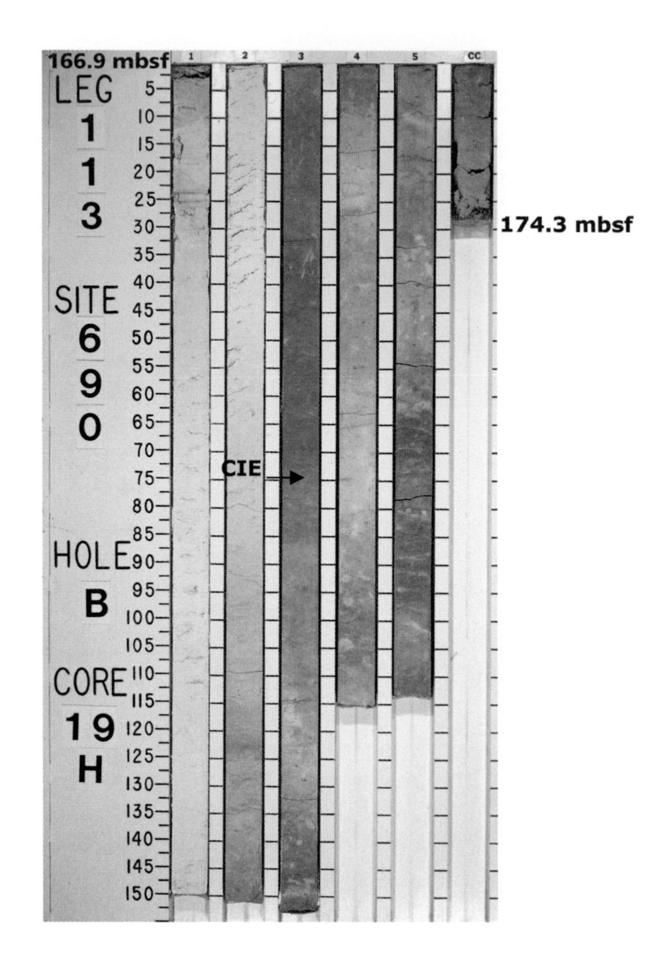

FIGURE 9.29. ODP Hole 690B-Core 19H. The **onset of the PETM** (**CIE** = carbon isotope excursion) occurs in Section 3, 75 cm (arrow). Courtesy of IODP.

6 How does the **visual record** (i.e., appearance of the sediment in the core photo) of the PETM at Site 690 (Figure 9.29) compare with other PETM records in the deep sea (e.g., ODP Leg 198 Shatsky Rise (Figure 9.7), ODP Leg 199 central Pacific (Figure 9.11), or ODP Leg 208, Walvis Ridge (Figure 9.15))?

____

____

____

Elemental proxies, such as Fe (iron) and Ca (calcium) are useful for exploring ancient climatic and oceanic processes (Figures 9.30 & 9.31). **Fe is a signal for a terrigenous material**, while **Ca is a signal for calcareous plankton** in the surface ocean. Therefore, **Fe–Ca cycles** could record multiple processes related to production, dilution, or dissolution. These could be:

- increased or decreased flux of terrigenous material to the ocean caused by climate changes (e.g., increased Fe flux could cause a **dilution** of the Ca signal),

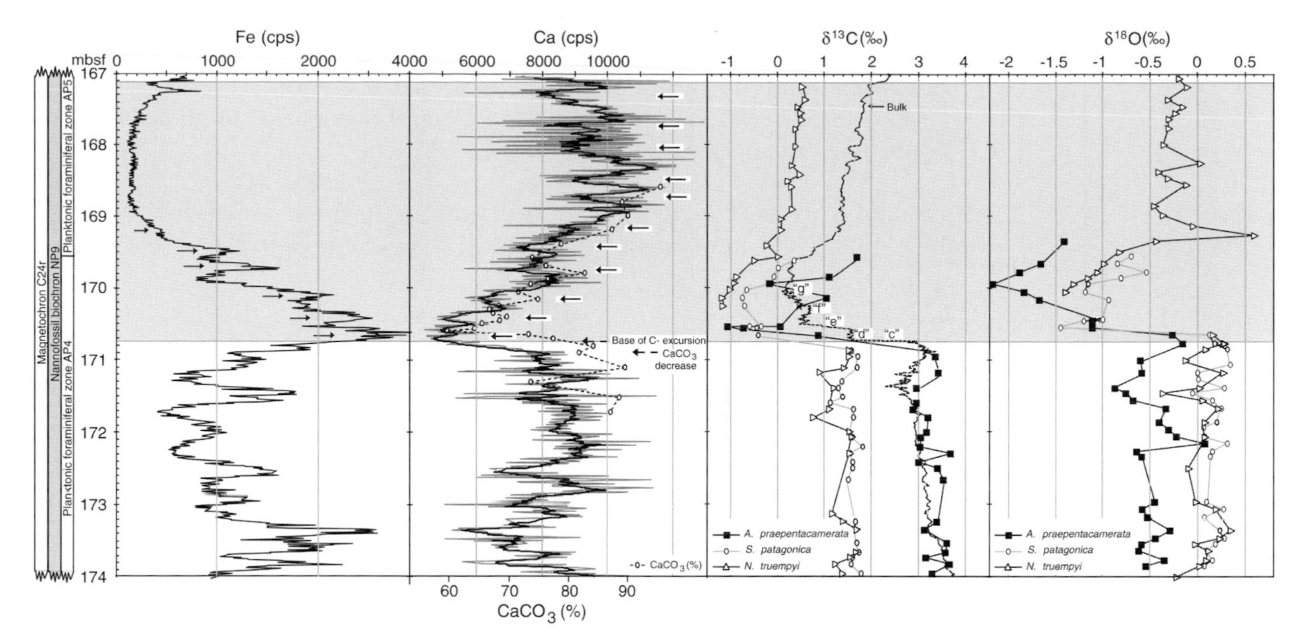

FIGURE 9.30. Geochemical data for ODP Hole 690B-Core 19H (167–174 mbsf). Vertical bars on far left: magnetostratigraphy and biostratigraphy. Graphs (left to right): Fe intensity curve (cps = counts per second), Ca intensity curve with five-point moving average (thick line) and %$CaCO_3$ (dashed line), multi-species carbon and oxygen isotope data [*Acarinina praepentacamerata* (mixed-layer planktic foram), black boxes; *Subbotina patagonica* (thermocline planktic foram), open circles; *Nuttalides truempyi* (benthic foram), open triangles. **Arrows** = cycles identified within the Fe and Ca records. **Shaded area** corresponds to the **PETM interval**. From Röhl et al., 2000.

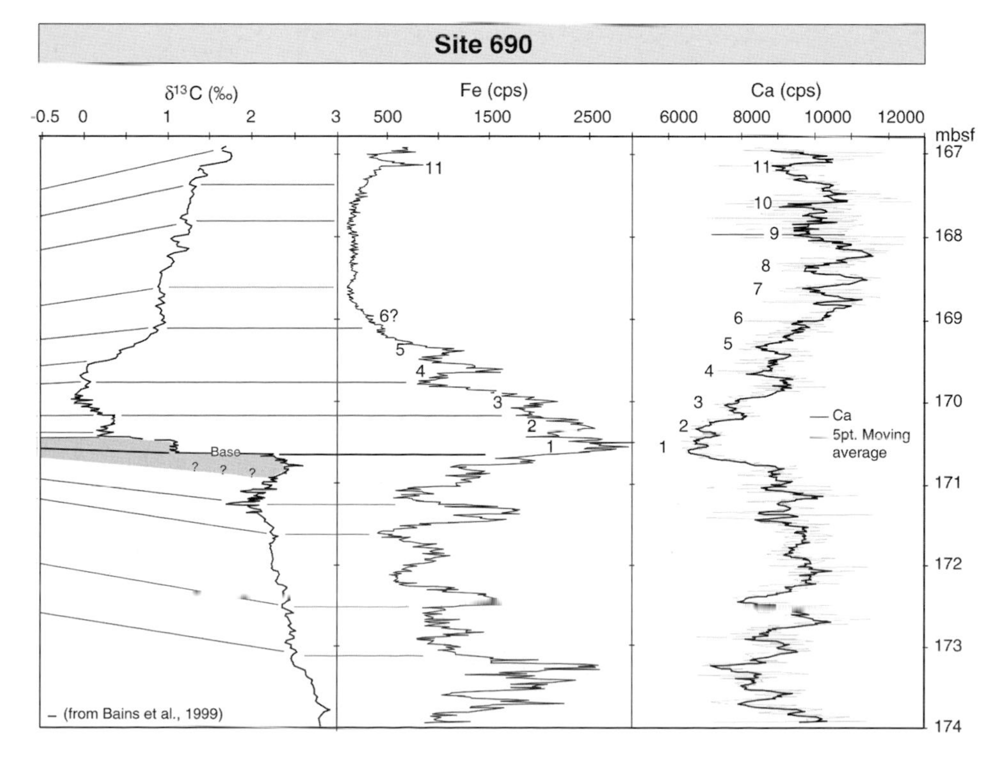

FIGURE 9.31. X-Ray fluorescence (iron, Fe, and calcium, Ca, in counts per second) and carbon isotope data from Hole 690B. Base = base of CIE. From Röhl et al., 2000.

NAME ___

- increased or decreased flux of microfossil carbonate caused by higher or lower **productivity** in the photic zone, respectively, or
- variable preservation and deposition of carbonate due to **dissolution** at depth (e.g., changing position of the CCD).

7 How many **"cycles"** (repetitions of a distinct pattern) are recorded in the Ca data in the PETM interval in Hole 690B (Figures 9.30 & 9.31)?

___

___

8 Iron (Fe) and calcium (Ca) concentrations in Hole 690B record orbital (i.e., Milankovich) cyclicity patterns (Figures 9.30 & 9.31; also see Chapter 8). Orbital **precessional cycles** have an average periodicity of approximately **21,000 years** and orbital **obliquity cycles** have a peridocity of approximately **41,000 years**. Assuming that the **onset of the PETM (CIE)** occurred approximately **54.95 Ma** at **170.75 mbsf** and the bottom of magnetic Chron 24n dated at **56.14 Ma** is recorded at approximately **154.62 mbsf**, what is the likely periodicity recorded in the geochemical cycles preserved in the PETM interval in Hole 690B: precession or obliquity? Show your work to support your interpretation.

___

___

___

___

9 Based on high-resolution analyses of the PETM in Hole 690B (Figures 9.30 & 9.31), does the onset of the PETM appear to be a single event or multiple events? Explain.

___

___

___

___

___

10 Which cycles (numbers 1–11 in the Ca data in Figure 9.31) correspond to the darkest, most clay-rich/carbonate-poor interval in Core 690B-19H (Figure 9.29)?

___

NAME

11 Refer back to the core photograph of Core 690B-19H (Figure 9.29). Describe how the geochemical records ($\delta^{13}C$, Fe, Ca) shown in Figures 9.30 & 9.31 change upwards through the core as the sediments change from darker color (orange brown) early in the PETM to lighter color (white) as the recovery progresses.

12 How long did the PETM last (i.e., from onset of the event to full recovery)?

13 Examine Figure 9.32 and describe how similar the Site 690 (Weddell Sea, Antarctica) and Site 1263 (Walvis Ridge, Southeast Atlantic) records are to each other.

14 Are there any significant differences between the two sites?

NAME

15 What conclusions can you draw about the PETM based on these two high-resolution studies of the PETM interval?

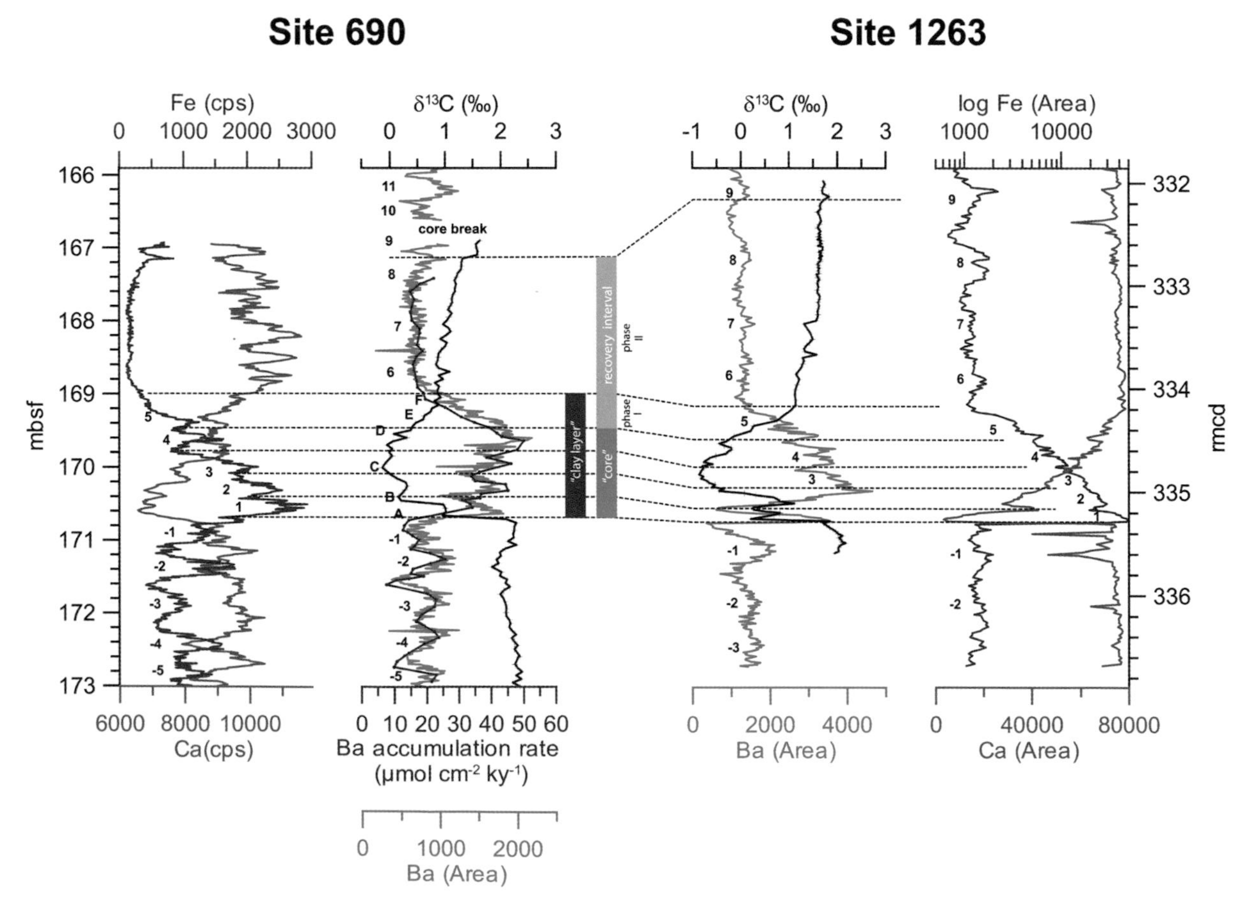

FIGURE 9.32. Comparison and correlation of **PETM interval at ODP Site 690**, Weddell Sea, Antarctica (left) and **ODP Site 1263**, Walvis Ridge, southeast Atlantic (right). Compilation of geochemical data from Sites 690 and 1263: **Fe** (iron; red line, in counts per second, cps, and area), **Ca** (calcium; blue line, in cps, and area), Ba (barium; green line, total counts), **Ba** accumulation rates (fushia line; from Bains et al., 2000), and **bulk carbon isotope data** from (black line) for Site 690 (Bains et al., 1999) and Site 1263 (Zachos et al., 2005). From Rohl et al., 2007.

NAME

The Paleocene–Eocene Thermal Maximum (PETM) Event

Part 9.5. Global Warming Today and Lessons from the PETM

The impact of human activity on global climate reached scientific consensus with the release of the report by the Intergovernmental Panel on Climate Change (IPCC, February 2007):

> "Most of the observed increase in globally averaged temperatures since the mid-20th century is *very likely* due to the observed increase in anthropogenic greenhouse gas concentrations."
> Note: According to the IPCC, "very likely" means there is a 90–99% probability of this conclusion.

While many people have long suspected this outcome, the **scientific method** is inherently conservative and it literally took decades of observation, data collection and analysis, debate, and testing of the results to reach this profound and sobering conclusion: human activity has unequivocally altered the composition of our atmosphere and is very likely to be responsible for the recent record of global warming primarily through the burning of fossil fuels.

The PETM occurred approximately 55 million years ago and records an abrupt change in the ocean–climate system related to global warming. How this ancient example of global change compares with present conditions of increasing concentrations of atmospheric carbon dioxide and rising global temperatures will be explored in this part of the PETM exercise.

50-Year Record of Increasing Atmospheric Carbon Dioxide

Figure 9.33 shows the instrument record of increasing atmospheric carbon dioxide (CO_2) as measured near the summit of Manua Loa, Hawaii since 1958.

1 How much has atmospheric CO_2 increased since 1958?

---

2 What was the concentration of CO_2 in 1990?

---

3 What was the concentration of CO_2 in 2000?

---

4 What was the concentration of CO_2 in 2010?

NAME

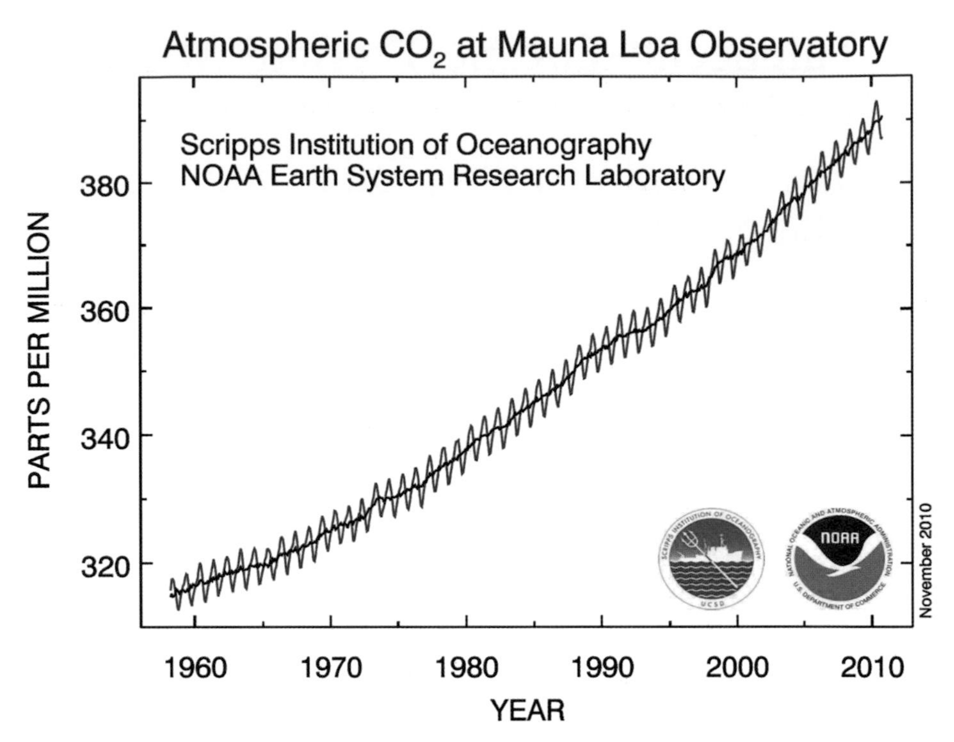

FIGURE 9.33. This figure shows the **recent history of atmospheric carbon dioxide (CO_2) concentrations** directly measured at Mauna Loa, Hawaii (ppmv = parts per million by volume). This curve is known as the **Keeling curve** and is an essential piece of evidence of the man-made increases in greenhouse gases that are believed to be the cause of global warming. The longest such record exists at Mauna Loa, but these measurements have been independently confirmed at many other sites around the world. The **annual cycle of atmospheric carbon dioxide variation** is derived from instrument data by taking the average concentration for each month across all measured years. The red curve shows the average monthly concentrations and the black curve is a 12-month moving average. Courtesy of NOAA.

5 Has the rate of increasing CO_2 been constant during the past 50 years? Explain.

6 Why does the CO_2 curve display a saw-tooth pattern?

NAME

Global Temperature Trends of the Past 150 Years

Figure 9.34 shows the **instrumental record** (i.e., measured with a thermometer rather than a proxy record) of average global temperatures since the mid-1800s.

7 Make a list of observations about the data trends in Figure 9.34.

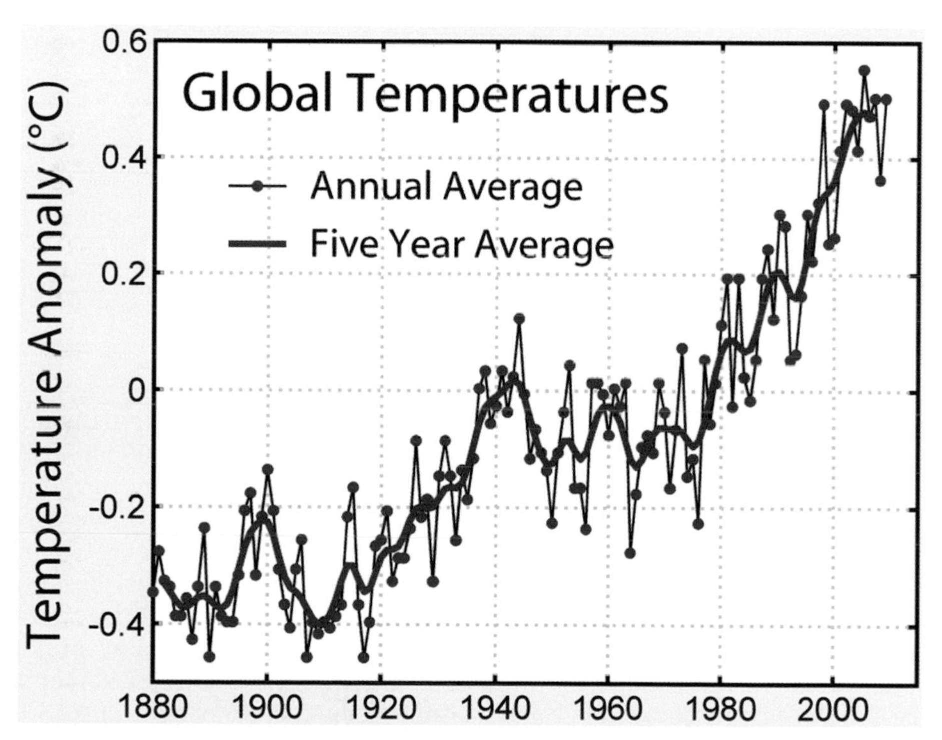

FIGURE 9.34. Image showing the **instrumental record of global average temperatures** as compiled by the Climatic Research Unit of the University of East Anglia and the Hadley Centre of the UK Meteorological Office. Data set HadCRUT3 was used. HadCRUT3 is **a record of surface temperatures collected from land and ocean-based stations**. The most recent documentation of this data set is Brohan et al., 2006. **Following the common practice of the IPCC, the zero on this figure is the mean temperature from 1961–1990**. This figure was originally prepared by Robert A. Rohde from publicly available data and is part of the Global Warming Art project (http://en.wikipedia.org/wiki/Image:Instrumental_Temperature_Record.png).

NAME ______________________________

8 How does this trend in global average temperatures (Figure 9.34) compare with the concentration of atmospheric CO_2 (Figure 9.33) during the past 50 years (approximately 1960–2010)?

____

____

____

____

9 What is the rate of global temperature rise for the 50 years between 1957 and 2007 (i.e., the interval for which we also have an instrument record of CO_2)? Report your results in **degrees Celsius (°C) per century**. Show your work.

____

____

____

____

____

Ice Core Records of Ancient Carbon Dioxide Concentrations

As we have seen, instrument records of changing atmospheric CO_2 are available dating back to 1958 (Figure 9.33). To determine pre-1958 concentrations of CO_2 we must rely on **"proxy" records** (i.e., indirect evidence). **Ice cores** collected in Greenland and Antarctica have provided a record of CO_2 for the past approximately 400,000 years (Figure 9.35). The snow that accumulates in the high latitudes becomes progressively compacted and recrystallized by subsequent years of snow accumulation to form **firn** and eventually ice. Tiny bubbles of air become trapped in the firn and ice (see Chapter 1). These bubbles can be sampled and measured for the concentration of CO_2 and other gases. The ice core record shown in Figure 9.35 spans the past four glacial–interglacial cycles, which have repeated with a cyclicity of approximately 100,000 years. This periodicity corresponds to predicted Milankovitch climate forcing related to the eccentricity of the Earth's orbit around the Sun.

10 What was the **natural range of variability of atmospheric CO_2** prior to the Industrial Revolution (circa 1800)?

____

NAME

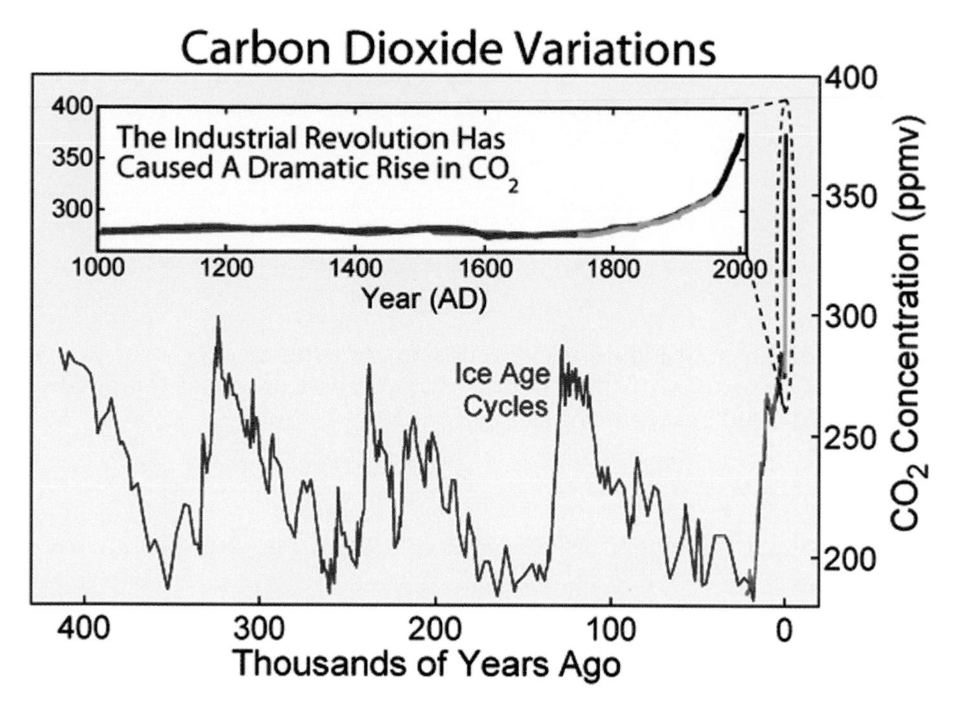

FIGURE 9.35. Variations in the **concentration of carbon dioxide (CO_2) in the atmosphere during the last 400 thousand years.** Data sources include: (blue) Vostok ice core, Antarctica (Fischer et al., 1999); (green) EPICA ice core, Antarctica (Monnin et al., 2004); (red) Law Dome ice core, Antarctica (Etheridge et al., 1998); (cyan) Siple Dome ice core, Antarctica (Neftel et al., 1994); (black) Mauna Loa Observatory, Hawaii (Keeling and Whorf, 2004). Throughout most of the record, the largest changes can be related to **glacial–interglacial cycles.** Although the glacial cycles are most directly caused by changes in the Earth's orbit (i.e., **Milankovitch cycles**), these changes also influence the carbon cycle, which in turn feeds back into the glacial system. This figure was originally prepared by Robert A. Rohde from publicly available data and is incorporated into the Global Warming Art project (http://en.wikipedia.org/wiki/File:Carbon_Dioxide_400kyr.png).

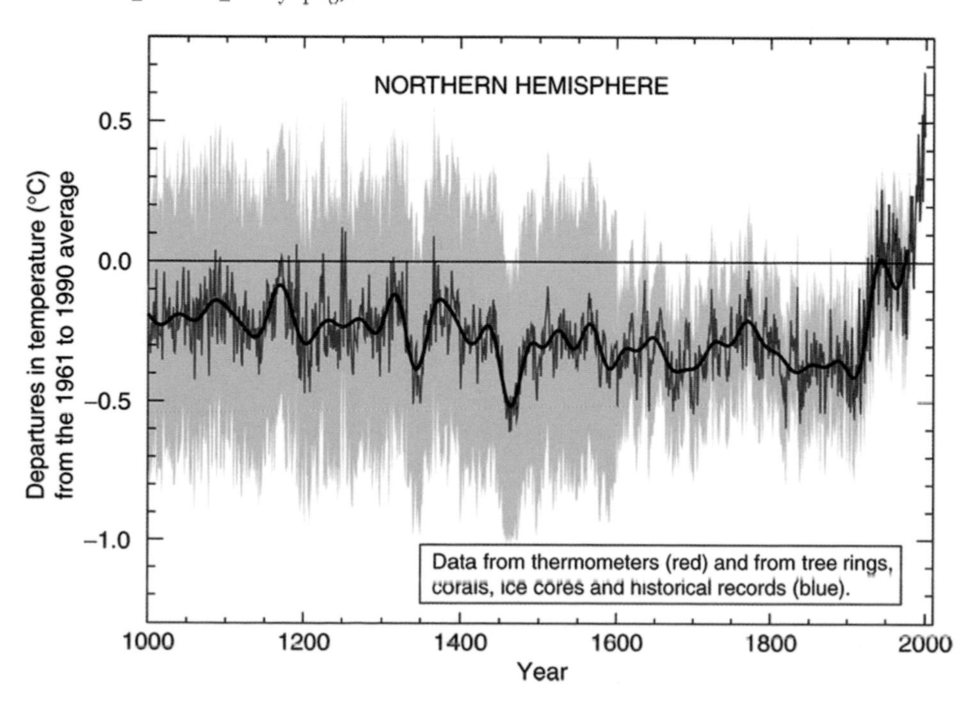

FIGURE 9.36. Northern hemisphere temperature changes over the past 1000 years. This data plot is commonly referred to as the **"hockey stick"** and is from the Intergovernmental Panel on Climate Change Fourth Assessment Report (2007). It is the IPCC's reproduction of the work of Mann et al. (1998, 1999). The colored lines are the reconstructed temperatures (blue = proxy data; red = instrument data), and the gray shaded region represents estimated error bars.

NAME ______________________________

11 The present concentration of CO_2 exceeds 390 ppmv (parts per million by volume). How does this compare with the natural range of variability of atmospheric CO_2 for the last 400,000 years?

____

____

12 How does the inset of Figure 9.35 (trends in atmospheric CO_2 over the past 1000 years) compare with the "hockey stick" trend of global temperatures over the past 1000 years shown in Figure 9.36?

____

____

____

____

____

How Does the PETM Compare with Modern Global Warming?

The PETM is an ancient example of how the Earth's ocean–climate system responded to very rapid global warming, probably triggered by the sudden release of greenhouse gases from sedimentary deposits. While the PETM was not related to human activity, it was a global event caused by changing greenhouse gas concentrations whose outcomes may foreshadow similar patterns of change in the near future. Reflect on what you learned about the PETM in this Chapter (Parts 9.1–9.5) and consider what lessons this past event has for us today.

13 What was the minimum (and maximum) increase in sea surface temperatures during the PETM?

____

____

14 How long did it take for this rapid rise of temperatures?

____

15 Knowing the magnitude of temperature increase and the amount of time required to increase global temperature, estimate a range of possible rates of temperature change in degrees Celsius (°C) per century during the onset of the PETM?

NAME

16 How does this estimate of PETM rate of temperature change compare to the modern rate of temperature change?

17 What can we learn from this comparison of PETM and modern rates of global warming?

NAME

References

Bains, S., et al., 1999, Mechanisms of climate warming at the end of the Paleocene. Science, **285**, 724–7.

Bains, S., et al., 2000, Termination of global warmth at the Paleocene/Eocene boundary through productivity feedbacks. Nature, **407**, 171–4.

Bralower, T.J., et al., 2002, Initial Reports of the Ocean Drilling Program, vol. 198, College Station, TX, Ocean Drilling Program; http://www-odp.tamu.edu/publications/198_IR/198ir.htm.

Brohan, P., et al., 2006, Uncertainty estimates in regional and global observed temperature changes: a new dataset from 1850. Journal Geophysical Research, **111**, D12106. doi:10.1029/2005JD006548.

Clechenko, E.R., et al., 2007, Terrestrial records of a regional weathering profile at the Paleocene–Eocene boundary in the Williston Basin of North Dakota. Geological Society of America Bulletin, **119**(3/4), 428–42.

Etheridge, D.M., Steele, L.P., Langenfelds, R.L., Francey, R.J., Barnola, J.M, and Morgan, V.I., 1998, Historical CO_2 records from the Law Dome DE08, DE08-2, and DSS ice cores, *in* Trends: A Compendium of Data on Global Change, Carbon Dioxide Information Analysis Center, Oak Ridge National Laboratory, U.S. Department of Energy, Oak Ridge, Tenn., U.S.A, http://cdiac.ornl.gov/trends/co2/lawdome.html.

Fischer, H., Wahlen, M., Smith, J., Mastroianni, D., and Deck, B., 1999, Ice core records of Atmospheric CO_2 around the last three glacial terminations: Science, v. **283**, 1712–1714.

IPCC, 2007, Summary for Policymakers, In Climate Change 2007: The Physical Science Basis. Contribution of Working Group I to the Fourth Assessment Report of the Intergovernmental Panel on Climate Change, Solomon, S., et al., (eds), Cambridge University Press, Cambridge, UK.

Kaiho, K., et al., 2006, Anomalous shifts in tropical Pacific planktonic and benthic foraminiferal test size during the Paleocene-Eocene thermal maximum. Palaeogeography, Palaeoclimatolology, Palaeoecology, **237**, 456–64.

Keeling, C.D., and Whorf, T.P., 2004, Atmospheric CO_2 records from sites in the SIO air sampling network, *in* Trends: A Compendium of Data on Global Change, Carbon Dioxide Information Analysis Center, Oak Ridge National Laboratory, U.S. Department of Energy, Oak Ridge, Tenn., U.S.A., http://cdiac.ornl.gov/trends/co2/sio-keel-flask/sio-keel-flaskmlo_c.html.

Kennett, J. and Stott, L.D., 1991, Abrupt deep sea warming, paleoceanographic changes and benthic extinctions at the end of the Palaeocene. Nature, **353**, 319–22.

Lyle, M., et al., 2002, Proceedings ODP, Initial Reports, vol. 199, College Station, TX, Ocean Drilling Program. http://www-odp.tamu.edu/publications/199_IR/199ir.htm

Mann, M.E., et al., 1998, Global-scale temperature patterns and climate forcing over the past six centuries, Nature, **392**, 779–787, doi:10.1038/33859.

Mann, M. E., et al., 1999, Northern hemisphere temperatures during the past millennium: Inferences, uncertainties, and limitations. Geophysical Research Letters, **26**(6), 759–62. doi:10.1029/1999GL900070

Miller, K.G., et al., 1991, Unlocking the ice house: Oligocene–Miocene oxygen isotopes, eustasy, and margin erosion. Journal of Geophysical Research, **96**, 6829–48.

Monnin, E., Steig, E.J., Siegenthaler, U., Kawamura, K., Schwander, J., Stauffer, B., Stocker, T.F., Morse, D.L., Barnola, J.M., Bellier, B., Raynaud, D., and Fischer, H.,

NAME

2004, Evidence for substantial accumulation rate variability in Antarctica during the Holocene, through synchronization of CO2 in the Taylor Dome, Dome C and DML ice cores: *Earth and Planetary Science Letters*, v. **224**, p. 45–54. doi:10.1016/j.epsl.2004.05.007

Neftel, A., Friedli, H., Moor, E., Lötscher, H., Oeschger, H., Siegenthaler, U., and Stauffer, B., 1994, Historical CO_2 record from the Siple Station ice core, *in* Trends: A Compendium of Data on Global Change, Carbon Dioxide Information Analysis Center, Oak Ridge National Laboratory, U.S. Department of Energy, Oak Ridge, Tenn., U.S.A., http://cdiac.ornl.gov/trends/co2/siple.html.

Röhl, U., et al., 2000, A new chronology for the late Paleocene thermal maximum and its environmental implications. Geology, **28**(10), 927–30.

Röhl, U., et al., 2007, On the duration of the Paleocene–Eocene thermal maximum (PETM). Geochemistry, Geophysics, Geosystems, **8**(12), Q12002, doi:10.1029/2007GC001784.

Ruddiman, W.F., 2008, Earth's Climate Past and Future, W.H. Freeman and Co., New York, 388 pp.

Schiermeier, Q., 2003, Gas leak! Nature, **423**, 681–2.

Sluijs, A., et al., and the Expedition 302 Scientists, 2006, Subtropical Arctic Ocean temperatures during the Palaeocene/Eocene thermal maximum. Nature, **441**, 610–13.

Sluijs, A., et al., 2007, Environmental precursors to rapid light carbon injection at the Palaeocene/Eocene boundary. Nature, **450**, 1218–21.

Storey, M., et al., 2007, Paleocene–Eocene thermal maximum and the opening of the northeast Atlantic. Science, **316**, 587–9.

Thomas, E., 2003, Extinction and food at the seafloor: A high-resolution benthic foraminiferal record across the Initial Eocene Thermal Maximum, Southern Ocean Site 690. In Causes and Consequences of Globally Warm Climates in the Early Paleogene, Wing, S.L., et al. (eds), Geological Society of America Special Paper 369, Boulder, Colorado, pp. 319–32.

Thomas, E. and N. J. Shackleton, 1996, The latest Paleocene benthic foraminiferal extinction and stable isotope anomalies. In Correlation of the Early Paleogene in Northwest Europe, R. W. Knox et al. (eds), Geological Society Special Publication, vol. 101, pp. 401–41.

Thomas, D.J., et al., 2002, Warming the fuel for the fire: Evidence for the thermal dissolution of methane hydrate during the Paleocene–Eocene thermal maximum. Geology, **30**(12), 1067–70.

Wing, S.L., et al., 2005, Transient floral change and rapid global warming at the Paleocene–Eocene boundary. Science, **310**, 993–6.

Zachos, J., et al., 2001, Trends, rhythms, and aberrations in global climate 65 Ma to present. Science, **292**, 686–93.

Zachos, J.C. et al., 2003, A transient rise in tropical sea surface temperature during the Paleocene–Eocene Thermal maximum. Science, **302**, 1551–4.

Zachos, J.C., et al., 2004, Proceedings ODP, Initial Reports, vol. 208, College Station, TX, Ocean Drilling Program, http://www-odp.tamu.edu/publications/208_IR.htm

Zachos, J.C., et al., 2005, Rapid acidification of the ocean during the Paleocene–Eocene thermal maximum. Science, **308**, 1611–15.

Chapter 10 Glaciation of Antarctica: The Oi1 Event

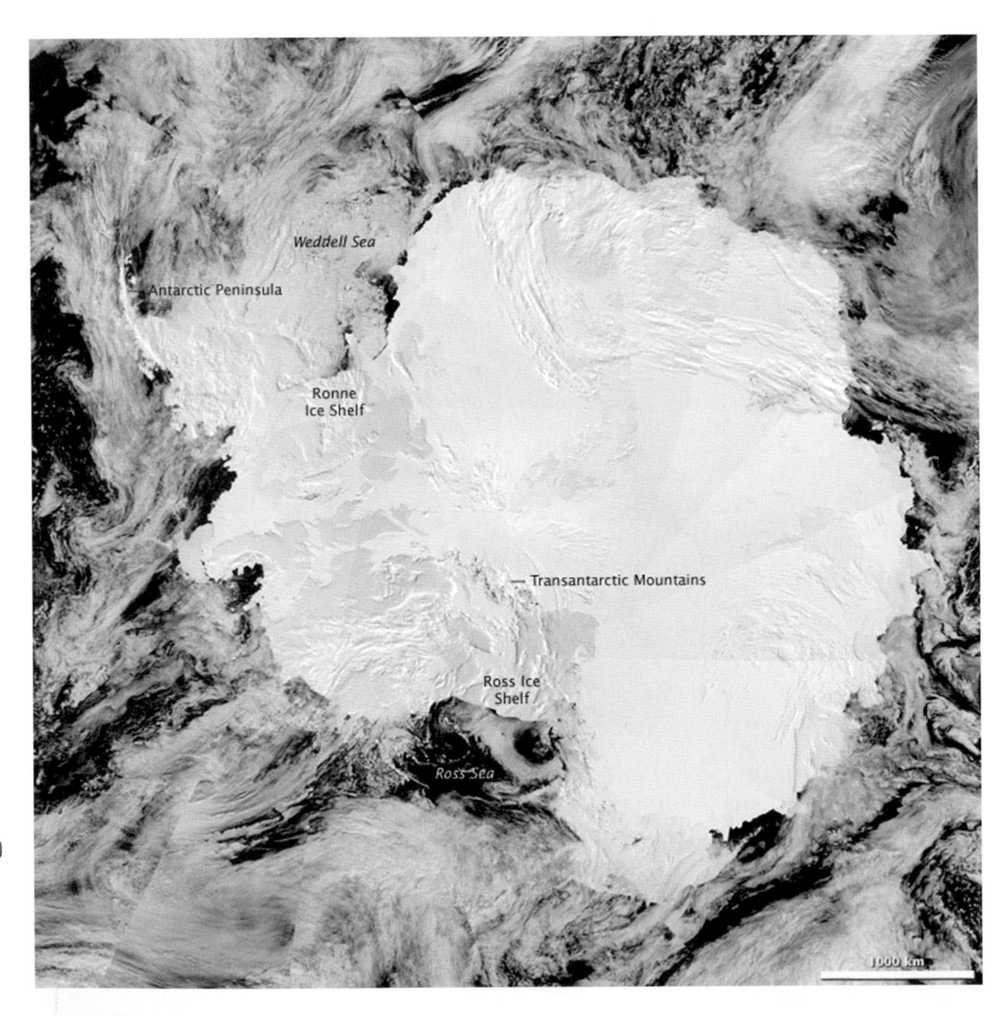

FIGURE 10.1. Moderate resolution imaging spectroradiometer (MODIS) on NASA's Aqua satellite captured this composite image of a summer day on Antarctica, January 27, 2009. Courtesy of NASA.

SUMMARY

The Oi1 event refers to an abrupt positive excursion of benthic foraminiferal oxygen isotope values across the Eocene–Oligocene boundary. It is attributed to the rapid glaciation of Antarctica approximately 33.5 Ma and

Reconstructing Earth's Climate History: Inquiry-Based Exercises for Lab and Class, First Edition. Kristen St John, R Mark Leckie, Kate Pound, Megan Jones and Lawrence Krissek.

NAME

is one of best examples of a **threshold** event in the ocean–climate system. In this chapter you will explore the Oi1 event in the context of Cenozoic climate change. In **Part 10.1**, you will evaluate some of the early evidence for the glaciation of Antarctica. In **Part 10.2**, you will evaluate direct and indirect evidence for glaciation and global climate change. You will make connections between different data sets and consider the possible influence of climate feedback in triggering the Oi1 event. Multiple hypotheses for the cause of the Oi1 event will be considered. In **Part 10.3**, you will explore the relationship between chemical weathering and atmospheric CO_2. You will learn about strontium isotopes as a proxy for chemical weathering and consider the role of climate feedback leading up to the glaciation of Antarctica. In **Part 10.4**, you will explore how the Oi1 event influenced the evolution of global ocean circulation. In particular you will investigate the formation of **psychrosphere**, which describes the cold deep water in all the ocean basins today.

Glaciation of Antarctica: The Oi1 Event

Part 10.1. Initial Evidence

You were introduced to **oxygen isotopes** as **a proxy for climate change** in Chapter 6. Consider again the composite **benthic foram $\delta^{18}O$** plot of Zachos et al. (2008; Figure 10.2). Recall, this is the record of changing ratios of $^{18}O/^{16}O$ in the calcium carbonate ($CaCO_3$) shells of Cenozoic deep-sea benthic foraminifera.

1 (a) What is the nature of the $\delta^{18}O$ signature that indicates cooling of the deep ocean and/or growth of large ice sheets? Specifically, is cooling/ice sheet growth indicated by more positive/greater values of $\delta^{18}O$ or by more negative/lesser values?

(b) Give a brief explanation of **why** the ratio of $^{18}O/^{16}O$ in a benthic foram shell changes as a function of cooling and/or increasing ice volume.

NAME

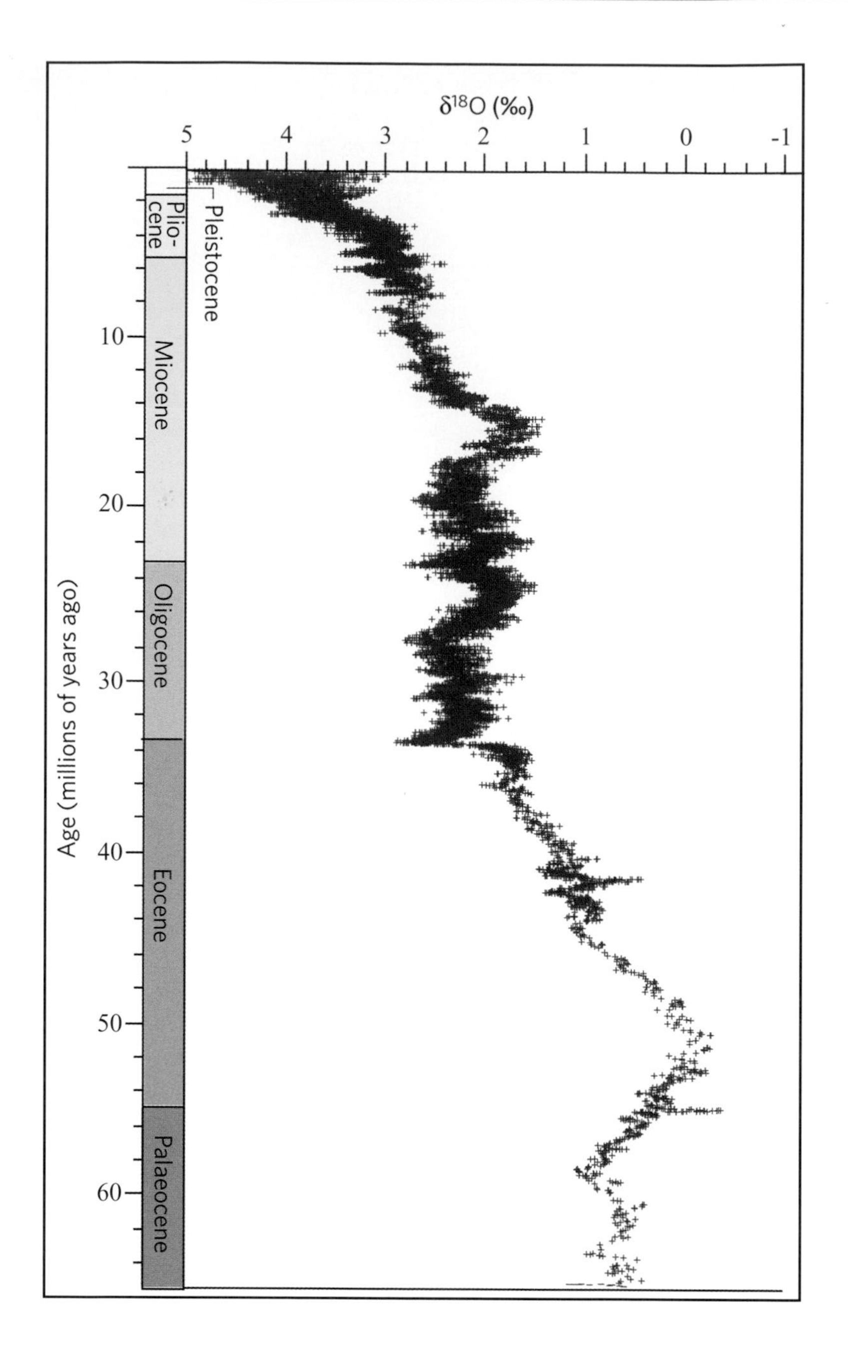

FIGURE 10.2. 65 million year **composite record** of benthic foraminiferal stable oxygen isotope values ($\delta^{18}O$) from the Deep Sea Drilling Project (DSDP) and Ocean Drilling Program (ODP) sediment cores. Modified from Zachos et al., 2008, which is an update of Zachos et al., 2001.

2 Based on the $\delta^{18}O$ data in Figure 10.2, **what intervals of the Cenozoic are likely to be times of significant continental-scale glaciation?** Make a list; give the approximate age (or age range) and the portion of the epoch (e.g., early, middle, or late Miocene) during which glaciation occurred.

NAME ______________________________

Approximate Age of Glaciation	Portion of Epoch

3 What **evidence** in the rock record (e.g., fossils, types of sedimentary deposits) would support the $\delta^{18}O$-based interpretation that glaciation occurred at these times?

____

____

____

____

4 Where **in the ocean** would you look for geological evidence of glaciation?

____

____

____

____

ODP Legs 119 and 120 drilled a number of sites along the Antarctic continental margin off Prydz Bay (Indian Ocean sector of Antarctica) and on the Kerguelen Plateau in the southern Indian Ocean, respectively (Figure 10.2). **ODP Site 748** is located far from Antarctica (58°26.45'S, 78°58.89'E) and is perched up on the top of the southern Kerguelen Plateau (1290 m water depth) away from any possible influence of down-slope turbidite flows (i.e., submarine debris flows).

Hole 748B contains a concentration of coarse sediment grains that have been interpreted as **ice rafted debris (IRD)** based on: (1) the isolation of the drill site away from land, on top of a bathymetric high and (2) the distinctive surface features and conchoidal fractures on the grain surfaces (Figure 10.4;

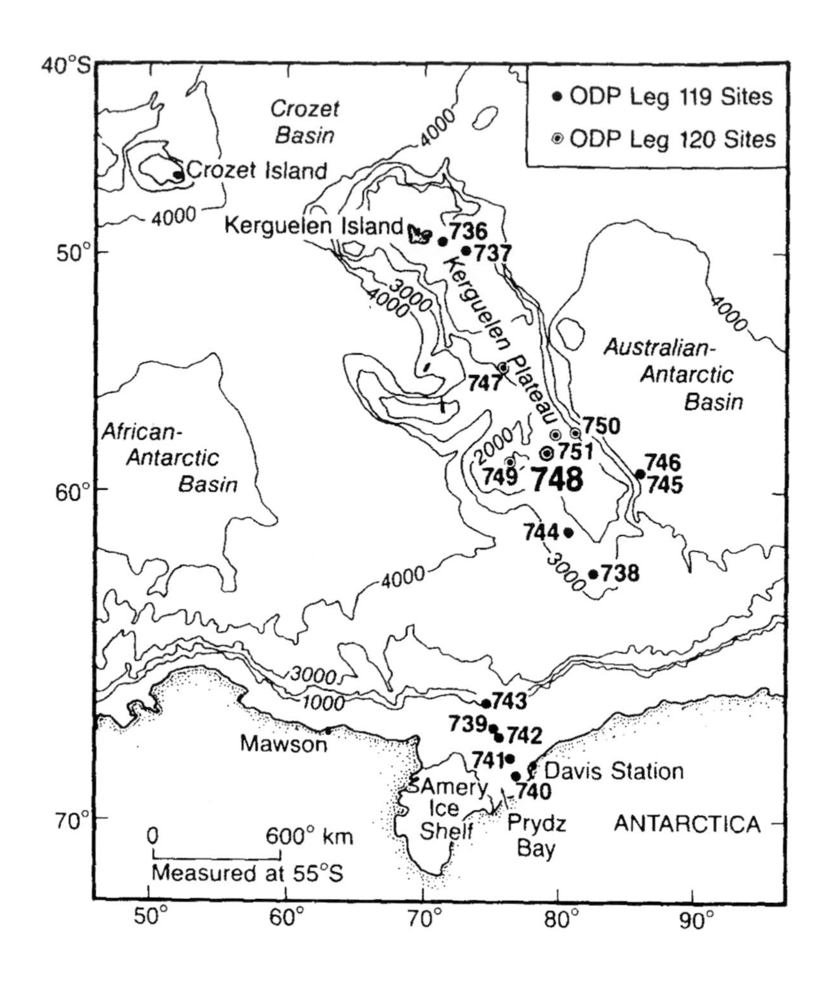

FIGURE 10.3. Map of the southern Indian Ocean showing sites cored during ODP Legs 119 and 120 on Kerguelen Plateau and north of Prydz Bay, Antarctica. From Zachos et al., 1992.

Zachos et al., 1992). In addition to the IRD, Zachos et al. analyzed the **oxygen and carbon isotopic composition of four benthic and planktic foraminiferal species** extracted from Hole 748B. In Figure 10.4, benthic forams are represented by solid triangles (*Cibicidoides* spp.) and solid circles (*Gyroidina*); planktic forams are represented by the open circles (*Subbotina angiporoides*) and 'plus' symbol (*Chiloguembelina cubensis*).

5 (a) What is the age of the **positive isotope excursion**? What is the age of the **IRD** concentration?

____

____

(b) What are the implications of these findings?

____

NAME

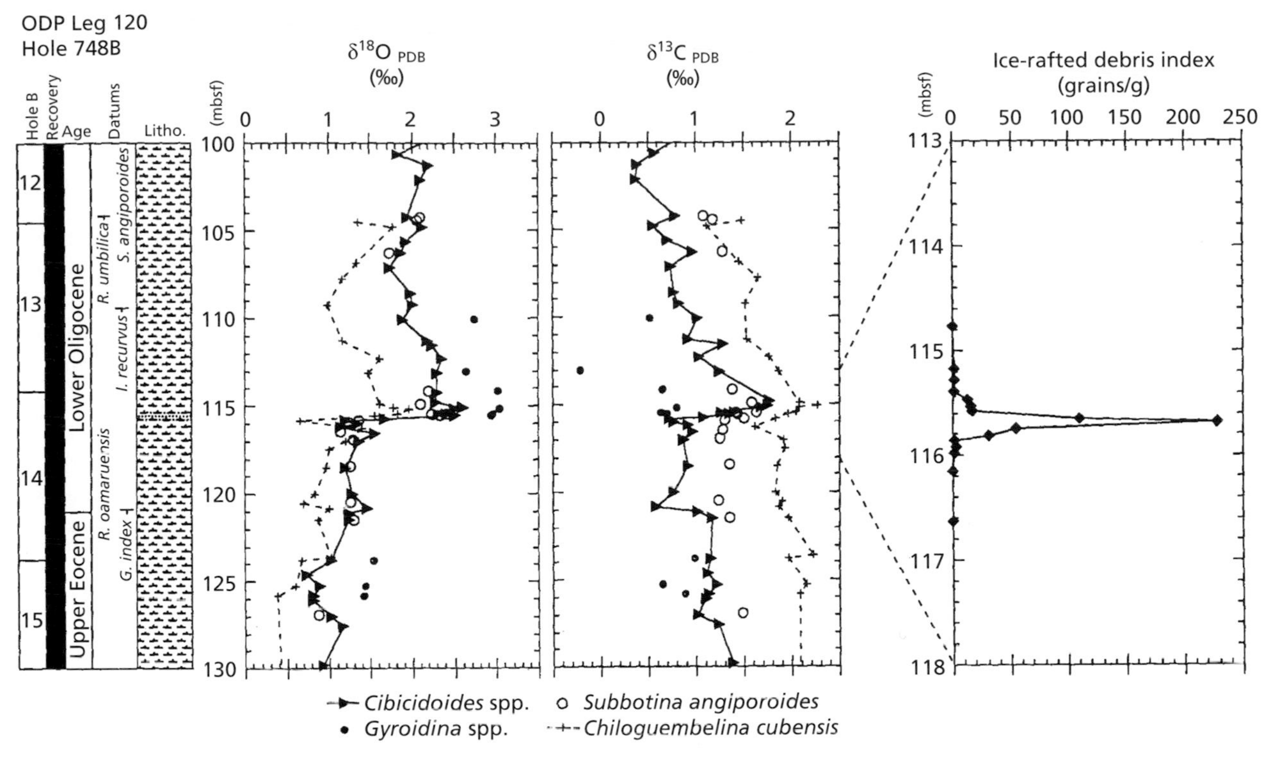

FIGURE 10.4. Upper Eocene–lower Oligocene stable isotope data and ice rafted debris record from Hole 748B. Core information, core recovery, biostratigraphic datum levels, and lithology are shown on the left. From Zachos et al., 1992.

NAME

Glaciation of Antarctica: The Oi1 Event

Part 10.2. Evidence for Global Change

In this part of the exercise we will consider a variety of data from a number of locations around the world (Figure 10.5). Students will assess data for **six localities** and report back to the class. The table at the end of this section will be used to compile the key findings from each location and build a spatial assessment of the **impact of the 'Oi1 event'**, the **positive isotope excursion in the earliest Oligocene**. The goal is to explore possible environmental consequences of the Oi1 event across the globe.

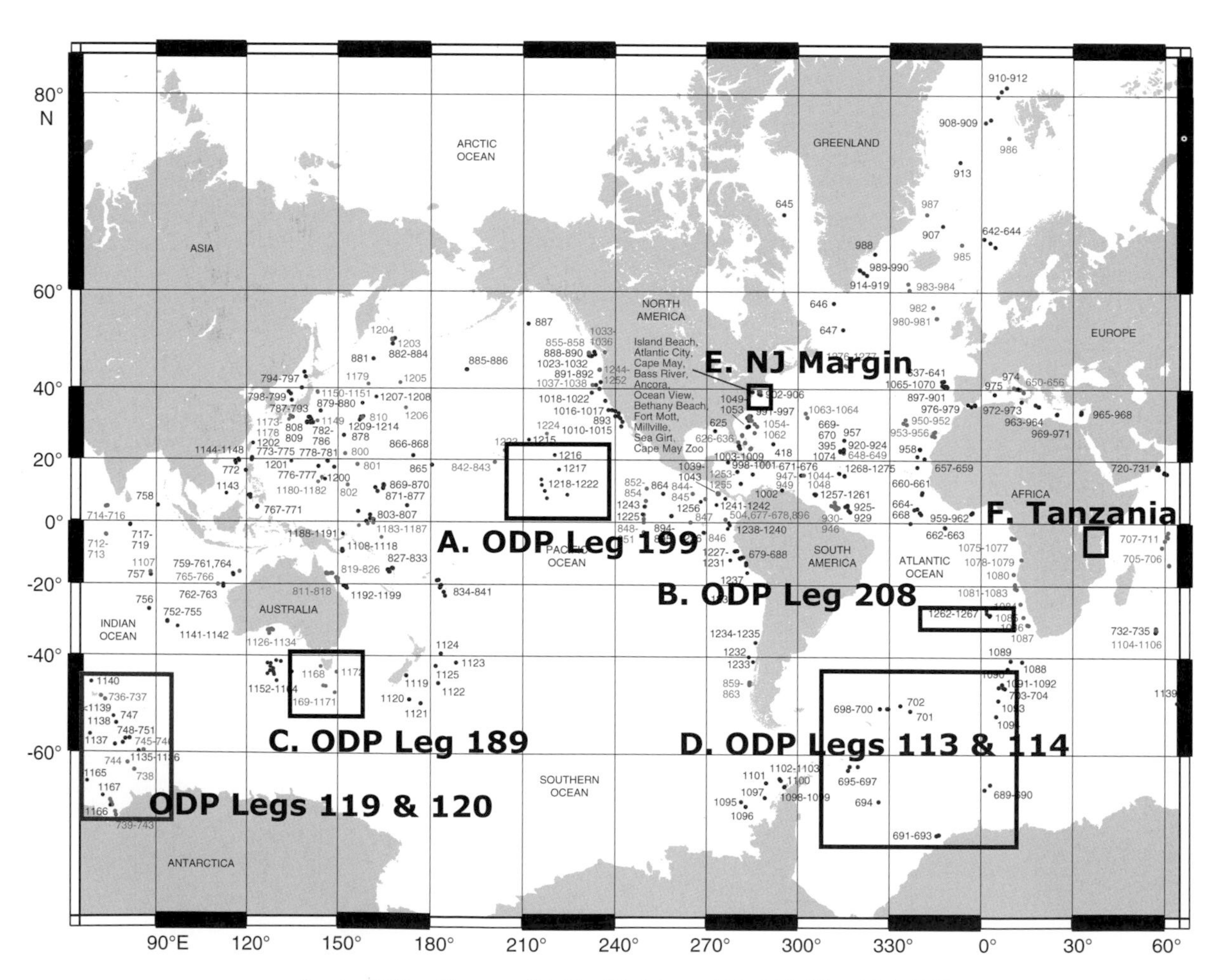

FIGURE 10.5. Map of all the Ocean Drilling Program (ODP) sites (1985–2003). Location of the Kerguelen Plateau and Prydz Bay, Antarctica drill sites from ODP Legs 119 and 120 (Figure 10.3) are shown in blue box. The **six localities (A–F)** used in this exercise, including land-based sections, are shown in black boxes. Courtesy of IODP.

NAME

A1. High-Resolution Record of the Oi1 Event at ODP Site 1218 (ODP Leg 199) in the Tropical Pacific

1 (a) Compare the **tropical $\delta^{18}O$ data** (Figure 10.6) with the **polar $\delta^{18}O$ data** (Figure 10.4). What similarities exist between these two records?

(b) What do these similarities suggest about the **geographic extent** of environmental or climatic change at this time?

2 What is the approximate **duration of the Oi1 event** as marked by the positive oxygen isotope excursion in Figure 10.6?

3 What happens to the **carbonate content** (reported as $\%CaCO_3$ and/or $CaCO_3$ mass accumulation rates, MARs) in the sediments across the Eocene–Oligocene transition and Oi1 event at ODP Site 1218? (**Hint:** think like a geologist; you're reconstructing a geological story so answer the question in chronological order, from older to younger in the core or in the data.)

Read the explanation for the changing CCD level in the box below and answer the following question.

NAME ______________________________

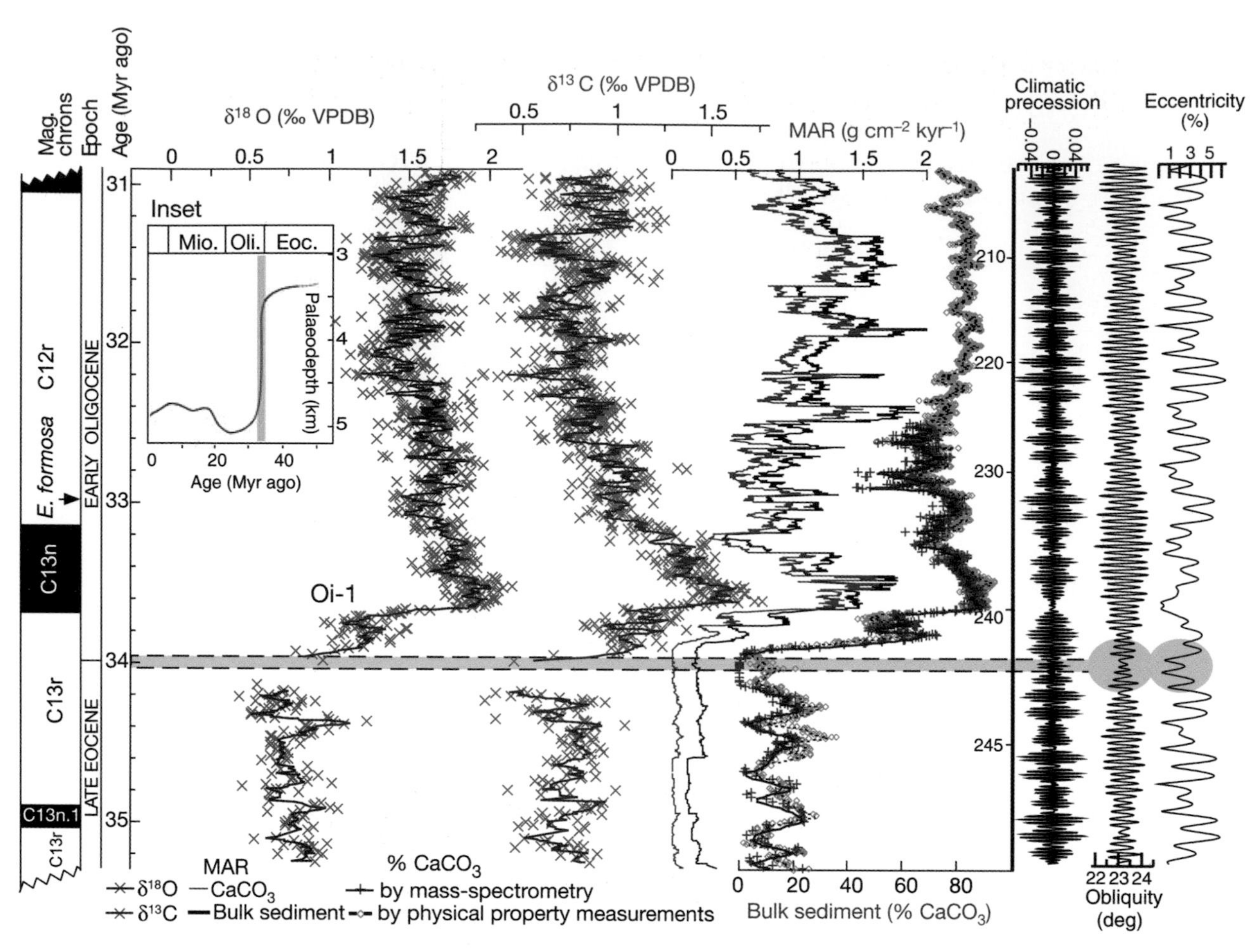

FIGURE 10.6. High-resolution paleoceanographic records from ODP Site 1218 (8°53.38'N, 135°22.00'W; water depth 4862 m) showing global changes in the ocean–climate system across the Eocene–Oligocene transition. Paleomagnetic reversal stratigraphy, calcareous nannofossil datum level, and age shown on the left. Benthic foram **δ^{18}O data** shown in blue; **δ^{13}C data** shown in purple. **Carbonate percentage** (bulk sediment %$CaCO_3$) and **carbonate mass accumulation rates** (MARs; g cm^{-2} kyr^{-1}) are each calculated by two methods, shown in red and black. Orbital forcing parameters shown on the right. The **Oi1 isotope excursion** coincides with a minimum in obliquity and eccentricity highlighted by gray circles. From Coxall et al., 2005.

4 Propose a hypothesis to account for the observed change in carbonate content at ODP Site 1218 (Figure 10.6).

____

____

____

____

____

NAME

DEEP-SEA CARBONATE AND THE CALCITE (OR CARBONATE) COMPENSATION DEPTH (THE 'CCD')

A change in carbonate content is a visual and striking aspect of many deep-sea records of the **PETM** (Chapter 9) and **Oi1** (this chapter) events. Carbonate content (percent carbonate, %$CaCO_3$) in marine sediments is influenced by: (1) **production** of carbonate by calcareous plankton in the surface ocean, (2) **dilution** of carbonate input by mud and other terrigenous sediments coming in from the weathering of land masses or volcanic islands, and (3) **dissolution** of carbonate by corrosive deep waters. The depth in the deep ocean marking the balance between carbonate rain (i.e., microscopic shells of calcareous plankton) from above and dissolution at depth is called the **calcite (carbonate) compensation** depth, or **CCD**. In general, no carbonate accumulates below the depth of the CCD.

The CCD is actually a dynamic surface that varies through space and time. The **depth of the CCD** is influenced by a number of factors, most important being **carbonate production in the overlying surface waters** (e.g., the CCD is deeper under areas with high influx of pelagic carbonate such as at the equator), and **deep water age** (i.e., deep waters originate as surface waters where they acquire their characteristics, such as temperature, salinity, nutrients, and dissolved gases, including oxygen and carbon dioxide, before sinking to become deep water masses). As deep waters move through the ocean basins and get older (i.e., time away from the surface increases), they acquire nutrients and dissolved CO_2 from the **respiration** of deep-sea organisms and by the **decomposition of organic matter** raining down from the surface ocean. This accumulating **CO_2 lowers the pH and makes the deep waters more corrosive to carbonate**.

The level of the CCD in the modern ocean is deeper in the Atlantic (approximately 5000 m) because deep waters form today in the Greenland-Norwegian Sea and in the Weddell Sea of Antarctica (i.e., at the north and south ends of the Atlantic); in other words, the **deep waters of the Atlantic today are 'young' and CO_2-poor**. In contrast, the level of the CCD in the Pacific, particularly the North Pacific, is much shallower (approximately 4000–4500 m) because the deep waters have been away from the surface for a long time: 1000 years or more. The **deep waters of the present-day Pacific are 'old', CO_2-rich, and corrosive to carbonate.** Another factor that controls the depth of the CCD is the **global sea level**. As the sea level rises, more carbonate is deposited across tropical shallow water shelves and epeiric seas (shallow seas on the continental crust); as a consequence, the level of the CCD rises to balance the loss of carbonate flux to the deep-sea. As the sea level falls, carbonate shelves and epeiric seas are reduced in size or drained and the level of the CCD falls. This **sea level control on the depth of the CCD is called shelf–basin partitioning**.

NAME

A2. The Eocene–Oligocene Transition and Oi1 Event in the Deep Tropical Pacific (ODP Leg 199)

5 Examine the core photos in Figure 10.7. Note that sediment color reflects the type of sediment (lithology) and carbonate content: lightest sediments are **nannofossil ooze**, darkest sediments are **radiolarian ooze**. Nannofossils are composed of carbonate, whereas radiolarians have shells made of silica. Make a list of observations about these cores from different sites in tropical Pacific. How are the sediments at these sites similar; how are they different?

Similarities	Differences

Note: Answer space continues on next page.

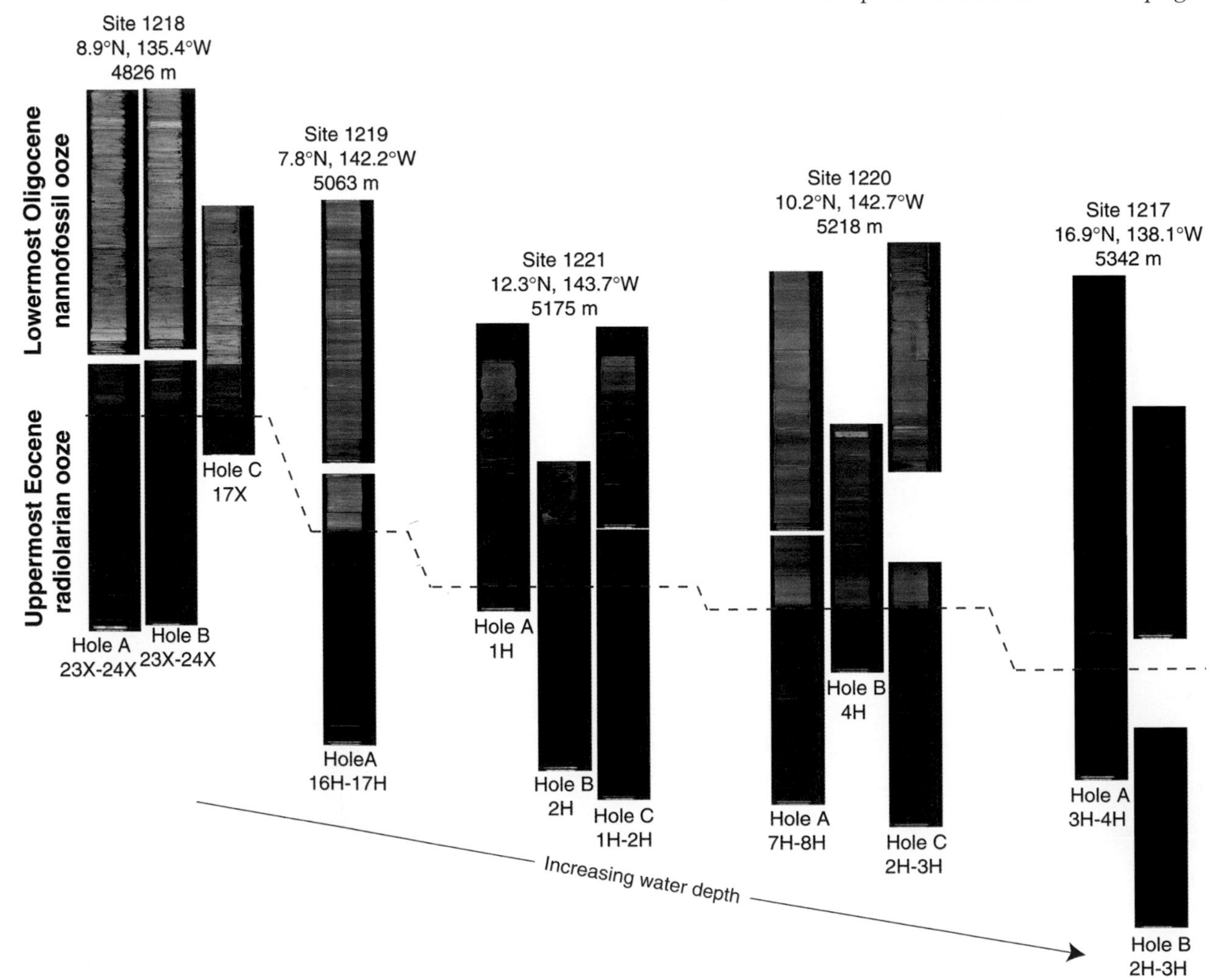

FIGURE 10.7. Composite digital mages of cores taken across the Eocene–Oligocene transition at ODP Leg 199 sites. Note the sites are arranged from shallowest (Site 1218; Figure 10.6) to deepest (Site 1217). Dashed line = base of lithologic transition. From: Shipboard Scientific Party, 2002.

NAME ___

6 Propose a hypothesis about a possible cause of **the change from radiolarian ooze to nannofossil ooze** at the ODP Leg 199 sites (Figure 10.7). Refer to the box above about the CCD, as well as Chapter 2 Seafloor Sediments.

___

___

___

___

___

B. The Eocene–Oligocene Transition and Oi1 Event in the Deep Southeast Atlantic along the Walvis Ridge (ODP Leg 208)

7 Examine the core photos in Figure 10.8. Make a list of observations. How are the sediments at these sites similar; how are they different?

Similarities	Differences

8 A **lithologic change from clay or clay-bearing nannofossil ooze to nannofossil ooze** occurs across the E–O transition and the Oi1 event at these Walvis Ridge sites. This lithologic change is associated with changes in the physical properties of the sediment, including decreasing magnetic susceptibility (MS) and gamma ray attenuation (GRA), and a lightening of the sediment color (L*). Collectively these data all indicate an upcore decrease in the relative proportion of clay to calcium carbonate ($CaCO_3$; nannofossils) in the sediment. What could this compositional change suggest about climate conditions, and why? (**Hint:** The box concerning the CCD may help formulate some ideas.)

___

___

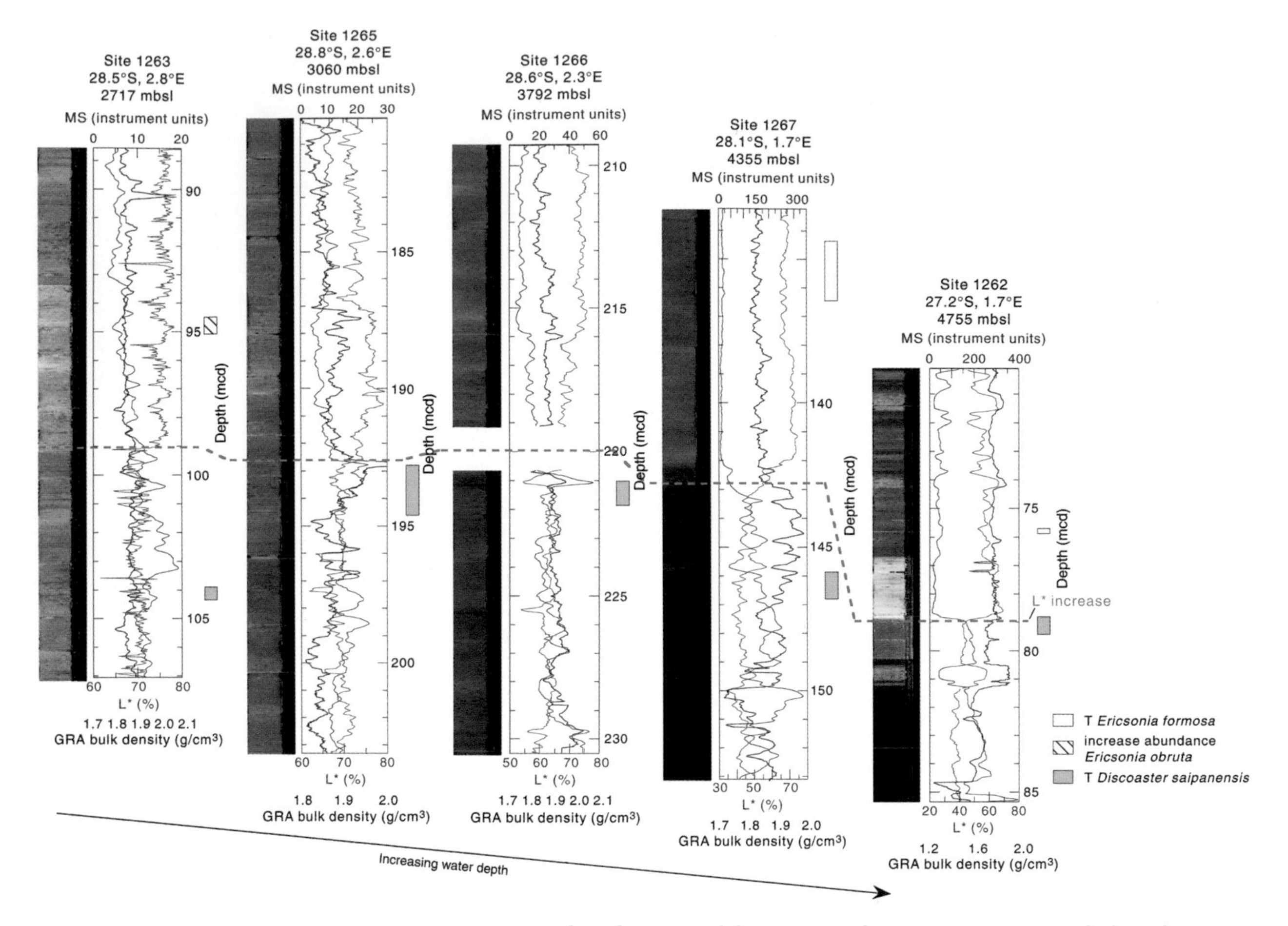

FIGURE 10.8. Composite digital images of the Eocene–Oligocene transition (including the Oi1 event) at ODP Leg 208 sites; the sites are arranged according to water depth down the Walvis Ridge. Also shown are magnetic susceptibility (MS in blue), color reflectance lightness (L* in red), and gamma ray attenuation bulk density data (GRA in black). The dashed line indicates a change in color reflectance that can be correlated from site to site. The location of calcareous nannofossil bioevents in the Eocene–Oligocene transition interval is shown as rectangles corresponding to the resolution of each event; the top (T) of *Ericsonia formosa* (open box symbol) occurs above the interval depicted at Sites 1263, 1265, and 1266. From: Shipboard Scientific Party, 2004.

C. The Eocene–Oligocene Transition on the South Tasman Rise (ODP Leg 189)

Note: the South Tasman Rise was once connected to Antarctica. ODP Leg 189 drilled a series of sites on the divergent margin of the South Tasman Rise designed to document the timing of the break-up between Australia–Tasmania and Antarctica, and to seek evidence for the opening of the **Tasman ocean gateway** between Tasmania and Antarctica (Figure 10.9).

NAME ______________________________

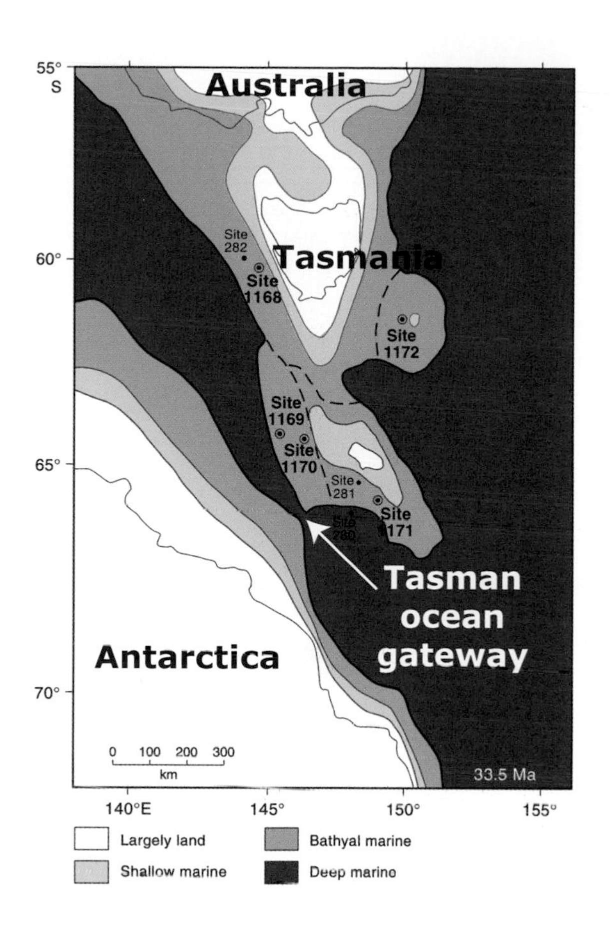

FIGURE 10.9. Paleogeographic map of the Tasmanian region **at the time of the Eocene/Oligocene boundary (approximately 33.5 Ma)** as the South Tasman Rise separated from Antarctica. Based partly on Royer and Rollet (1997) and Leg 189 results. Dark blue = deep water (>2000 m), light blue and green = shallow water (<2000 m). Solid circles = DSDP Leg 29 sites. Open circles with dots = ODP Leg 189 sites. From: Shipboard Scientific Party, 2001.

9 Based on Figure 10.10, describe how the **lithology on the South Tasman Rise changed during Eocene–Oligocene time**:

____

____

____

____

____

____

10 Tasmania and Antarctica separated during the time that these sediments were deposited. Describe how the lithostratigraphy in these cores tell the story of the break-up of Tasmania and Antarctica.

____

____

NAME

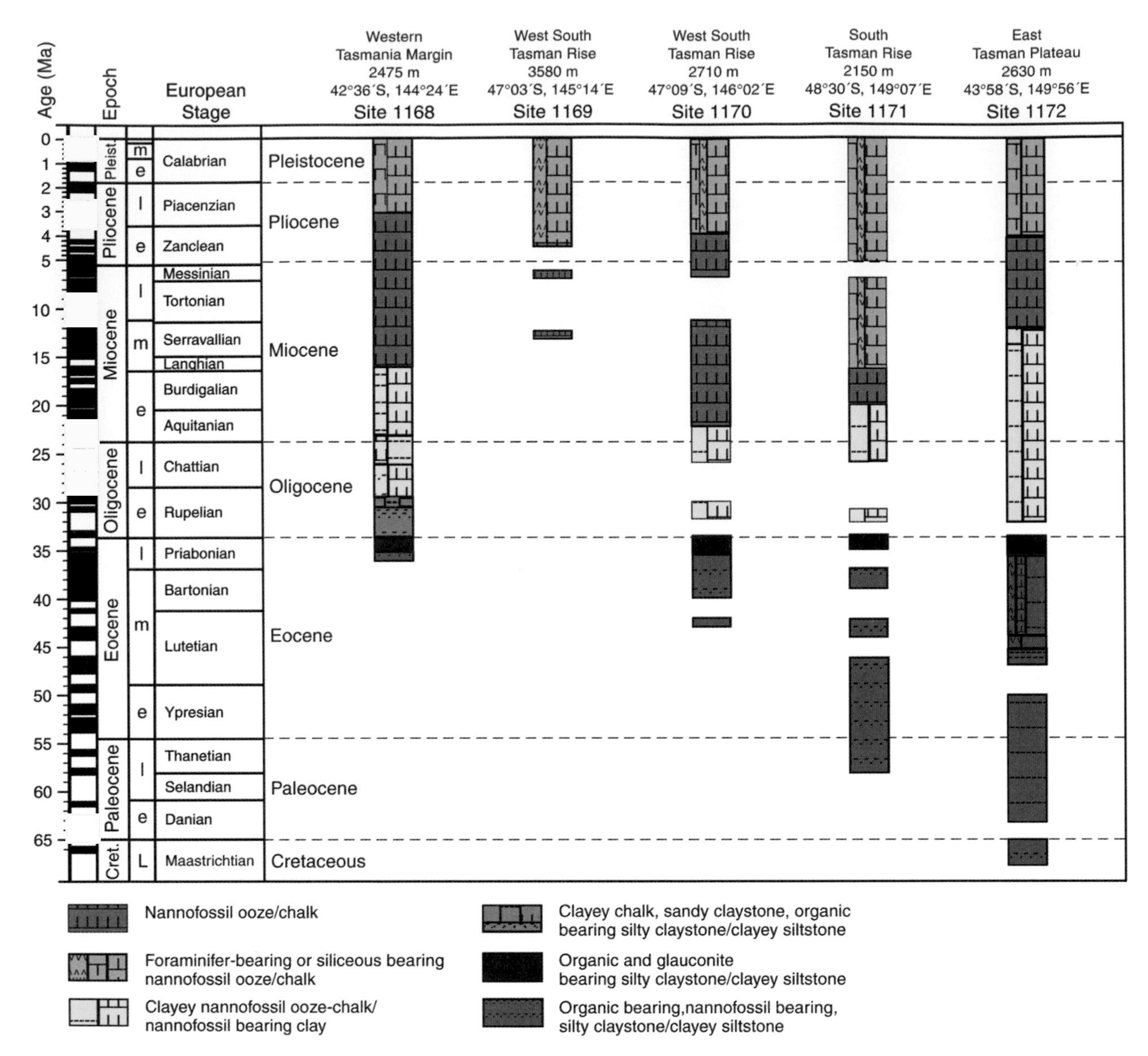

FIGURE 10.10. Summary of the **sediment lithogies** for ODP Leg 189 Sites 1168 through 1172, arranged west to east on the South Tasman Rise (Figure 10.9). These major lithologies are plotted against time to show major hiatuses (erosional breaks in each section). Silty claystone and clayey siltstone lithologies record **shallow neritic (shelf) depositional environments**; clayey nannofossil ooze-chalk and nannofossil bearing clay record **deeper water bathyal (slope) depositional environments**. **Glauconite** is a clay mineral that forms in marine environments, especially shelf environments characterized by slow sedimentation rates due in part to rising sea level and/or current-swept areas of the continental margin. From Shipboard Scientific Party, 2001.

NAME ____________________

11 Propose a hypothesis about how the break-up of Tasmania and Antarctica may be related to the Oi1 event and the glaciation of Antarctica.

____

____

____

____

____

____

____

D. The Eocene–Oligocene Transition in the Deep Southern South Atlantic (Southern Ocean, ODP Legs 113 & 114)

D1. The Eocene–Oligocene Transition at Site 699 (ODP Leg 114)

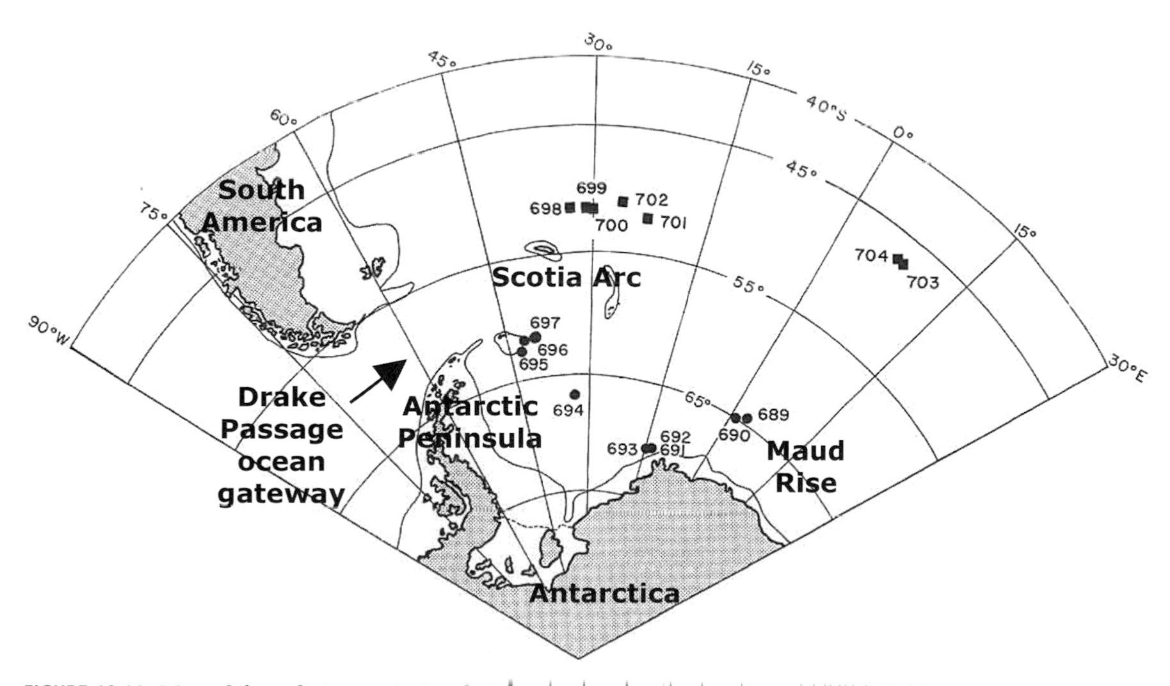

FIGURE 10.11. Map of the **sub-Antarctic South Atlantic** showing the location of ODP Leg 113 Sites 689–697 (circles) and Leg 114 Sites 688–704 (squares) in the Atlantic sector of the **Southern Ocean**. The **Drake Passage** is the **ocean gateway** between the southern tip of South America and the northern tip of the Antarctic Peninsula. ODP Sites 689, 699, and 701 are shown below. From Shipboard Scientific Party, 1988.

NAME ______________________________

12 Use the data in Figure 10.12 to describe the **changes in microfossil assemblages** (i.e., the relative abundances of calcareous nannofossil and siliceous microfossils) in Hole 699A (Figure 10.11).

____

____

____

____

____

____

____

13 Does the change in the dominant type of microfossil through the Eocene–Oligocene transition appear to be **abrupt or gradual**?

____

14 How can the **changing microfossil assemblages** be used to infer changes in ocean circulation or climate? Recall from Chapter 3 that both calcareous nannofossils and diatoms (a siliceous microfossil) are phytoplankton (i.e., primary producers).

____

____

____

____

15 Propose a hypothesis about how this change in type of phytoplankton may be related to the Oi1 event and the glaciation of Antarctica:

____

____

____

____

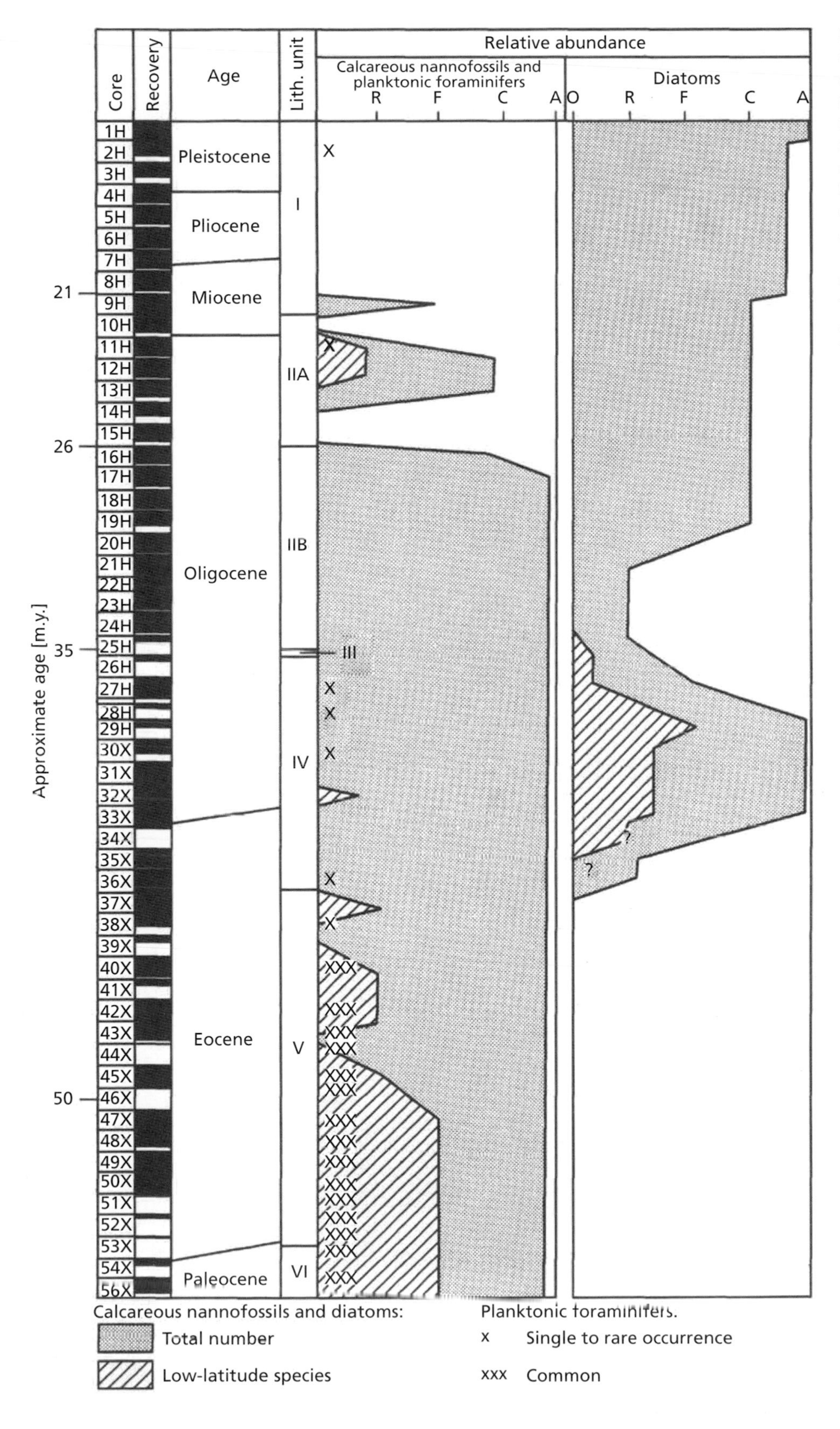

FIGURE 10.12. Changes in the **relative abundance of calcareous microfossils (primarily calcareous nannofossils) and siliceous microfossils (primarily diatoms)** found in the Cenozoic age sediments cored in **Hole 699A** in the Southern Ocean (Figure 10.11). Core numbers and core recovery are plotted on the left. Also shown are the sediment age and lithologic units (I–VI). From Shipboard Scientific Party, 1988, Site 699.

NAME

D2. The Eocene–Oligocene Transition in the ODP Leg 114 Sites

16 What are the **dominant lithologies (i.e., rock and sediment types) in the Late Cretaceous, Paleocene, and Eocene** (Figure 10.13)?

Age	Dominant Lithologies
Eocene	
Paleocene	
Late Cretaceous	

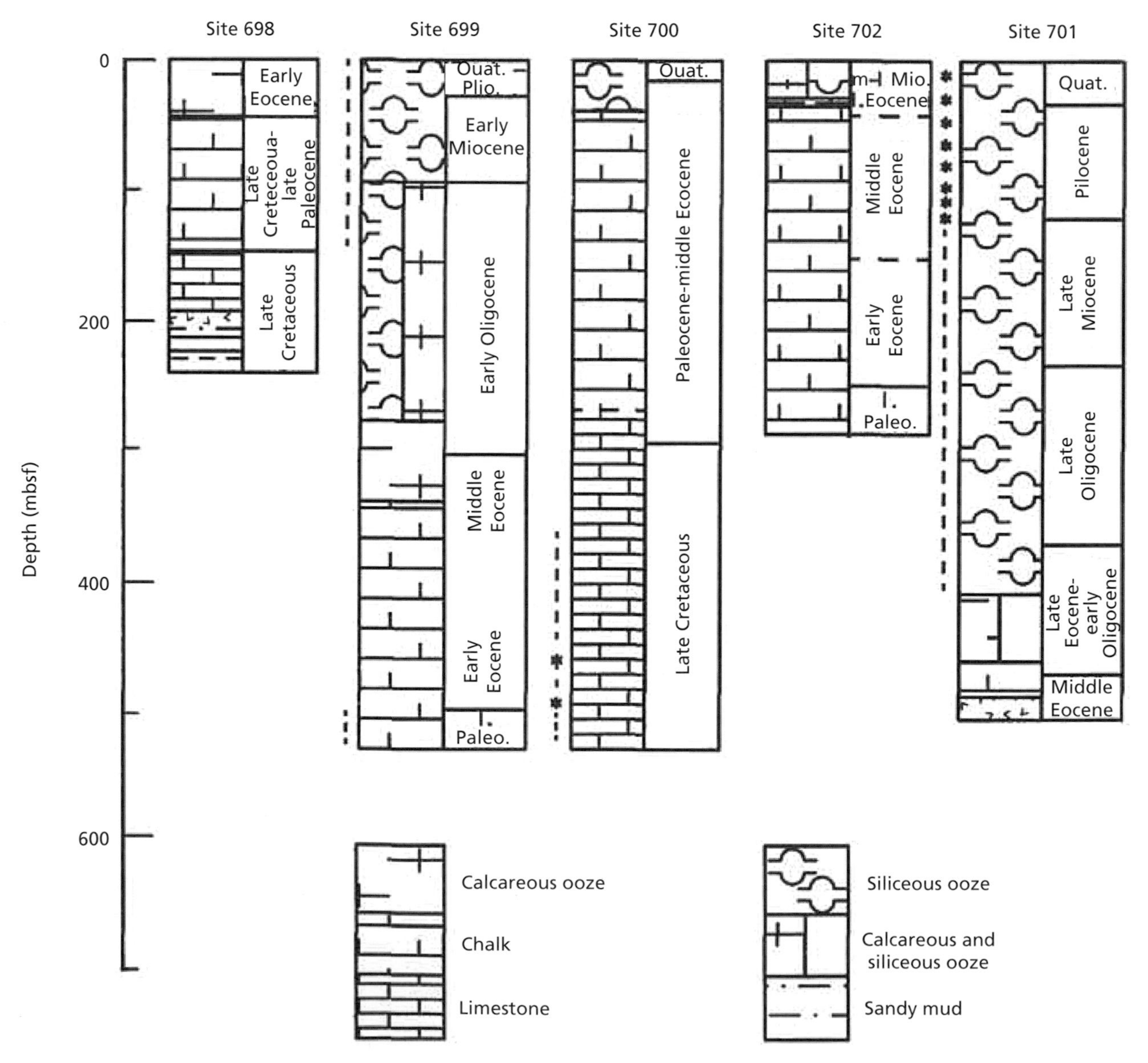

FIGURE 10.13. Lithostratigraphy of ODP Leg 114 sites. Here Site 699 (Figure 10.12) is compared with other Leg 114 sites (Figure 10.11), including Site 701. From Shipboard Scientific Party, 1988.

NAME ______________________________

17 What are the **dominant lithologies in the Oligocene, Miocene, Pliocene, and Quaternary**?

Age	Dominant Lithologies
Quaternary	
Pliocene	
Miocene	
Oligocene	

18 (a) Based on the data presented in Figure 10.13, what can you infer about **changes in climate or changes in the ocean environment** during the Cenozoic?

(b) Propose a hypothesis about how this change may be related to the Oi1 event and the glaciation of Antarctica:

D3. The Eocene–Oligocene Transition at Site 689 (ODP Leg 113)

19 Examine Figure 10.14. What type of sediment does **diatom ooze–siliceous ooze** replace at Site 689? When (at what age) did the change in lithology begin?

NAME ______________________

20 Does the change in lithology appear to be **gradual or abrupt**?

____

21 How does the **Maud Rise lithologic record near Antarctica** (Figure 10.14) compare with the Southern Ocean sites north of the Scotia Arc (ODP Leg 114 Sites 699 and 701; Figure 10.13)? Refer to map (Figure 10.11).

____

____

____

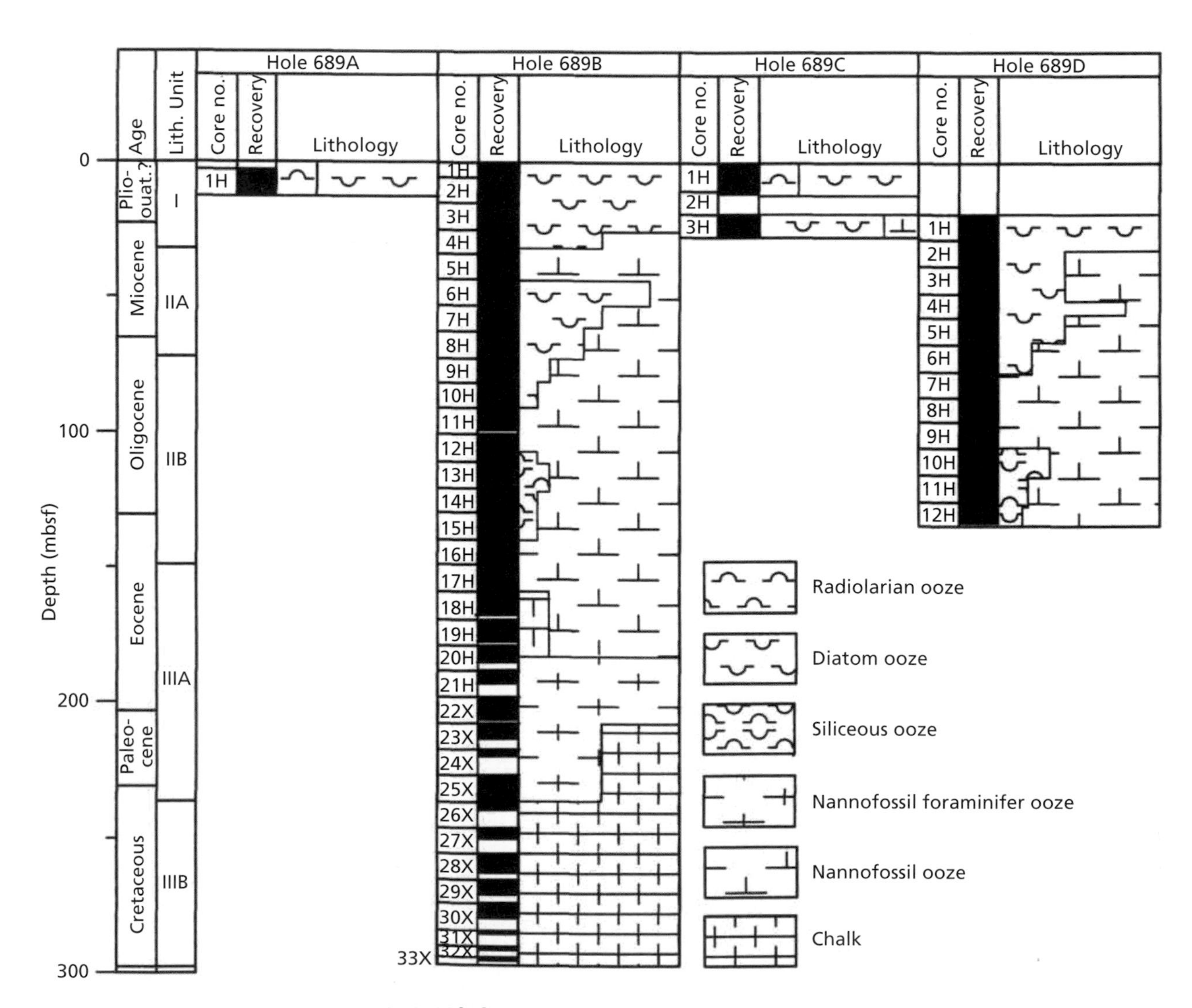

FIGURE 10.14. Lithologic summary of Site 689 (Holes 689A–D) on Maud Rise in the Southern Ocean off Antarctica (Figure 10.11). From Shipboard Scientific Party, 1988, Site 689.

NAME

E. New Jersey Coastal Plain and Continental Margin Sea Level Reconstruction

22 Examine Figure 10.15. Based on the New Jersey margin record of sea level change, how much did **sea level change** during the Eocene–Oligocene transition?

23 What is the **likely cause** of this change in sea level?

24 Refer back to Figure 10.6. What is the **magnitude of the Oi1 $\delta^{18}O$ positive excursion**, in permil (i.e., use the ‰ scale on the graph)?

25 $\delta^{18}O$ of deep-sea benthic forams is controlled primarily by temperature and ice volume. **If an increase in ice volume produced a 10 meter drop in global sea level and resulted in a 0.1‰ increase in $\delta^{18}O$**, how much of the Oi1 excursion was due to changing ice volume (and therefore sea level) at the Eocene–Oligocene transition? Show your calculation.

26 How does this estimate compare with the **New Jersey margin record** of sea level change (Figure 10.15)?

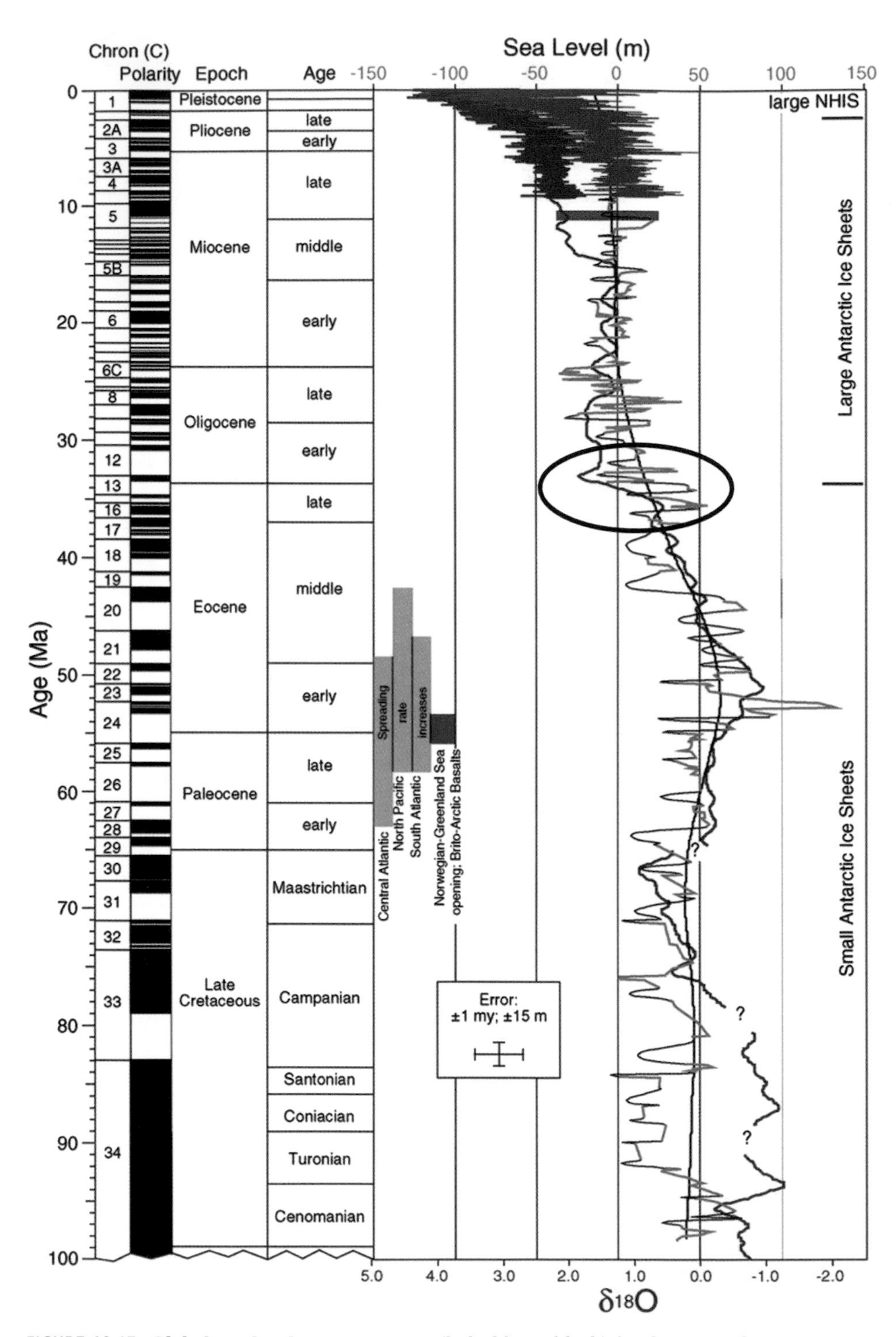

FIGURE 10.15. **Global sea level reconstruction** (light blue + black) for the interval 100–7 Ma based on stratigraphic, paleontological, and geophysical data from the NJ margin (Figure 10.5). The benthic foram $\delta^{18}O$ synthesis plot is shown for comparison (red). Modified from Miller et al., 2005.

NAME

27 (a) If **a 1°C decrease in temperature produces a 0.25‰ increase in $\delta^{18}O$**, how much may deep-sea temperatures have changed during the Oi1 event? Show your work.

(b) Based on your response to Questions 25–27(a), do you interpret the **positive oxygen isotope excursion** of the Oi1 event to represent a **purely ice volume increase** (sea level fall) signal or a **mix of ice volume and temperature decrease**? Explain.

F. Tanzania Coastal Plain Biotic Turnover

28 Examine Figure 10.16. The authors of this study attribute the extinctions and biotic turnover of planktic foraminifers during the Eocene–Oligiocene transition to a **major ecological disturbance**. Speculate about ecological changes that may have affected near-surface dwelling plankton.

29 The **global extinction** of planktic forams occurs in **two steps** as observed in Tanzania (Figure 10.16) and elsewhere. Are there other data from this exercise

NAME

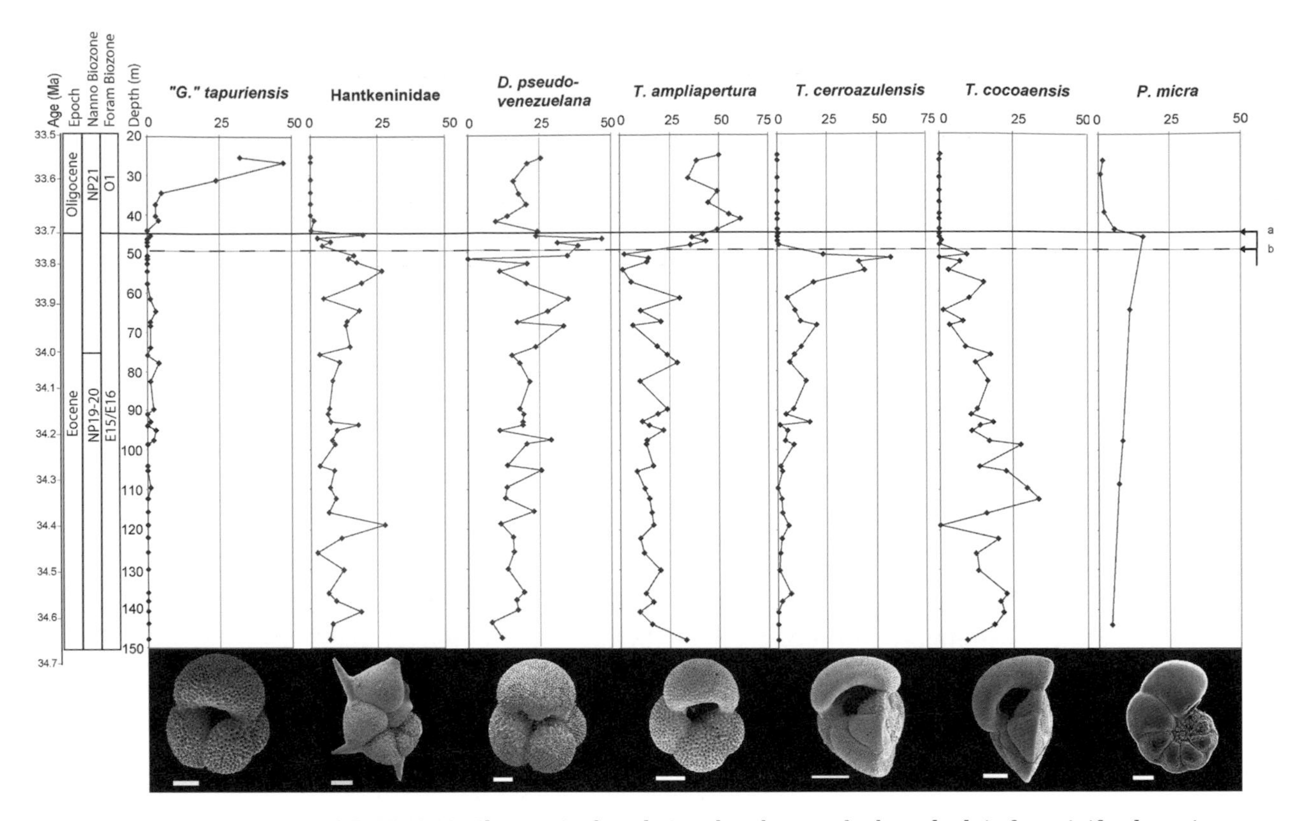

FIGURE 10.16. Changes in the relative abundances of select **planktic foraminiferal species** through the Eocene–Oligocene transition in Tanzania. Scale bars = 100 μm (=0.1 mm), except far right specimen of *Pseudohastigerina micra* = 50 μm. The solid horizontal line is the Eocene–Oligocene boundary marked by the **global extinction of five species of *Hantkenina***. The dashed line marks the global extinction of *Turborotalia cerroazulensis* and *T. cocoaensis*. From Wade and Pearson, 2008.

(i.e., previous data plots in Part 10.2) suggesting that environmental changes across the Eocene–Oligocene transition may have occurred in two steps?

30 This marine record comes from a drill core taken on the tectonically stable coastal plain (i.e., on shore) of Tanzania. What does this imply about **global sea level during the late Eocene**?

Synthesis and Discussion

31 Use Table 10.1 to summarize the **key observational data** from each of the deep-sea and land-based records presented above. List any questions that you have about what you have seen in the data.

32 Use the data summarized in Table 10.1 to identify **possible connections** (i.e., cause and effect, timing) among the following:

Glaciation:

Ocean circulation:

Global cooling:

NAME ____________________

TABLE 10.1. Synthesis of observations.

Figure	Location (Figure 10.5)	Key Observations	Your Questions
10.6	**A1.** ODP Site 1218 (Leg 199) **Tropical Pacific**		
10.7	**A2.** ODP Leg 199 **Tropical Pacific**		
10.8	**B.** ODP Leg 208 **SE Atlantic** (Walvis Ridge)		
10.10	**C.** ODP Leg 189 **S. Tasman Rise**		
10.12	**D1.** ODP Site 699 (Leg 114) **Southern Ocean**		
10.13	**D2.** ODP Leg 114 **Southern Ocean**		
10.14	**D3.** ODP Site 689 (Leg 113) Southern Ocean		
10.15	**E. New Jersey margin cores**		
10.16	**F. Tanzania coastal plain cores**		

NAME

Productivity:

Sea level:

Biotic evolution and extinction:

33 Write a paragraph in which you hypothesize about the cause and effect of glaciation on Antarctica. Be sure to support your hypothesis with evidence from this exercise.

NAME

NAME ______________________

Glaciation of Antarctica: The Oi1 Event

Part 10.3. Mountain Building, Weathering, CO_2 and Climate

Chemical weathering of rocks and minerals at the Earth's surface can have an important impact on the long-term climate (see Chapter 5, Part 5.4). The chemical weathering of **silicate** rocks and minerals **consumes atmospheric CO_2**. The weathering reaction (shown below) produces dissolved cations, bicarbonate, and silicic acid. The **dissolved by-products** of chemical weathering are transported by rivers to the ocean and provide the **raw materials** for organisms to precipitate hard parts of **calcium carbonate** or **opaline silica** and provide important **nutrients** to drive photosynthesis and **primary productivity**.

$$Mg_2SiO_4 + 4CO_2 + 4H_2O \rightarrow 2Mg^{2+} + 4HCO_3^- + H_4SiO_4$$

olivine (a silicate mineral) + carbon dioxide + water →
magnesium and bicarbonate ions in solution + silicic acid in solution

The chemical weathering of **aluminosilicates**, such as feldspar, also **consumes CO_2** from the atmosphere. This weathering reaction produces **clay minerals** together with dissolved cations, bicarbonate, and silicic acid.

$$2KAlSi_3O_8 + 2H_2CO_3 + 9H_2O \rightarrow Al_2Si_2O_5(OH)_4 + 4H_4SiO_4 + 2K^+ + 2HCO_3^-$$

orthoclase feldspar + carbonic acid + water →
kaolinite (a clay mineral) + silicic acid in solution +
potassium and bicarbonate ions in solution

1 (a) Based on the reactions above, explain how the **chemical weathering** processes described above can affect **climate**.

____

____

____

____

____

(b) If the rate of chemical weathering were to increase, what type of **climate response** would you predict?

____

NAME ___________________________________

____

____

2 If the rate of chemical weathering were to increase, what response might we expect to see in **seafloor sediments**?

____

____

____

____

STRONTIUM ISOTOPES AND CONTINENTAL WEATHERING

A useful **proxy for long-term chemical weathering of continental rocks** is the **strontium isotope record of marine carbonates**. Stable isotopes of strontium are incorporated in trace (but measurable) abundances in the calcium carbonate shells of marine organisms, such as foraminifera. Changes in the **strontium isotopic ratio ($^{87}Sr/^{86}Sr$)** of foram shells faithfully track the **Sr isotopic composition of seawater over time** (Figure 10.17), which in turn tracks the balance between relatively higher $^{87}Sr/^{86}Sr$ input from weathering of continental crust and relatively lower $^{87}Sr/^{86}Sr$ from hydrothermal reactions during oceanic crust formation at divergent plate boundaries. Thus, the **$^{87}Sr/^{86}Sr$ proxy** can be affected by mountain building, sea level, and changes in the rate of seafloor spreading. For example, the subcontinent of India began to collide with Asia approximately 50 Ma. A consequence of this continent–continent collision has been the creation of the most massive uplift on the planet: the Himalaya Mountains and the Tibetan Plateau. This major tectonic event is recorded in the $^{87}Sr/^{86}Sr$ ratio of seawater (Figure 10.17).

3 Read the box above on strontium isotopes and continental weathering and then answer the following questions:

(a) As $^{87}Sr/^{86}Sr$ values have risen over the past 40 million years, do you predict that atmospheric CO_2 values have risen or fallen?

____

____

(b) What impact might the chemical weathering-induced change in CO_2 have had on global climate over the past 40 million years?

NAME

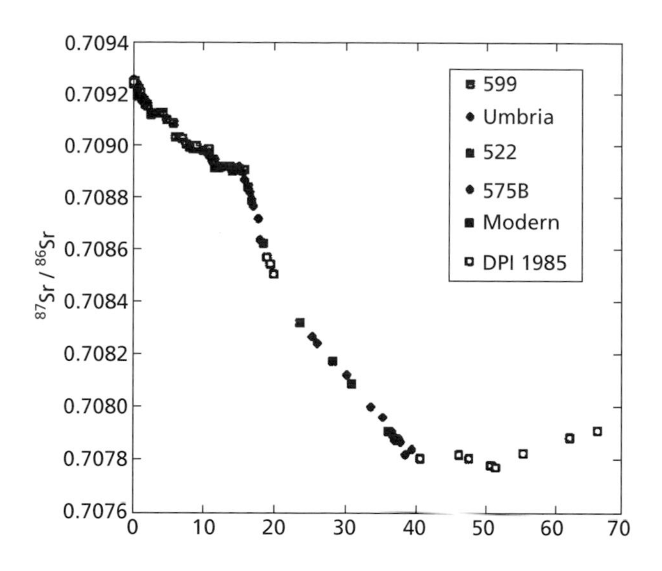

FIGURE 10.17. **Strontium isotopic composition of seawater** for the past 70 million years. **Continental rocks** are relatively radiogenic and therefore **more enriched in $^{87}Sr/^{86}Sr$** than the non-radiogenic hydrothermal activity associated with ocean crust production at spreading centers. Note: collision of India with Asia and the uplift of the Himalaya Mountains and Tibetan Plateau began after 50 Ma. From Raymo and Ruddiman, 1992.

4 Summarize the relationship between the strontium isotope composition of seawater, the chemical weathering of continental rocks, and atmospheric CO_2.

5 What other processes, besides chemical weathering of continental rocks, might affect atmospheric CO_2 concentrations, and therefore, global climate? (**Hint:** think back to the long-term carbon cycle introduced in Chapter 5, Part 5.4.)

NAME

Putting it All Together

In Parts 10.1 and 10.2, you made a number of observations from geologic records that are coincident with the glaciation of Antarctica (the so-called 'Oi1 event'). It is clear that **the glaciation of Antarctica in the earliest Oligocene had global consequences.** As you have seen in Part 10.3, chemical weathering is a natural "sink" for atmospheric CO_2. If this process occurs on a large scale, the loss of CO_2 from the atmosphere would promote **global cooling**. This process of **CO_2 drawdown** can be considered to be a **"reverse greenhouse"** effect.

6 Make a list of **possible causal mechanisms for the Oi1 event** (i.e., the rapid glaciation of Antarctica as evidenced by the positive excursion of $\delta^{18}O$ values in planktic and benthic forams). Explain how each would lead to the growth of continental-scale ice sheets.

7 **Positive feedback** amplifies climate change, while **negative feedback** diminishes climate change (see Chapter 5, Part 5.2).

(a) What feedbacks associated with the global environmental changes of the Eocene–Oligocene transition (e.g., CO_2, climate, glaciation, sea level, weathering, productivity, ocean circulation) may have contributed to the rapid glaciation of Antarctica approximately 33.7 Ma? A sketch may help.

NAME

(b) Could multiple feedbacks be involved? Explain.

Climate models can inform us about dynamic processes in the climate system. Large-scale computer models that simulate the Earth's atmosphere are called **general circulation models** or **GCMs** (See Chapter 6, Part 6.1). Figure 10.18 shows the results of a modeled simulation driven by progressively declining atmospheric CO_2 that resulted in rapid growth of a continental scale ice sheet on Antarctica.

8 Based on the many ocean–climate system relationships you learned about in this exercise (Parts 10.1, 10.2, and 10.3), **suggest at least two different processes that may have caused CO_2 to fall during the Eocene–Oligocene transition**. List your ideas and compare with others.

9 Read the box below about alternative hypotheses for the glaciation of Antarctica and examine Figures 10.19 and 10.20. Do any of the data we have considered thus far in this exercise (Parts 1, 2, and 3) support either of these **alternative hypotheses**? If so, how?

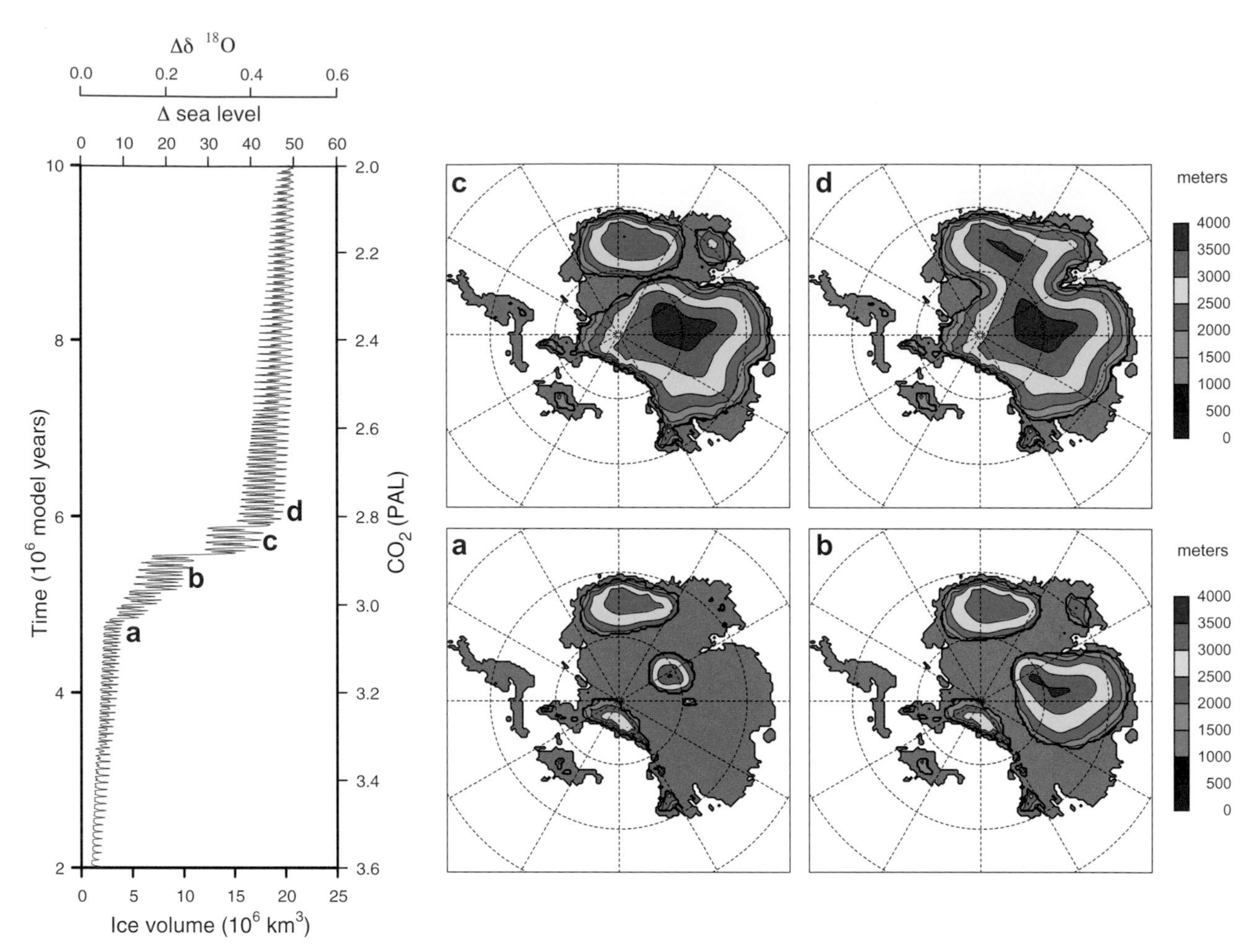

FIGURE 10.18. This composite figure depicts a **simulated initiation of East Antarctic glaciation in the earliest Oligocene using a coupled global climate model (GCM)-ice sheet model.** From DeConto and Pollard, 2003; figure modified by DeConto, 2011. According to this model the main triggering mechanism for initial inception and development of the East Antarctic ice sheet was **decreasing levels of CO_2 concentration in the atmosphere coupled with favorable orbital parameters** (Milankovitch forcing; refer to Chapter 8). Once a certain **threshold** of CO_2 was reached, the continent glaciated very rapidly within two eccentricity cycles (<200 kyr; a–b–c–d). Note this model shows that initiation of glaciation was principally a "two-step" process.

ALTERNATE HYPOTHESES FOR THE GLACIATION OF ANTARCTICA

Decreasing CO_2 is one hypothesis that can explain how and why the continent of Antarctica glaciated in the earliest Oligocene. Other hypotheses of the cause of the initial glaciation of Antarctica are as follows:

(1) Plate tectonics and thermal isolation of Antarctica (Figure 10.19): Opening of the Tasman Rise and Drake Passage ocean gateways allowed the development of the **Antarctic Circumpolar Current** (e.g., Kennett, 1977, 1978).

(2) Explosive volcanism (Figure 10.20): concentration of explosive volcanism (ignimbrite flare-ups) in the circum-Pacific during the latest Eocene and earliest Oligocene may have contributed to **global cooling** and may have been a contributing factor leading to rapid glaciation on Antarctica (e.g., Sigurdsson et al., 2000; Jicha et al., 2009).

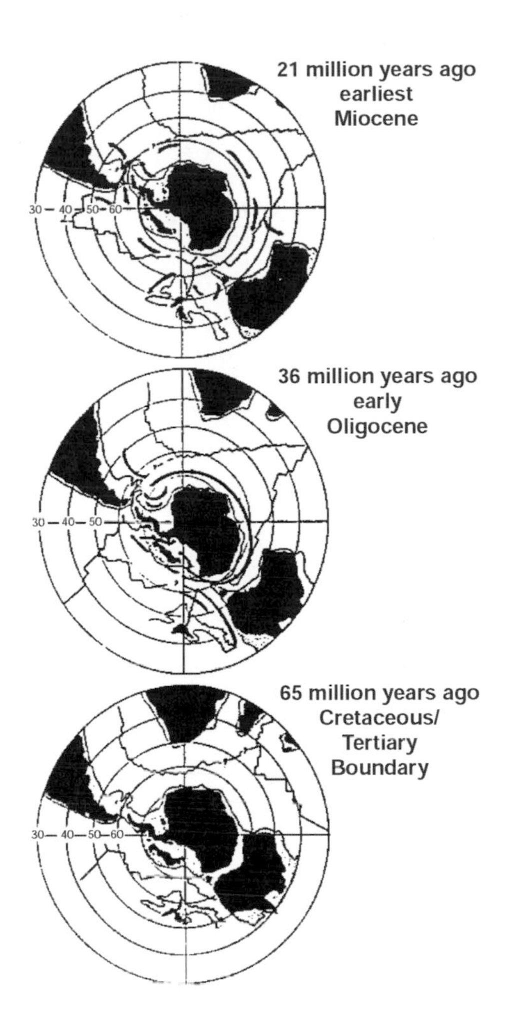

FIGURE 10.19. Antarctica and surrounding continents showing the development of the **Antarctic Circumpolar Current** during the Cenozoic. Modified from Kennett, 1978.

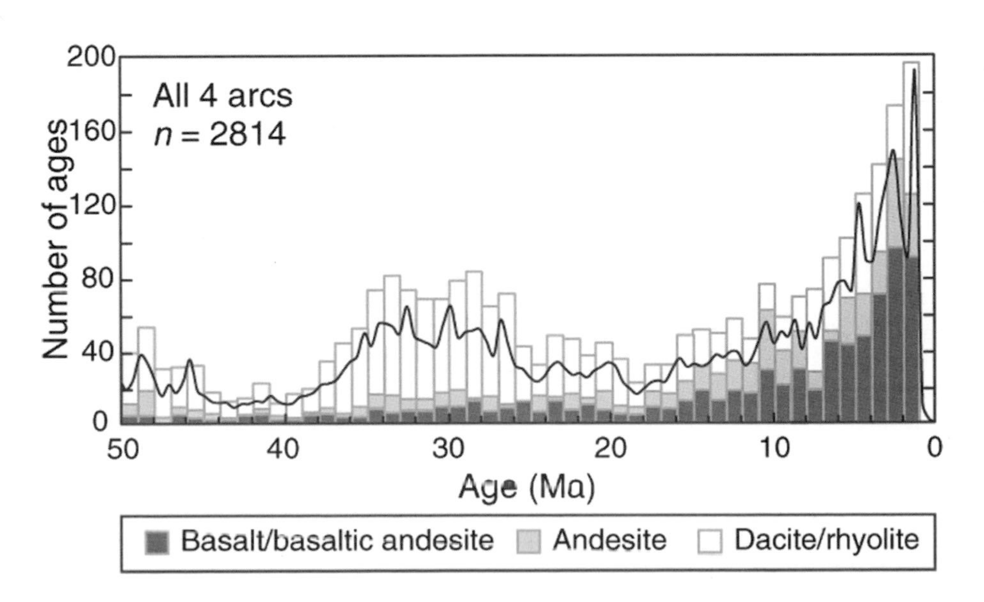

FIGURE 10.20. Histogram of the number of K–Ar and 40Ar/39Ar ages from four circum-Pacific arcs shown per 1 million years. Note the concentration of dacite/rhyolite eruption ages in the Eocene–Oligocene transition approximately 34–33 Ma. From Jicha et al., 2009.

NAME ______________________

Glaciation of Antarctica: The Oi1 Event

Part 10.4. Legacy of the Oi1 Event: The Development of the Psychrosphere

The following two plots (Figure 10.21) are low-resolution summaries of planktic foram (open symbols) and benthic foram (closed symbols) oxygen isotope data and paleotemperature estimates from Deep Sea Drilling Project sites in the low latitudes and the high latitudes of the Pacific.

1 In Figure 10.21, **circle** the Oi1 event as recorded by foram data in the North Pacific and the Southern Ocean. How are these data similar and how are they different?

Similarities	Differences

2 Which forams tell us about the character of the deep waters?

____

3 Which forams tell us about the character of the surface waters?

____

4 The North Pacific sites (Figure 10.21, left) are located in the low latitudes. What does the difference in $\delta^{18}O$ between planktic and benthic forams in the low latitude sites tell you about:

(a) the character of the water column (i.e., the surface ocean compared to the deep ocean)

____

____

____

NAME

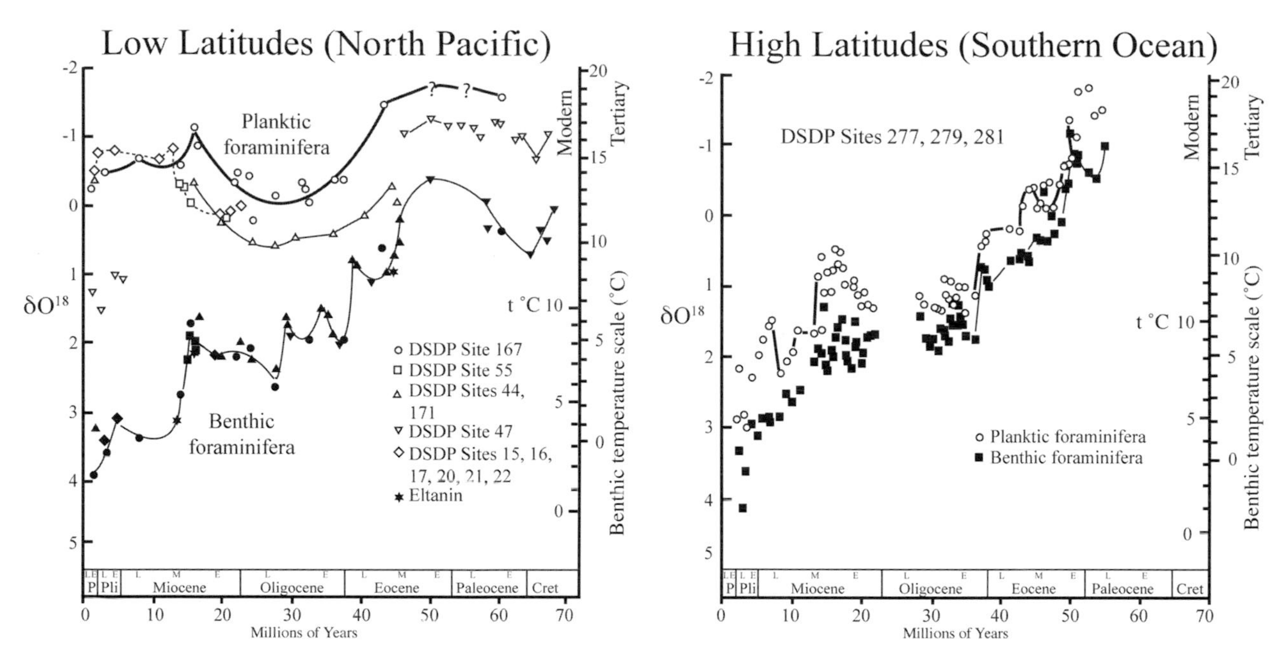

FIGURE 10.21. Cenozoic deep-sea oxygen isotope data for DSDP sites in the **North Pacific** (left) and **Southern Ocean** (right). **Note the Modern and Tertiary benthic temperature scale on both plots.** Modified from Savin, 1977.

(b) the timing of changes in the low latitude water column?

WATER MASSES AND WATER COLUMN STRATIFICATION

In the **modern** ocean (last 5 million years in Figure 10.21), **the temperature of deep waters in the low and high latitudes are approximately the same.** Furthermore, the **temperature of high latitude deep waters is very similar to high latitude surface waters.** This is because deep waters in the modern ocean form in the high latitudes; winter cooling and annual sea ice formation cause surface waters to become more dense and sink. This process of deep water formation yields the **stratified** ocean we have

NAME ______________________________

today with warm waters in the low latitudes overlying cool intermediate waters, and very cold and dense deep and bottom water masses as shown in Figure 10.22. A **water mass** is a parcel or volume of water with distinct temperature, salinity, and chemical characteristics.

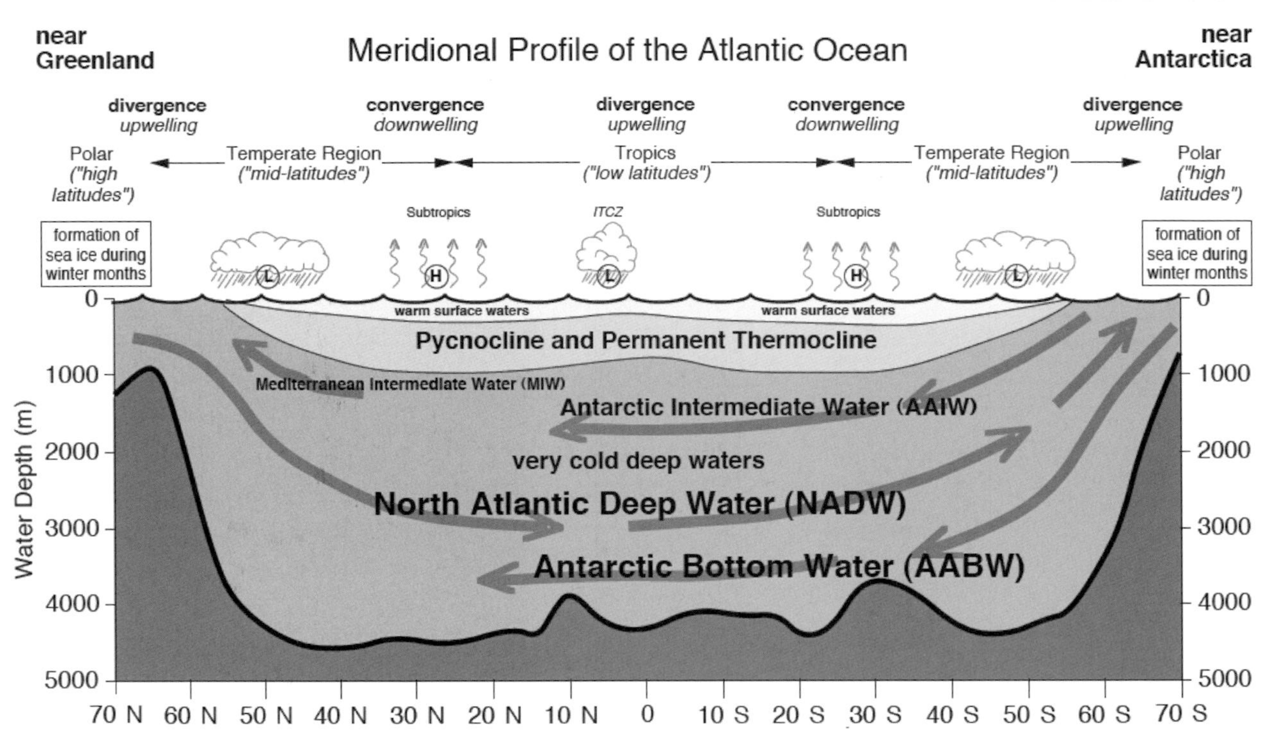

FIGURE 10.22. North–South (meridional) profile of the Atlantic Ocean showing the distribution and sources of intermediate, deep, and bottom water masses. Figure drawn by Mark Leckie.

Read the box above on Water Masses and Water Column Stratification and answer Question 5.

5 (a) The changes in surface ocean and deep ocean temperatures track each other in the southern high latitudes (Southern Ocean), but this is not what is observed in the low latitudes (North Pacific). Using data from Figure 10.21, predict **when** the **deep water masses of the world ocean** became more similar to the modern ocean.

____

____

(b) What evidence supports your answer?

____

____

NAME

6 Where might deep waters have formed in Late Cretaceous, Paleocene, or Eocene time? Explain the reasoning for your answer. (Note the Tertiary benthic temperature scale on the right vertical axes of both plots in Figure 10.21.)

THE PSYCHROSPHERE, DEEP WATER MASSES, AND THERMOHALINE CIRCULATION

The term **psychrosphere** refers to the very cold deep waters that fill all the ocean basins of the world ocean. Today, deep waters form in the northern and southern high latitudes of the Atlantic Ocean, specifically in the Greenland-Norwegian Sea and in the Weddell Sea (Figures 10.22 and 10.23). Deep (and bottom) waters form at the surface in the high latitudes where they acquire their physical and chemical characteristics (temperature, salinity, dissolved gases) before sinking. Surface waters sink when they become more dense owing to winter cooling and sea ice formation, which makes the waters saltier. This **downwelling** process typically occurs during the dark and cold winter months. **Thermohaline circulation** describes the process of deep-water mass formation and movement through the world ocean.

The deep waters that form in the Greenland-Norwegian Sea are called **North Atlantic Deep Water (NADW)** and the deep waters that form in the Weddell Sea are called **Antarctic Bottom Water (AABW)**. NADW and AABW mix in the Southern Ocean adjacent to Antarctica and form a new water mass called **Circumpolar Water (CPW)**. CPW is the deep-water mass that fills the deep Indian and Pacific Ocean basins. By the time deep waters reach the North Pacific they are approximately 1000 years old. In other words, these deep North Pacific waters have been away from the ocean surface for a millennium and are the oldest waters in the world ocean. The **global conveyor** (Figure 10.22) describes the movement of waters through the entire world ocean, including thermohaline circulation of deep waters through the ocean basins, upwelling of deep and intermediate waters at various places, and the wind-driven circulation of surface water masses. Because deep waters only form in the Atlantic today, **the Atlantic exports deep waters to the world ocean and imports surface waters** thereby completing the conveyor.

NAME

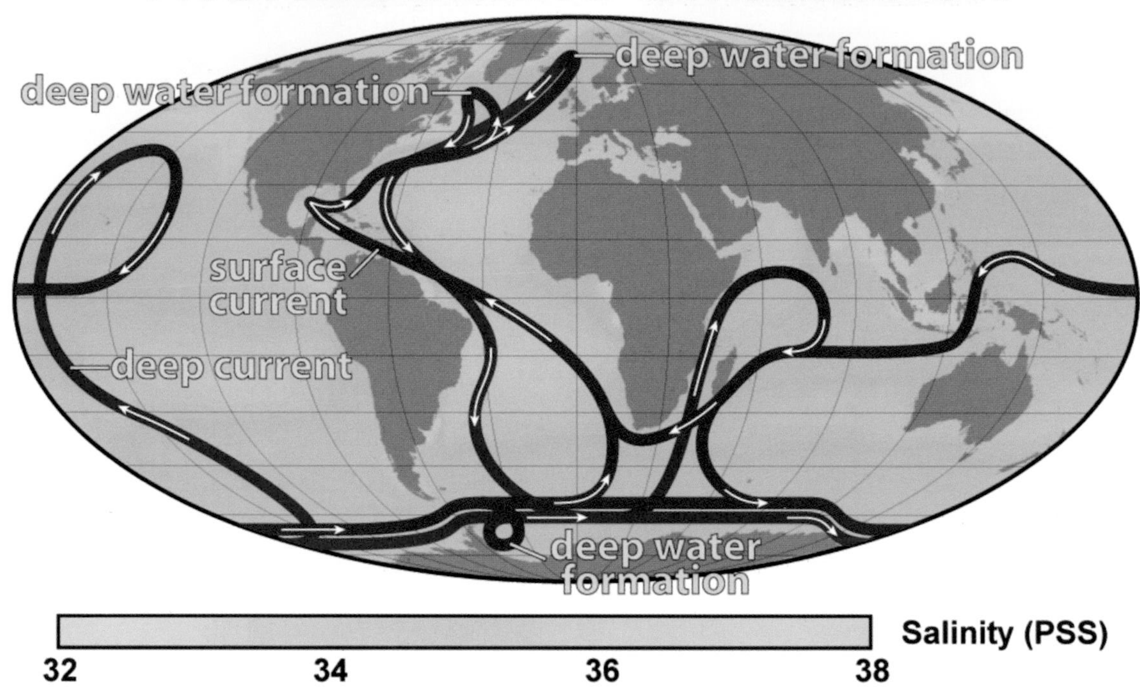

FIGURE 10.23. Conceptual diagram of the **global conveyor** of ocean circulation linking wind-driven surface currents (red) with thermohaline driven deep and bottom water currents (blue). From http://en.wikipedia.org/wiki/File:Thermohaline_Circulation_2.png. Figure originally drawn by Robert Simmon, NASA.

7 Read the box above describing the "Psychrosphere, Deep Water Masses, and Thermohaline Circulation". Write a paragraph explaining why **the locations of deep water formation** may have changed over the past 65 million years. Cite evidence presented in Figure 10.21 suggesting that **thermohaline circulation** was different in the geologic past.

NAME

NAME ______________________

References

Coxall, H.K., et al., 2005, Rapid stepwise onset of Antarctic glaciation and deeper calcite compensation in the Pacific Ocean. Nature, **433**, 53–7.

DeConto, R.M. and Pollard, D., 2003, Rapid Cenozoic glaciation of Antarctica induced by declining atmospheric CO_2. Nature, **421**, 245–9.

Jicha, B.R., et al., 2009, Circum-Pacific arc flare-ups and global cooling near the Eocene–Oligocene boundary. Geology, **37**, 303–6.

Kennett, J.P., 1977, Cenozoic evolution of Antarctic glaciations, the circum-Antarctic ocean, and their impact on global paleoceanography. Journal of Geophysical Research, **82**, 3843–60.

Kennett, J.P., 1978, The development of planktonic biogeography in the Southern Ocean during the Cenozoic. Marine Micropaleontology, **3**, 301–45.

Miller, K.G., et al., 2005, The Phanerozoic record of global sea-level change. Science, **310**, 1293–8.

Raymo, M.E. and Ruddiman, W.F., 1992, Tectonic forcing of late Cenozoic climate. Nature, **359**, 117–22.

Royer, J.-Y. and Rollet, N., 1997. Plate-tectonic setting of the Tasmanian region. In West Tasmanian Margin and Offshore Plateaus: Geology, Tectonic and Climatic History, and Resource Potential. Exon, N.F. and Crawford, A.J. (eds), Australian Journal of Earth Science, **44**, 543–60.

Savin, S.M., 1977, The history of the Earth's surface temperature during the past 100 million years. Annual Reviews of Earth and Planetary Science, **5**, 319–55.

Shipboard Scientific Party, 1988. Preliminary Results of Subantarctic South Atlantic Leg 114 of the Ocean Drilling Program. In Proceedings ODP, Initial Reports, vol. 114, Ciesielski, P.F., et al., (eds), College Station, TX Ocean Drilling Program, pp. 797–804. http://www-odp.tamu.edu/publications/114_IR/114TOC.HTM

Shipboard Scientific Party, 1988, Site 689. In Proceedings ODP, Initial Reports, vol. 113, Barker, P.E, et al. (eds), College Station, TX, Ocean Drilling Program, pp. 89–181. http://www-odp.tamu.edu/publications/113_IR/VOLUME/CHAPTERS/ir113_05.pdf.

Shipboard Scientific Party, 1988. Site 699. In Proceedings ODP, Initial Reports, vol. 114, Ciesielski, P.F., et al., (eds), College Station, TX, Ocean Drilling Program, pp. 151–254. http://www-odp.tamu.edu/publications/114_IR/VOLUME/CHAPTERS/ir114_06.pdf.

Shipboard Scientific Party, 2001. Leg 189 summary. In Proceedings ODP, Initial Reports,vol. 189, Exon, N.F., et al., (eds), College Station, TX Ocean Drilling Program, pp. 1–98. http://www-odp.tamu.edu/publications/189_IR/chap_01/c1_f7.htm#50522

Shipboard Scientific Party, 2002. Leg 199 summary. In Proceedings ODP, Initial Reports, vol. 199, Lyle, M., et al., (eds), College Station, TX Ocean Drilling Program, pp. 1–87. http://www-odp.tamu.edu/publications/199_IR/chap_01/c1_f11.htm#539014

Shipboard Scientific Party, 2004. Leg 208 summary. In Proceedings ODP, Initial Reports, vol. 208, Zachos, J.C., et al., (eds), College Station, TX Ocean Drilling Program, pp. 1–112. http://www-odp.tamu.edu/publications/208_IR/chap_01/chap_01.htm

Sigurdsson, H., et al., 2000, History of circum-Caribbean explosive volcanism: $^{40}Ar/^{39}Ar$ dating of tephra layers. ODP Scientific Results, **165**, 299–314.

Wade, B.S. and Pearson, P.N., 2008, Planktonic foraminiferal turnover, diversity fluctuations and geochemical signals across the Eocene/Oligocene boundary in Tanzania. Marine Micropaleontology, **68**, 244–55.

Zachos, J.C., et al., 1992, Early Oligocene ice-sheet expansion on Antarctica: Stable isotope and sedimentological evidence from Kerguelen Plateau, southern Indian Ocean. Geology, **20**, 569–73.

Zachos, J., et al., 2001, Trends, rhythms, and aberrations in global climate 65 Ma to present. Science, **292**, 686–93.

Zachos, J., et al., 2008. An early Cenozoic perspective on greenhouse warming and carbon-cycle dymanics. Nature, **451**, 279–83.

Chapter 11 Antarctica and Neogene Global Climate Change

FIGURE 11.1. The ANDRILL drill site on the McMurdo Ice Shelf during the 2006/07 field season. From http://antarcticsun.usap.gov/science/contenthandler.cfm?id=2092. Photo provided by the ANDRILL program.

SUMMARY

This investigation introduces you to the status and role of Antarctica in Cenozoic (specifically Neogene) climate change and sets the stage for evaluating the two sediment cores retrieved from the floor of McMurdo Sound by the Antarctic Geologic Drilling Project (ANDRILL) in 2006 and

Reconstructing Earth's Climate History: Inquiry-Based Exercises for Lab and Class,
First Edition. Kristen St John, R Mark Leckie, Kate Pound, Megan Jones and Lawrence Krissek.

NAME

2007 (Figure 11.1). The cores are introduced in Chapter 12 (Interpreting Antarctic Sediment Cores). In this chapter you will build basic geographic and geologic knowledge of Antarctica and use geologic reasoning. In **Part 11.1,** you will review your understanding of the oxygen isotope curve, interpret global climate conditions from this curve, and assess the validity of your global interpretations based on the global distribution of sediment cores. In **Part 11.2,** you will become familiar with the geography and geologic units of the Ross Sea region of Antarctica and review or build your knowledge of southern hemisphere seasons, sea-ice, ice-shelves, and the challenges associated with obtaining a sediment core from the floor of McMurdo Sound. You will also build and use your understanding of simple geologic maps, cross sections, and the geologic time scale, so you can explain the reasons for selecting drill sites in McMurdo Sound. In **Part 11.3,** you will review the existing data from sediment cores in the Ross Sea region of Antarctica and use the knowledge gained in Parts 11.1 and 11.2 to identify a target stratigraphic interval and select two drill sites. Evaluation of the ANDRILL core is undertaken in Chapter 12 "Interpreting Antarctic Sediment Cores".

Antarctica and Neogene Global Climate Change

Part 11.1. What Do We Think We Know About the History of Antarctic Climate?

Introduction

Over the last 40 years, numerous studies have used sediment cores recovered from the ocean floor to examine the history of the Earth's climate during the Cenozoic (the last 65 million years). Many of these studies have identified changes in the Earth's climate during the Cenozoic and have invoked conditions in Antarctica as a major influence on climate and environments elsewhere around the world. In particular, the steps in the Antarctic climate that have been interpreted from these studies include those shown in Table 11.1.

Table 11.1. General interpretation of Oligocene–Holocene Antarctic climate from the late 1970s (Kennett and Shackleton, 1976 a, b; Shackleton and Kennett, 1975; Kennett, 1978).

Approximately 34 million years ago	The first major cooling in Antarctica and the first development of large ice sheets in Antarctica
Approximately 15 million years ago	A second additional cooling step and development of major ice sheets in Antarctica
Approximately 15 million years ago to present time	Persistence of a cold polar climate and large stable ice sheets in Antarctica

NAME ____

1 The oxygen isotope curve (Figure 11.2) was constructed using data from multiple ocean drilling sites and is used to interpret Antarctic ice volume. Explain how and why this curve supports the three major steps listed in Table 11.1.

____

____

____

____

____

____

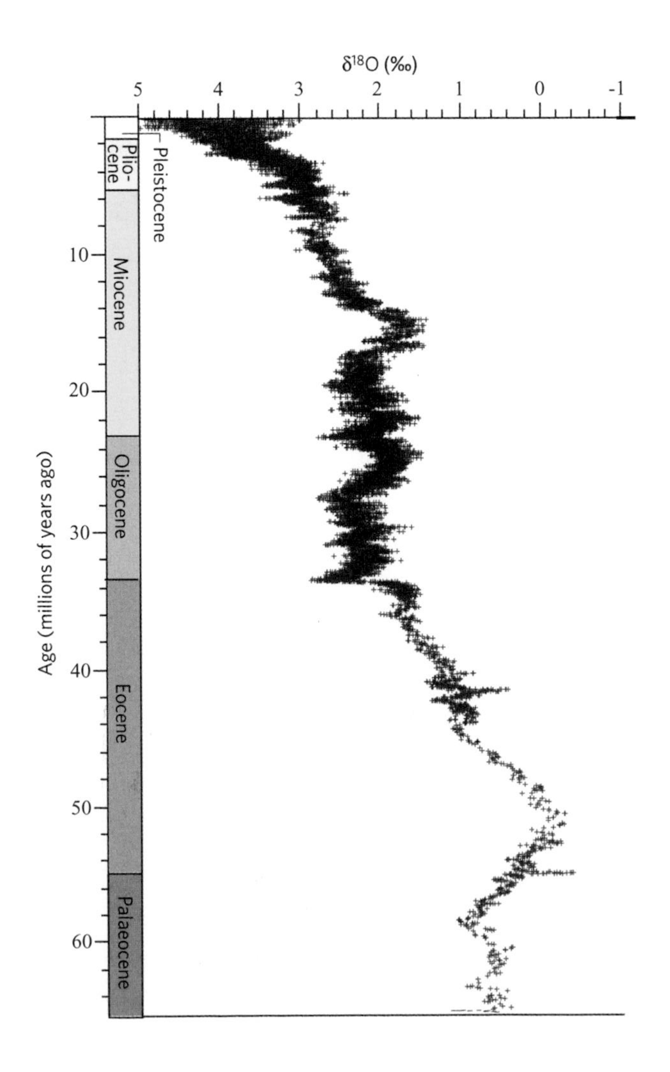

FIGURE 11.2. Composite marine record for $\delta^{18}O$ from benthic foraminiferal calcite. Modified from Zachos et al., 2008.

NAME

2 What do you notice about the distribution of deep sea drilling sites (Figure 11.3), some of which were used to build the composite oxygen isotope curve in Figure 11.2?

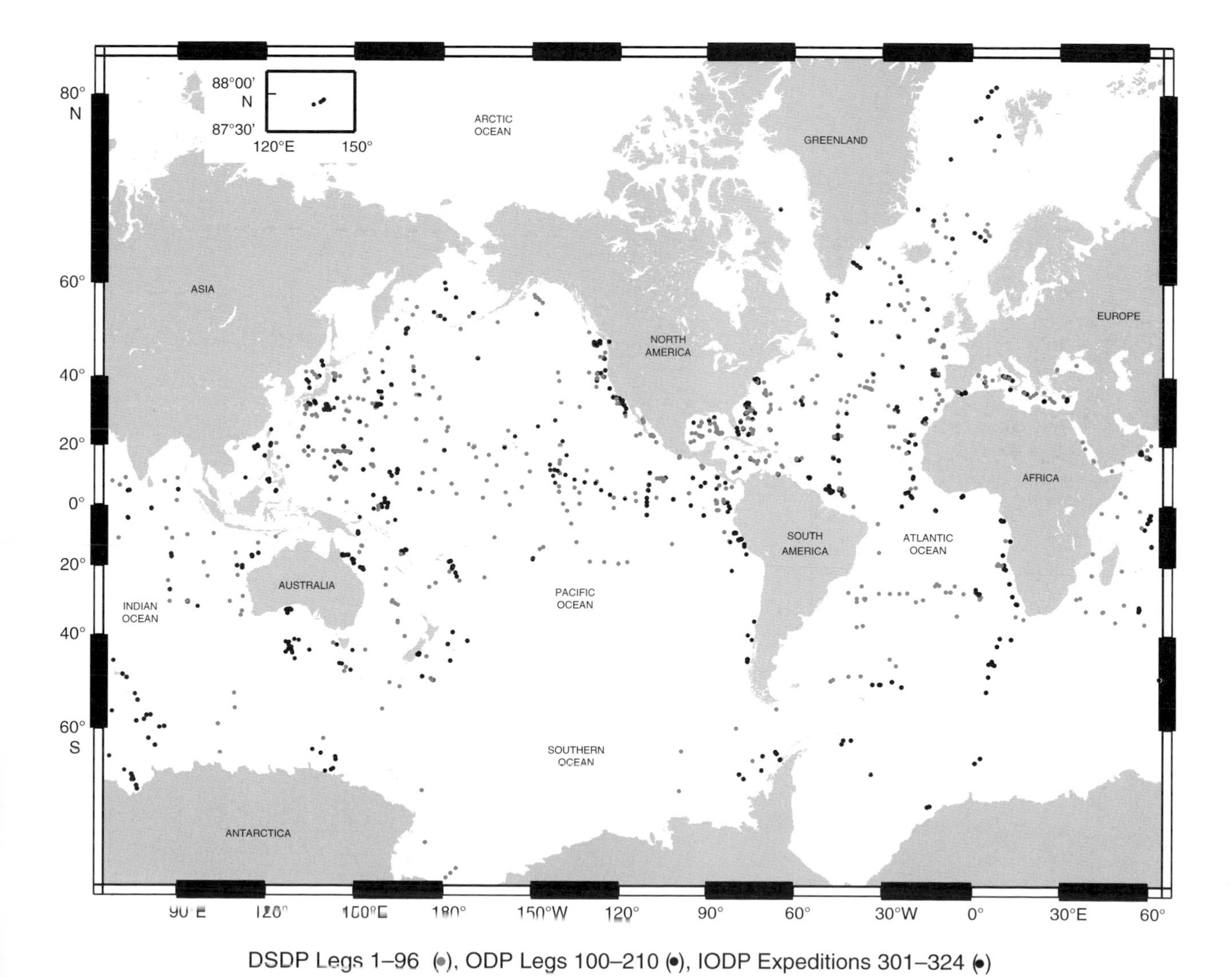

FIGURE 11.3. Map showing distribution of DSDP, ODP, and IODP core locations. Map courtesy of IODP.

NAME ______________________________

3 Does the distribution of sites in Figure 11.3 provide a relatively direct indicator of climatic conditions near the Antarctic continent? Explain your answer.

____

____

____

____

4 Based on the distribution of drill site locations, where would you go to test the interpretations in Table 11.1 further? Put three X's on the map to indicate your selection of future drill sites. Explain the reasoning for your site selections.

____

____

____

____

____

____

5 Based on what you have learned from other exercises in this book that you have done thus far, what observations and data from sediment cores can be used to help interpret past climates?

____

____

____

____

____

____

____

6 Thinking back to Chapter 2, what type(s) of sediment would you expect to be deposited (a) adjacent to a glaciated Antarctica and (b) to an unglaciated Antarctica?

NAME ______________________________

(a) ______________________________

____

(b) ______________________________

____

7 How might changes in global climate, such as an increase in global average temperature or a change in global precipitation patterns, have an impact on Antarctica?

____

____

____

____

____

____

8 Not only can changes in global climate affect regional climate (Question 7), but changes in regional climate can in turn affect global climate. Use your understanding of the Earth's system feedbacks (Chapter 5), of thermohaline circulation (Chapter 10), and sea level to explain how melting of the Antarctic ice sheets might affect global climate?

____

____

____

____

____

____

____

____

____

NAME ____________________

Antarctica and Neogene Global Climate Change

Part 11.2. What is Antarctica's Geographic and Geologic Context?

1 Antarctica has been described as the coldest, windiest, driest, and harshest place on earth. Find out what the **current weather** is at McMurdo Station, Antarctica. http://www.wunderground.com/global/AA.html.

Date/Time:	
Precipitation:	
Windspeed and Direction:	
Temperature (°F):	

2 What is the temperature (from question 1) in °C? ________
Conversion °F to °C = ([F]–32) × 0.55556) Show your calculations below.

____

____

____

Watch the videos and peruse the websites introduced in Questions 3–5 to learn about the Antarctic geography as well as logistical, technological, and scientific challenges associated with scientific research in Antarctica, then answer Questions 3 to 8.

3 Watch "A Tour of the Cryosphere: Earth's Frozen Assets" produced by NASA at: http://www.nasa.gov/vision/earth/environment/cryosphere.html (select the version with narration). What are the main types of ice body that make up the cryosphere in the Antarctic region?

____

____

____

____

____

NAME

4 Learn about Antarctic sea ice and its seasonal variability at: http://earthobservatory.nasa.gov/Features/WorldOfChange/sea_ice_south.php.

(a) On each of the maps below (Figure 11.4) indicate which specific month is represented.

(b) Explain how you used the reasons for seasonal variability in sea-ice extent to select your answer to 4(a).

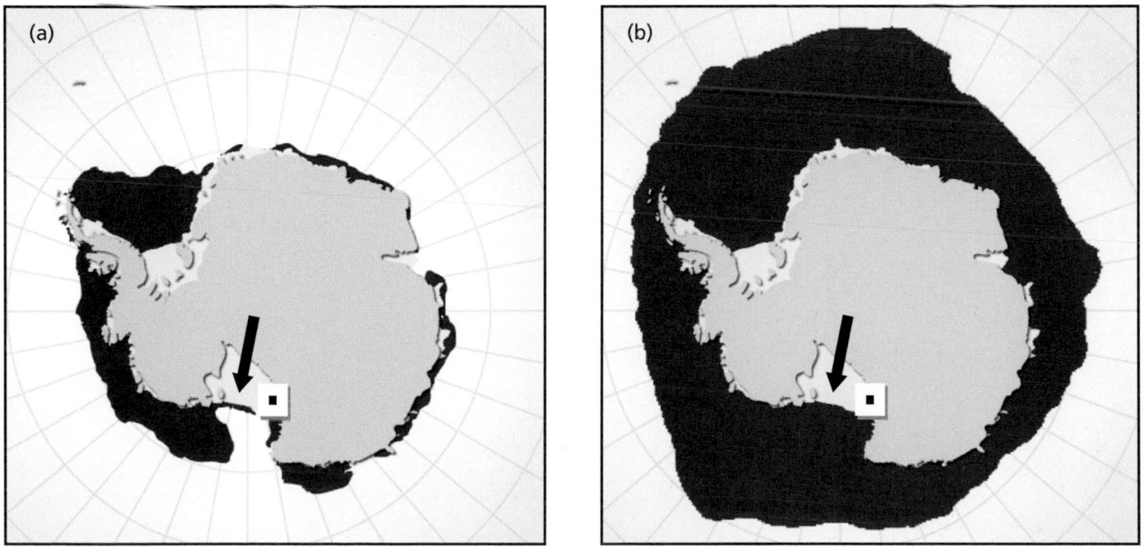

(a)

Month:

(b)

Month:

FIGURE 11.4. Maps of Antarctica and the Southern Ocean; extent of sea ice shown in dark blue. Ross Sea region indicated by an arrow and McMurdo Station by a black square in a white box. Maps courtesy of the Australian Antarctic Division (http://www.classroom.antarctica.gov.au/6-climate/6-3-annual-ice-cycle).

NAME ______________________

5 Watch: "Southbound" (ANDRILL, 2006) at: http://www.andrill.org/iceberg/videos/2006/index.html, and "Antarctica Today" (ANDRILL, 2007), "Antarctic Geology" (ANDRILL, 2007), and "Historical Journey" (ANDRILL, 2007) at: http://www.andrill.org/iceberg/videos/2007/index.html.

Use the information from these online videos to summarize the technical, logistical, and scientific challenges of doing research in Antarctica:

____

____

____

____

____

6 Based on the information conveyed in the videos above (Questions 3–5), outline the climatic history of Antarctica for the past 350 million years. Record your responses in the table below. (For more information on icehouse vs. greenhouse conditions see Chapter 5.)

Name and Age (in millions of years) of Time Block	Climatic Condition: Icehouse or Greenhouse?	Proxy Evidence for the Climatic Condition

NAME ____________________

7 In Questions 1 and 2 you examined data on current weather at McMurdo Station. In Question 6 you summarized data on Antarctic paleoclimates. Use your knowledge and resources to:

(a) Explain the difference between weather and climate at a given location (e.g. Antarctica):

(b) Explain the difference between "local climate" and "global climate":

In the past 50 years much research has taken place near McMurdo Station, the largest US research base in Antarctica (Figure 11.5). McMurdo Station is located on Ross Island, an island in the western Ross Sea built from active and recently active volcanoes. The Ross Island volcanoes and other volcanic features in the region form the Erebus Volcanic Province, where volcanic activity has occurred throughout the past approximately 15 million years. The Trans Antarctic Mountains (TAM) mark the eastern margin of East Antarctica, and are well exposed approximately 60 km to the west of Ross Island. The TAM are composed of older rocks (Figures 11.6 & 11.7). Note that on average only 2% of the surface of Antarctica today is exposed rock; the rest is covered by ice.

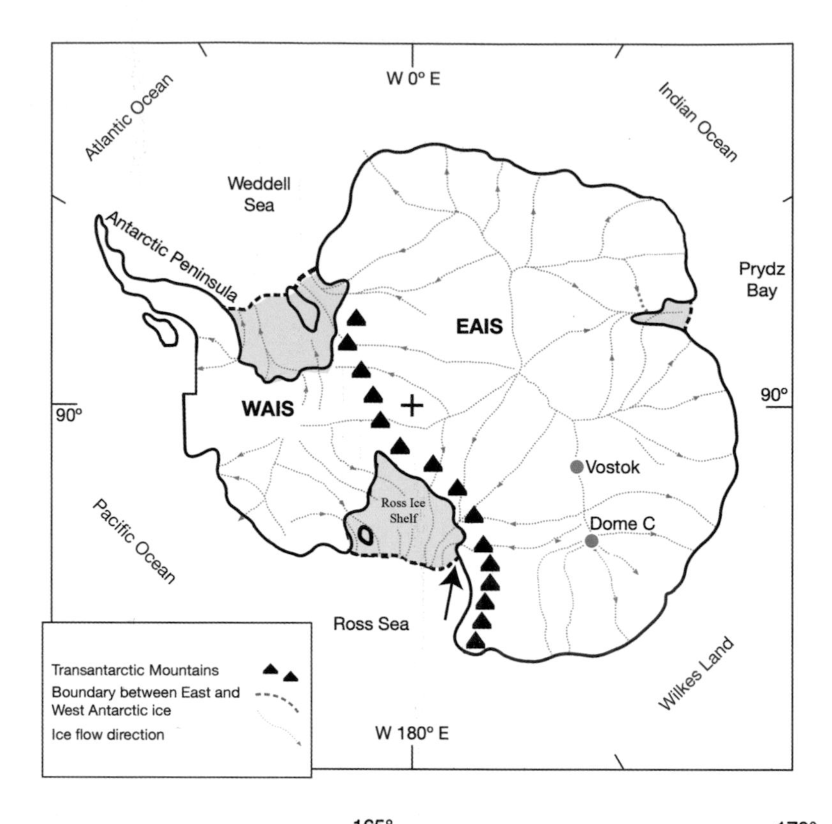

FIGURE 11.5. Map of Antarctica showing location of the East Antarctic Ice Sheet (EAIS), the West Antarctic Ice Sheet (WAIS), the Trans Antarctic Mountains (TAM) and the Ross Ice Shelf. The map also shows the main ice divides. The arrow points to Ross Island. Modified from Hambrey et al, 2003.

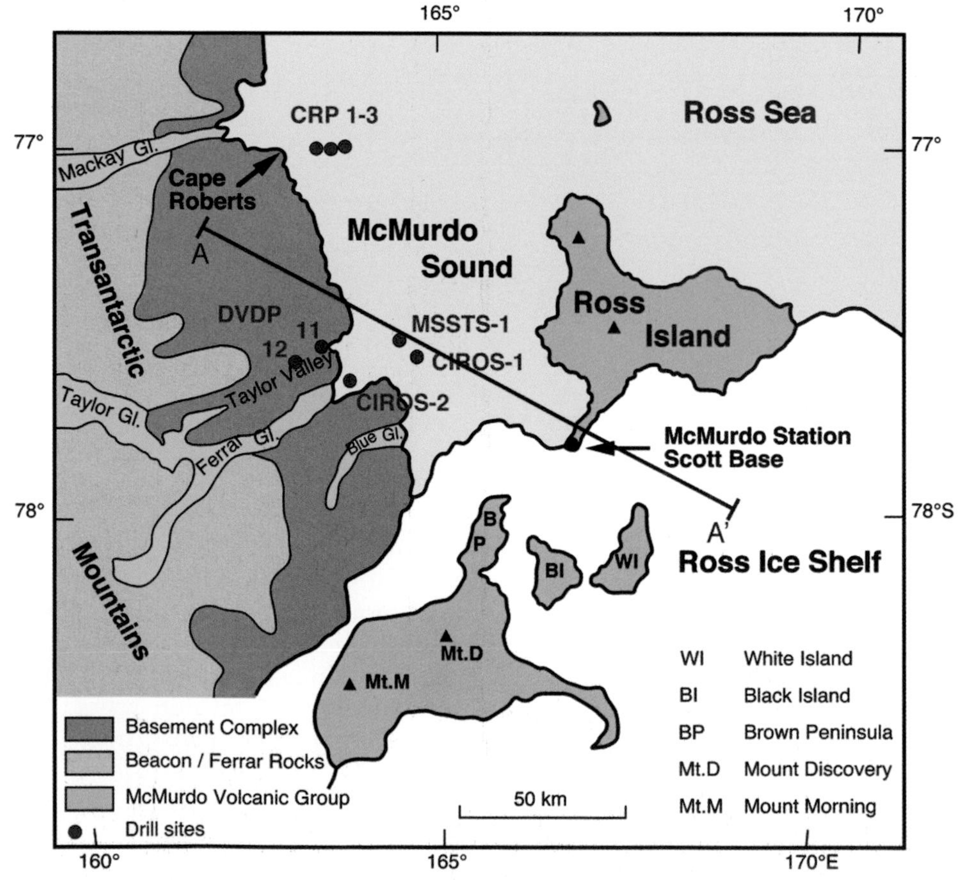

FIGURE 11.6. Simplified geologic map of McMurdo Sound region showing Ross Island and the Trans Antarctic Mountains, as well as location of selected drill sites (CRP, DVDP, CIROS and MSSTS). Rocks identified as "Basement Complex" and "Beacon/ Ferrar" on this map constitute "Bedrock" and are detailed in Figure 11.7, as are the McMurdo Volcanics. The line of cross section in Figure 11.8 (A–A') is also shown. **The Ross Sea (ocean) is blue** in this image, and **the Ross Ice Shelf is white.** Outlet glaciers shown in bright blue. Map drawn by Werner Ehrmann, Universitat Leipzig.

NAME ___

McMurdo Sound is a southward extension of the seasonally open-water portion of the Ross Sea; it lies between Ross Island and the East Antarctic mainland (Figure 11.6), and is covered during the southern hemisphere winter by sea ice approximately 2–6 m thick. During the southern hemisphere summer, most of this sea ice melts and open-water conditions are present. The McMurdo Ice Shelf lies to the south of Ross Island and connects to the east and south with the larger Ross Ice Shelf. Both of these ice shelves are fed by glacial ice flowing from the large West Antarctic Ice Sheet and the much larger East Antarctic Ice Sheet and cover the remainder of the Ross Sea. Water depths beneath the ice shelves are as much as approximately 1000 m.

8 Based on the geologic map and related key for the western Ross Sea area near Ross Island (Figure 11.7), **estimate the percentage of land area exposed in the western Ross Sea area**. Consider only areas above sea level.

___

___

___

9 Is this more or less than is exposed in Antarctica in general? (Note: In Figure 11.5 areas of white are covered by ice and gray is ice shelf.)

___

___

___

___

10 Based on the summary information provided and that presented in Figures 11.6 & 11.7, what are the general rock types and ages exposed in the **Trans Antarctic Mountains**?

Trans Antarctic Mountain Rock Types	Trans Antarctic Mountain Rock Ages

NAME

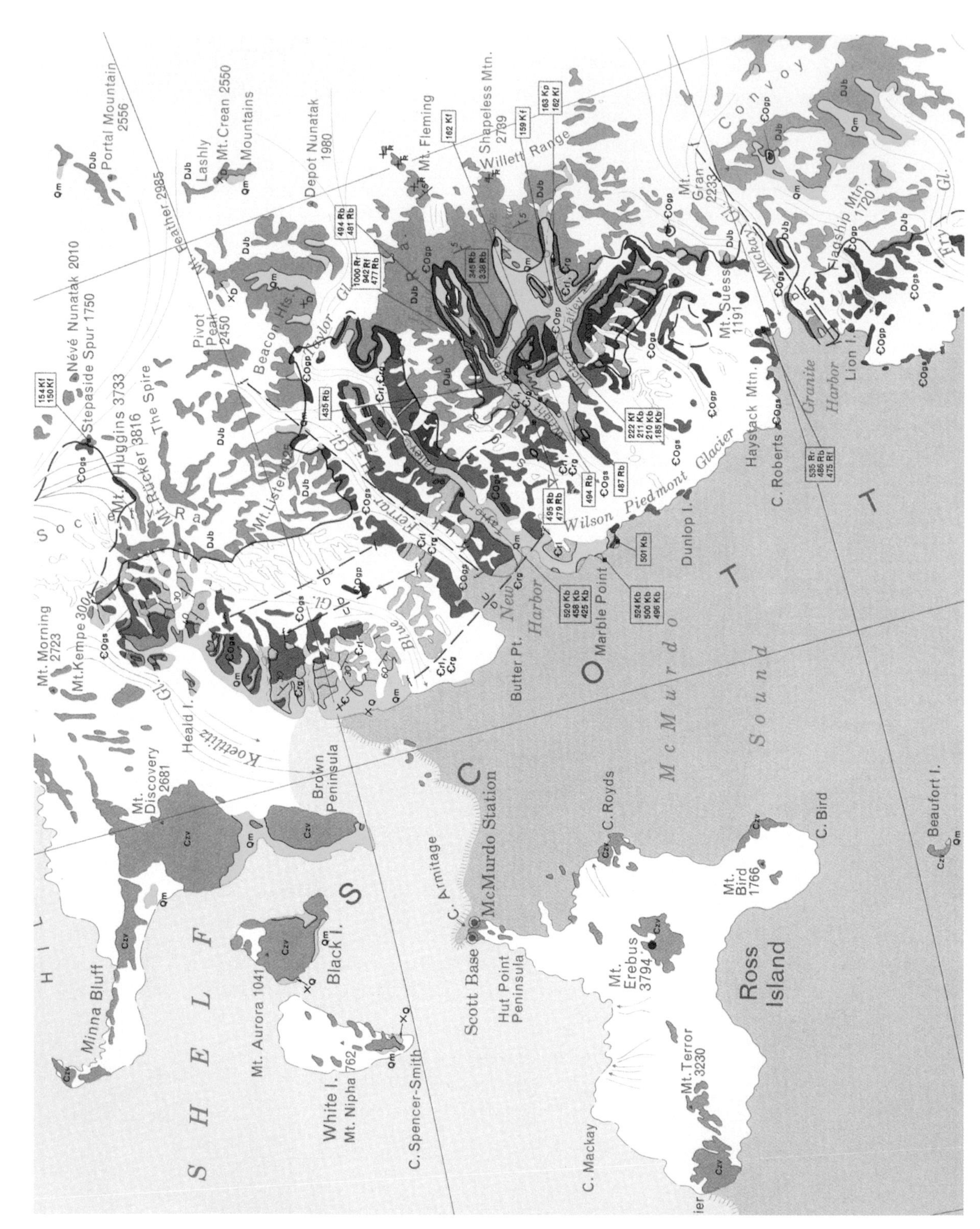

FIGURE 11.7. (a) Portion of geologic map and (b) legend from Geologic Map of Antarctica, Sheet 14, Terra Nova Bay – McMurdo Sound Area, Victoria Land, Note that south is towards the top of the map, and that white regions on the map indicate ice (and therefore no exposed bedrock). From Warren, 1969.

NAME

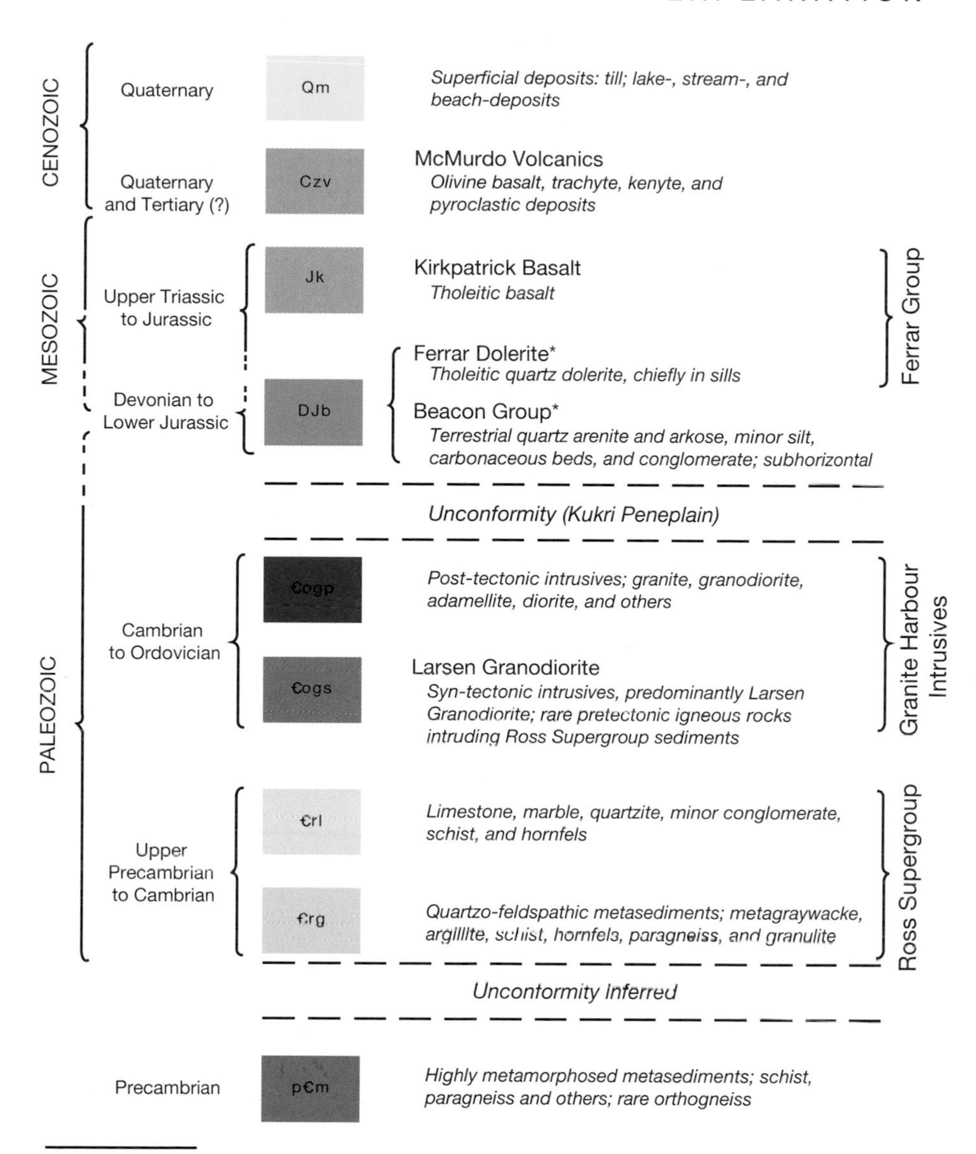

FIGURE 11.7. *Continued*

NAME ______________________________

11 Based on information given above and on the maps (Figures 11.6 & 11.7) what is the general rock types and ages on **Ross Island** (the location of McMurdo Station & Scott Base)?

Ross Island Rock Types	Ross Island Rock Ages

12 How much information does the type of rock you identified in Question 11 give you about climate at the time the rocks formed? Explain.

____

____

____

Sedimentary layers are somewhat like wall posts on your facebook wall. Each wall post tells another step or part of a story. The oldest wall posts are at the bottom and the youngest (newest, most recent) wall posts are at the top. Presumably, the more wall posts there are, the more complete the story (or record) is. This general analogy also applies to a book or journal; the more pages there are, the **more complete**, or **more detailed**, the story is.

13 Using the wall post or journal analogy, consider a complete sequence of sediments (i.e. a lifetime of wall posts!). Broadly speaking, does a **thicker** or **thinner** sedimentary sequence provide a more complete record of geologic events affecting the area adjacent to the basin in which the sediments are accumulating? Explain your reasoning.

____

____

____

14 Based on the maps (Figures 11.6–11.7) and the cross section (Figure 11.8), where **in general** are you most likely to find a complete Cenozoic sedimentary sequence?

____

NAME

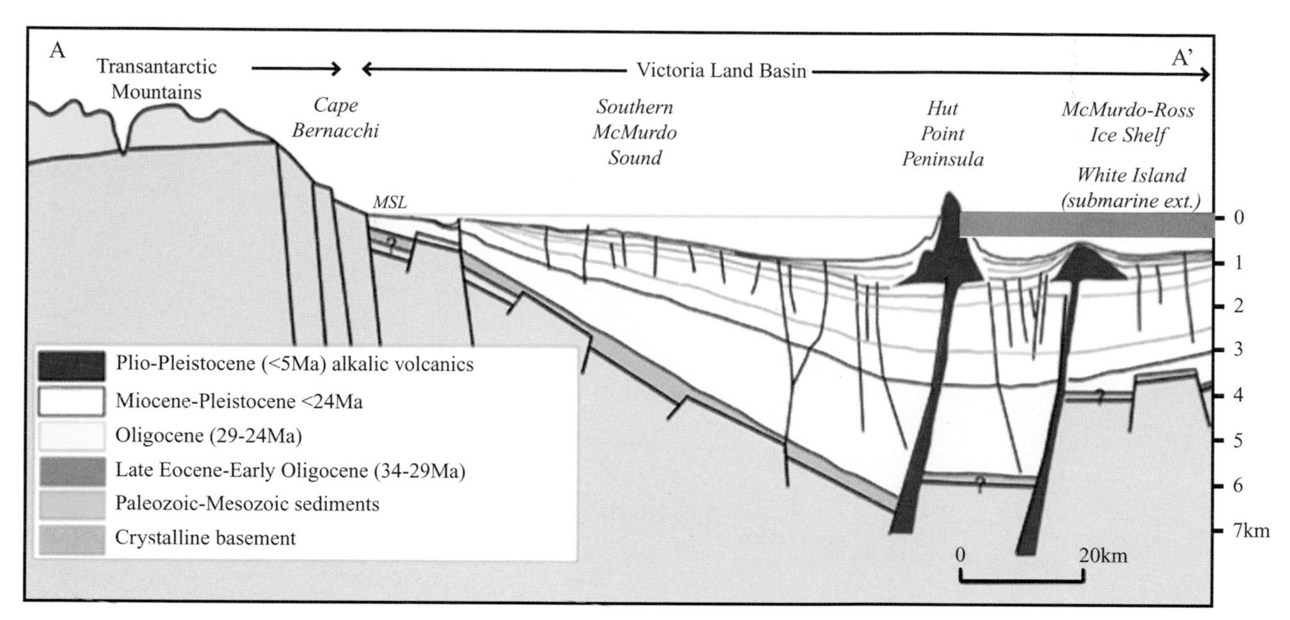

FIGURE 11.8. Simplified geologic cross-section of the western Ross Sea region. Adapted from Naish et al., 2005, ANDRILL Contribution 4 http://andrill.org/publications). The line of cross-section is shown on Figure 11.6. The colored lines in the Miocene–Pleistocene sediments are marker horizons (or "reflectors") based on seismic surveys. Hut Point Peninsula is the south-southwest projection of Ross Island. The gray area at sea level is the Ross Ice Shelf; the pale blue line at sea level represents the region covered by varying amounts of sea-ice. MSL = mean sea level. Note that this is essentially the same as Figure 11.11 (which has additional information), however they are used for different purposes. From Harwood et al., 2008–2009.

15 On Figure 11.8 draw a distinct vertical line where the Cenozoic sedimentary sequence is thickest and is undisrupted by faults.

(a) What was the probable source of the terrigenous Cenozoic sediments?

NAME ______________________________

(b) Considering that Southern McMurdo Sound and the McMurdo-Ross Ice Shelf are covered with sea ice or ice shelves, how might you obtain a drill core in the location you have selected?

____

____

____

____

____

____

Antarctica and Neogene Global Climate Change

Part 11.3. Selecting The Best Drill Sites for the Science Objectives

In the past 35 years, several drilling projects have taken place in the McMurdo Sound region. The sites drilled by these projects are shown in Figure 11.9 and the ages of the sediments recovered are shown in Figure 11.10. For most of these projects, a drilling rig was placed on the sea ice or the ice shelf at the drilling location during the southern hemisphere spring and early summer, in other words, the ice (rather than a ship) acted as the "drilling platform".

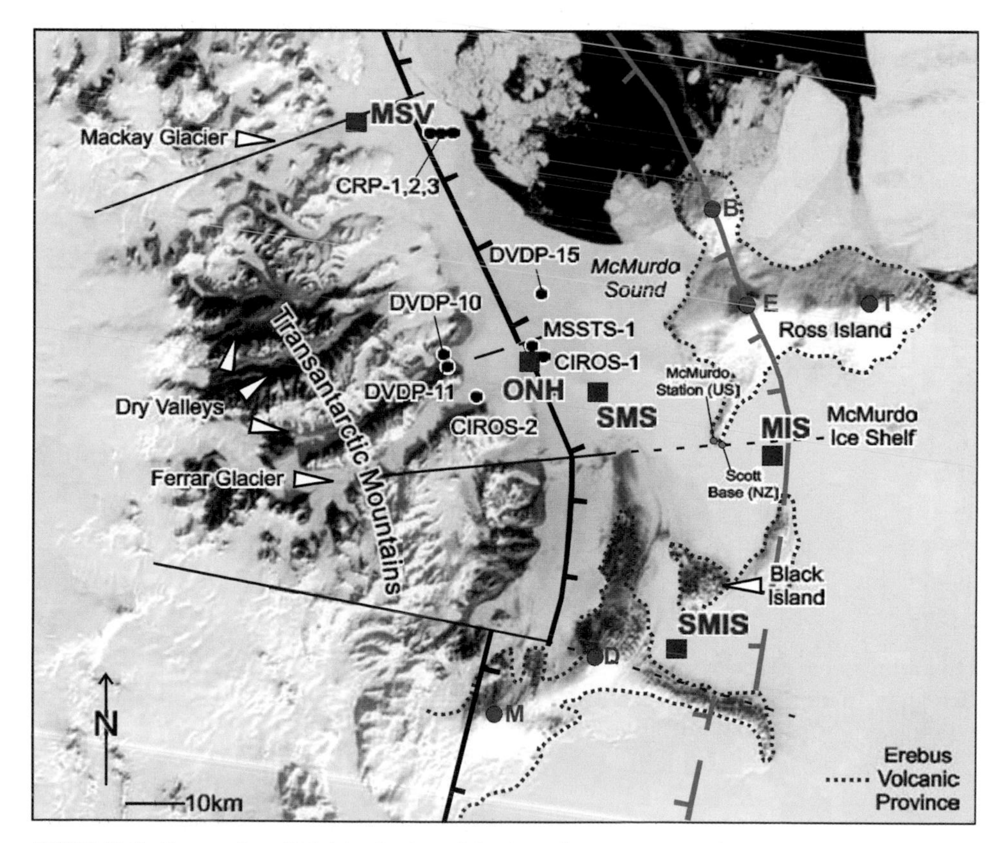

FIGURE 11.9. Geography of McMurdo Sound Region, showing geographic and tectonic features and location of drill cores. Red dotted line outlines extent of the Erebus Volcanic Province (Kyle and Cole, 1974). Volcanic centers of Erebus (E), Terror (T), Bird (B), Discovery (D) and Morning (M) are marked. Also shown are locations of previous stratigraphic drill holes (DVDP, CIROS, MSSTS, and CRP) in McMurdo Sound. Completed and Potential ANDRILL sites shown by red squares. From Harwood et al., 2008–2009.

NAME

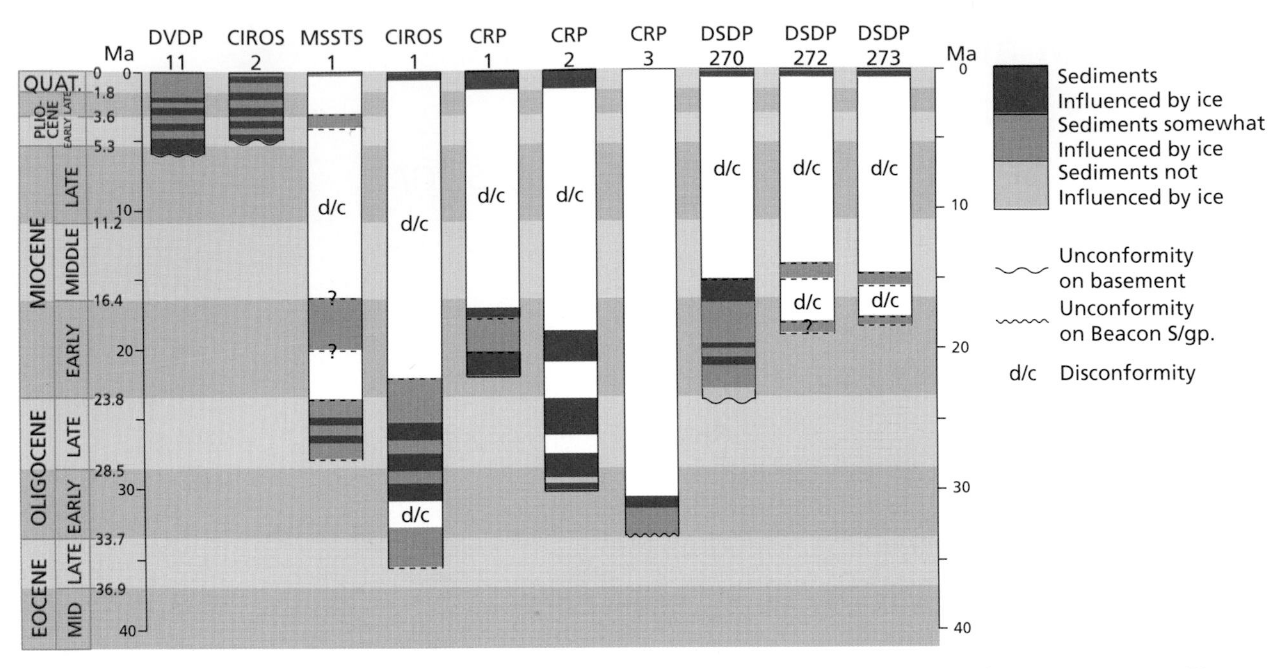

FIGURE 11.10. Age of sediments retrieved from drilling projects in the Ross Sea region. Figure adapted from Harwood et al., 2002. The location of the Deep Sea Drilling Project (DSDP) cores is not shown in Figure 11.9.

1 Do the datasets from the Ross Sea region (Figure 11.10) provide adequate data to evaluate the history of environmental and climatic conditions postulated (Table 11.1) for Antarctica **throughout** the past 34 million years? Explain your answer.

2 How does the drill core data summarized in Figure 11.10 **support or refute** the hypotheses about Antarctic Ice volume and climate that have been proposed (see Table 11.1)? Explain your answers.

NAME

Imagine that you have been assigned the job of developing the next drilling program for the Ross Sea region. Consider the following logistical and financial constraints:

- The project is funded for two drill sites.
- Funding is US$20 million.
- The drill rig is limited to a maximum 1500 m of penetration below the seafloor at any site drilled.

3 Examine Figure 11.10 and identify the **age interval** within the Cenozoic that should be your primary drilling target. In other words, what age interval has been least sampled by previous drilling programs, so that you would obtain the most scientific "bang for the buck" by sampling it? Explain your choice of an age interval.

4 Assuming that your project is enormously successful and recovers 3000 m of core, how much does 1 m of core cost? Show your calculations.

The sediments younger than approximately 5 Ma that have been sampled previously in the Ross Sea region form relatively thin layers and are poorly dated. As a result, the environmental and climatic history of Antarctica for the last 5 million years is poorly known.

NAME ____________________

5 Why might you be particularly interested in developing a detailed history of Antarctica for the past 5 million years? Explain.

____

____

____

____

____

Figure 11.11 is a generalized geologic cross section of the western Ross Sea region. The colored lines beneath the seafloor of the Victoria Land Basin indicate the positions of distinctive seismic reflectors (or distinctive sediment layers). Where known, the interpreted ages of these reflectors are listed at the bottom of Figure 11.11.

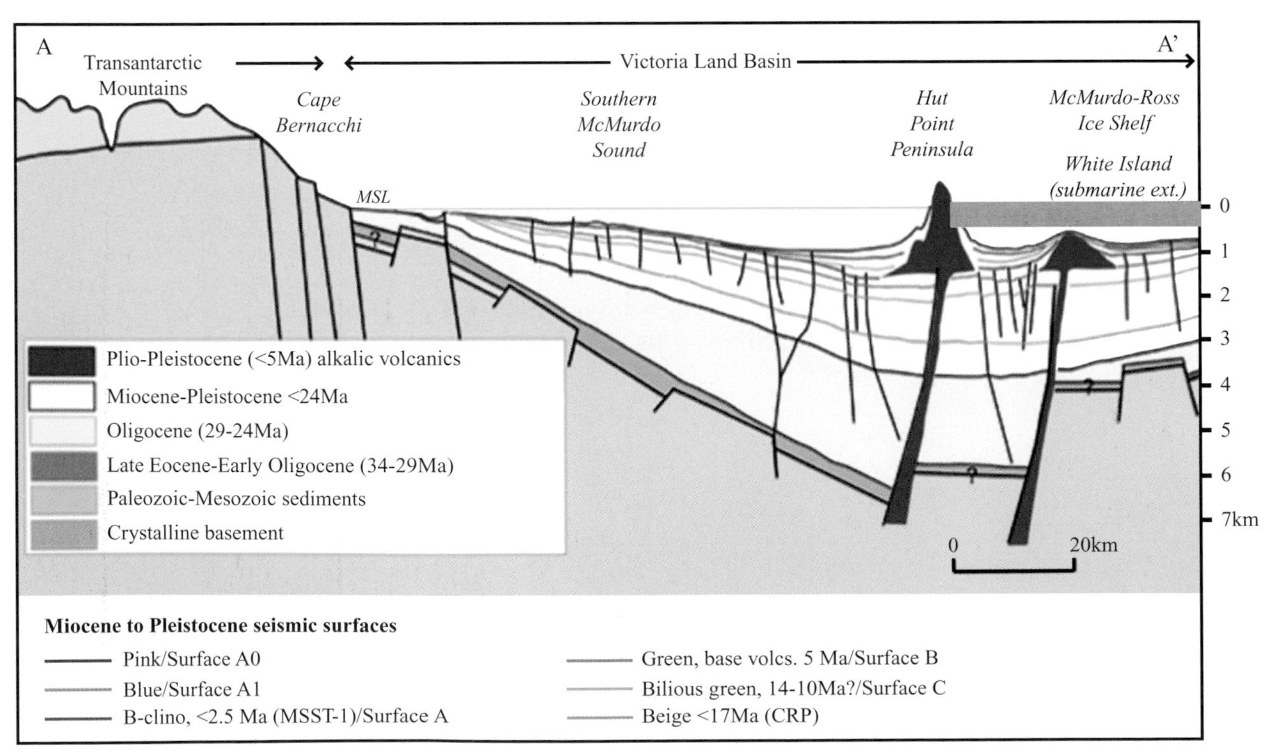

FIGURE 11.11. Cross-section of the Victoria Land Basin of western Ross Sea region (adapted from Naish et al., 2005); line of cross-section shown on Figure 11.8; Hut Point Peninsula is part of Ross Island; MSL = mean sea level; gray area to right of Hut Point Peninsula is the Ross Ice Shelf. This figure is similar to Figure 11.8, but it contains more information on the seismic reflectors. From Harwood et al., 2008–2009.

NAME ____

6 Use Figure 11.11 to **pick locations for two drill sites** – one to sample the age range you identified as your primary drilling target in Question 3 above and the second to sample sediments younger than approximately 5 Ma. In each case, you want to core through the target interval, but are limited to total penetration into the seafloor of 1500 m or less; you also need to make sure the core does not intersect any faults. Clearly show your drill sites on Figure 11.11 and **explain/justify your choice of drill sites in the space below**.

____

____

____

____

____

NAME ___________________

References

Hambrey, M.J., Webb, P/-N., Harwood, D.M., et al., 2003. Neogene glacial record from the Sirius Group of the Shackleton Glacier region, central Transantarctic Mountains, Antarctica. *Geological Society of America Bulletin*, **115** (8), 994–1015.

Harwood, D.M., Florindo, F., Levy, R., Fielding, C.R., Pekar, S.F., Speece, M.A. and the SMS Science Team, 2005. ANDRILL Southern McMurdo Sound Project, Scientific Prospectus. ANDRILL Contribution 5, ANDRILL Science Management Office, University of Nebraska-Lincoln, Lincoln, NE, 29 pp. Available at http://andrill.org/publications

Harwood, D.M., et al. (eds), 2002. Future Antarctic Margin Drilling: Developing a Science Plan for McMurdo Sound. ANDRILL Science Management Office (SMO). Contribution 1. University of Nebraska–Lincoln, Lincoln, NE, vol. vi, 301 pp.

Harwood, D.M., Florindo, F., Talarico, F., et al., 2008–2009. Background to the ANDRILL Southern McMurdo Sound Project, Antarctica," from *Studies from the ANDRILL, Southern McMurdo Sound Project, Antarctica, Initial Science Report on AND-2A* **15**:1. Available at http://www.mna.it/english/Publications/TAP/volume14.html

Kennett, J.P., 1978. The development of planktonic biogeography in the Southern Ocean during the Cenozoic. Marine Micropaleontology, **3**, 301–45.

Kennett, J.P. and Shackleton, N.J., 1976a, Oxygen isotopic evidence for the development of the psychrosphere 38 Myr ago. Nature, **260** (5551), 513–15.

Kennett, J.P. and Shackleton, N.J., 1976b, Critical development in evolution of deep-sea waters 38 m.y. ago; oxygen isotopic evidence from deep-sea sediments. EOS, Transactions of the American Geophysical Union, **57** (4), 256.

Kyle, P.R. and Cole, J.W., 1974, Structural control of volcanism in the McMurdo Volcanic Group, Antarctica. Bulletin of Volcanology, **38**, 16–25.

Naish, T.R., Powell, R., Barrett, P., Horgan, H., Dunbar, G., Wilson, G., Levy, R., Robinson, N., Carter, L., Pyne, A., Neissen, F., Bannister, S., Balfour, N., Damaske, D., Henrys, S., Kyle, P. and Wilson, T., 2005: ANDRILL McMurdo Ice Shelf Project, Scientific Prospectus. ANDRILL Contribution 4, ANDRILL Science Management Office, University of Nebraska-Lincoln, Lincoln, NE, 18 pp. Available at http://andrill.org/publications

Shackleton, N.J. and J.P. Kennett, 1975. Paleotemperature history of the Cenozoic and the initiation of Antarctic glaciation: Oxygen and Carbon isotope analyses in DSDP Sites 277, 279 and 281. In Proceedings ODP, Initial Reports of the Deep Sea Drilling Project, vol. 29, J.P. Kennett and R.E. Houtz et al. (eds), US Government Printing Office, Washington, pp.743–56.

Warren, G., 1969. Geology of the Terra Nova Bay-McMurdo Sound Area, Victoria Land. Antarctic Map Folio Series 12, Geology, Sheet 14, American Geographical Society, New York.

Zachos, J.C., et al., 2008. An early Cenozoic perspective on greenhouse warming and carbon-cycle dynamics. Nature, **451**, 279–83.

Chapter 12 Interpreting Antarctic Sediment Cores: A Record of Dynamic Neogene Climate

FIGURE 12.1. ANDRILL Scientists discussing a part of the ANDRILL-1B core in the Core Lab, McMurdo Station Season 2006/7. Courtesy of the ANDRILL Program.

Reconstructing Earth's Climate History: Inquiry-Based Exercises for Lab and Class,
First Edition. Kristen St John, R Mark Leckie, Kate Pound, Megan Jones and Lawrence Krissek.

NAME

SUMMARY

This set of investigations focuses on the **use of sedimentary facies** (lithologies interpreted to record particular depositional environments) **to interpret paleoenvironmental and paleoclimatic changes in Neogene sediment cores from the Antarctic margin**. Particular attention will be given to characteristics of settings close to the ice (ice-proximal) and far from the ice (ice-distal) in high-latitude settings. In **Part 12.1,** you will build your knowledge of polar sediment lithologies and the corresponding facies through conceptual diagrams, geological reasoning, and use of core images and core logs (a graphical summary of the sediments). In **Part 12.2,** the core log for the entire 1285 m ANDRILL 1-B core is presented. You will characterize each of the key lithostratigraphic subdivisions and use your knowledge of depositional facies to write a brief history of the Neogene climatic and environmental conditions in the Ross Sea region. In **Part 12.3,** you will use your core log reading skills and facies knowledge to evaluate patterns in the Pliocene sediments from ANDRILL 1-B. You will correlate quantitatively patterns in your dataset with cycles in insolation, influenced by changes in the Earth's orbit during the Pliocene.

Interpreting Antarctic Sediment Cores: A Record of Dynamic Neogene Climate

Part 12.1. What Sediment Facies are Common on the Antarctic Margin?

Figures 12.2 & 12.3 illustrate the variety of **depositional environments** and **sediment types** possible at the margin of a glacially influenced land mass, such as in the Ross Sea region, Antarctica. Figure 12.2 shows outlet glaciers from the East Antarctic Ice Sheet (EAIS, upper left) cutting through the rocks of the Trans Antarctic Mountains to the Ross Sea (lower right). Note that *Nothofagus* trees would have only been present during a period of temperate or warmer conditions. In Figure 12.3 the glacier (stippled) is flowing from right to left, into Antarctic Ocean waters. The different sediment lithologies are labeled (e.g., diamict).

NAME

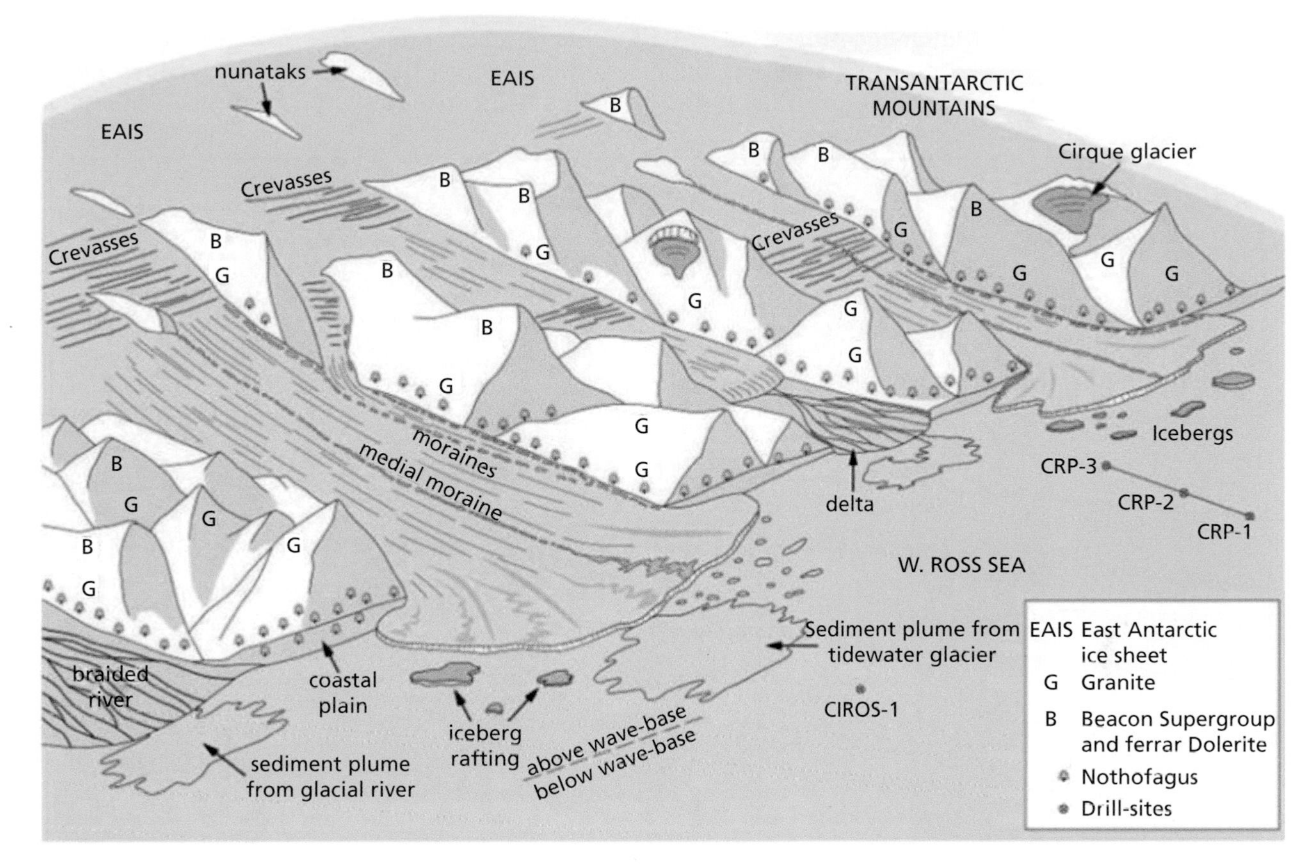

FIGURE 12.2. General paleoenvironmental setting for sedimentation along the flanks of the Trans Antarctic Mountains in late Oligocene time. From Hambrey et al., 2002. *Nothofagus* refers to southern beeches (Genus *Nothofagus*, Family Nothofagaceae) which are native to temperate oceanic to tropical regions in the southern hemisphere.

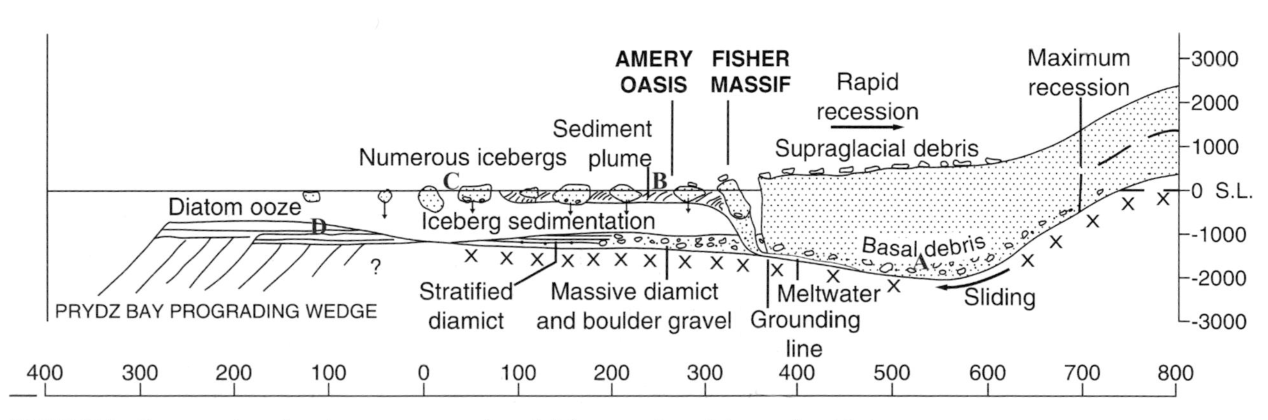

FIGURE 12.3. Cross section showing a conceptual model for growth and decay of a tidewater glacier in Lambert Graben Fjord, Lambert Glacier Region, Antarctica. From Hambrey and McKelvey, 2000. The vertical scale is in meters and the horizontal scale is in kilometers. Letters A–D are explained in Question 1.

NAME

1 Match locations A–D in the conceptual model (Figure 12.3) to locations within the paleonvironmental setting (Figure 12.2). Do this by **writing letters A–D on Figure 12.2**. In the following table, **explain** why you placed each letter where you did.

Location	Your Reasoning
A. On the surface of, and beneath, grounded outlet glacier	
B. Region where a braided river, or subglacial meltwater, enters ocean waters	
C. Marine region characterized by icebergs	
D. Marine region below wave base and seaward of melting icebergs	

The following section introduces several important lithologies that are diagnostic of ice-distal to ice-proximal depositional settings in polar regions. Note that these lithologies are more specific than the general marine lithologies (e.g., siliceous ooze, glaciomarine) of the global ocean that were introduced in Chapter 2.

Answer Questions 2–5, based on:

- The **information on sediment type** in the boxes below
- Figures 12.2 & 12.3
- The short videos '5:Telling Time' (2007) and '6:Cenozoic Global Climate' (2007) from http://www.andrill.org/iceberg/videos/2007/index.html

SEDIMENT TYPE: DIATOMITE

Diatoms are an important group of single-celled free-floating photosynthetic protists (i.e., phytoplankton) that precipitate an opaline silica shell. They are particularly important in areas of open water around Antarctica. Some species of diatoms thrive in the ocean waters under ice shelves. Sediment made up of diatoms is called **diatom ooze** or **siliceous ooze** if it is unlithified, and **diatomite** if it is lithified. These sediments are typically a pale yellowish brown or greenish grayish yellow in color and may be bedded or laminated; sometimes they are bioturbated.

NAME ___

2 If you are interpreting the history of depositional conditions on the continental margin of Antarctica, what important piece of **environmental information about a location** would you gain from the presence of diatom-rich sediments at that site?

___

___

___

___

SEDIMENT TYPE: SEDIMENT CONTAINING ICEBERG-RAFTED DEBRIS

When a glacier terminates in the ocean, icebergs can break off (or 'calve') from the front of that glacier or ice shelf, and drift out to sea. The icebergs carry sediment particles that were eroded by and embedded in the glacier as it moved across the land. The particles range from clay- to gravel-sized. As the iceberg melts, the sediment particles are released and settle on the seafloor. The most noticeable particles deposited by this process are the larger gravel-sized particles, which are called **iceberg-rafted debris (IRD)**; these particles are particularly noticeable as isolated large grains within finer grained sediment (terrigenous sands, silts, and muds). The finer grained sediment containing the IRD may also contain a low concentration of diatoms, particularly at locations further from the end of the glacier as open-marine conditions become more dominant.

3 If you are interpreting the history of depositional conditions on the continental margin of Antarctica, what important piece of **environmental information about a location** would you gain from the presence of a mud that contains a few percent of diatoms, as well as isolated gravel-sized (or larger) grains?

___

___

___

___

___

NAME ___________

SEDIMENT TYPE: DIAMICTITE

A diamictite is a deposit of poorly sorted clastic sediments – in other words, it contains a mixture of particles ranging in size from very small ("clay") to very large ("gravel"). In addition, diamictites contain little or no internal pattern of layering (i.e., they are unstratified or poorly stratified). Diamictites can be deposited by several processes, including glacial activity and landslides. A **diamictite** that was deposited directly from glacial ice is called **till**. Most tills are deposited beneath a glacier, rather than along the glacier's sides.

4 If you are interpreting the history of depositional conditions on the continental margin of Antarctica, what important **environmental information about a location** would you gain from the presence of a till at that location?

____

____

____

____

SEDIMENT TYPE: WELL-SORTED SANDS AND/OR GRAVELS

In some places at the base of a glacier, large quantities of meltwater form subglacial streams. Where these subglacial streams flow out into the ocean from the glacier's front, the flowing water can remove the smaller sediment particles from the glacial till. This process of removing the smaller particles is called "winnowing" and leaves a deposit of **well-sorted sands and/or gravels** close to the end of the glacier. The smaller particles are carried away from the glacier front as clouds of muddy water (also known as "sediment plumes") and are deposited at more distant locations where turbidity current deposits, IRD, and diatoms may be present.

5 If you are interpreting the history of environmental conditions on the continental margin of Antarctica, what important **environmental information about a location** would you gain from the presence of well-sorted sands and gravels at that location?

____

____

NAME

6 In the table below, **name and describe the sediment type (e.g., diamictite) expected for each depositional environment listed**. Use the information in Figures 12.2 & 12.3, the lithologic information in the text boxes, and your answers to Questions 1–5. Succinctly **explain the environmental processes** that produce those sediment characteristics.

Depositional Environment	Sediment Type and Description	Environmental Processes Producing the Lithologies
Open ocean, beyond iceberg influence		
Open ocean, within iceberg influence		
Glacial front, near the exit of a subglacial stream		
Subglacial (i.e., underneath the glacier)		

7 Images of four core intervals are shown in Figure 12.4. Compare them with your summary of sediment characteristics (Question 6) and the information presented in the boxes describing **Sediment types**. In the table on the next page match each of the sediment images/descriptions to one of the depositional environments you described in Question 6, explain your reasoning and ask question(s) about features you observe that do not seem to fit the "model".

NAME

Sediment Core	Sediment Name and Depositional Environment	Reasoning and Questions
ANDRILL 1-B 424.67–425.67 mbsf	Name: Environment:	
ANDRILL 1-B 133.07–133.88 mbsf	Name: Environment:	
ANDRILL 1-B 792.25–793.25 mbsf	Name: Environment:	
ANDRILL 1-B 80.50–81.50 mbsf	Name: Environment:	

NAME

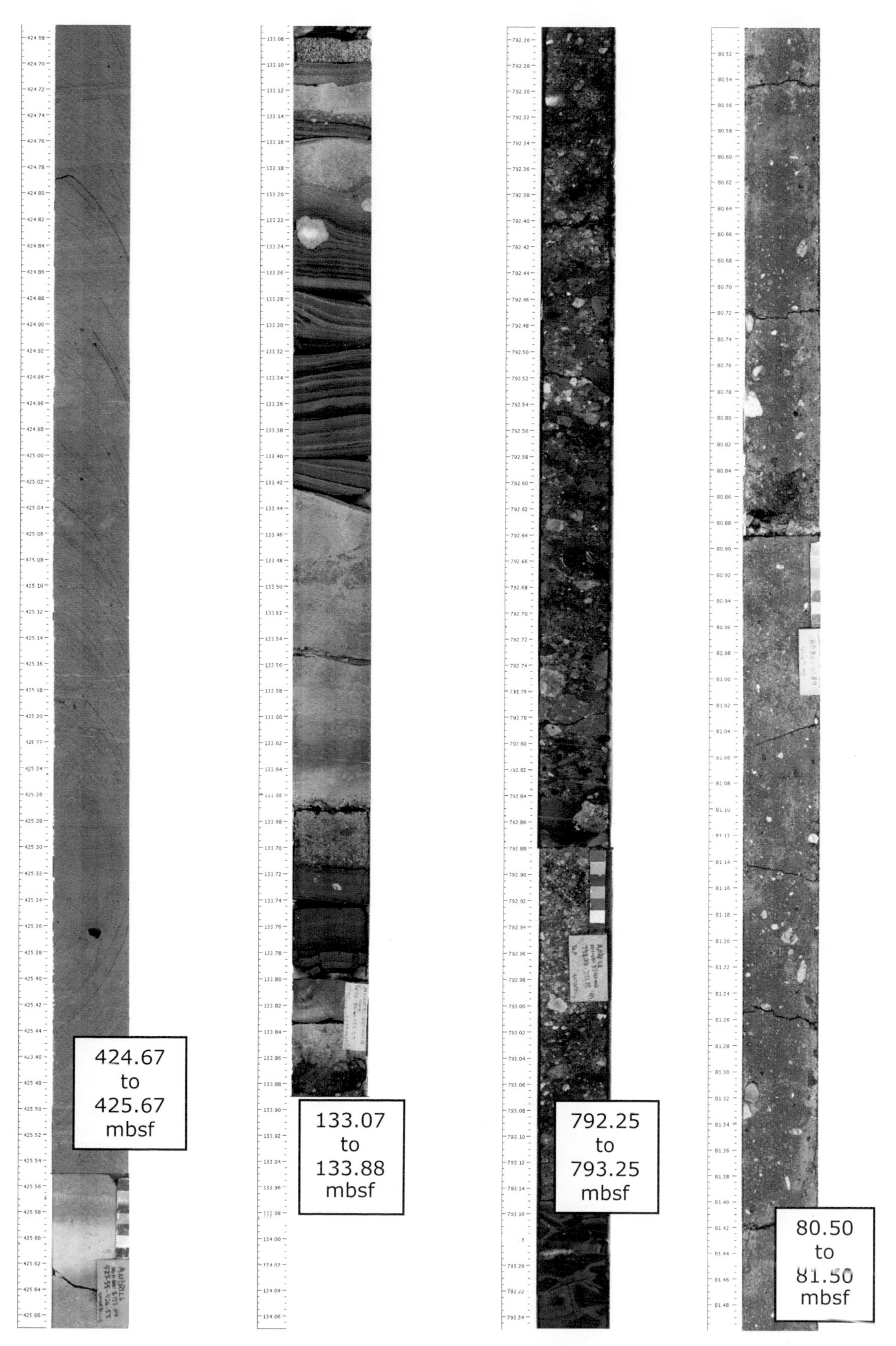

FIGURE 12.4. Four core intervals from ANDRILL 1-B. Photos courtesy of the ANDRILL Program.

NAME ______________________________

SEDIMENTARY FACIES

When sedimentologists describe a core or outcrop, they identify distinctive sediment lithologies based on a suite of objective observable properties, such as grain size, sorting, composition, and/or color (see Chapter 2). When lithologies are interpreted in terms of their location within a complex three-dimensional conceptual model of the region's depositional setting, lithologic **facies** are defined. The term facies can also be used in conjunction with the interpreted environment e.g. "ice-proximal facies" or "ice-distal facies".

Each of the sediment types you described and interpreted is one of the facies recognized by sedimentologists as they describe and interpret sediment cores from the Antarctic margin. Because the environment that exists at one location can change through time, the facies being deposited at that location can also change. In other words, **depositional environments (and the facies deposited in those environments) migrate laterally as glacial ice advances and retreats across the area**. Over time, such changes produce a vertical stack of different sedimentary facies; this is known as **Walther's Law**.

8 Examine Figure 12.5, and imagine that you are stuck underneath the glacier at the point labeled "**A**". Over time, as the glacier retreats toward the south (the right side of the diagram), all of the other depositional settings shown in Figure 12.5 will also shift toward the south. When the glacier has retreated furthest to the south, open-marine conditions extend to "**A**". Demonstrate your understanding of Walther's Law (see box on Sedimentary Facies) by making a list of the sedimentary facies you would expect to be deposited at location A in this scenario. Be sure to put the oldest deposit at the bottom and more recent deposits at the top.

____

____

____

____

____

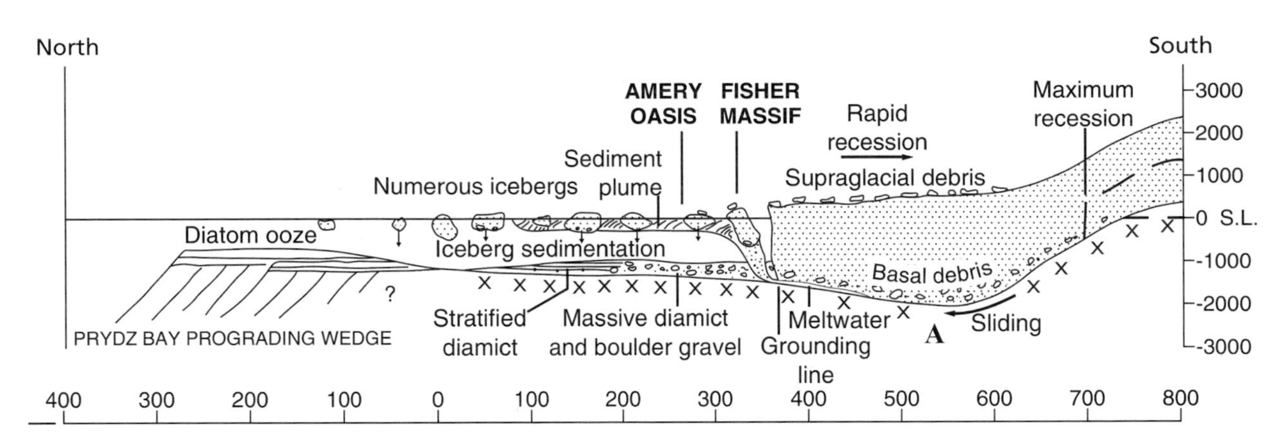

FIGURE 12.5. Cross section showing part of conceptual model for growth and decay of tidewater glacier in Lambert Graben Fjord, Lambert Glacier Region, Antarctica. Scales are in meters (vertical) and kilometers (horizontal). Vertical scale is in meters above or below sea level. Horizontal scale is in kilometers. This figure is nearly identical to Figure 12.3. From Hambrey & McKelvey, 2000.

9 Use the core logging sheet on the next page to **draw a simple stratigraphic column** that shows the vertical stack of sediments you would expect to be deposited at "A" as the glacier "retreats" landward (i.e. to the south) and then "advances" seaward (i.e. to the north). Note that you will not be able to fill in the columns labeled color, sketch, and depth, so leave these columns blank. In the "comments" section (a) label each facies in the vertical sequence, and (b) indicate the depositional environment present at "A" when that facies was being deposited. For help with development of your predictive model go to the simulation at http://andrill.org/system/files/web/images/edu/iceshelfadvanceretreat.swf

NAME ______________________________

Scale	Clay	Silt	Sand: VFS	Sand: FS	Sand: MS	Sand: CS	Sand: VCS	Gravel	Color	Sketch	Depth	Comments (Fossils,bioturbation, sed structures, etc)
											S D	

NAME

SEQUENCE MOTIFS

Sedimentologists working on the Antarctic continental margin have recognized several typical sedimentary sequences that develop during the advance and retreat of the ice. One example of a typical sequence, called a **"sequence motif"** by the ANDRILL sedimentologists, is shown in Figure 12.6. Note that the phrase "ice-proximal" is used to mean sediments deposited closer to the ice and "ice-distal" to mean sediments deposited further away from the ice.

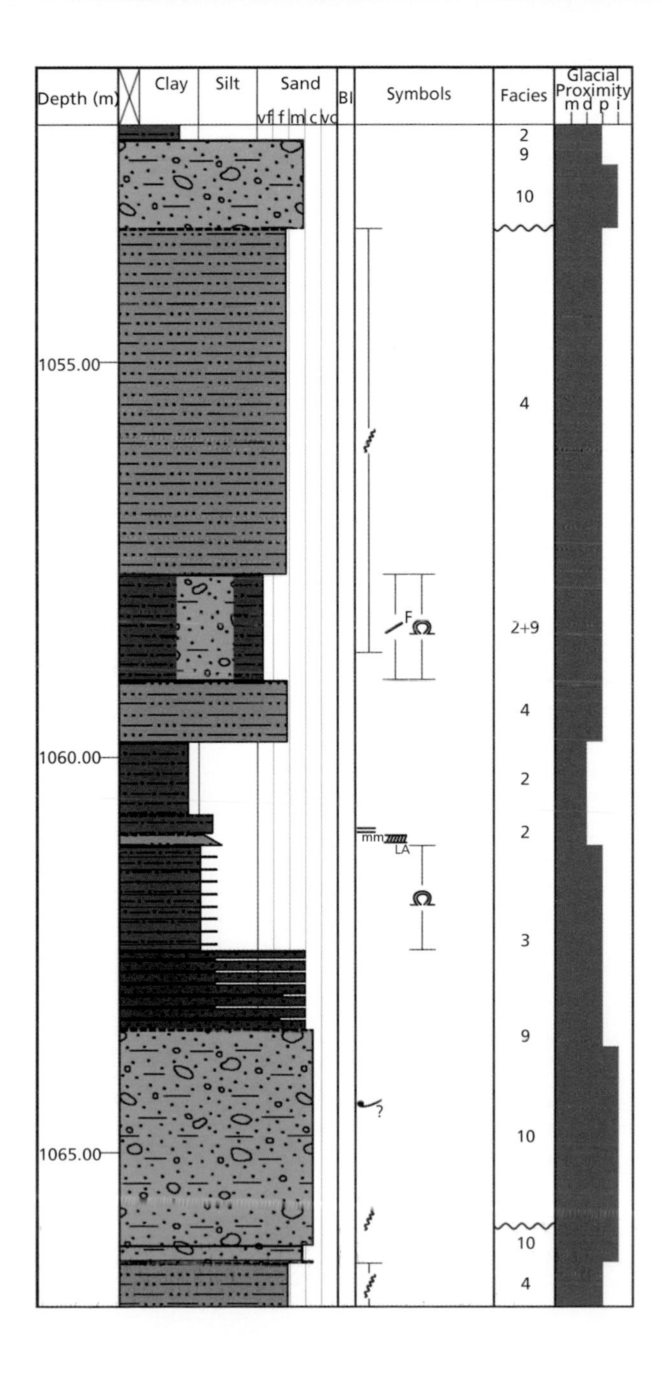

FIGURE 12.6. Example of a **sequence motif** from ANDRILL core 1-B, 1053.3–1066.4 mbsf. Green is diamictite, brown is conglomerate or breccias, gray is terrigenous siltstone and sandstone. The columns on the right of the graphic log show the main sedimentary features by symbols, facies number, and a glacial proximity graph. In the glacial proximity column, the width of the dark bar indicates ice proximity (wider = more proximal). From Krissek et al., 2007.

NAME

10 How does the vertical sequence you proposed for "**A**" in Question 9 compare with the vertical sequence shown in Figure 12.6? What is similar between the two? What is different? What explanations can you propose for any differences?

Similarities:

Differences:

Explanation:

NAME ____

11 In the next section (Part 11.2), you are going to be presented with the entire 1285 m core log for ANDRILL 1-B. Recall that a core log is a graphical summary of the sediments recovered from one location. How could you use the information on depositional environments, facies, and sequence motifs that you learned in this investigation to aid in your interpretation of the 1285 m core log for ANDRILL 1-B? List your strategies.

____

____

____

____

____

____

NAME ______________________________

Interpreting Antarctic Sediment Cores: A Record of Dynamic Neogene Climate
Part 12.2. ANDRILL 1-B The BIG Picture

ABOUT ANDRILL

ANDRILL (ANtarctic geological DRILLing) is an international program involving scientists, students, engineers, technicians, drillers, and educators from the USA, New Zealand, Italy, and Germany. ANDRILL's goal is to drill and recover sediment cores from the ocean floor beneath the Antarctic ice shelf and sea-ice, where the most complete sedimentary records of Antarctica's glacial, climatic, and environmental history for the past 65 million years are most likely to be found. In order to drill and recover these cores, a new drilling and coring system was developed that could be placed on top of the ice shelf and sea-ice.

Two ANDRILL projects have been completed successfully: the **McMurdo Ice Shelf Project (MIS)** in late 2006 and the **Southern McMurdo Sound Project (SMS)** in late 2007. Studies of the cores recovered during these projects are presently underway and will continue for many years to come.

This investigation will focus on sediments recovered during the MIS Project. As an introduction to ANDRILL operations and sediment core description, watch the short ANDRILL video journals listed below (downloadable from http://www.andrill.org/iceberg/videos/index.html):

ANDRILL Video Journals

4: *Selecting Where to Drill* (2007)
6: *The Drill Rig* (2006)
7: *Physical Properties and Logging* (2006)
8: *Core Curation* (2006)
9: *Sedimentology Team* (2006)
10: *Paleontology* (2006)
11: *Petrology* (2006)
12: *Paleomagnetism* (2006).

NAME ______________________________

1 Use the information in the video journals to write a paragraph that summarizes the types of observation and data that are collected from the sediment cores before they are shipped to the Antarctic Core Repository in the United States.

____

____

____

____

____

____

____

____

____

____

LITHOSTRATIGRAPHIC UNITS (LSUs)

Figure 12.7 is a graphic summary (i.e., core log) of the sediments recovered in the 1285-m long ANDRILL MIS core (officially named "AND-1B"). In this graphic summary, glacial tills are shown in green, mudstones (generally containing IRD) are shown in gray, well-sorted sands and gravel are shown in brown, and diatom-rich sediments ("diatomites") are shown in yellow. Volcanic rocks and sediments rich in volcanic material are shown in orange, but will not be considered further here.

As the core was described, the sedimentologists identified "**lithostratigraphic units**" (**LSUs**). The LSUs are intervals of the core that are either:

- Dominated by a single sediment type (i.e., facies) or
- Exhibit a relatively consistent pattern of interbedding of two or more sediment types (i.e., two or more facies).

The LSUs are numbered in **increasing order down-core** and some LSUs are further subdivided (e.g., LSU 2.1, 2.2, 2.3) to highlight more subtle differences in the relative abundance of sediment types.

2 In order to examine the differences between these LSUs, use Figure 12.7 to **complete the table (located after Figure 12.7)** by estimating the contribution of each sedimentary facies (i.e., diatomite, mudstone, till, and sand and gravel) to each LSU. To make each estimate, first identify the LSU number

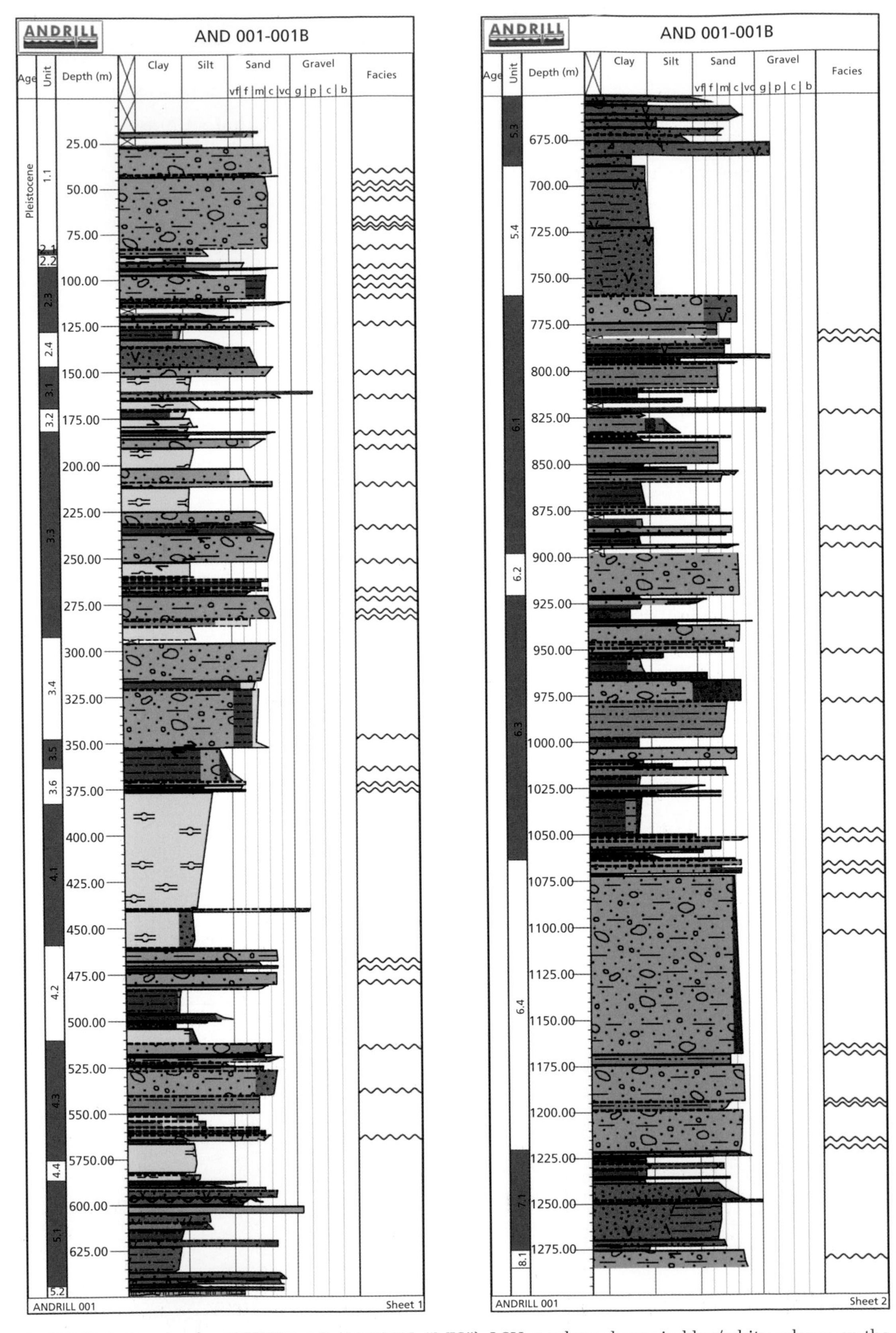

FIGURE 12.7. Core log for ANDRILL 1-B (ANDRILL "MIS"), LSU numbers shown in blue/white column on the left. The wavy lines represent breaks or gaps in the sedimentary sequence caused by erosion by grounded ice. These gaps in the record are called unconformities. Green = diamictite, yellow = diatomite, brown = conglomerate or breccias, grey = terrigenous siltstone and sandstone, orange = volcaniclastic sediment. From Krissek et al., 2007.

NAME ______________________________

from the blue/white column on the left-hand side; then sum the thicknesses of the occurrences of each of the four facies within that LSU and divide by the total vertical thickness of that LSU.

LSU	% Diatomite	% Mudstone	% Sand and Gravel	% Till
LSU 1				
LSU 2				
LSU 3				
LSU 4				
LSU 6.1 + 6.2 + 6.3				
LSU 6.4				

3 The presence (100%) or absence (0%) of diatomite (open marine) compared with the till (sub-glacial) is clearly a key criterion for identifying the LSUs. However, abundance variations of as little as 5–10% in one of the four facies can help distinguish between LSUs. Based on your estimates in Question 2, are there significant differences in facies abundances in the LSUs identified by ANDRILL scientists? What are the significant differences?

____

____

____

____

NAME ______________________________

4 Use your answer from Question 2 and Figure 12.7 to summarize the major compositional changes that take place as one moves "up-core" from LSU 6.4 to LSU 1.

____

____

____

____

____

____

____

As you learned in Part 12.1, the environmental conditions that existed at a location during times in the past can be interpreted from the characteristics of the **sedimentary facies** deposited during those times.

5 In order to develop a general environmental history for the ANDRILL-1B site, use your data from Question 2 (Part 12.2) and notes from Part 12.1 of this chapter to complete the table below.

LSU	Most Abundant Facies	2nd Most Abundant Facies	Most Common Environment	2nd most Common Environment
LSU 1				
LSU 2				
LSU 3				
LSU 4				
LSU 6.1 + 6.2 + 6.3				
LSU 6.4				

NAME ______________________________

6 Use the data from the table you have just completed to **write a history of climatic and environmental conditions at the site of ANDRILL-1B**, from the time of deposition of LSU 6.4 to the time of deposition of LSU 1. Describe the environments present during deposition of each LSU, the stability or variability of conditions, and whether you would classify that interval as a time dominated by ice (glacials), a time dominated by the absence of ice (interglacials), or a time of repeated glacial–interglacial cycles.

NAME

Interpreting Antarctic Sediment Cores: A Record of Dynamic Neogene Climate

Part 12.3. Pliocene Sedimentary Patterns in the ANDRILL 1-B Core

Sedimentary sequences that show some kind of repetition or pattern are generally referred to as being **cyclic**. The cyclicity may take place over a regularly repeated time interval. This time interval is referred to as the periodicity of the cycles (see Chapter 8). In this investigation we will analyze the 135–250 mbsf interval of the 1285-m long sediment core recovered by the ANDRILL McMurdo Ice Shelf project in 2006/07 (Figure 12.8). **Facies patterns in this part of the sedimentary sequence will be evaluated qualitatively and quantitatively to identify possible climate cycles.**

In Figure 12.8 the narrow yellow intervals are **diatomites** and the wider intervals in green, brown, and gray are **terrigenous** sediments that contain common to abundant gravel-sized clasts. These terrigenous sediments can be generalized in this investigation as **diamictites** (or tills).

Note that for Question 1, you should consider the dominant lithologic facies in each interval. Sedimentologists use the term facies or lithology somewhat interchangeably for distinctive "packets" of sediment. Recognition of each sedimentary lithology or sediment type is based on a suite of objective, observable properties such as grain size, sorting, composition, and color (see Part 12.2).

1 Starting from 227 mbsf and **working upwards** to 150 mbsf, list the vertical succession of lithologic facies (either diamictite or diatomite) in the table below. Start the list at the bottom of the table, so the oldest intervals are at the bottom and the youngest at the top. List the core interval, the thickness of each interval, and the lithologic facies.

Core Interval (in mbsf)	Interval thickness	Lithologic Facies

Note: The works space for Question 1 continues on page 424.

NAME

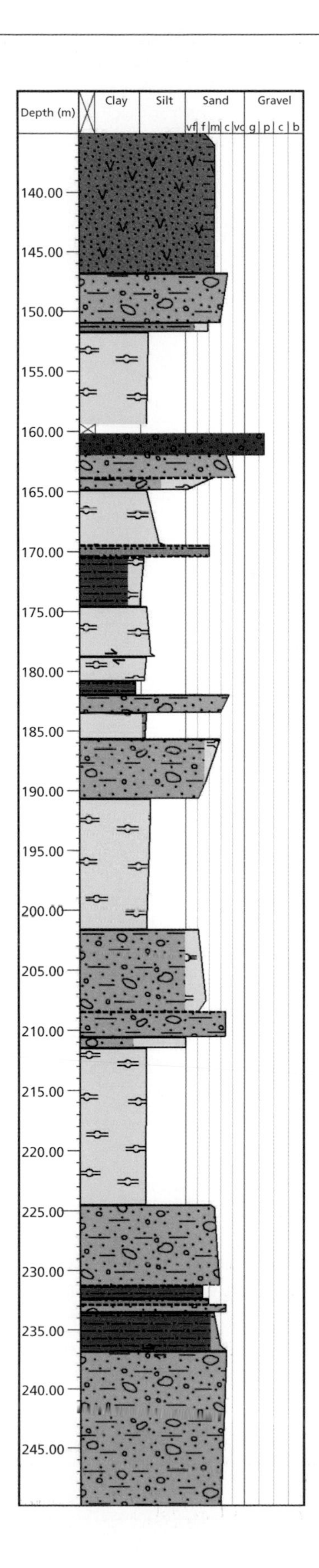

FIGURE 12.8. Graphical log of ANDRILL Core 1-B showing interval from 135–250 mbsf. Yellow intervals are diatomites. Green, brown, and gray intervals are terrigenous sediments. The scale at the top of the column gives grain size. Adapted from Naish et al., 2009.

NAME ____

2 Describe the general pattern of vertical arrangement of lithologic facies that you listed in Question 1. For example, is the stratigraphic pattern regular or random? Provide some specific details.

____

____

____

____

3 Let's define a "couplet" of these sediments as an interval of diamictite (till and other terrigenous sediments) overlain by an interval of diatomite. Starting at 227 mbsf and working up to 150 mbsf, how many "couplets" of diamictite overlain by diatomite can you identify?

____

____

4 Based on this vertical arrangement of lithologic facies, how did the environmental conditions at this site change as these sediments were deposited?

____

____

____

____

NAME ______________________________

5 Based on a combination of biostratigraphic, paleomagnetic, and radiometric data, the estimated age at 150 mbsf is between 1.94 Ma and 2.58 Ma, and the age at 227 mbsf is approximately 2.95 Ma.

(a) Given these data, what is the **fastest sedimentation rate** possible for this interval? Show all your calculations, give the rate in m/myr and cm/yr.

______________ m/myr

______________ cm/yr

(b) Given these data, what is the **slowest sedimentation rate** possible for the 150–227 mbsf interval? Show all your calculations, give the rate in m/myr and cm/yr.

______________ m/myr

______________ cm/yr

6 Assuming that the 150–227 mbsf interval represents 1.01 million years, what is the average duration of each couplet? This is the maximum average duration of the couplets. Show all your calculations; give the answer in myr, kyr, and yr.

______________ myr

______________ kyr

______________ yr

7 Assuming that the 150–227 mbsf interval represents 0.37 million years, what is the average duration of each couplet? This is the minimum average duration of the couplets. Show all your calculations; give the answer in myr, kyr, and yr.

______________ myr

______________ kyr

______________ yr

NAME ____

8 When data are being used to assess large-scale connections (such as links to global-scale patterns of climate variability), averages are often most useful. In this case, use your answers to Questions 6 and 7 to calculate the average duration of each couplet; in other words, calculate the average of the "maximum average duration" from Question 6 and the "minimum average duration" from Question 7. Show all your calculations; give your answer in my, ky, and yr.

_______________myr

_______________kyr

_______________yr

9 Compare the couplet durations you have calculated in Questions 6, 7, and 8 to the periodicities of eccentricity, obliquity, and precession (see Chapter 8). Does the Pliocene sedimentation in the Ross Sea, Antarctica, match any of these global-scale patterns of climate variability? Describe.

____

____

____

____

____

10 In order to evaluate the couplet durations that you have calculated, it is important to have a sense of the **completeness of the stratigraphic record**.

(a) Is it likely that all the sediments that initially were deposited at this location have been preserved? Why or why not?

____

____

____

____

____

____

NAME ______________________________________

(b) Use your knowledge about age determination (Chapter 3 and 4) to explain how you might you test whether the stratigraphic record is complete.

____

____

____

____

____

____

11 You have already calculated that the sediments presently at 227 mbsf were deposited 0.37–1.01 million years before the sediments present at 150 mbsf. Geoscientists working on the core, however, have estimated that only 36% of that time—0.37 to 1.01 million years—is actually represented by the sediments in this interval; the other 64% of the elapsed time is not represented by sediments. Using this estimate of the time actually represented by sediments (36%) and your answers to Questions 6 and 7, recalculate how much time (in **years**) is actually represented by each couplet if one considers:

(a) The **minimum average** couplet duration for the interval 150–227 mbsf. Show your calculations.

______________ yr

(b) The **maximum average** couplet duration for the interval 150–227 mbsf. Show your calculations.

______________ yr

(c) Use the results from (a) and (b) to determine the **overall average couplet duration** for the interval 150–227 mbsf. Show your calculations.

______________ yr

NAME ______________________________

12 Do any of the time intervals calculated in Question 11 match global-scale patterns of climate variability (eccentricity, obliquity, precession; see Ch. 8)? If yes, **which global-scale pattern and what is its periodicity?**

____

____

____

13 If you look back to the core log and the table you completed for Question 1 (Part 12.3), you will note the variability in thickness of the individual lithologic facies and couplets. Also look back at the rates that you calculated. If each couplet was deposited during the same length of time, controlled by a global-scale pattern of climate variability (eccentricity, obliquity, or precession), what could account for the variations in couplet thickness observed in this core?

____

____

____

____

____

14 What additional datasets from this core might you look at or obtain to test your hypotheses regarding cyclicity? (Hint: think about datasets explored in Ch. 8.)

____

____

____

____

____

NAME

References

Hambrey, M.J. and McKelvey, 2000. Major Neogene fluctuations of the East Antarctic ice sheet: Stratigraphic evidence from the Lambert Glacier region. Geology, **28** (10), 887–90.

Hambrey, M.J., et al., 2002. Late Oligocene and early Miocene glacimarine sedimentation in the SW Ross Sea, Antarctica: the record from offshore drilling. In Glacier-Influenced Sedimentation on High Latitude Continental Margins, 2002. Dowdeswell, J.A. and C.O. Cofaigh (eds.), Geological Society, London, Special Publications 203, pp. 105–08, 392 pp.

Krissek, L., et al., and the ANDRILL-MIS Science Team, 2007. Sedimentology and stratigraphy of the AND-1B Core, ANDRILL McMurdo Ice Shelf Project, Antarctica. Terra Antarctica, **14** (3), 185–222.

Naish, T., et al., and The ANDRILL-MIS Science Team, 2007. Synthesis of the Initial Scientific Results of the MIS Project (AND-1B Core), Victoria Land Basin, Antarctica. Terra Antarctica, **14** (3), 317–27.

Naish, T., et al., 2009. Obliquity-paced Pliocene West Antarctic ice sheet oscillations. Nature, **458**, 322–28. doi: 10.1038/nature07867.

Chapter 13 Pliocene Warmth: Are We Seeing Our Future?

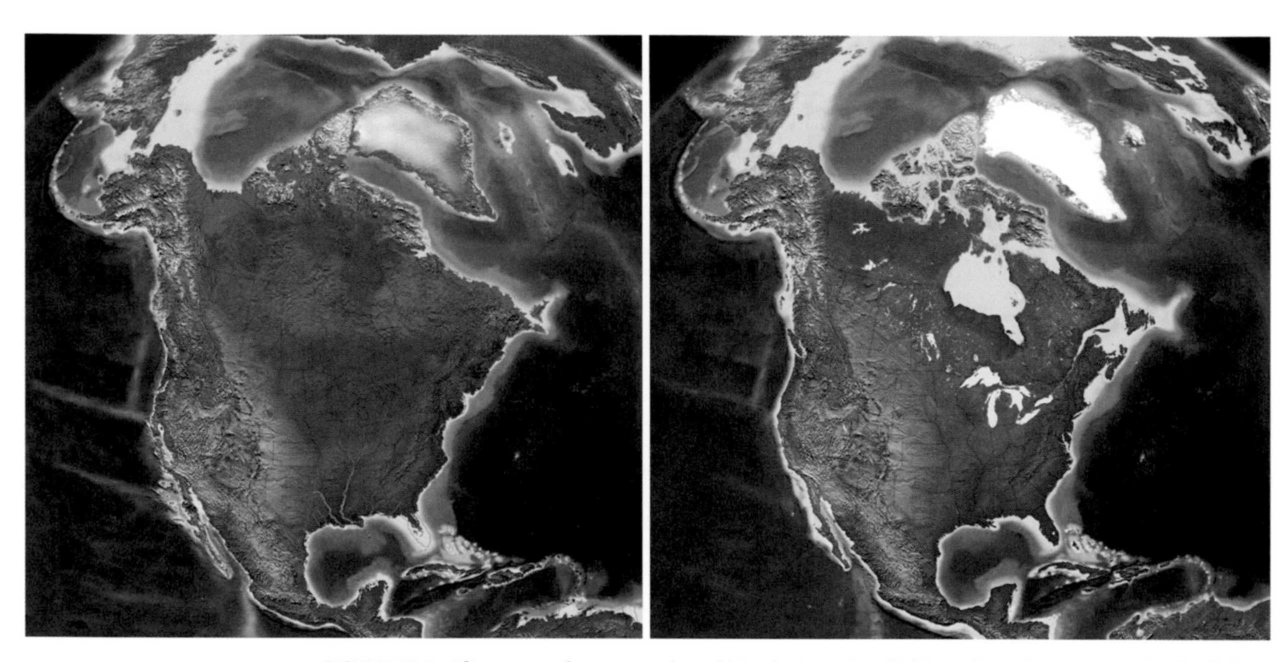

FIGURE 13.1. Pliocene paleogeography of North America (**left**), and modern geography (**right**). From Dr. Ron Blakey: http://jan.ucc.nau.edu/~rcb7/

SUMMARY

The Pliocene Epoch (5.3–2.6 Ma) represents a time interval immediately preceding the growth of large continental ice sheets in the Northern Hemisphere and cyclical glacial–interglacial cycles. There is abundant evidence that early to mid-Pliocene time (approximately 5–3 Ma) was warmer and global sea level was higher than today. In addition, plate configurations, mountain belts, and ocean circulation were more like today than any other time in the Earth's history. For these reasons, paleoclimatologists are very interested in studying the Pliocene as a possible analog for

Reconstructing Earth's Climate History: Inquiry-Based Exercises for Lab and Class, First Edition. Kristen St John, R Mark Leckie, Kate Pound, Megan Jones and Lawrence Krissek.

NAME

the evolving trends of global warming today. In **Part 13.1**, you will consider evidence for global warmth and the role of greenhouse gases in causing this warmth, specifically carbon dioxide (CO_2). In **Part 13.2**, you will consider the magnitude of sea level change during the early to mid-Pliocene, and compare this with ongoing sea level rise today.

Pliocene Warmth: Are We Seeing Our Future?

Part 13.1. The Last 5 Million Years

1 What's different about the Pliocene paleogeographic reconstruction (Figure 13.1, left panel), compared with the modern geography (Figure 13.1, right panel)? List ten things that are different.

Ten differences:

1.

2.

3.

4.

5.

6.

7.

8.

9.

10.

The Deep-Sea Benthic Foraminiferal Oxygen Isotope Record

In Chapter 6, you were introduced to the $\delta^{18}O$ record from benthic foraminifera as a global proxy for seawater temperature and ice volume. In Chapter 8, you identified orbitally driven cycles in these (and other) data and calculated the resulting periodicities. Explore the benthic foraminiferal $\delta^{18}O$ record further by examining Figure 13.2 below and answering Questions 2–5.

NAME ______________________

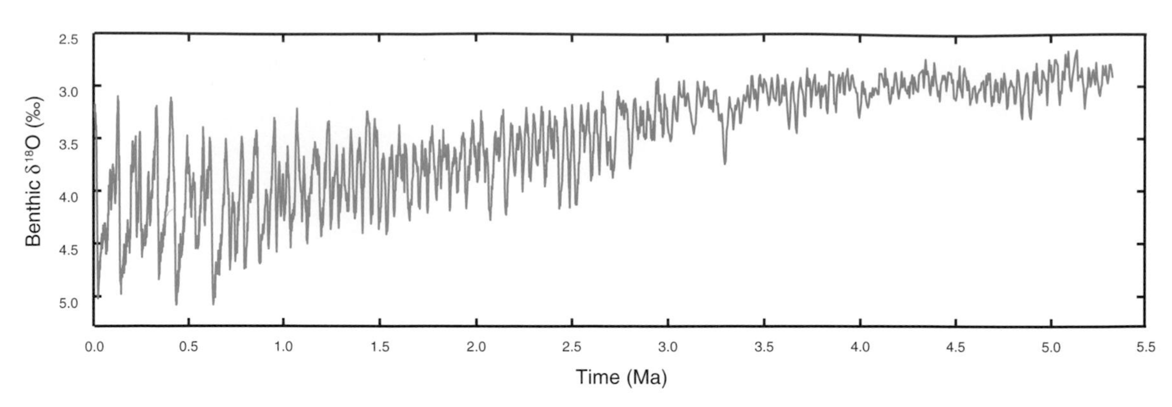

FIGURE 13.2. Composite deep-sea benthic foraminiferal oxygen isotope curve for the past 5.3 myr. This composite plot is derived from a "stack" of 57 globally distributed Pliocene–Pleistocene records, which have been aligned by an automated graphic correlation algorithm. From Lisiecki and Raymo, 2005.

2 In general terms, describe the **trend** of changing deep-sea oxygen isotope values over the past 5.3 million years (Pliocene to present).

____

____

____

____

____

____

____

____

3 The **pattern** of deep-sea oxygen isotope values shows some distinct differences in frequency and amplitude during the past 5.3 million years. **Frequency** refers to the number of repeated events (cycles) in a certain period of time, whereas **amplitude** refers to the magnitude of the changing $\delta^{18}O$ values. Characterize the frequency and amplitude of the $\delta^{18}O$ data for the following times. Record your answers in the table below.

Time	Frequency	Amplitude
approximately 1 Ma to the present		
3 Ma to approximately 1 Ma		
Approximately 5 Ma to 3 Ma		

NAME

4 If climate change influenced both the deep-sea benthic foram oxygen isotope record (Figure 13.2) and the Antarctic sedimentary record from the McMurdo Sound area (Chapter 12, Figure 12.7), these two datasets should show some similarities. Compare these two records.

(a) How does the up-core general change in lithology in ANDRILL 1-B during the Pliocene and Pleistocene compare with the global trend in $\delta^{18}O$ during this time?

(b) The late Pliocene and Pleistocene sediments in ANDRILL 1-B are characterized by couplets of glacially influenced sediment (diamictite) overlain by a layer of diatomite (Chapter 12, Figure 12.7). How does the average duration of each couplet (Part 12.3, Question 8) compare with the periodicity of the $\delta^{18}O$ data (Figure 13.2) for this same time period? (Remember, **periodicity** (time/cycle) is the **inverse of frequency** (number of cycles/time).)

NAME ______________________________

____

____

____

THE PLIOCENE RESEARCH, INTERPRETATION AND SYNOPTIC MAPPING (PRISM) PROJECT

Data analysis and climate modeling are used to estimate past temperatures. Estimates of global warming during the mid-Pliocene Epoch, early Piacenzian Age (Figure 13.3, shaded interval: 3.264–3.025 Ma) suggest that **global mean annual temperatures were 2°C warmer than today**. The **Intergovernmental Panel on Climate Change (IPCC)** report of 2007 has predicted that a 2°C warming during the 21st century (i.e., this century) falls within the range of likely global warming in the coming decades (see Chapter 7, Part 7.1). If true, in this century, global average temperatures could reach values, which have not been experienced since the mid-Pliocene. The **Pliocene Research, Interpretation and Synoptic Mapping** group (**PRISM**) of the US Geological Survey is a project to understand global warming in the Earth's recent past under conditions with similar plate configurations, paleogeography, and ocean circulation. Numerical simulations are designed to understand the sensitivity, climatic impact, and feedbacks forced by future global warming. Data thus far suggest that Pliocene warmth was the **consequence of elevated greenhouse gases and increased ocean heat transport**, together with associated, but yet unresolved, feedback in the ocean–climate system. From: http://geology.er.usgs.gov/eespteam/prism/index.html

5 The vertical line at approximately 3.2‰ (per mil) $\delta^{18}O$ represents present day oxygen isotope values in the deep-sea. What can we infer about temperature and ice volume during early to mid-Pliocene time compared to today? Are they warmer or colder? Is there less ice or more ice? Explain.

____

____

____

____

____

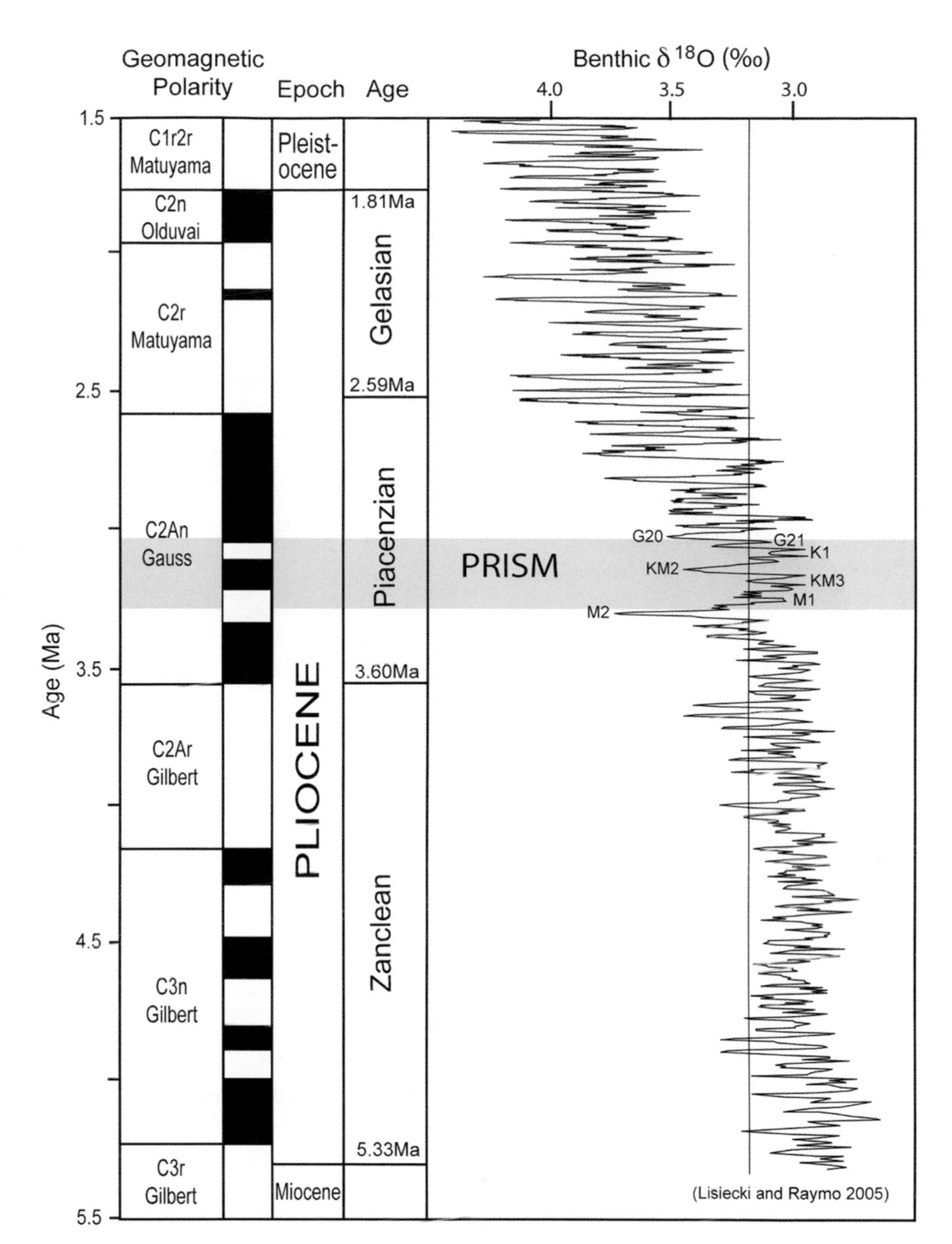

FIGURE 13.3. Geomagnetic polarity time scale and deep-sea benthic foraminiferal oxygen isotope data for the Pliocene Epoch from Lisiecki and Raymo (2005). The shaded interval in the mid-Pliocene is being studied in detail by the Pliocene Research, Interpretation and Synoptic Mapping group (**PRISM**) of the US Geological Survey (see box above). Note: the International Commission on Stratigraphy has recently moved the **Pliocene/Pleistocene Epoch boundary** down to the Piacenzian/Gelasian Age boundary (2.6 Ma). The **vertical line at approximately 3.2‰ (per mil) $\delta^{18}O$ represents present-day oxygen isotope values in the deep-sea.**

6 Compare **modern mean annual SST** (Figure 13.4) with estimated **mid-Pliocene mean annual SST** (Figure 13.5). List five differences between the two SST maps:

1.

2.

3.

4.

5.

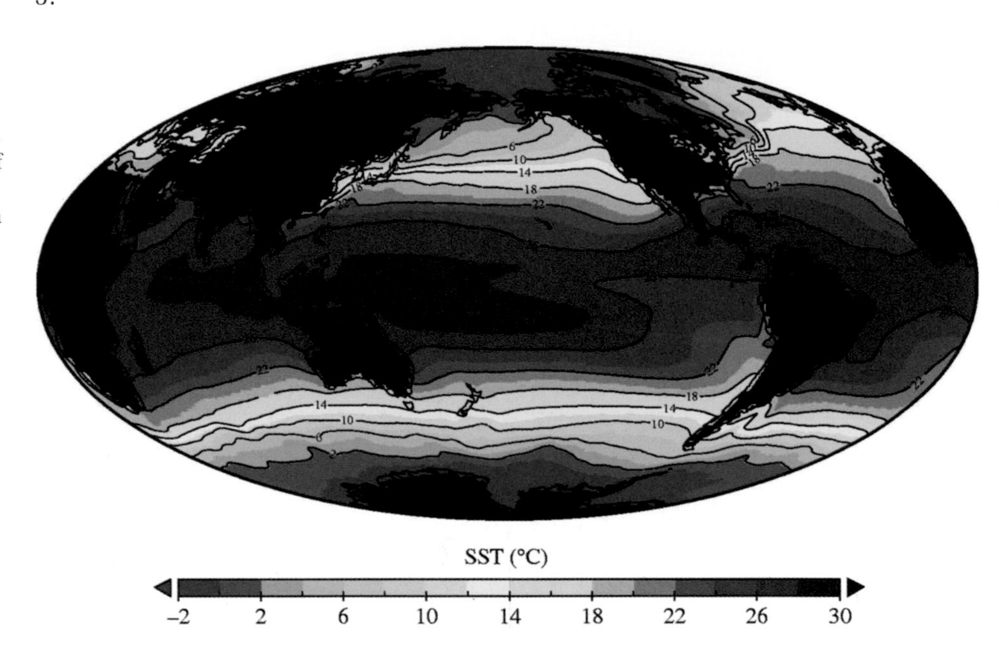

FIGURE 13.4. Modern mean annual sea surface temperature **(SST).** The Western Pacific Warm Pool (WPWP) is the large pod of 30°C water in the tropical western Pacific and eastern Indian Ocean. From Dowsett and Robinson, 2009.

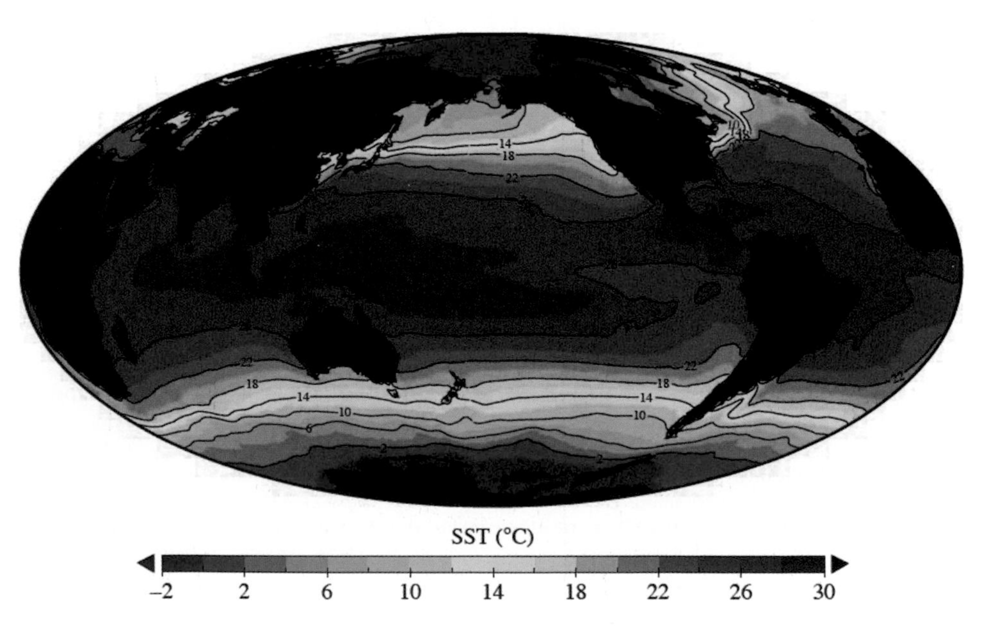

FIGURE 13.5. Pliocene reconstruction map (PRISM3) of mean annual sea surface temperature **(SST)** based on numerical data and climate modeling. This study focused on the tropical Pacific. From Dowsett and Robinson, 2009.

NAME

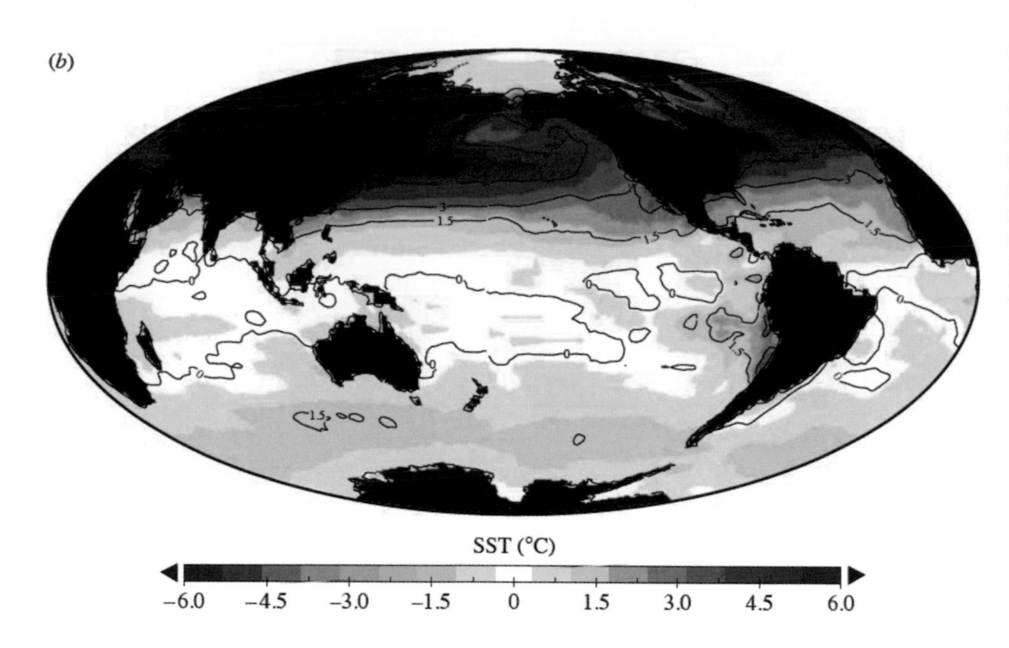

FIGURE 13.6. Mid-Pliocene PRISM3 SST anomaly map: these **sea surface temperature anomalies** are **the difference between mid-Pliocene modeled mean** annual SST values **and present day values.** From Dowsett and Robinson, 2009.

7 Based on the SST anomaly data depicted in Figure 13.6, where do SSTs increase the most (tropical zone, temperate zone, polar zone; northern hemisphere, southern hemisphere)?

____

____

8 Where do SSTs show the least amount of change?

____

____

The above study (Dowsett and Robinson, 2009; Figures 13.4–13.6) focused on sea surface temperatures (SSTs) during the mid-Pliocene. Next, we will consider data from an **early Pliocene** peat deposit in the high Canadian Arctic. Peat is a sedimentary deposit rich in decaying terrestrial organic matter and forms in bogs, marshes, and wetlands. Ballantyne et al. (2010; Figure 13.7) estimated **mean annual (air) temperature (MAT)** for the early Pliocene using multiple proxies for the early Pliocene: (a) oxygen isotopes of fossil wood cellulose and annual tree ring widths, (b) coexistence of paleovegetation, and (c) bacterial tetraether composition in paleosols (ancient soils). The MAT estimates from the three proxies are statistically indistinguishable and are plotted together in Figure 13.7 (black filled circles within red oval). The peat deposit provides the highest latitude terrestrial estimates of paleotemperature for the early Pliocene and anchors the northern end of this latitudinal (i.e., meridional) MAT gradient.

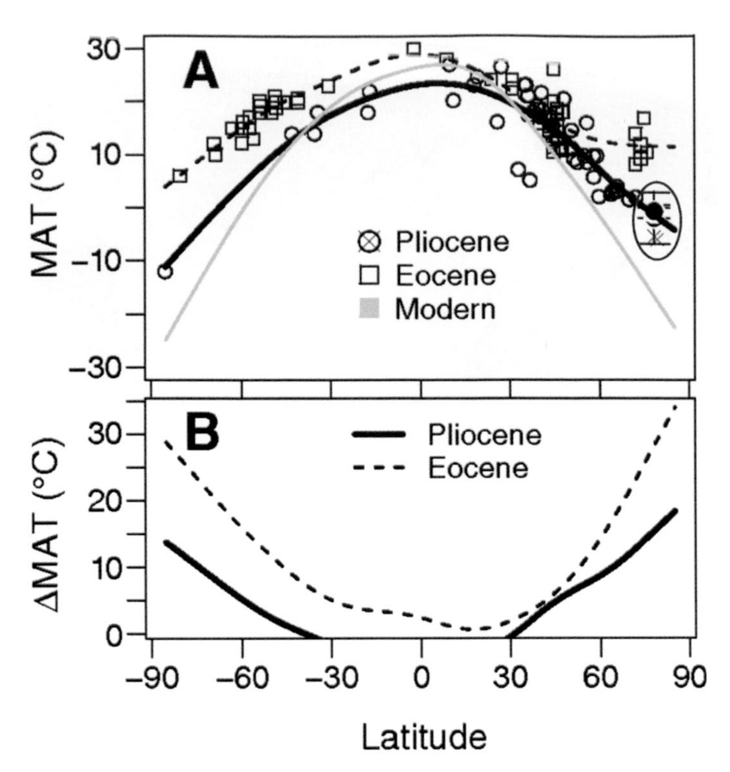

FIGURE 13.7. (a) **Latitudinal mean annual temperature (MAT) gradients** of the past (early Pliocene, solid black line; and early Eocene, dashed black line) compared with the present (Modern; gray line). The three independent temperature estimates from the high Canadian Arctic peat deposit are shown as filled black circles with standard error bars (within red oval). (b) Difference between present day MAT gradient and early Eocene (dashed black line) and early Pliocene (solid black line). 90 = North Pole, –90 = South Pole, 0 = Equator. From Ballantyne et al., 2010.

9 The consequence of global warming of the past can be seen by examining the early Pliocene latitudinal temperature gradient (solid black line; Figure 13.7). **During the early Pliocene, what latitudes experienced the greatest deviation from modern MAT temperatures** (i.e., what regions of the planet were most sensitive to rising temperatures during the early Pliocene)?

10 How much warmer were **mean annual temperatures** in the Arctic during the early to mid-Pliocene compared with today?

NAME

11 The **Early Eocene Climatic Optimum (EECO; approximately 53–50 Ma)** is an example of **extreme warmth** (refer back to Chapter 6; Figure 6.2). How does the early Eocene latitudinal temperature gradient compare with the early Pliocene latitudinal temperature gradient?

12 The mid-Pliocene SST data (Figures 13.4–13.6) and the early Pliocene and early Eocene MAT data (Figure 13.7), provide clues to the distribution of temperatures across the planet during times of global warming. What can you conclude about differential temperature effects across the planet during times of global warmth?

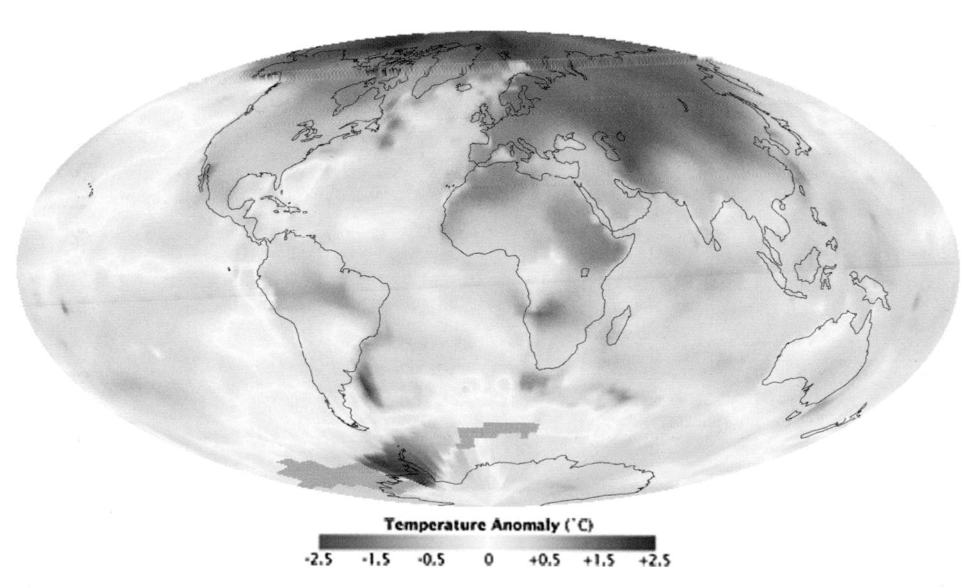

FIGURE 13.8. Map showing how much warmer **mean annual temperatures (MATs)** were in the most recent decade (2000–2009; the warmest decade on record) compared with average MATs for the period 1951–1980 (i.e., the map shows **temperature anomalies**). The data are based on temperatures recorded at weather stations around the world and satellite data over the ocean. Gray blocks in the Southern Ocean adjacent to Antarctica lack data. From Goddard Institute for Space Studies (GISS), NASA Earth Observatory: http://earthobservatory.nasa.gov/IOTD/view.php?id=42392

NAME ____

13 How do these ancient trends of temperature change (early Pliocene and early Eocene) compare with the modern trends of global warming shown in Figure 13.8?

____

____

____

____

Atmospheric Carbon Dioxide and Global Climate

As we have seen in previous exercises, the concentration of greenhouse gases in our atmosphere, such as carbon dioxide (CO_2) and methane (CH_4), play a primary role in controlling mean annual temperatures across the planet (see Chapters 5, 9, and 10).

14 What is the concentration of CO_2 in the atmosphere today? A good source for up-to-date trends in atmospheric CO_2 is http://www.esrl.noaa.gov/gmd/ccgg/trends/

____

15 As a first approximation, what do you predict the concentration of CO_2 was like during the early to mid-Pliocene relative to today: much higher, somewhat higher, the same, somewhat lower, much lower?

____

Review the data presented in Figure 13.9 below and answer the questions that follow.

16 Describe the general trend of atmospheric CO_2 from the early Pliocene to the present (Figure 13.9).

____

____

____

____

17 Draw a horizontal line on each of the six plots in Figure 13.9 showing present day pCO_2 values (see your answer to Question 14). How do early and mid-Pliocene estimated values of pCO_2 compare with modern values?

____

NAME

____

____

____

18 What take-away messages do you glean from these Pliocene data relative to present day pCO_2 values?

____

____

____

____

____

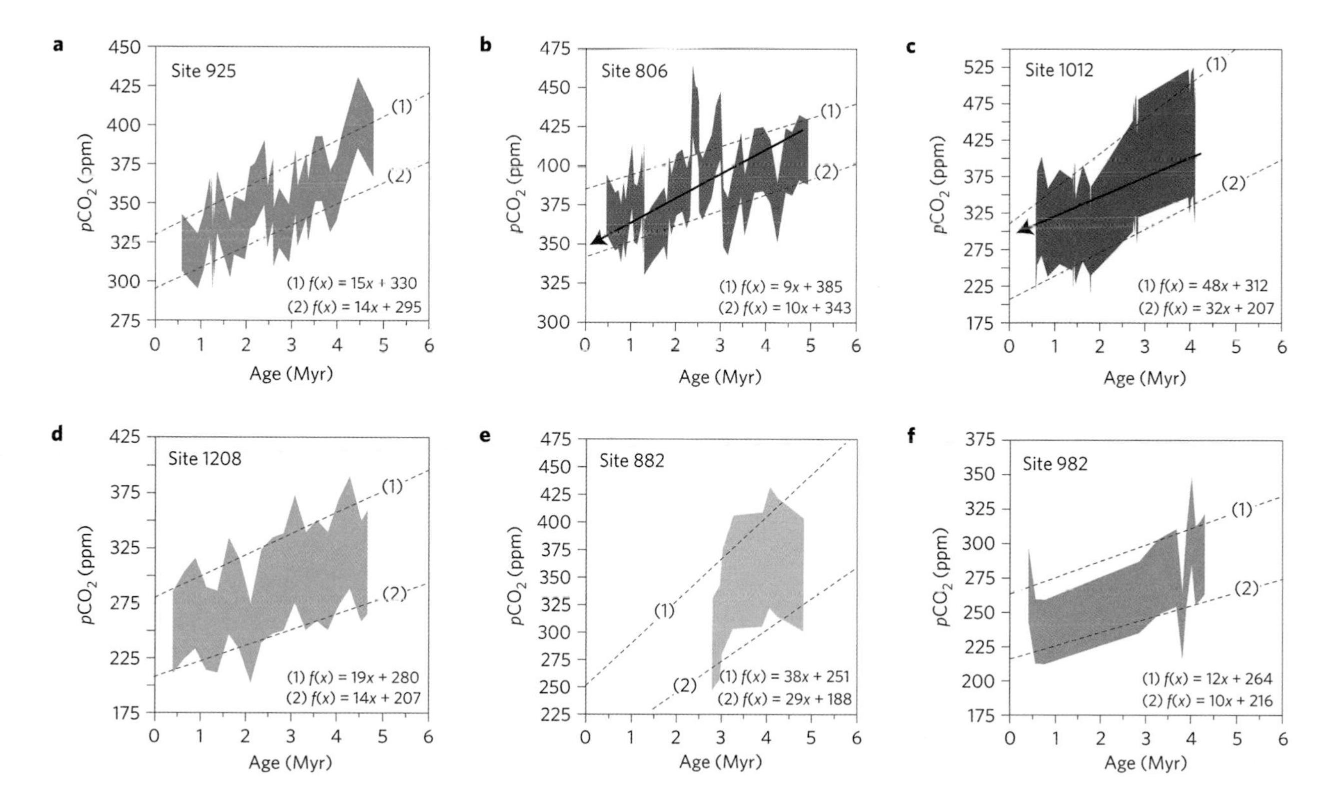

FIGURE 13.9. Estimates of **Pliocene–Pleistocene atmospheric concentration of carbon dioxide [pCO_2 (ppm)]** at six different ODP sites (a–f) calculated from alkenone carbon isotope fractionation (ε_p) and phosphate (PO_4^{3-}) proxies. Upper and lower pCO_2 estimates show a calculated range of values based on depth in photic zone (0 to 75 m) of the coccolithophorids (phytoplankton) responsible for the alkenone biomarkers used in this study. The dashed lines are linear regressions of maximum and minimum CO_2 estimates. From Pagani et al., 2009.

NAME

Pliocene Warmth: Are We Seeing Our Future?

Part 13.2. Sea Level Past, Present, and Future

Onset of Northern Hemisphere Glaciation

1 Based on the Lisiecki and Raymo (2005) benthic foram isotope values shown in Figure 13.10, over what million year interval did Pliocene warmth end and cooling begin?

---

2 On the right-hand side of the oxygen isotope plot above (Figure 13.10, *y*-axis), add the terms "more ice" and "less ice" corresponding to the higher and lower $\delta^{18}O$ values, respectively.

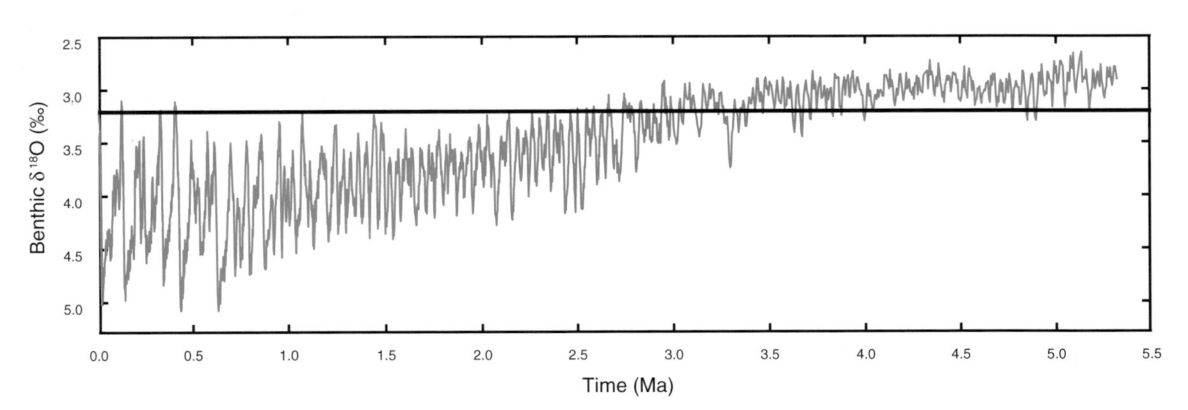

FIGURE 13.10. Composite deep-sea benthic foraminiferal oxygen isotope curve for the past 5.3 myr. This composite plot is derived from a "stack" of 57 globally distributed Pliocene–Pleistocene records, which have been aligned by an automated graphic correlation algorithm. **The horizonal line represents present day deep-sea oxygen isotope values (approximately 3.2‰).** From Lisiecki and Raymo, 2005.

3 The onset of northern hemisphere glaciation is believed to have begun approximately 2.6 Ma based on the occurrence of ice-rafted debris in the deep-sea of the northern high latitudes, the deep-sea benthic foram record, such as that of Lisiecki and Raymo (2005), and other proxy records (see Chapter 14). **With a vertical arrow, mark** the point on the Lisiecki and Raymo oxygen isotope curve (Figure 13.10) corresponding to the onset of northern hemisphere glaciation.

NAME ___

4 Changes in ice volume and water temperature can affect the oxygen isotopic composition of deep-sea benthic forams. If the difference between early to mid-Pliocene deep-sea benthic foram $\delta^{18}O$ values and those of today (approximately 3.2‰; Figure 13.10) is based solely on ice volume differences and not on changes in deep water temperatures, we can estimate the corresponding change in sea level (since changes in ice volume directly affect sea level). Approximately how different was early to mid-Pliocene sea level compared with today, if 0.11‰ = 10 m of sea level? Show your work.

___

___

___

___

___

___

___

5 Based on Figure 13.11, describe the nature of global sea level change since the Pliocene.

___

___

___

___

___

6 Is sea level expected to remain static over time?

___

___

7 What is the principal mechanism that controls global sea level?

___

___

___

NAME

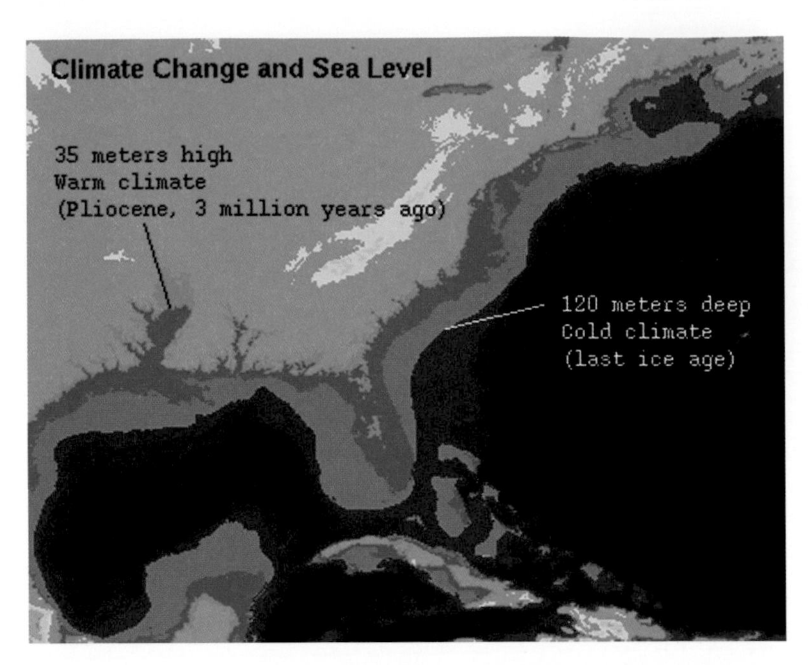

FIGURE 13.11. Map of the US East Coast, Gulf Coast, Gulf of Mexico and northern Caribbean depicting the changes in **global (eustatic) sea level** since the Pliocene. The light blue color shows the area exposed by lowered sea level during the **Last Glacial Maximum (LGM)**, about 18,000 years ago (18 ka). During the LGM, the continental shelves were completely exposed as sea level fell approximately 125 m (approximately 410 ft.). **The dark green color shows the approximate area flooded during the global warmth and related sea level rise of the mid-Pliocene (+35 m),** about 3 million years ago. From http://geochange.er.usgs.gov/data/sea_level/ofr96000.html). Reproduced with the permission of Peter Schweitzer and Robert Thompson.

8 Compare the data presented in Tables 13.1 and 13.2, and the maps in Figures 13.12 & 13.13. What coastal cities in the USA would flood if the Greenland Ice Sheet and the West Antarctic Ice Sheet (the most vulnerable large ice sheets today) were to both partially melt, each losing a half of their ice volume?

9 If the rate of sea level rise along the eastern seaboard is approximately 3.5 mm/yr (Figure 13.14), how many years will it take for sea level to rise 1 m? Show your work.

NAME ______________________________

____

____

TABLE 13.1. Estimated **potential maximum sea level rise** from the total melting of present day glaciers.

Location	Volume (km^3)	Potential sea level rise (m)
East Antarctic ice sheet	26,039,200	64.80
West Antarctic ice sheet	3,262,000	8.06
Antarctic Peninsula	227,100	0.46
Greenland	2,620,000	6.55
All other ice caps, ice fields, valley glaciers	180,000	0.45
TOTAL	**32,328,300**	**80.32**

Modified from Williams and Hall, 1993. From USGS fact sheet: Sea Level and Climate, http://pubs.usgs.gov/fs/fs2-00/

Table 13.2. Elevations of major coastal cities in the USA (based on major airport elevation).

City, State	Elevation (feet)	Elevation (m)
San Diego, CA	15	4.6
San Francisco, CA	11	3.3
Washington, DC	16	4.9
Miami, FL	11	3.3
Honolulu, HI	13	4.0
New Orleans, LA	4	1.2
Boston, MA	20	6.1
New York, NY	13	4.0
Newark, NJ	18	5.5
Portland, OR	27	8.2
Philadelphia, PA	21	6.4
Providence, RI	55	16.8
Houston, TX	98	29.9
Norfolk, VA	27	8.2

From: http://www.altimeters.net/cityaltitudes2.html

NAME ______________________________

Sea Level Rise

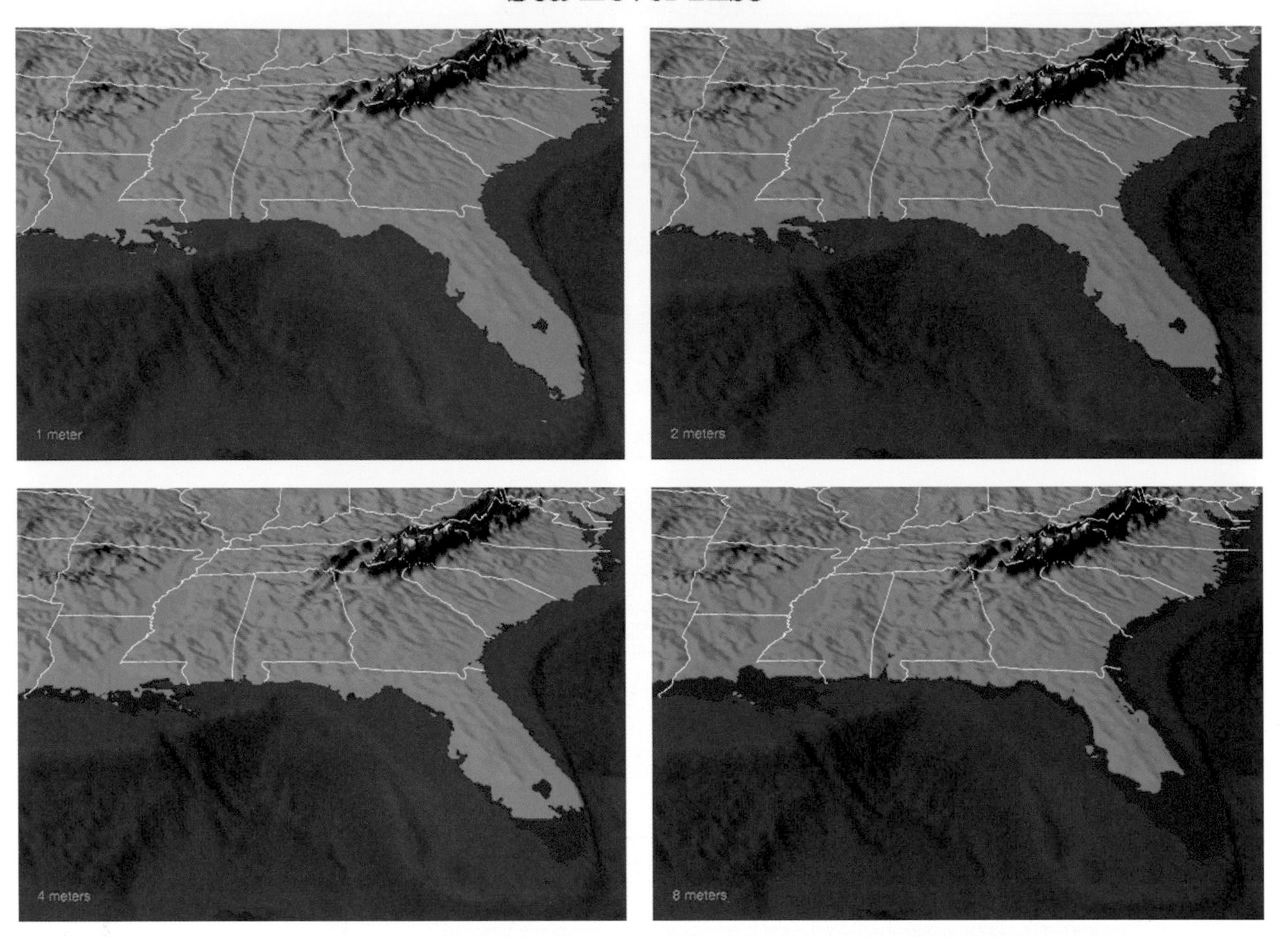

FIGURE 13.12. Series of maps depicting the extent of flooding (red areas) of the southeast USA coastal plain caused by sea level rise: upper left = 1 m, upper right = 2 m, lower left = 4 m, and lower right = 8 m. Courtesy of the National Geophysical Fluid Dynamics Laboratory: http://www.gfdl.noaa.gov/climate-impact-of-quadrupling-co2.

10 The sea level is rising at a rate of approximately 3–4 mm/year along the New Jersey coastline (Figure 13.14). It is rising at a rate of approximately 6 mm/year along the Virginia coastline (Zervas, 2009), and approximately 9 mm/year along the periphery of the Mississippi River delta in Louisiana (Miner et al., 2009). The 2007 IPCC Report states that the average global rate of sea level rise for the period of 1950–2000 was approximately 1.8 mm/yr. Speculate about why the rate of sea level rise is so much higher along the eastern and Gulf seaboards of the USA compared to the global average.

____

____

____

NAME

11 Note that sea level is relative to the land surface and, in some places, the land is also subsiding (sinking) in addition to being submerged by the rising global sea level. Speculate about what might cause subsidence in the eastern and Gulf coasts of the USA?

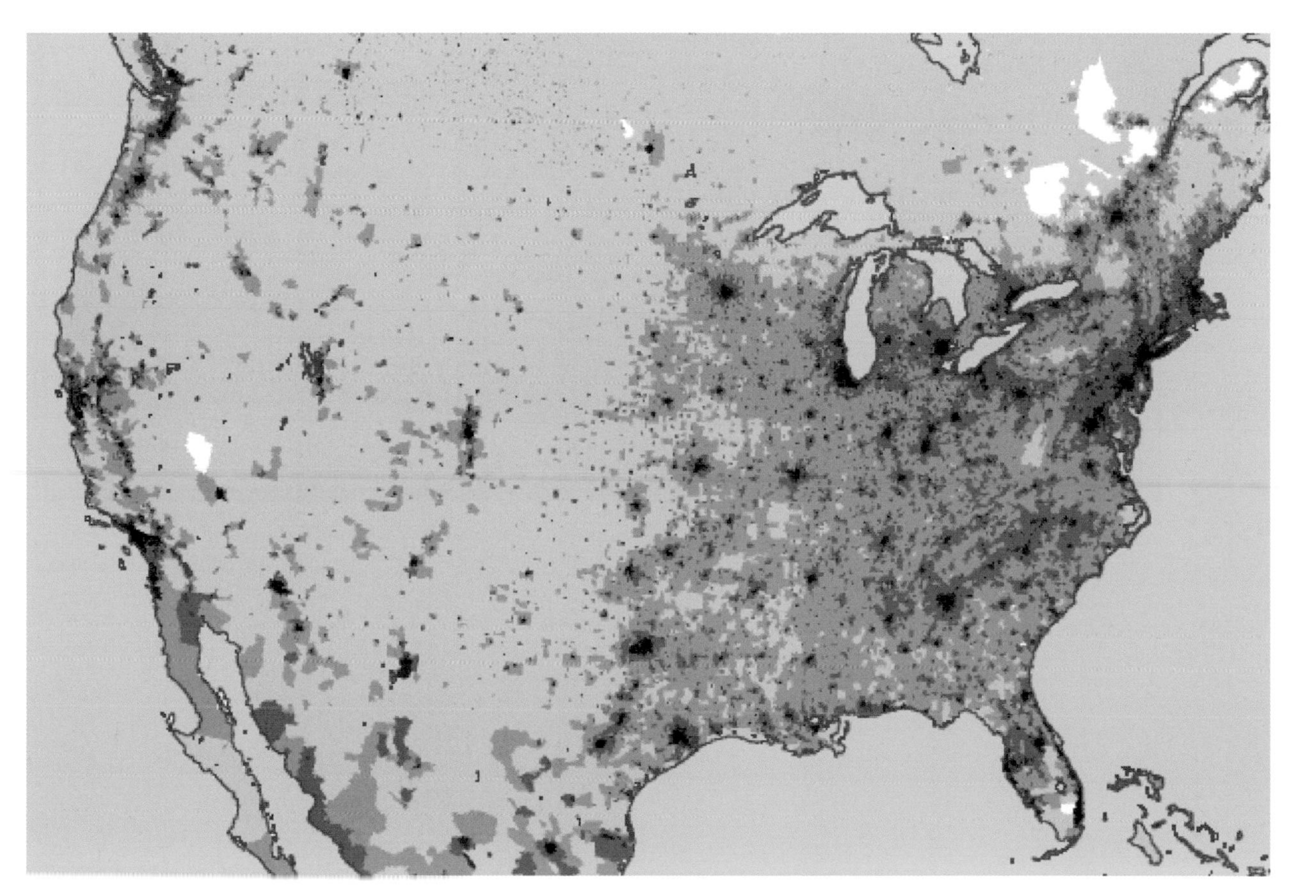

FIGURE 13.13. Population density across the lower 48 states, northern Mexico and southern Canada. From Gridded Population of the World, version 3 (GPWv3), produced by the Center for International Earth Science Information Network (CIESIN) of the Earth Institute at Columbia University. http://sedac.ciesin.columbia.edu/gpw/

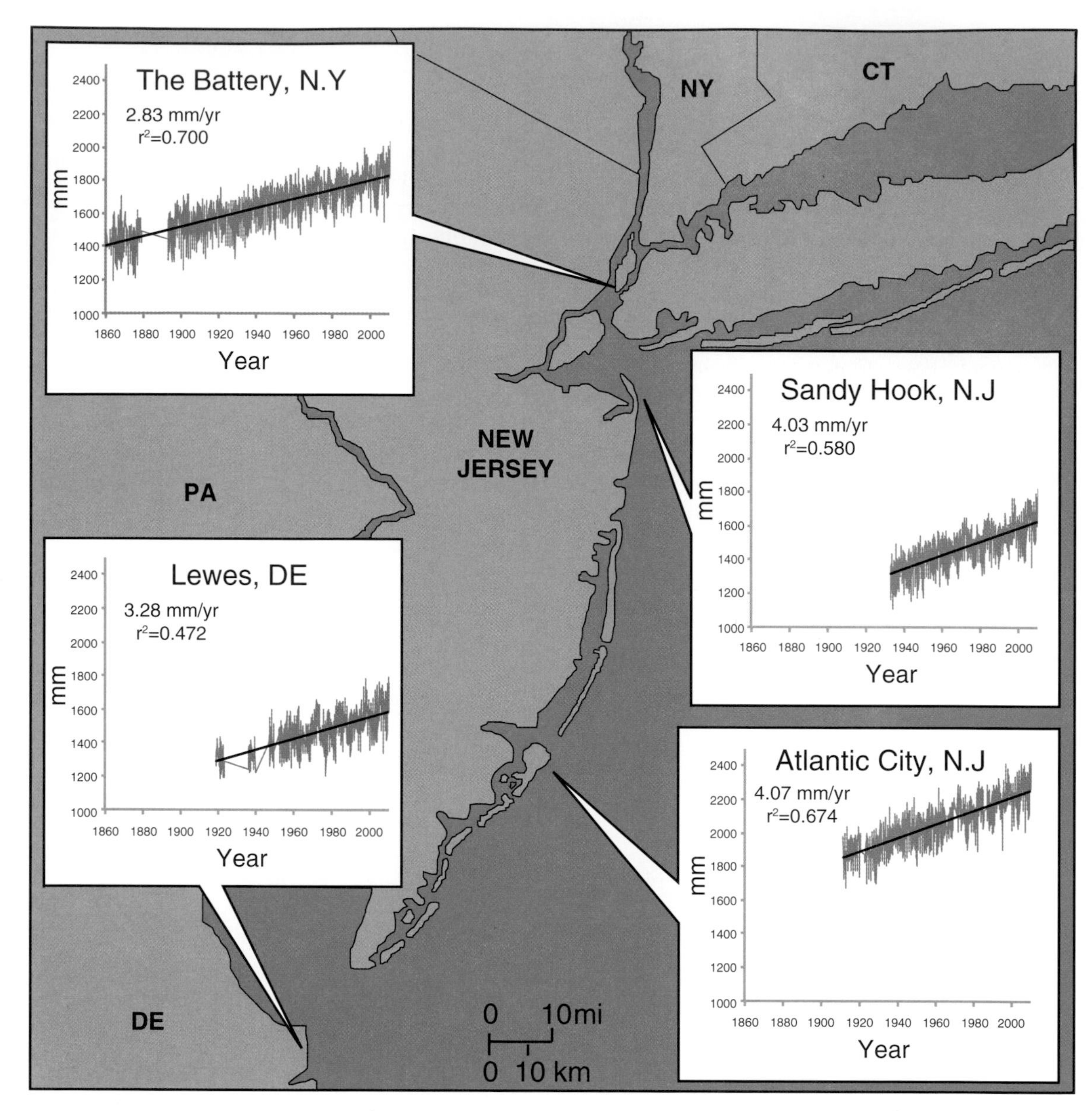

FIGURE 13.14. Sea level is rising at a rate of approximately 3–4 mm/year along the New Jersey coastline based on tide gauge stations. Figure adapted from http://climatechange.rutgers.edu/images/nj_sl_rise_map.jpg. Graphs made by Serena Dameron from historical tide station data from http://tidesandcurrents.noaa.gov

NAME

References

Ballantyne, A.P., et al., 2010, Significantly warmer Arctic surface temperatures during the Pliocene indicated by multiple independent proxies. Geology, **38**(7), 603–6.

Dowsett, H.J. and Robinson, M.M., 2009, Mid-Pliocene equatorial Pacific sea surface temperature reconstruction: A multi-proxy perspective. Philosophical Transactions of the Royal Society A, **367**, 109–25.

IPCC, 2007, Summary for Policymakers, In Climate Change 2007: The Physical Science Basis. Solomon, S., et al. (eds). Contribution of Working Group I to the Fourth Assessment Report of the Intergovernmental Panel on Climate Change, Cambridge University Press, Cambridge, UK and New York, USA.

Lisiecki, L.E. and Raymo, M.E., 2005, A Pliocene–Pleistocene stack of 57 globally distributed benthic $\delta^{18}O$ records. Paleoceanography, **20**, doi:10.1029/2004PA001071.

Miner, M.D., et al., 2009, Delta lobe degradation and hurricane impacts governing large-scale coastal behavior, south-central Louisiana. Geo-Marine Letters, **29**, 441–53.

Pagani, M., et al., 2009, High Earth-system climate sensitivity determined from Pliocene carbon dioxide concentrations. Nature Geoscience, **3**, 27–30, doi:10.1038/ngeo724.

Williams, R.S. and Hall, D.K., 1993, Glaciers, in Chapter on the cryo-sphere. In Atlas of Earth Observations Related to Global Change. Gurney, R.J., et al. (eds), Cambridge University Press, Cambridge, UK, pp. 401–22.

Zervas, C., 2009, Sea level variations of the United States 1854–2006. NOAA Technical Report, NOS CO-OPS, US Department of Commerce, National Oceanic and Atmospheric Administration, National Ocean Service., Silver Spring, MD.

Chapter 14 Northern Hemisphere Glaciation

FIGURE 14.1. A gibbous moon glows over the ice near Qaanaaq in northwestern Greenland. From the National Snow and Ice Data Center, Photo and Image Gallery.

SUMMARY

This investigation introduces the **characteristics and possible causes of the northern hemisphere glaciation during the Cenozoic**. In **Part 14.1**, you will make predictions about where and why continental ice sheets form. In **Part 14.2**, you will examine geological, geochemical, and paleontological data to infer the spatial extent and temporal history of this glaciation. In **Part 14.3**, you will read abstracts critically from seven peer-reviewed papers to decipher proposed mechanisms for the expansion of northern hemisphere glaciation at approximately 2.6 Ma and reflect on the scientific value of multiple working hypotheses.

Reconstructing Earth's Climate History: Inquiry-Based Exercises for Lab and Class,
First Edition. Kristen St John, R Mark Leckie, Kate Pound, Megan Jones and Lawrence Krissek.

NAME

Northern Hemisphere Glaciation
Part 14.1. Concepts and Predictions

Introduction

This exercise examines the expansion and contraction of large land-based ice sheets, such as the Greenland Ice Sheet, which is the last vestige of the great northern hemisphere ice sheets of the Pleistocene.

1 Predict where on the Earth's surface you would expect perennial (year round) ice to form. Why there?

Where:

Why:

2 Predict what might **cause** a continental-size ice sheet to form? Explain your reasoning.

3 Predict **what types of records** might provide clues to the existence and dynamic history (e.g., expansion and contraction) of northern hemisphere continental ice sheets? Make a list.

NAME

NAME

Northern Hemisphere Glaciation
Part 14.2. What is the Evidence?

Introduction

The temporal and spatial history of Cenozoic northern hemisphere glaciation is derived from **geological**, **geochemical**, and **paleontological** sources. Here you will assess a subset of evidence from each of these sources to develop a conceptual model of the timing and extent of Cenozoic northern hemisphere glaciation.

To do:

Read the introductory information on geological, geochemical, and paleontological sources of evidence on the following pages. Then **use the various data in the figures to answer the eight questions in Table 14.1**.

When examining the various data, pay close attention to the type of proxies used, the site locations, the age ranges of the data, and the data resolutions (both temporal and spatial). All these factors influence how data are useful in addressing hypotheses and answering scientific questions. Note that the data presented here do not represent all the geological, geochemical, and paleontological information published on the northern hemisphere glaciations, but are only a small representative sampling of the wealth of peer-reviewed scientific literature. Not every question in Table 14.1 will have a clear-cut answer, so formulate your best answer based on the available data. Support your answer with specific evidence from the information in the maps and graphs (Figures 14.4–14.17).

FIGURE 14.2. Receding glacier in the Swiss Alps. Notice the brown **lateral moraines** along the sides of the glacier and the **ground moraine** and **pro-glacial lake** at the foot of the glacier. Photo from Kristen St. John.

NAME ______________________

Table 14.1 Data interpretations

Question	Answer	Evidence
1 How far south did ice sheets extend in North America during the Pleistocene? (see Figures 14.4 & 14.5)		
2 **When** in the Cenozoic (last 65.5 million years) did glacial ice first form in the northern hemisphere? (see Figures 14.7–14.13)		
3 **Where** did the first Cenozoic northern hemisphere ice form? (see Figures 14.7–14.13)		
4 **When** in the Cenozoic did northern hemisphere glaciation become widespread? (see Figures 14.7–14.13)		
5 Once northern hemisphere glaciation was widespread, did the northern hemisphere: (a) remain persistently icy and cold, (b) cycle regularly between icy/cold and less icy/warm, or (c) irregularly switch from icy/cold to less icy/warm? (see Figures 14.4–14.17)		

Question	Answer and Interpretation
6 What challenges might there be for reconstructing a glacial history from land-based deposits compared to reconstructing a glacial history from marine deposits?	
7 What is the value in investigating northern hemisphere glaciation from a muiltproxy approach?	
8 What other types of information would you like to examine to compare with the data here? List and explain.	

FIGURE 14.3. Close-up of the leading edge of the glacier shown in Figure 14.2. Notice the sediment in and on the ice. Also notice the **ground moraine** at the bottom of the photo. Photo from Kristen St. John.

Geological Evidence: Moraine Landforms and Till Sediments

Glaciers are powerful agents of erosion, incorporating sedimentary debris into the ice as they slide through mountain valleys and across the land surface. Debris frozen into the base of a glacier will abrade the underlying bedrock as the glacier moves. Ice-incorporated sediment will eventually be deposited as the glacial ice melts. The most common type of land-based glacial deposit is a **moraine** (Figures 14.2 and 14.3). Moraines are landforms composed of **glacial tills**, a mixture of sediment of different grain sizes that was transported by the glacier and deposited below, in front of, or along the sides of that glacier. Maps showing the local and global extent of northern hemisphere ice sheets, based largely on the distribution of glacial moraine landforms, are shown in Figures 14.4 & 14.5.

The map in Figure 14.4 is representative of the detailed mapping of different types and ages of **moraines** and other glacial deposits. A **ground moraine** (Figure 14.3) is a thin till layer deposited during a time of steady ice retreat. A **ridge moraine** is a pile of till deposited at the edge of the glacier during a pause in ice retreat. Note the widespread distribution of Wisconsin deposits relative to older deposits, which were eroded by the advance of Wisconsin ice sheets (from the Last Glacial Maximum; LGM).

NAME

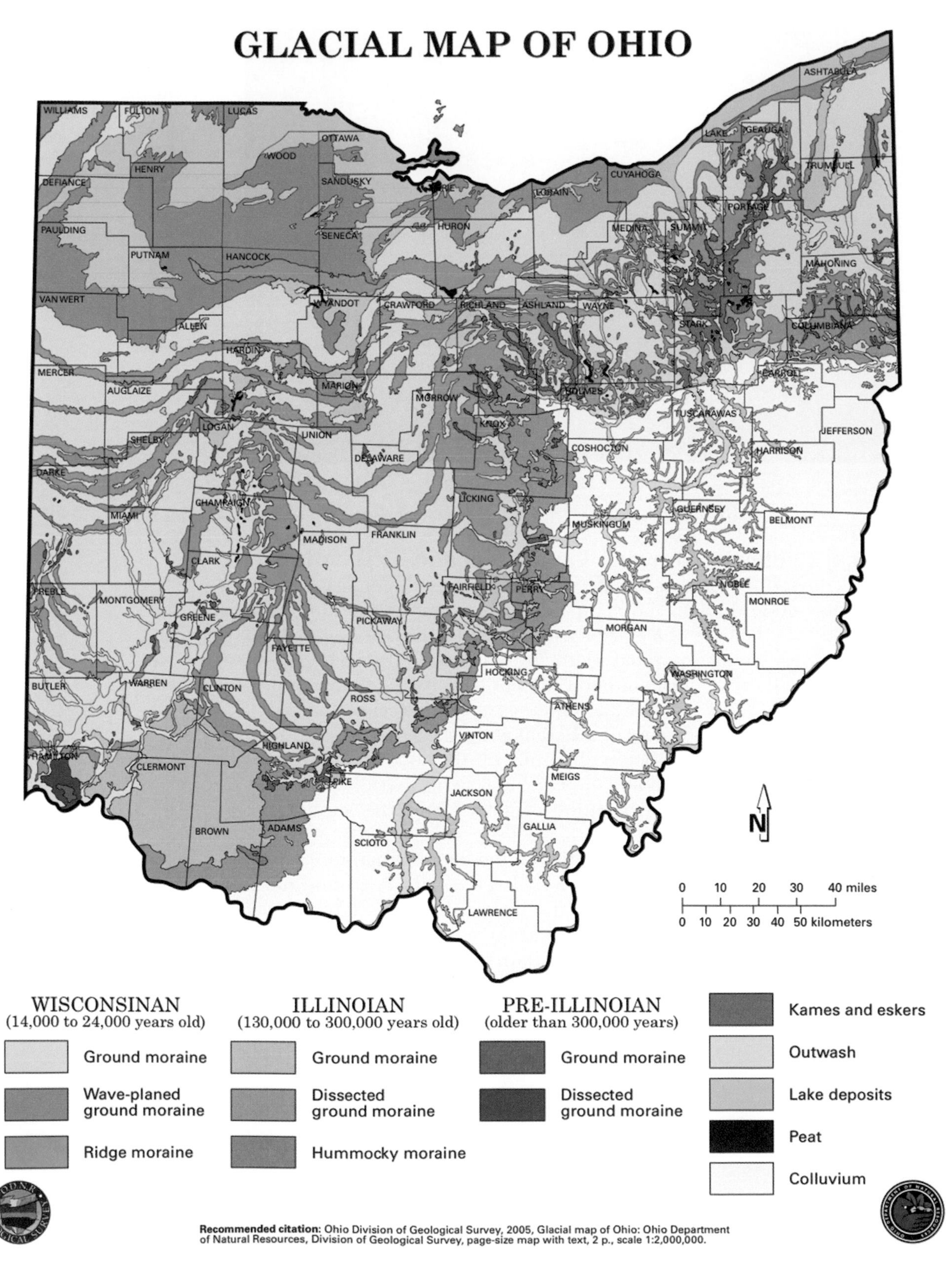

FIGURE 14.4. Map showing the distribution of **glacial deposits** in Ohio. Map courtesy of the Geological Survey, Ohio Division of Natural Resources.

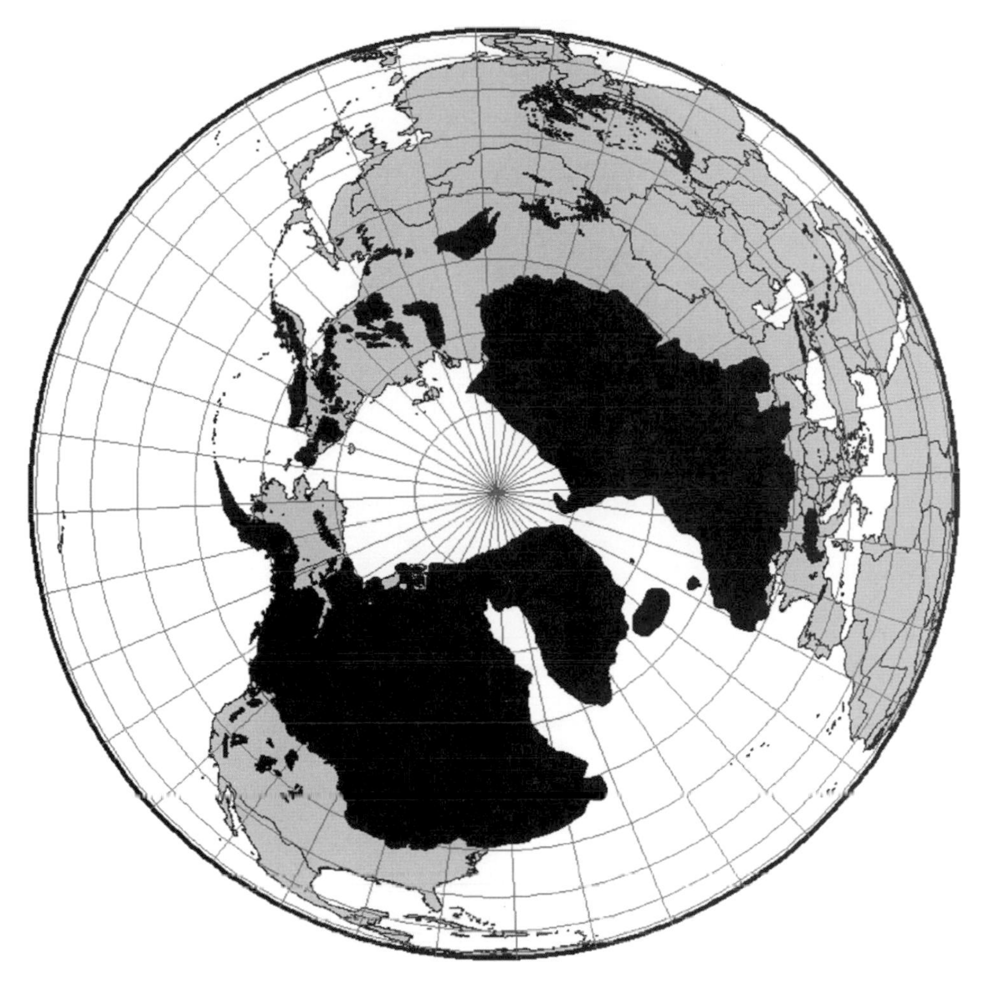

FIGURE 14.5. This map shows a reconstruction of the maximum extent of Pleistocene ice sheets in the northern hemisphere (Ehlers and Gibbard, 2007). It is based on geologic evidence of glaciations from the work of over 200 scientists working in 60 countries and territories as part of the International Union for Quaternary Research (INQUA) project "Extent and Chronology of Glaciations". Evidence is derived from field work, general stratigraphic correlation, age-dating, and modeling. From Elhers and Gibbard, 2003.

Geological Evidence: Ice-Rafted Debris (IRD)

As glaciers expand and reach sea level, their edges "calve" off into the ocean as debris-laden icebergs. As these icebergs melt within the relatively warmer seawater, their "**ice-rafted debris**" (**IRD**) will settle through the water column and accumulate on the seafloor. The composition (minerals and rock fragments) and grain size of the IRD will depend on the local geology in the glaciated source area. Clays to cobbles (and even boulders) can be transported to the sea by icebergs. The most obvious IRD records are those of isolated pebbles in a finer grained matrix (Figure 14.6); these pebbles are appropriately called "**dropstones**".

In general, the presence of IRD is interpreted to record the presence of glacially calved icebergs at the time that these sediments were deposited, and the abundance of IRD is interpreted as a general indicator of the relative abundance of icebergs and glacial ice (i.e., more IRD = more icebergs = more glacial ice at sea level = larger ice sheets on land). Note that sea ice that forms along shallow shorelines can also transport clay to sand-sized IRD. This can complicate interpretations of IRD records from oceans that are covered with sea ice, such as the Arctic.

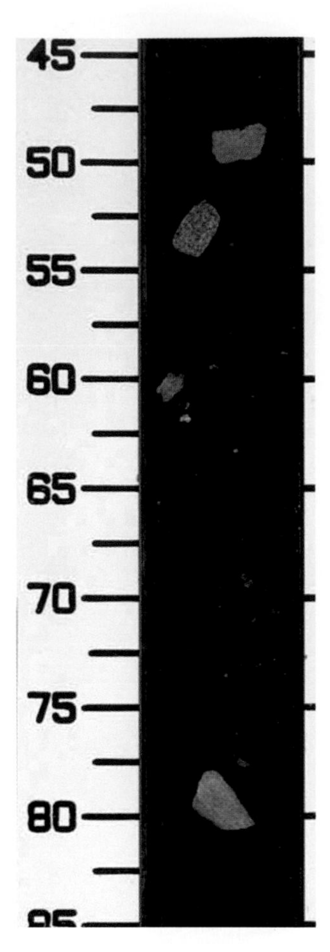

FIGURE 14.6. Photo of a split core from the Norwegian-Greenland Sea showing abundant dropstones of sedimentary and igneous lithologies in late Pliocene age black clayey mud (151-913A-9X-1, 45–85 cm). Age was determined using biostratigraphy and magnetostratigraphy. Courtesy of IODP.

Representative IRD records from the Arctic, North Atlantic, and North Pacific Oceans are shown in Figures 14.7 to 14.13. These IRD data are displayed in one of three ways: (1) as the IRD weight percent (wt%), (2) as the number of IRD grains per gram of sediment (#IRD/g), or (3) as the IRD mass accumulation rate ($g/cm^2/kyr$). Each of these is a measure of the relative abundance, or relative importance, of IRD compared to the bulk sediment sample.

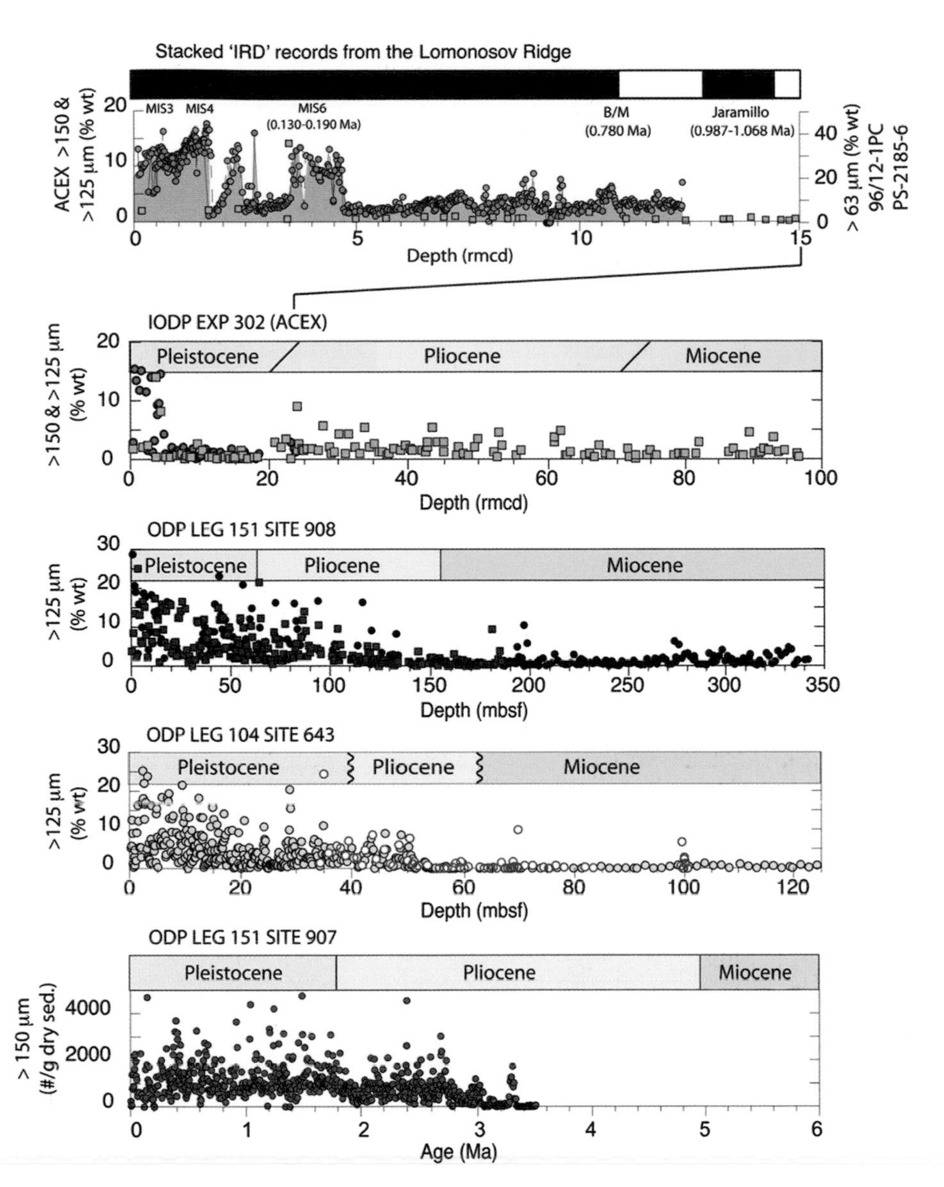

FIGURE 14.7. Compilation of **Miocene to Pleistocene IRD** records from the **Norwegian-Greenland Seas (Site 908 is west of Svalbard, Site 907 is north of Iceland, Site 643 is northeast of Iceland) and Arctic Ocean (site 302 is on the Lomonosov Ridge in the central Arctic).** (originally drawn by Matt O'Regan; from Polyak et al., 2010). Sediments at Exp. 302, Site 908, and Site 643 have the same age range as the Pliocene–Pleistocene sediments at Site 907. Note that Ma = million years ago.

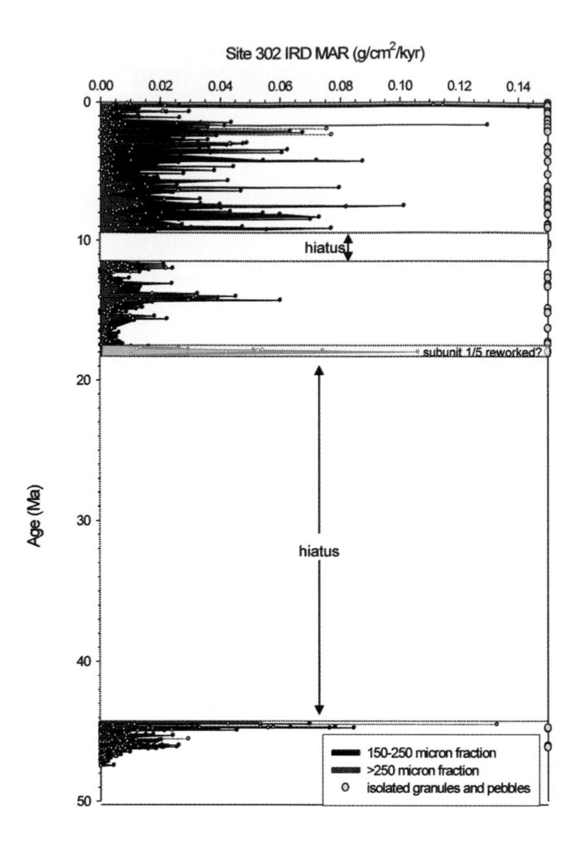

FIGURE 14.8. Long-term record of IRD abundance (i.e., mass accumulation rates, MAR) from the Lomonosov Ridge, central **Arctic Ocean** (Site d in Figure 14.7; modified from St. John, 2008). A **hiatus** represents a period of erosion or non-deposition in the sedimentary sequence. Biostratigraphy and magnetostratigraphy provide the basis for age control at this site. Note that Ma = million years ago.

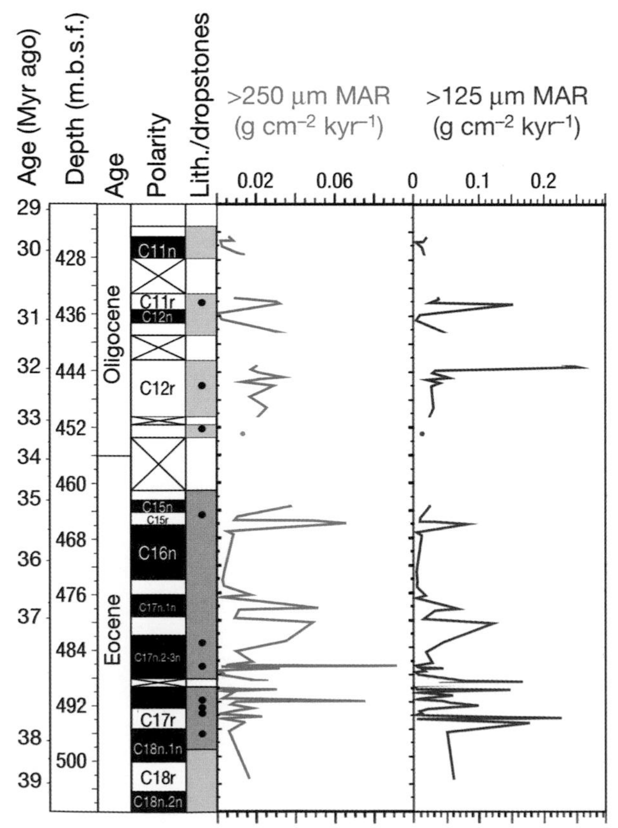

FIGURE 14.9. IRD abundance (mass accumulation rate) data from ODP site 913, **Greenland Basin.** Modified from Eldrett et al., 2007. Intervals of the "Polarity" column marked with an X are intervals not recovered by coring. Magnetostratigraphy and biostratigraphy provide the basis for age control at this site. Note that Myr = million years.

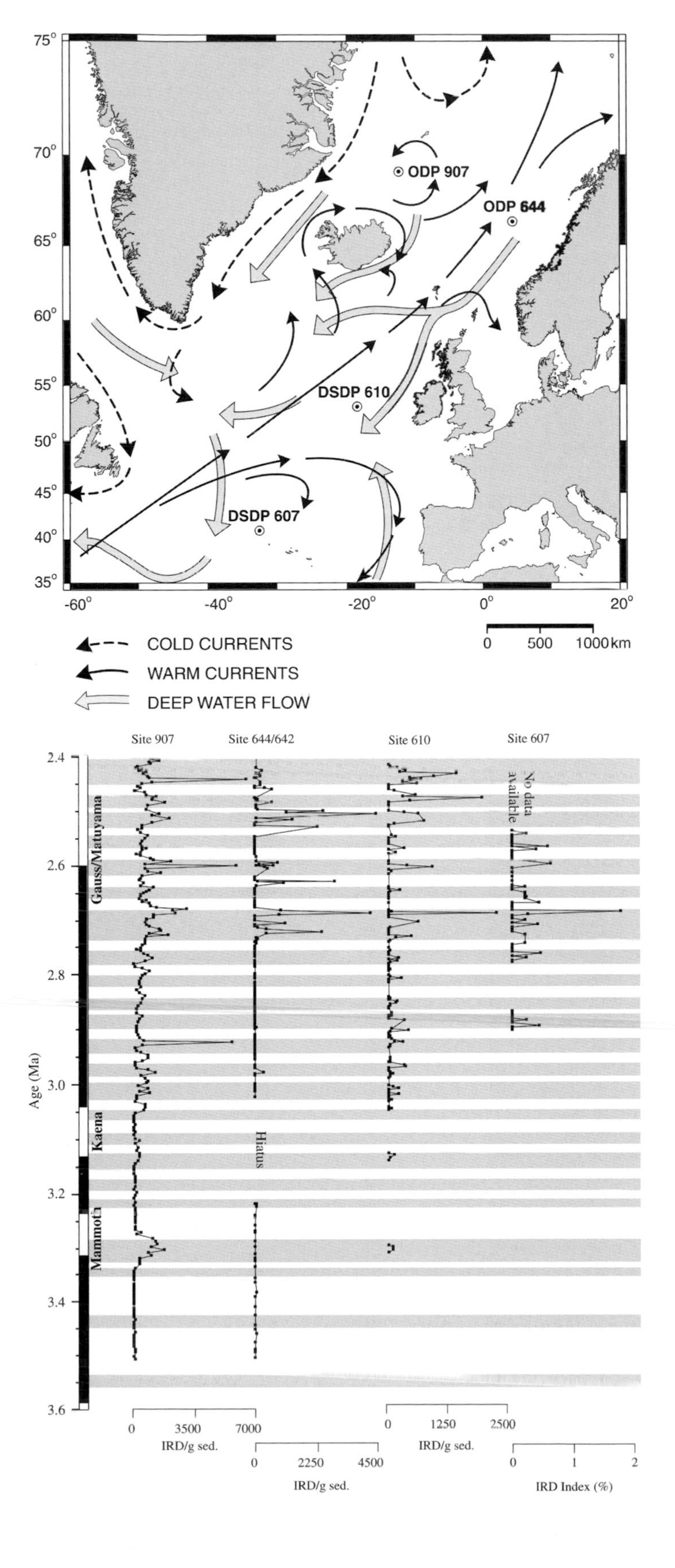

FIGURE 14.10. Map showing the main ocean currents of the present northern **North Atlantic Ocean and Norwegian-Greenland Sea**, and the location of ODP and DSDP sites used in the IRD study shown in Figure 14.11. From Flesche-Kleiven et al., 2002.

FIGURE 14.11. IRD abundance records from Sites 907, 644/642, 610 and 607 (see Figure 14.10; modified from Flesche-Kleiven et al., 2002). In this study, IRD is defined as mineralogical grains with a diameter >0.125 mm. The number of these grains per gram of dry sediment represents the IRD abundance. The horizontal shaded bars mark the stratigraphic positions of glacial oxygen isotope stages. Magnetostratigraphy is shown on the left. Magnetostratigraphy and oxygen-isotope geochemistry provide the basis for age control and correlation between these sites. Note that Ma = million years ago.

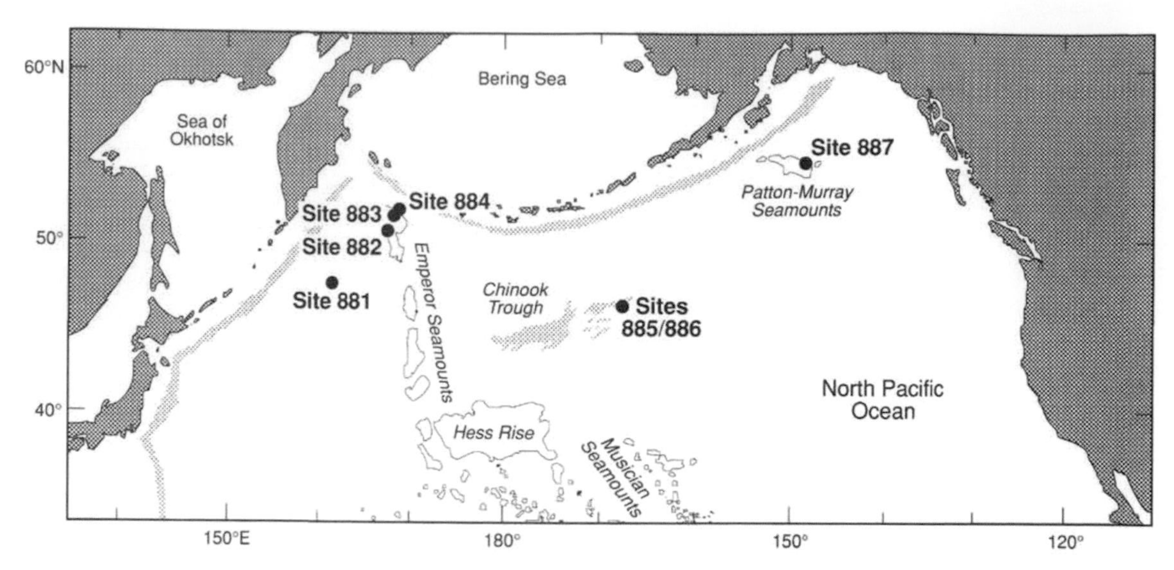

FIGURE 14.12. Map showing the location of **North Pacific Ocean** sediment cores included in the IRD study by Krissek (1995). The data from this study is shown in Figure 14.13. Map from Shipboard Scientific Party, 1993.

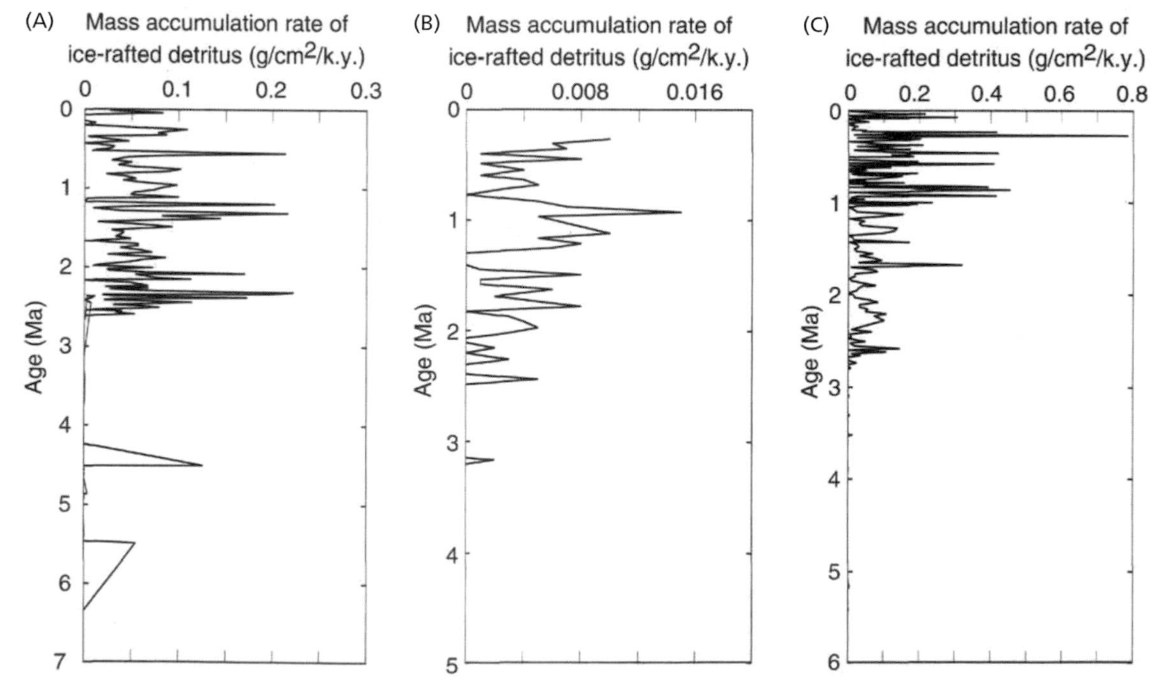

FIGURE 14.13. IRD abundance (mass accumulation rate) records from the North Pacific. In this study IRD is defined as terrigenous grains between 0.25 and 2.00 mm diameter. Note change in scales between plots. (A) Data for Site 881. (B) Data for Site 883. (C) Data for Site 887. From Krissek, 1995. Magnetostratigraphy and biostratigraphy provide the basis for age control at these sites. Note Ma = million years.

NAME

Geochemical Evidence: Stable Oxygen Isotopes in Marine Microfossils and in Glacial Ice

As global atmospheric temperatures vary, precipitation patterns change, and as ice sheets expand and contract, the geochemistry of both ocean water and glacial ice will be affected. **Stable oxygen isotope** records (Figures 14.14–14.16) from marine microfossils and from glacial ice itself are indirect indicators of these climatic changes.

Note that the oldest ice core in Greenland extends back approximately 250 kyr, with a robust record back to approximately 120 kyr. The records in Figure 14.16 show oxygen-isotope variability for the last 110 kyr measured from ice cores. **For glacial ice records, more negative δ ^{18}O values indicate colder conditions and less negative δ ^{18}O values indicate warmer conditions.**

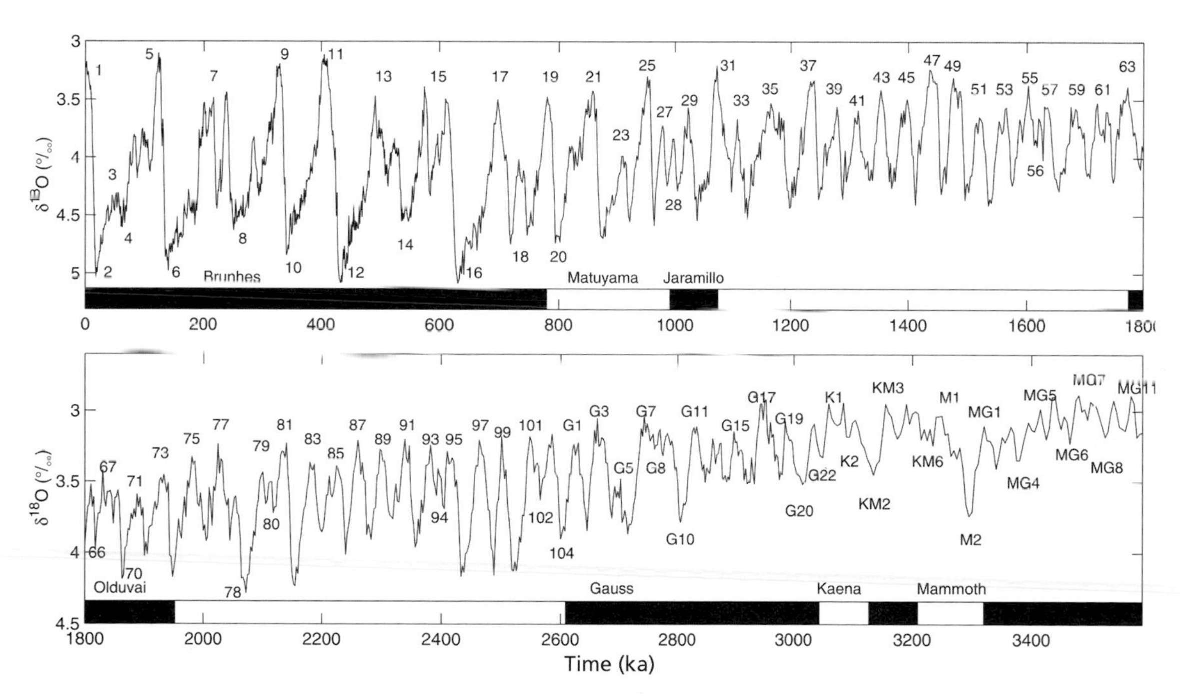

FIGURE 14.14. A "stacked" **marine oxygen isotope** record constructed by graphically correlating 57 globally distributed benthic foraminifera δ^{18}O records. Modified from Lisiecki and Raymo, 2005. The magnetic polarity is shown at the base of each panel in the figure. **For marine oxygen isotope records, greater δ^{18}O values indicate times of expanded ice and smaller δ^{18}O values indicate times of less ice (see Chapter 6).** Note that ka = thousand years ago.

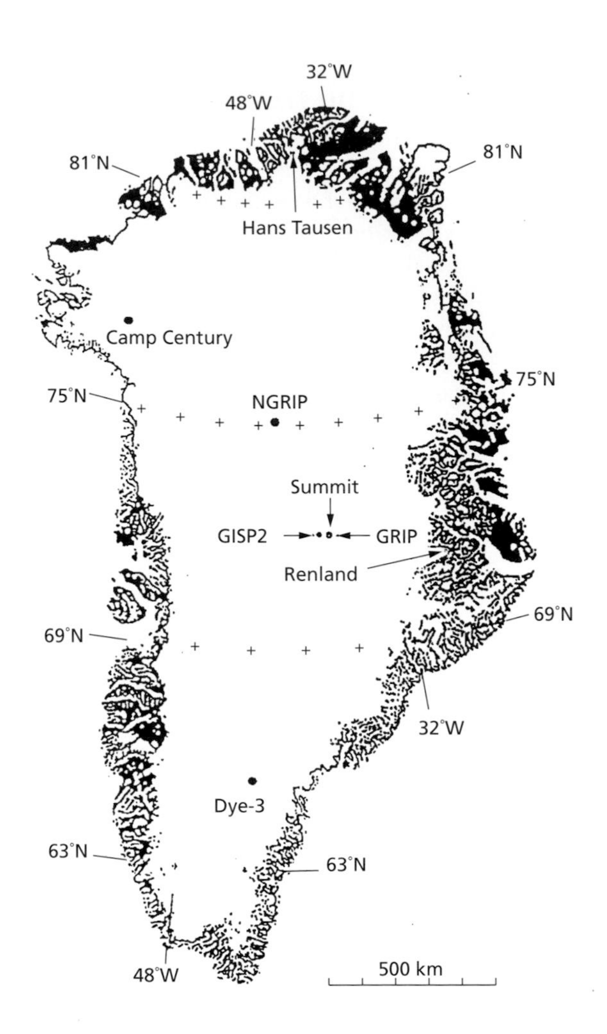

FIGURE 14.15. Map showing the locations of ice core sites in Greenland. From Johnsen et al., 2001.

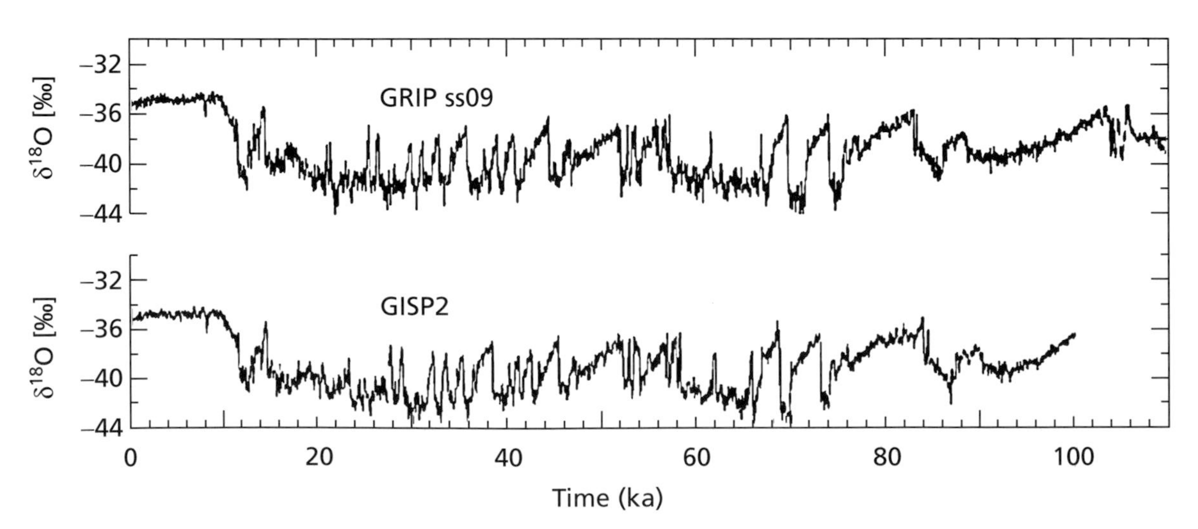

FIGURE 14.16. Stable oxygen isotope profiles ($\delta\ ^{18}O$) from the two central Greenland **ice cores** (GRIP and GISP2, shown in Figure 14.15. Modified from Johnsen et al., 2001. Note ka = thousand years ago.

NAME

Paleontological Evidence: Pollen from Plants

Just as changing temperatures and precipitation result in ice sheet expansion and contraction, these variations in climate also affect the distribution and abundance of plants and animals. The fossil record (e.g., Figure 14.17) of plants and animals are indirect indicators of paleoclimate conditions.

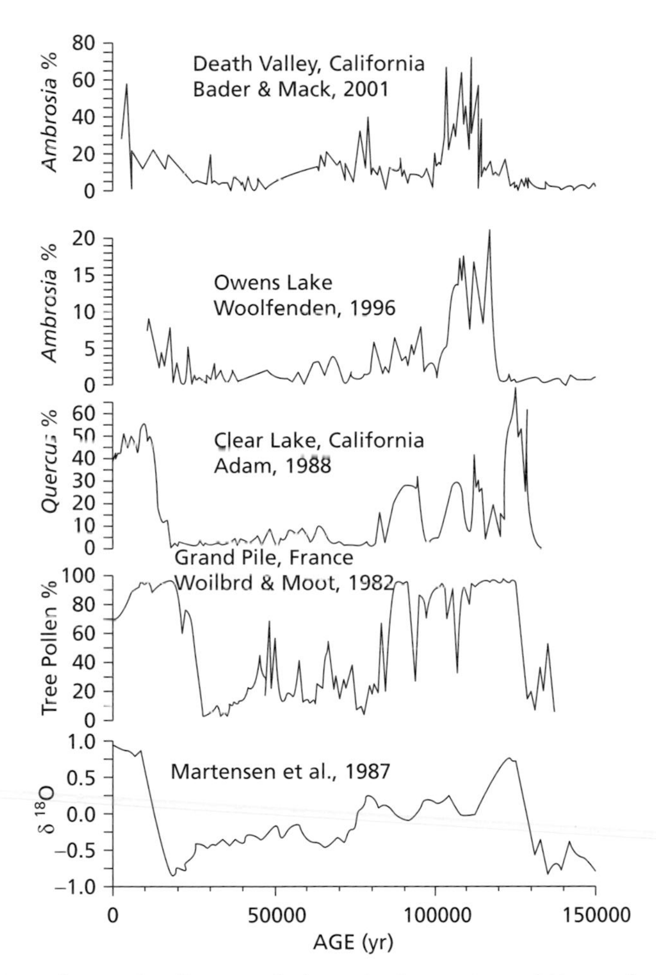

FIGURE 14.17. Compilation of pollen records from North America and Europe for the last 150,000 years correlated with the marine stable oxygen-isotope record ($\delta^{18}O$). Data compiled by O. Davis, University of Arizona.

NAME

Northern Hemisphere Glaciation
Part 14.3. What Caused It?

Introduction
What caused the northern hemisphere glaciation? This is a simple question, but like a challenging detective case, not necessarily easy to answer. To investigate the cause of the widespread expansion of northern hemisphere ice between 3–2.5 Ma you will read several abstracts from published peer-reviewed journal articles. The abstracts are summaries of the main points in the paper written by the authors.

1 **Use what you have learned from each abstract to complete Table 14.2 below. Then use your completed Table 14.2 to synthesize the information on the expansion of northern hemisphere glaciation.** Since these abstracts are from the primary literature they are written for a scientific audience and may contain terms that are unfamiliar to you. When you come across such terms, look them up. Highlight, underline, make notes in the margins, too, if that helps you decipher the abstract.

Abstract 1. Influence of Late Cenozoic Mountain Building on Ocean Geochemical Cycles (Raymo et al., 1988)
In a steady-state ocean, input fluxes of dissolved salts to the sea must be balanced in mass and isotopic value by output fluxes. For the elements strontium, calcium, and carbon, rivers provide the primary input, whereas marine biogenic sedimentation dominates removal. Dissolved fluxes in rivers are related to rates of continental weathering, which in turn are strongly dependent on rates of uplift. The largest dissolved fluxes today arise in the Himalayan and Andean mountain ranges and the Tibetan Plateau. During the past 5 myr, uplift rates in these areas have increased significantly; this suggests that weathering rates and river fluxes may also have increased. The oceanic records of carbonate sedimentation, level of the calcite compensation depth, and $\delta^{13}C$ and $\delta^{87}Sr$ in biogenic sediments are consistent with a global increase in river fluxes since the late Miocene. The cooling of global climate over the past few million years may be linked to a decrease in atmospheric CO_2 driven by enhanced continental weathering in these tectonically active regions.

NAME ____________________

Table 14.2 Summary of abstracts

Abstract number	Abstract Title and Date of Publication	Proposed Mechanism (Hypothesis) for the Onset of Widespread Northern Hemisphere Glaciation	Parts of the Earth System This Mechanism Would Involve	Types of Data and/or Method Used to Support This Hypothesis	Location of Data Set(s)	Terms used in Abstract That You Needed to Look Up
1	Influence of late Cenozoic mountain building on ocean geochemical cycles (1988)					
2	Forcing of Late Cenozoic northern hemisphere climate by plateau uplift in southern Asia and the American West (1989)					
3	Volcanic triggering of late Pliocene glaciation: evidence from the flux of volcanic glass and ice-rafted debris to the North Pacific Ocean (2001)					
4	Initiation of northern hemisphere glaciation and strengthening of the northeast Indian monsoon: Ocean Drilling Program Site 758, eastern equatorial Indian Ocean (2003)					
5	Regional climate shifts caused by gradual global cooling in the Pliocene epoch. (2004)					
6	North Pacific seasonality and the glaciation of North America 2.7 million years ago (2005)					
7	Closure of the Panama Seaway during the Pliocene: implications for climate and northern hemisphere glaciation (2008)					

Abstract 2. Forcing of Late Cenozoic Northern Hemisphere Climate by Plateau Uplift in Southern Asia and the American West (Ruddiman and Kutzbach, 1989)

Geologic evidence indicates that net vertical uplift occurred on a large (kilometer) scale and at accelerating rates during the middle and late Cenozoic in the plateaus of southern Asia and the American west. Based on this evidence, general circulation model sensitivity tests were run to isolate the unique effects of plateau uplift on climate. The experiments simulated significant climatic changes in many places, some far from the uplifted regions. The basic direction of most of these simulated responses to progressive uplift is borne out by changes found in the geologic record: winter cooling in North America, northern Europe, northern Asia, and the Arctic Ocean; summer drying in the North American west coast, the Eurasian interior, and the Mediterranean; winter drying in the North American northern plains and the interior of Asia; and changes over the North Atlantic Ocean conducive to increased formation of deep water. The modeled changes result from increased orographic diversion of westerly winds, from cyclonic and anticyclonic surface flow induced by summer heating and winter cooling of the uplifted plateaus, and from the intensification of vertical circulation cells in the atmosphere caused by exchanges of mass between the summer-heated (and winter-cooled) plateaus and the mid-latitude oceans. Disagreements between the geologic record and the model simulations in Alaska and the Southern Rockies and plains may be related mainly to the lack of narrow mountain barriers in the model orography. Taken together, the observed regional trends comprise much of the pattern of "late Cenozoic climatic deterioration" in the northern hemisphere that culminated in the Pliocene–Pleistocene ice ages. The success of the uplift sensitivity experiment in simulating the correct pattern and sign of most of the observed regional climatic trends points to uplift as an important forcing function of late Cenozoic climatic change in the northern hemisphere at time scales longer than orbital variations; however, the modest amplitude of the uplift-induced cooling simulated at high latitudes indicates a probable need for additional climatic forcing.

Abstract 3. Volcanic Triggering of Late Pliocene Glaciation: Evidence from the Flux of Volcanic Glass and Ice-Rafted Debris to the North Pacific Ocean (Prueher and Rea, 2001)

Mass accumulation rates (MAR) of different components of North Pacific deep-sea sediment provide detailed information about the timing of the onset of major northern hemisphere glaciation that occurred at 2.65 Ma. An increase in explosive volcanism in the Kamchatka–Kurile and Aleutian arcs occurred at this same time, suggesting a link between volcanism and glaciation. Sediments recovered by piston-coring techniques during ODP Leg 145 provide a unique opportunity to undertake a detailed test of this possibility. Here we use volcanic glass as a proxy for explosive volcanism and ice-rafted debris (IRD) as a proxy for glaciation. The MAR of both glass and IRD increase markedly at 2.65 Ma. Further, the flux of the volcanic glass increased just prior to the flux of ice-rafted material, suggesting that the cooling resulting from explosive volcanic eruptions may have been the ultimate trigger for the mid-Pliocene glacial intensification.

NAME

Abstract 4. Initiation of Northern Hemisphere Glaciation and Strengthening of the Northeast Indian Monsoon: Ocean Drilling Program Site 758, Eastern Equatorial Indian Ocean

(Gupta and Thomas, 2003)

The Indian monsoon system, as recorded by ocean-floor biota (benthic foraminifera) at Ocean Drilling Program Site 758 in the eastern equatorial Indian Ocean, has varied dramatically over the past 5.5 myr, long after the onset of the monsoons at 10–8 Ma. Benthic foraminifera that thrive with high productivity year round were common before the formation of northern hemisphere continental ice sheets approximately 3.1–2.5 Ma, indicating that the summer (southwest) monsoon had high intensity and long seasonal duration. Ca. 2.8 Ma benthic faunas became dominated by taxa that flourish with a seasonally strongly fluctuating food supply, indicating that the northeast (winter) monsoon, during which primary productivity is relatively low, increased in duration and strength to form a system similar to that of today. The change occurred coeval with the initiation of the northern hemisphere glaciation, documenting a close link between the development of the Indian monsoon and northern hemisphere glaciation.

Abstract 5. Regional Climate Shifts Caused by Gradual Global Cooling in the Pliocene Epoch

(Ravelo et al., 2004)

The Earth's climate has undergone a global transition over the past four million years, from warm conditions with global surface temperatures about 3°C warmer than today, smaller ice sheets and higher sea levels to the current cooler conditions. Tectonic changes and their influence on ocean heat transport have been suggested as forcing factors for that transition, including the onset of significant northern hemisphere glaciation approximately 2.75 million years ago, but the ultimate causes of the climatic changes are still under debate. Here we compare climate records from high latitudes, subtropical regions and the tropics, indicating that the onset of large glacial/interglacial cycles did not coincide with a specific climate reorganization event at lower latitudes. The regional differences in the timing of cooling imply that global cooling was a gradual process, rather than the response to a single threshold or episodic event as previously suggested. We also find that high-latitude climate sensitivity to variations in solar heating increased gradually, culminating after cool tropical and subtropical upwelling conditions were established two million years ago. Our results suggest that mean low-latitude climate conditions can significantly influence global climate feedback.

Abstract 6. North Pacific Seasonality and the Glaciation of North America 2.7 Million Years Ago (Haug et al., 2005)

In the context of gradual Cenozoic cooling, the timing of the onset of significant northern hemisphere glaciation 2.7 million years ago is consistent with Milankovitch's orbital theory, which posited that ice sheets grow when polar summertime insolation and temperature are low. However, the role of moisture supply in the initiation of large northern hemisphere ice sheets has

NAME

Abstract 6. (*Continued*)

remained unclear. The subarctic Pacific Ocean represents a significant source of water vapor to boreal North America, but it has been largely overlooked in efforts to explain northern hemisphere glaciation. Here we present alkenone unsaturation ratios and diatom oxygen isotope ratios from a sediment core in the western subarctic Pacific Ocean, indicating that 2.7 million years ago late summer sea surface temperatures in this ocean region rose in response to an increase in stratification. At the same time, winter sea surface temperatures cooled, winter floating ice became more abundant and global climate descended into glacial conditions. We suggest that the observed summer warming extended into the autumn, providing water vapor to northern North America, where it precipitated and accumulated as snow and thus allowed the initiation of northern hemisphere glaciation.

Abstract 7. Closure of the Panama Seaway During the Pliocene: Implications for Climate and Northern Hemisphere Glaciation (Lunt et al., 2008)

The "Panama Hypothesis" states that the gradual closure of the Panama Seaway, between 13 million years ago (13 Ma) and 2.6 Ma, led to decreased mixing of Atlantic and Pacific water Masses, the formation of North Atlantic deep water and strengthening of the Atlantic thermohaline circulation, increased temperatures and evaporation in the North Atlantic, and increased precipitation in northern hemisphere (NH) high latitudes, culminating in the intensification of northern hemisphere Glaciation (NHG) during the Pliocene, 3.2–2.7 Ma. Here we test this hypothesis using a fully coupled, fully dynamic ocean atmosphere general circulation model (GCM), with boundary conditions specific to the Pliocene, and a high resolution dynamic ice sheet model. We carry out two GCM simulations with "closed" and "open" Panama Seaways and use the simulated climatologies to force the ice sheet model. We find that the models support the "Panama Hypothesis" in as much as the closure of the seaway results in a more intense Atlantic thermohaline circulation, enhanced precipitation over Greenland and North America, and ultimately larger ice sheets. However, the volume difference between the ice sheets in the "closed" and "open" configurations is small, equivalent to about 5 cm from sea level. We conclude that although the closure of the Panama Seaway may have slightly enhanced or advanced the onset of NHG, it was not a major forcing mechanism. Future work must fully couple the ice sheet model and GCM and investigate the role of orbital and CO^2 effects in controlling NHG.

Synthesis

2 Do any of the proposed mechanisms involve only one part of the Earth system? Why do you think this is?

NAME

3 Can these mechanisms ever truly be proved, or can they only be disproved?

4 Do you view any of the proposed mechanisms as mutually exclusive? Why or why not?

5 Several of these hypotheses arise because of a concurrent timing of some change in the Earth system and northern hemisphere glacial expansion. Does concurrent timing imply causal relationships? Explain.

NAME

6 The growth of ice sheets requires sufficient temperature and adequate moisture supply. Which of these mechanisms aim to satisfy the temperature requirement and which aim to satisfy the moisture requirement?

7 Are the data used to support the hypotheses proposed in the abstracts largely marine or non-marine (land-based)? Why do you think that is?

NAME

NAME

References

Adam, D.P., 1988, Correlations of the Clear Lake, California, Core CL-73-4 pollen sequence with other long climate records. Special Paper. Geological Society of America, **214**, 81–95.

Eldrett, J.S., et al., 2007, Continental ice in Greenland during the Eocene and Oligocene. Nature, **446**, 176–9.

Elhers, J. and Gibbard, P.L., 2003, Extent and chronology of glaciations. Quaternary Science Reviews, **22**, 1561–8.

Elhers, J. and Gibbard, P.L., 2007, The extent and chronology of Cenozoic global glaciation. Quaternary International, **164–165**, 6–20.

Flesche-Kleiven, H., et al., 2002, Intensification of northern hemisphere glaciations in the circum Atlantic region (3.5–2.4 Ma) – ice-rafted detritus evidence. Palaeogeography, Palaeoclimatology, Palaeoecology, **184**, 213–23.

Gupta, A.K. and Thomas, E, 2003, Initiation of northern hemisphere glaciation and strengthening of the northeast Indian monsoon. Ocean Drilling Program Site 758, eastern equatorial Indian Ocean. Geology, **31**, 47–50.

Haug, G.H., et al., 2005, North Pacific seasonality and the glaciation of North America 2.7 million years ago. Nature, **433**, 821–5.

Johnsen, S. J., et al., 2001, Oxygen isotope and palaeotemperature records from six Greenland ice-core stations: Camp Century, Dye-3, GRIP, GISP2, Renland and North-GRIP. Journal of Quaternary Science, **16**, 299–307.

Krissek, L.A., 1995, Late Cenozoic ice-rafting records from Leg 145 sites in the North Pacific: late Miocene onset, late Pliocene intensification, and Pliocene–Pleistocene events. In Proceedings of the Ocean Drilling Program, Scientific Results, vol. 145, Rea, D.K., et al., (eds), College Station, TX, Ocean Drilling Program, pp. 179–94. doi:10.2973/odp.proc.sr.145.118.1995.

Lisiecki L. E. and Raymo, M.E., 2005, A Pliocene–Pleistocene stack of 57 globally distributed benthic $\delta^{18}O$ records. Paleoceanography, **20**, PA1003, doi:10.1029/2004PA001071.

Lunt, D. J., et al., 2008, Closure of the panama seaway during the pliocene: Implications for climate and northern hemisphere glaciation. Climate Dynamics, **30** (1), 1–18.

Martinson, D.G., et al., 1987, Age dating and the orbital theory of the ice ages: Development of a high-resolution 0–300,000 year chronostratigraphy. Quaternary Research, **27**, 1–29.

Polyak, L., et al., 2010, History of Sea Ice in the Arctic. Quaternary Science Review (QSR) Special Issue: Past Climate Variability and Change in the Arctic, (in press). doi:10.1016/j.quascirev.2010.02.010.

Prueher, L.M. and Rea, D.K., 2001, Volcanic triggering of late Pliocene glaciation: evidence from the flux of volcanic glass and ice-rafted debris to the North Pacific Ocean. Palaeogeography, Palaoeclimatology, Palaeoecology, **173**, 215–30.

Ravelo, A.C., et al., 2004, Regional climate shifts caused by gradual global cooling in the Pliocene epoch. Nature, **429**, 263–7.

Raymo, M.E., et al., 1988, Influence of late Cenozoic mountain building on ocean geochemical cycles. Geology, **16**, 649–53.

NAME

Ruddiman, W.F. and Kutzbach, J.E., 1989, Forcing of Late Cenozoic northern hemisphere Climate by Plateau Uplift in Southern Asia and the American West. Journal of Geophysical Research, **94**(D15), 18409–27.

Shipboard Scientific Party, 1993, Introduction to Leg 145: North Pacific Transect, In Proceeding of the Ocean Drilling Program, Initial Reports, vol. 145, Rea, D.K., et al. (eds), College Station, TX, Ocean Drilling Program, pp. 5–7.

St. John, K., 2008, Cenozoic History of Ice-Rafting in the Central Arctic: Terrigenous Sands on the Lomonosov Ridge. Paleoceanography, **23**, PA1S05.

Woillard G.M. and Mook, W.G., 1982, Carbon-14 dates at Grande Pile: correlation of land and sea chronologies. Science, **215**, 159–61.

Woolfenden, W.B., 1996, 180,000-year pollen record from Owens Lake, CA: terrestrial vegetation change on orbital scales. Quaternary Research, **59**, 430–44.

Index

Reconstructing Earth's Climate History: Inquiry-Based Exercises for Lab and Class, First Edition. Kristen St John, R Mark Leckie, Kate Pound, Megan Jones and Lawrence Krissek.